CONVERSION FACTORS*

D0118727

Length
1 m = 3.281 ft
1 m = 3.937 × 10 in.

Area
1 m² = 1.550 × 10³ in.²
1 m² = 1.076 × 10 ft²

Volume
1 m³ = 6.102 × 10⁴ in.³
1 m³ = 3.532 × 10 ft³
1 m³ = 2.642 × 10² U.S. gallons

Mass
1 kg = 2.205 lbm

Force
1 N = 2.248 × 10⁻¹ lbf

Energy
1 J = 9.478 × 10⁻⁴ Btu
    = 7.376 × 10⁻¹ ft-lbf
1 kW-hr = 3.412 × 10³ Btu
    = 2.655 × 10⁶ ft-lbf

Power
1 W = 3.412 Btu/hr
1 W = 1.341 × 10⁻³ hp
1 W = 2.844 × 10⁻⁴ tons of
    refrigeration

Pressure
1 Pa = 1.450 × 10⁻⁴ lbf/in.²
1 Pa = 2.088 × 10⁻² lbf/ft²
1 Pa = 9.869 × 10⁻⁶ std atm
1 Pa = 2.961 × 10⁻⁴ in. mercury
1 Pa = 4.019 × 10⁻³ in. water

Temperature
1 deg R difference = 1 deg F
    difference = 5/9 deg C difference
    = 5/9 deg K difference
deg F = 9/5 (deg C) + 32

$$1 \text{ m/s}^2 = 3.281 \text{ ft/sec}^2$$

Mass Density
1 kg/m³ = 6.243 × 10⁻² lbm/ft³

Mass Flow Rate
1 kg/s = 2.205 lbm/sec
1 kg/s = 7.937 × 10³ lbm/hr

Volume Flow Rate
1 m³/s = 2.119 × 10³ ft³/min
1 m³/s = 1.585 × 10⁴ gal/min

Thermal Conductivity
$$1 \frac{W}{m\text{-}C} = 5.778 \times 10^{-1} \frac{Btu}{hr\text{-}ft\text{-}F}$$
$$1 \frac{W}{M\text{-}C} = 6.934 \frac{Btu\text{-}in}{hr\text{-}ft^2\text{-}F}$$

Heat-transfer Coefficient
$$1 \frac{W}{M^2\text{-}C} = 1.761 \times 10^{-1} \frac{Btu}{hr/ft^2\text{-}F}$$

Specific Heat
$$1 \frac{J}{kg\text{-}C} = 2.389 \times 10^{-4} \frac{Btu}{lbm\text{-}F}$$

Viscosity, Absolute
$$1 \frac{N\text{-}s}{m^2} = 10^3 \text{ centipoise} = 1 \text{ Pa-s}$$
$$1 \frac{N\text{-}s}{m^2} = 6.720 \times 10^{-1} \frac{lbm}{ft\text{-}sec}$$

Viscosity, Kinematic
1 m²/s = 1.076 × 10 ft²/sec
1 m²/s = 10⁶ centistoke

*All factors have been rounded off to four significant figures.

Heating, Ventilating and Air Conditioning
ANALYSIS AND DESIGN

Heating, Ventilating, and Air Conditioning

ANALYSIS and DESIGN

THIRD EDITION

FAYE C. McQUISTON
JERALD D. PARKER
Oklahoma State University

WILEY
JOHN WILEY & SONS
New York • Chichester • Brisbane • Toronto • Singapore

Library of Congress Cataloging-in-Publication Data

McQuiston, Faye C.
  Heating, ventilation, and air conditioning.

  Bibliography: p.
  Includes index.
    1. Heating.   2. Ventilation.   3. Air conditioning.
I. Parker, Jerald D.   II. Title.
TH7222.M38   1988   697   87-34604
ISBN 0-471-63757-2

Printed in the United States of America

10 9 8 7 6 5 4 3 2 1

Preface

The normal evolution of analysis and design procedures as published in the *ASHRAE Handbook* and in other books of a similar nature has made it desirable to revise the second edition of this text. The original objective was to prepare an up-to-date book, based on ASHRAE literature, that would be a convenient classroom teaching aid. This objective has not changed. One of the principal motivations for development of this book was our desire to contribute to the continued growth of the HVAC industry through the education of both beginning and practicing engineers. Following a study of the complete text, a student should be capable of participating in the design of all types of HVAC systems. The book is intended for use at the undergraduate and beginning graduate levels by students who have taken the basic engineering science courses, including thermodynamics, heat transfer, and fluid mechanics. It can be used in two regular semester courses, with a great deal of latitude in the makeup of those courses. Even though the book is designed primarily as a teaching device, it is also useful to practicing engineers as a ready source of background material and as a valuable aid in studying new procedures.

In response to a number of suggestions for the improvement of the second edition, we have made several revisions. New problems have been written, including types that can be solved by hand or with available computer software. A complete solutions manual for these problems is available from the publisher. The instructor is encouraged to provide examples and problems that emphasize the requirements of the local geographical region and the instructor's design philosophies.

More emphasis has been placed on automatic control with an introduction in Chapter 2, and discussion at appropriate places in other chapters. Automatic Control is treated with the objective of showing where and how it is used rather than how to design a control system.

Chapter 3 has been revised to include more practical applications and to clarify certain areas. The latest and most precise moist air properties are given in Table A-2. Increased emphasis has been placed on the effects of barometric pressure in Chapter 3 and other appropriate places. Psychometric charts for high altitudes have been added in the pocket inside the rear cover.

Recent new data for floor slabs and underground walls and floors have been included in Chapter 6. Chapters 7 and 8 have been revised to cover energy calculation by the bin method.

Chapter 9 has been revised to cover modern practices in hydronic system design to include variable flow with more emphasis placed on large systems. New data from the *ASHRAE Handbook* for duct fittings have been placed in Chapter 11. Emphasis has also been placed on the overall design of the air-distribution system with the concept of *energy grade line* introduced as an aid to understanding in this area. More discussion and data related to fans and *system effect* have also been introduced.

Chapter 15 has been revised to emphasize the real refrigeration system by using actual compressor data. Control of the compressor and the system is now discussed.

The personal computer has become common in the teaching environment and is particularly useful in the design and analysis of HVAC systems. The instructor using this text is encouraged to involve students in the use of the personal computer and the many programs available for them. Moreover we may be contacted about software covering the areas of psychrometrics, load calculations, pipe and duct design, and coil simulation.

The conversion from English to the international system of units (SI) in the United States is proceeding more slowly than was anticipated. However, engineers should be able to use both systems of units when they enter practice. Therefore, this book continues to use a dual system of units. Although emphasis is placed on the English system, all of the equations are written to be dimensionally correct in either system of units. Extensive data, examples, and problems are provided in SI units. Instructors may blend the two systems of units as they see fit.

We are deeply indebted to ASHRAE for providing a great deal of support. Many other companies and individuals contributed suggestions, material, and photographs. Our thanks to everyone.

Faye C. McQuiston
Jerald D. Parker

About the Authors

Faye C. McQuiston is a Professor of Mechanical and Aerospace Engineering at Oklahoma State University, Stillwater, Oklahoma. He received B. S. and M. S. degrees in mechanical engineering from Oklahoma State University in 1958 and 1959 and a Ph. D. in mechanical engineering from Purdue University in 1970. Dr. McQuiston joined the Oklahoma State faculty in 1962 after a three-year period in industry. He was a National Science Foundation Faculty Fellow from 1967 to 1969. Dr. McQuiston is an active member of the American Society of Heating, Refrigerating and Air Conditioning Engineers (ASHRAE), presently a member of the Technology Council and several technical committees. He is a past member of the Education, Research and Technical and Standards Committees and was honored with the Best Paper Award in 1979, the Region VIII Award of Merit in 1981, the Distinguished Service Award in 1984, and the E. K. Campbell Award in 1986. He was also elected to the grade of Fellow in 1986. He is a registered professional engineer and a consultant to several system design and equipment manufacturing firms. Dr. McQuiston is active in research related to the design of heating and air conditioning systems and particularly in the areas of heat exchanger design and simulation, load calculations, and energy calculations. He has written many papers in the general area of heating and air conditioning and is the coauthor of a basic fluid mechanics and heat transfer text.

Jerald D. Parker is Professor Emeritus of Mechanical and Aerospace Engineering at Oklahoma State University and Professor of Engineering at Oklahoma Christian college. He received B. S. and M. S. degrees in mechanical engineering from Oklahoma State University in 1955 and 1958 and a Ph. D. in mechanical engineering from Purdue University in 1961. He was at Oklahoma State University from 1955 to 1988 with the exception of one year spent in the engineering department at duPont in Newark, Delaware. He has been active in both the ASME, where he is a Fellow, and ASHRAE where he has served as Chairman of the Technical Committee on Fluid Mechanics and Heat Transfer. He has also been an active local member in the Americal Nuclear Society. He is a registered Professional Engineer and a member of the Oklahoma and National Societies of Professional Engineers. He is coauthor of a basic text in fluid mechanics and heat transfer and has contributed articles for handbooks, technical journals, and magazines. His research has been involved with ground coupled heat pumps, solar heated asphalt storage systems, and chilled water storage and distribution. He has served as a consultant in cases involving performance, accidents, fires, and explosions in heating, cooling, and process systems.

Contents

Symbols

English Letter Symbols

A	area, ft² or m²
A	apparent solar irradiation for zero air mass, Btu/(hr-ft²) or W/m²
a	light fixture classification, Table 8-19
ADPI	air distribution performance index, dimensionless
B	wet bulb coefficient, Eq. 12-27, lbmv/(lbma-F) or kgv/(kga-C)
B	atmospheric extinction coefficient.
b	transfer function coefficient, Btu/(hr-ft²-F) or W/(m²-C)
b	light fixture classification, Table 8-20
b	bypass factor, dimensionless
C	concentration, lbm/ft³ or kg/m³
C	unit thermal conductance, Btu/(hr-ft²-F) or W/(m²/C)
C	discharge coefficient, dimensionless
C	constant in Eq. 10-3, dimensionless
C	loss coefficient, dimensionless
C	fluid capacity rate, Btu/(hr-F) or W/C
C	clearance factor, dimensionless
C_d	overall flow coefficient, dimensionless
C_d	draft coefficient, dimensionless
C_p	pressure coefficient, dimensionless
C_v	flow coefficient, dimensionless
CLD	cooling load factor, dimensionless
CLTD	cooling load temperature difference, F or C
COP	coefficient of performance, dimensionless
c	specific heat, Btu/(lbm-F) or J/(kg-C)
c	transfer function coefficient, Btu/(hr-ft²-F) or W/(m²-C)
cfm	volume flow rate, ft³/min
clo	clothing thermal resistance, (ft²-hr-F)/Btu or (m²-C)/W
D	diameter, ft or m
D	diffusion coefficient, ft²/sec or m²/s
DD	degree days, F-day or C-day

db	dry bulb temperature, F or C
DR	daily range of temperature, F or C
d	bulb diameter, ft or m
d	sun's declination, degrees
d	transfer function coefficient, dimensionless
E	effective emittance, dimensionless
EDT	effective draft temperature, or C
ET	effective temperature, F or C
F	configuration factor, dimensionless
F	quantity of fuel, ft^3 or m^3
F_c	fraction of heat gain, dimensionless
F(s)	wet surface function, Eq. 14-60, dimensionless
f	friction factor, dimensionless
f_t	Darcy friction factor with fully turbulent flow, dimensionless
FP	correlating parameter, Eq. 14-51, dimensionless
G	irradiation, Btu/(hr-ft^2) or W/m^2
g	local acceleration due to gravity, ft/sec^2 or m/s^2
g	transfer function coefficient, Btu/(hr-ft) or W/C
g_c	dimensional constant, 32.17 (lbm-ft)/(lbf/sec^2) or 1.0 (kg-m)/(N-s^2)
H	heating value of fuel, Btu or J per unit of volume
H	head, ft or m
h	height or length, ft or m
h	heat-transfer coefficient, Btu/(hr-ft^2-F) or W/(m^2-C) (also used for mass transfer coefficient with subscripts *m, d,* and *i*)
h	hour angle, degrees
hp	horsepower
i	enthalpy, Btu/lbm or J/kg
J	Joule's equivalent, 778.28 (ft-lbf)/Btu
JP	correlating parameter, Eq. 14-46, dimensionless
J(s)	wet surface function, Eq. 14-58, dimensionless
$J_i(s)$	wet surface function, Eq. 14-59, dimensionless
j	Colburn *j*-factor, Eq. 12-18, dimensionless
K	color correction factor, dimensionless
K	resistance coefficient, dimensionless
K	constant in Eq. 10-2, dimensionless
K_t	unit length conductance, Btu/(ft-hr-F) or W/(m-C)
k	thermal conductivity, (Btu-ft)/(ft^2-hr-F), (Btu-in.)/(ft^2-hr-F) or (W-m)/(m^2-C)
k	isentropic exponent, c_p/c_v, dimensionless
L	fin dimension, Figs. 14-13 and 14-14, ft or m
L	total length, ft or m
Le	Lewis number, Sc/Pr, dimensionless
LM	latitude-month correction, F or C
LMTD	log mean temperature difference, F or C
l	latitude, degrees
l	lost head, ft or m
M	molecular weight, lbm/(lb mole) or kg/(kg mole)
M	constant in Eq. 10-1, F-min/ft or (C-s)/m
M	fin dimension, Figs. 14-13 and 14-14, ft or m
MRT	mean radiant temperature, F or C

m	mass, lbm or kg
$\dot{m}$	mass flow rate or mass transfer rate, lbm/sec or kg/s
N	number of hours or other integer
Nu	Nusselt number, hx/k, dimensionless
NC	noise criteria, dimensionless
NTU	number of transfer units, dimensionless
P	pressure, lb/ft² or psia or N/m² or Pa
P	heat exchanger parameter, Eq. 14-4, dimensionless
P	circumference, ft or m
Pr	Prandtl number, $\mu c_p/k$, dimensionless
PD	piston displacement, ft³/min or m³/s
p	partial pressure, lbf/ft² or psia or Pa
p	transfer function coefficient, dimensionless
	volume flow rate, ft³/sec or m³/s
q	heat transfer, Btu/lbm or J/kg
$\dot{q}$	heat transfer rate, Btu/hr or W
R	gas constant, (ft-lbf)/(lbm-R) or J/(kg-K)
R	unit thermal resistance, (ft²-hr-F)/Btu or (m²-C)/W
R	heat exchanger parameter, Eq. 14-4, dimensionless
R	fin radius, ft or m
R'	thermal resistance, (hr-F)/Btu or C/W
$\overline{R}$	gas constant, (ft-lbf)/(lb mole-R) or J/(kg mole-K)
Re	Rcynolds number $\rho \overline{V} D/\mu$, dimensionless
R_f	unit fouling resistance, (hr-ft²-F)/Btu, or (m²-C)/W
r	radius, ft or m
rpm	revolutions per minute
S	fin spacing, ft or m
S	equipment characteristic, Btu/(hr-F) or W/C
Sc	Schmidt number, v/D, dimensionless
Sh	Sherwood number, $h_m x/D$, dimensionless
SC	shading coefficient, dimensionless
SHF	sensible heat factor, dimensionless
SHGF	solar heat gain factor, Btu/(hr-ft²) or W/m²
s	entropy, Btu/(lbm-R) or J/(kg-K)
T	absolute temperature, R or K
t	temperature, F or C
t^*	thermodynamic wet bulb temperature, F or C
U	overall heat transfer coefficient, Btu/(hr-ft²-F) or W/(m²-C)
u	velocity in x direction, ft/sec or m/s
V	volume, ft³ or m³
$\overline{V}$	velocity, ft/sec or m/s
v	specific volume, ft³/lbm or m³/kg
v	transfer function coefficient, dimensionless
v	velocity in y direction, ft/sec or m/s
W	humidity ratio, lbmv/lbma or kgv/kga
W	equipment characteristics, Btu/hr or W
$\dot{W}$	power, Btu/hr or W
$WBGT$	wet bulb globe temperature, F or C
w	skin wettedness, dimensionless

w	work, Btu, or ft-lbf, or J
w	transfer function coefficient, dimensionless
X	normalized input, dimensionless
X	fraction of daily range
x	mole fraction
x	quality, lbmv/lbm or kgv/kg
x	throw, ft or m
x, y, z	length, ft or m
Y	normalized capacity, dimensionless

Subscripts

a	transverse dimension
a	air
a	average
a	attic
as	adiabatic saturation
as	denotes change from dry air to saturated air
avg	average
B	barometric
b	branch
b	longitudinal dimension
b	base
c	cool or coil
c	convection
c	ceiling
c	cross section or minimum free area
c	cold
c	condenser
c	Carnot
c	collector
cl	center line
D	direct
D	diameter
d	dew point
d	total heat
d	diffuse
d	design
d	downstream
dry	dry surface
e	equivalent
e	sol-air
e	equipment
e	evaporator
f	film
f	friction
f	fin
fg	refers to change from saturated liquid to saturated vapor
fl	fluorescent light

fl	floor
fr	frontal
g	refers to saturated vapor
g	globe
H	horizontal
HD	on horizontal surface, daily
h	heat
h	hydraulic
h	head
h	heat transfer
h	hot
i	j-factor for total heat transfer
i	inside
i	instantaneous
l	latent
l	liquid
m	mean
m	mass transfer
m	mechanical
ND	direct normal
n	integer
o	outside
o	total or stagnation
o	initial condition
oh	humid operative
P	pressure
p	constant pressure
p	pump
R	reflected
R	refrigerating
r	radiation
r	room air
s	stack effect
s	sensible
s	saturated vapor or saturated air
s	supply air
s	shaft
s	static
s	surface
sc	solar constant
sh	shade
sl	sunlit
t	temperature
t	total
t	contact
t	tube
u	unheated
u	upstream

V	vertical
v	vapor
v	ventilation
v	velocity
w	wind
w	wall
w	liquid water
wet	wet surface
x	length
x	extraction
1, 2, 3	state of substance at boundary of a control volume
1, 2, 3	a constituent in a mixture
∞	free stream condition

Greek Letter Symbols

α	angle of tilt from horizontal, degrees
α	absorptivity or absorptance, dimensionless
α	total heat transfer area over total volume, ft^{-1} or m^{-1}
α	thermal diffusivity, ft^2/sec or m^2/s
β	parameter in Eq. 14-23, dimensionless
β	altitude angle, degrees
γ	wall solar azimuth angle, degrees
Δ	change in a quantity or property
δ	boundary layer thickness, ft or m
ε	heat exchanger effectiveness, dimensionless
ε	emittance or emissivity, dimensionless
η	efficiency, dimensionless
θ	angle of incidence, degrees
θ	time, sec
μ	degree of saturation, percent or fraction
μ	dynamic viscosity, $\text{lbm}/(\text{ft-sec})$ or $(\text{N-s})/\text{m}^2$
v	kinematic viscosity, ft^2/sec or m^2/s
ρ	mass density, lbm/ft^3 or kg/m^3
ρ	reflectivity or reflectance, dimensionless
Σ	angle of tilt from horizontal, degrees
σ	Stefan-Boltzmann constant, $\text{Btu}/(\text{hr-ft}^2\text{-R}^4)$ or $\text{J}/(\text{s-m}^2\text{-K}^4)$
σ	free flow over frontal area, dimensionless
τ	transmissivity or transmittance, dimensionless
ϕ	fin parameter, Eq. 14-22, dimensionless
ϕ	solar azimuth angle, degrees
ϕ	relative humidity, percent or fraction
ψ	wall azimuth angle, degrees
Ψ	parameter in Eq. 14-23, dimensionless
Ψ	zenith angle, degrees

List of Charts*

*For the convenience of the reader all charts have been folded into a pocket located on the inside back cover.

Heating, Ventilating and Air Conditioning

ANALYSIS AND DESIGN

CHAPTER ONE

Introduction

The development and industrialization of the United States, especially the Southern states, would never have been possible without year-round control of the indoor environment. One has only to look for a manufacturing or printing plant, electronics laboratory or other high-technology facility and the vast office complexes associated with our economy to understand the truth of that statement. Indeed, virtually every residential, commercial, industrial, and institutional building in the United States, Canada, and other industrial countries of the world have controlled environments the year round. Therefore, the heating, ventilating, air conditioning and the associated refrigeration industry have become a large progressive element of our economy. During the period of about 1950 to 1970 when the modern HVAC industry was rapidly developing, major emphasis was placed on keeping a building cool or warm as required and comfort as we know it today received little attention. Also the cost of energy in the form of oil, coal, natural gas, or electricity was very low and led to little consideration of operating costs. With escalation of energy costs, the need for closely controlled environments, and better understanding of the parameters that define comfort and indoor air quality, the arts and sciences of HVAC system design and simulation have been stimulated and the industry is itself moving into high-technology areas. Even so, the basics of good system design have not changed much except in the vast array of tools and equipment available to consummate the design. In the chapters to follow emphasis is placed on the basic elements of HVAC system design taking into account recent developments and procedures to arrive at a functional, well-controlled energy-efficient system.

1-1 HISTORICAL NOTES

Human beings have always struggled to make their lives more comfortable through control of the immediate environment. Because of both necessity and natural laws the first human beings were most successful in situations requiring

1

heat. Although there is evidence of the use of evaporative effects and ice for cooling in very early times, it was not until the middle of the nineteenth century that a practical refrigerating machine was built.

By the end of the nineteenth century the concept of central heating was fairly well developed, and early in the twentieth century cooling for comfort got its start. Developments since that time have been rapid. The greatest advance has been in the change from rather artful design practices to the detailed analytical methods necessitated by complex building structures, high construction costs, and energy shortages. More detailed discussions of the history of heating, refrigeration, and air conditioning are given in articles by Willis R. Woolrich, John W. James, Walter A. Grant, and William L. McGrath (1). In these articles only the names of outstanding contributors appear. Because of the wide scope and diverse nature of the Heating, Ventilating and Air Conditioning (HVAC) field, literally thousands of engineers, their names too numerous to list, have developed the industry. The accomplishments of all these unnamed persons are summarized in the *ASHRAE* Handbook* consisting of four volumes entitled: *Fundamentals, Refrigeration, Equipment,* and *Systems.* These volumes are constantly being revised by ASHRAE members. Considerable research designed to improve the handbooks is also sponsored by ASHRAE and monitored by ASHRAE members. Unquestionably the ASHRAE Handbooks are the encyclopedias of the industry throughout the world. The principles presented in this textbook follow the handbooks closely.

1-2   AIR CONDITIONING AND ENVIRONMENTAL CONTROL

Historically *air conditioning* has implied cooling or otherwise improving the indoor environment during the warm months of the year. In modern times the term has taken on a more literal meaning that can be applied to year-round environmental situations. That is, air conditioning refers to the control of temperature, moisture content, cleanliness, air quality, and air circulation, as required by occupants, a process, or a product in the space. This general definition has lead to the alternate term *environmental control.* The functions required to air condition a space completely are briefly described next.

Heating

Heating is the transfer of energy to a space or to the air in a space by virtue of a difference in temperature between the source and the space or air. This process may take different forms such as direct radiation to the space, direct heating of the circulated air, or through heating of water that is circulated to the vicinity of the space and used to heat the circulated air. This type of heat transfer, which is manifested in a rise in temperature of the air, is called *sensible heat transfer.*

---

*ASHRAE is an abbreviation for the American Society of Heating, Refrigerating and Air-Conditioning Engineers, Incorporated.

Humidifying

The transfer of water to atmospheric air is referred to as humidification. Heat transfer is associated with this mass transfer process; however, the transfer of mass and energy are manifested in an increase in the concentration of water in the air–water vapor mixture. Here the term *latent heat transfer* is used. This process is usually accomplished by introducing water vapor or by spraying fine droplets of water into the circulating air stream. Wetted mats or plates may also be used.

Cooling

Cooling is the transfer of energy from the space or air supplied to the space by virtue of a difference in temperature between the source and the space or air. In the usual cooling process air is circulated over a surface maintained at a low temperature. The surface may be in the space to be cooled or at some remote location from it, the air being ducted to and from the space. Usually water or a volatile refrigerant is the cooling medium. Cooling usually denotes sensible heat transfer.

Dehumidifying

The transfer of water vapor from atmospheric air is called dehumidification. Latent heat transfer is also associated with this process. The transfer of energy is from the air; as a consequence, the concentration of water in the air–water vapor mixture is lowered. This process is most often accomplished by circulating the air over a surface maintained at a sufficiently low temperature to cause the condensation of water vapor from the mixture. It is also possible to dehumidify by spraying cold water into the air stream.

Cleaning

The cleaning of air usually implies filtering; however, it is sometimes necessary to deodorize the air. Filtering is most often done by a process in which dirt particles are captured in a porous medium. Electrostatic cleaners are also used, especially to remove very small particles; in some cases water sprays may be used. Odors are removed by activated charcoal or by exhaust and makeup air, for example.

Air Motion

The motion of air in the vicinity of the occupant should be sufficiently strong to remove energy generated by the body but gentle enough to be unnoticed. The desired air motion is achieved by the proper placement of air inlets to the space and by the use of various air-distributing devices. The importance of air motion especially where occupant comfort is required cannot be underestimated.

All of the functions of the air-conditioning system discussed before may not be active all of the time. Residences and many commercial establishments have inactive cooling and dehumidifying sections during the winter months and keep the heating and humidifying sections inactive during the summer. In large commercial installations, however, it is not uncommon to have all of the functions under simultaneous control for the entire year. Obviously this requires elaborate controls and sensing devices. Precise control of the moisture content of the air is difficult, and humidifying and dehumidifying are usually not done during both winter and summer unless they are absolutely necessary for process control or product preservation, even though the heating and cooling functions may be active all of the time. Because cleaning of the air and good air circulation is always necessary, these functions are used continuously.

1-3   UNITS AND DIMENSIONS

In HVAC computations as in all engineering work consistent units must be employed. A *unit* is a specific quantitative measure of a physical characteristic in reference to a standard. An example of a unit is the foot, which is used to measure the physical characteristic length. A physical characteristic, such as length, is called a *dimension*. Other dimensions of interest in HVAC computations are force, time, temperature, and mass.

In this text two systems of units will be employed. The first is called the English Engineering System and is most commonly used in HVAC work in the United States with some modification such as use of inches instead of feet. The second is the International System (or SI for Système International d'Unites) and is the system in use in engineering practice throughout most of the world and is being widely adopted in the United States.

Equipment designed using U.S. conventional units will be operational for years and even decades. For the foreseeable future, then, it will be necessary for the engineer to work in either system of units and to be able to make conversions from one system to another. The base SI units important in HVAC work are given in Table 1-1 with the comparable English Engineering unit.

Table 1-1

	Units	
Dimensions	**English Engineering System**	**International System**
Mass	pound mass (lbm)	kilogram (kg)
Length	foot (ft)	meter (m)
Time	second (sec)	second (s)
Temperature	degree Fahrenheit (F)	degree Kelvin (K)

The SI system is described in several documents: (2), (3), and (4). The system consists of seven base units (the four base units given in Table 1-1 plus units of electric current, amount of substance, and luminous intensity). In addition there are two supplementary units for the plane angle (rad), the solid angle (r), and a list of derived units. The derived units in the SI system, which are frequently used in HVAC, are given in Table 1-2.

Because there is only one unit for each dimension in the SI system, there is a carefully defined set of multiple and submultiple prefixes to avoid the use of a large number of zeroes for very large or very small quantities. These prefix names and symbols are given in Table 1-3 (4).

Conversion factors for changing to and from SI units are given in Appendix E.

Example 1-1. Describe the quantity of pressure equal to 1000 N/m² in suggested SI terminology.

Solution. 1000 can be described by the prefix kilo (k). The N/m² has the special name of pascal.
Therefore,

$$1000 \text{ N/m}^2 = 1 \text{ kilo pascal} = 1 \text{ kPa}$$

Table 1-2 DERIVED SI UNITS FREQUENTLY
USED IN HVAC

Dimension	Unit	Special Name and Symbol
Acceleration, angular	rad/s²	
Acceleration, linear	m/s²	
Area	m²	
Density	kg/m³	
Energy	N-m	joule (J)
Force	(kg-m)/s²	newton (N)
Frequency	l/s	hertz (Hz)
Power	J/s	watt (W)
Pressure	N/m²	pascal (Pa)
Specific heat capacity	J/(kg-C)	
Stress	N/m²	pascal (Pa)
Thermal conductivity	W/(m-C)	
Thermal flux density	W/m²	
Velocity, angular	rad/s	
Velocity, linear	m/s	
Viscosity, dynamic	(N-s)/m²	
Viscosity, kinematic	m/s²	
Volume	m³	

Table 1-3   MULTIPLE AND
                SUBMULTIPLE SI PREFIXES

Multiplying Factor	Prefix	Symbol
$1\ 000\ 000\ 000\ 000 = 10^{12}$	tera	T
$1\ 000\ 000\ 000 = 10^{9}$	giga	G
$1\ 000\ 000 = 10^{6}$	mega	M
$1\ 000 = 10^{3}$	kilo	k
$100 = 10^{2}$	hecto[a]	h
$10 = 10^{1}$	deka[a]	da
$0.1 = 10^{-1}$	deci	d
$0.01 = 10^{-2}$	centi[a]	c
$0.001 = 10^{-3}$	milli	m
$0.000\ 001 = 10^{-6}$	micro	μ
$0.000\ 000\ 001 = 10^{-9}$	nano	n
$0.000\ 000\ 000\ 001 = 10^{-12}$	pico	p

[a]These prefixes are to be avoided if possible.

Example 1-2. A certain corkboard has a thermal conductivity of 0.025 Btu/(hr-ft-F). Convert this quantity to the equivalent SI value.

Solution. According to the conversion factor given in Appendix E, to convert from Btu/(hr-ft-F) to W/(m-C) you must multiply the given value by 1.7307.

$$(0.025)\ \frac{\text{Btu}}{\text{hr-ft-F}}\ (1.7307)\ \frac{\text{W/(m-C)}}{\text{Btu/(hr-ft-F)}} = 0.043\ \text{W/(m-C)}$$

Because the prefix centi is to be avoided if possible (Table 1-3), the quantity is expressed as shown. Notice that the final quantity is not expressed to any more significant figures than the original quantity, in this case two.

In the SI system the temperature is given in degrees Kelvin (K), which is a thermodynamic absolute temperature. It is quite common to specify temperature in degrees Celsius or Centigrade even where all other units might be SI units. One degree change on the Celsius scale is identical to one degree change on the Kelvin scale. For this reason the value of thermal conductivity obtained in Example 1-1 would be no different if it had been expressed in the units of W/(m-K). In this text quantities involving temperature *change* may be expressed in either absolute or nonabsolute units.

The relationship between temperature on the Kelvin scale and temperature on the Celsius scale is

$$K = C + 273.15 \tag{1-1}$$

For example, 100 C is identical to 373.15 K.

In the English Engineering system the unit of temperature is the degree Fahrenheit. When the thermodynamic absolute temperature is needed, the temperature is specified in degrees Rankine (R). The relationship between the Fahrenheit scale and the Rankine scale is

$$R = F + 459.67 \qquad (1\text{-}2)$$

Since both the Rankine and Kelvin scales are absolute scales, absolute zero is identical on each scale. The relationship between the Rankine scale and the Kelvin scale is

$$K = \tfrac{5}{9} R \qquad (1\text{-}3)$$

For example 900 R is identical in temperature to 500 K. The relationship between the four temperature scales is given in Fig. 1-1.

Consistent units must always be employed in physical computations. For example, thermal and mechanical energy may be interchangeable in a given situation. Thermal energy traditionally has been specified in terms of British Thermal Units (Btu) in engineering work in English-speaking countries. Mechanical energy is frequently specified in the units of foot pounds force (ft-lbf). If the net energy (the sum of the thermal and mechanical energy) is to be computed, it is necessary to specify both types of energy in the same unit. The relationship between the Btu and the ft-lbf is

$$1\ Btu = 778.28\ ft\text{-}lbf \qquad (1\text{-}4)$$

Example 1-3.   A system is known to contain 100 Btu of thermal energy and 30,000 ft-lbf of mechanical energy. What is the total energy (mechanical plus thermal) contained by the system in Btu?

Solution.   Use Eq. 1-4 as a conversion factor.

$$100\ Btu + \frac{(30{,}000)\ ft\text{-}lbf}{(778.28)\ ft\text{-}lbf/Btu} = (100 + 38.5)\ Btu = 138.5\ Btu$$

In this example notice that an equation (Eq. 1-4) was changed to a conversion factor by simple algebra.

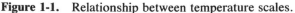

Figure 1-1.   Relationship between temperature scales.

$$1 \text{ Btu} = 778.28 \text{ ft-lbf}$$
$$1 = 778.28 \text{ ft-lbf/Btu}$$

Since 778.28 (ft-lbf)/Btu is equivalent to unity, it can be placed in the appropriate term in such a way as to cancel the undesired units, and yet not change the true value of the physical quantity represented by the term. In Example 1-3, 30,000 ft-lbf of energy is identical to 38.5 Btu of energy.

From Table 1-2 we see that the derived unit of energy in the SI system is the joule, which is equivalent to one newton-meter. The derived unit of power in the SI system is the watt, which is equivalent to one joule per second.

Example 1-4.   The specific heat of air at normal conditions is approximately equal to 0.241 Btu/(lbm-F). Express this value of the specific heat of air in SI units, joule per kilogram, and degree.

Solution.   From the conversion factor given in Appendix E the following relationship is true

$$(0.241) \frac{\text{Btu}}{\text{lbm-F}} (4.1868 \times 10^3) \frac{\text{J/(kg-C)}}{\text{Btu/(lbm-F)}} = 1010 \text{ J/(kg-C)}$$
$$= 1.01 \text{ kJ/(kg-C)}$$

1-4 FUNDAMENTAL CONCEPTS

Generally speaking, background preparation for a study of HVAC system design is covered in the typical thermodynamics, fluid mechanics, and heat transfer curriculum. The general concepts taught in dynamic systems are also important to understanding and analyzing any HVAC system.

Thermodynamics

The most important concept in this area is the First Law of Thermodynamics or energy balance. Associated with the energy balance are property data for the pure substance as well as mixtures of gases and vapor. The Second Law of Thermodynamics is also very important; however, its direct application is less frequent in actual system design.

Fluid Mechanics

The principles dealing with the flow of liquids and gases in pipes and ducts are most important here. The concepts are closely related and used in conjunction with thermodynamic concepts. Emphasis will be placed on complete fluid distributing systems instead of a single element that is a significant extension of basic fluid mechanic concepts. Most problems will be of a steady-flow nature even though changes in flow rate and fluid properties may occur from time to time.

Heat Transfer

Generally the most simple concepts dealing with conduction, convection, and radiation are used in typical system design. Again the concepts of thermodynamics and fluid mechanics are intertwined with most heat-transfer processes. In most cases, steady state can be assumed for design purposes. In those cases where transient effects are important, computer routines are usually used to obtain the required results. In this connection a basic understanding of a dynamic system is essential.

At appropriate places in the chapters that follow, the concepts discussed herein will be introduced and reviewed in a context useful in HVAC system design. In the next chapter, HVAC systems are described and their general functions discussed. This will give an indication of how all the basic concepts enter the design problem.

REFERENCES

1. "Environmental Control Principles," American Society of Heating, Refrigerating and Air-Conditioning Engineers. New York, 1985.

2. "ASHRAE SI for HVAC and R," 6th ed., American Society of Heating, Refrigerating and Air-Conditioning Engineers, New York, 1986.

3. W. F. Stoecker, "Using SI Units in Heating, Air-Conditioning, and Refrigeration," Business News Publishing Co., Birmingham, Mich., 1975.

4. "Metric Practice Guide E380-72," American Society for Testing and Materials, Philadelphia, 1972 (also ANSI Standard Z210.1).

5. "ASME Orientation and Guide for Use of Metric Units," 3rd ed., The American Society of Mechanical Engineers, New York, 1973.

6. "SI Units and Recommendations for the Use of Their Multiples and of Certain Other Units," International Standard ISO 1000, International Organization for Standardization, New York, 1973.

PROBLEMS

1-1 Write the following quantities in the suggested SI terminology.
 (a) 11,000 newton per square meter
 (b) 12,000 watts
 (c) 1600 joule per kilogram
 (d) 101,101 pascal
 (e) 12,000 kilowatt
 (f) 0.012 meter
 (g) 0.0000012 second

1-2 Convert the following quantities from English to SI units.
 (a) 98 Btu/(hr-ft-F)
 (b) 0.24 Btu/(lbm-F)

    (c) 0.04 lbm/(ft-hr)
    (d) 1050 Btu/lbm
    (e) 12000 Btu/hr
    (f) 14.7 lbf/in.^{2}

1-3 Convert the following quantities from SI to English units.
    (a) 120 kPa
    (b) 100 W/(m-C)
    (c) 0.8 W/(m^2-C)
    (d) 10^{-6} (N-s)/m^2
    (e) 1200 kW
    (f) 1000 kJ/kg

1-4 The kinetic energy of a flowing fluid is proportional to the velocity squared divided by two. Compute the kinetic energy per unit mass for the following velocities in the units indicated.
    (a) velocity of 100 ft/sec; English units
    (b) velocity of 600 m/s; SI units
    (c) velocity of 400 ft/sec; SI units
    (d) velocity of 500 m/s; English units

1-5 The potential energy of a fluid is proportional to the elevation of the fluid. Compute the potential energy for the elevations given below per unit mass.
    (a) elevation of 200 ft; English units
    (b) elevation of 70 m; SI units
    (c) elevation of 120 ft; SI units
    (d) elevation of 38 m; English units

1-6 A gas is contained in a vertical cylinder with a frictionless piston. Compute the pressure in the cylinder for the following cases.
    (a) piston mass of 20 lbm and area of 7 in.^{2}
    (b) piston mass of 10 kg and diameter of 100 mm

1-7 A pump develops a total head of 50 ft of water under a given operating condition. What pressure is the pump developing in SI units and terminology?

1-8 A fan is observed to operate with a pressure difference of 4 in. of water. What is the pressure difference in SI units and terminology?

1-9 Compute the Reynolds number (Re $= \rho \bar{V} D / \mu$) for 21 C water flowing in a standard 2 in. pipe at a velocity of one m/s using SI units. Tables B-3a and D-1 will be useful. What is the Reynolds number in English units?

1-10 Compute the thermal diffusivity ($\alpha = k/\rho c_p$) of the following substances in SI units.
    (a) saturated liquid water at 38 C
    (b) air at 47 C and 101.3 kPa
    (c) saturated liquid refrigerant 22 at 38 C
    (d) saturated liquid refrigerant 12 at 38 C

1-11 Compute the Prandtl number (Pr $= \mu c_p/k$) for the following substances in SI units.
    (a) saturated liquid water at 10 C
    (b) saturated liquid water at 60 C

(c) air at 20 C

(d) saturated liquid refrigerant 22 at 38 C

1-12 Compute the heat transferred from water as it flows through a heat exchanger at a steady rate of one m³/s. The decrease in temperature of the water is 5 C and the mean bulk temperature is 60 C. Use SI units.

1-13 Make the following volume and mass flow rate calculations in SI units. (a) Water flowing at an average velocity of 2 m/s in nominal 2½-in., type L copper tubing. (b) Standard air flowing at an average velocity of 4 m/s in a 30 cm diameter duct.

1-14 A room with dimensions of 3 m × 10 m × 20 m is estimated to have an infiltration rate of ¼ volume change per hour. Determine the infiltration rate in m³/s.

1-15 Air enters a heat exchanger at a rate of 5000 cubic feet per minute at a temperature of 50 F and pressure of 14.7 psia. The air is heated by hot water flowing in the same exchanger at a rate of 11,200 pounds per hour with a decrease in temperature of 10 F. At what temperature does the air leave the heat exchanger?

1-16 Water flowing at a rate of 1.5 kg/s through a heat exchanger heats air from 20 C to 30 C flowing at a rate of 2.4 m³/s. The water enters at a temperature of 90 C and the air is at 0.1 MPa. At what temperature does the water leave the exchanger?

1-17 Air at a mean temperature of 50 F flows over a thin-wall 1-in. O.D. tube, 10 feet in length, which has condensing water vapor flowing inside at a pressure of 14.7 psia. Compute the heat transfer rate if the average heat transfer coefficient between the air and tube surface is 10 Btu/(hr-ft²-F).

1-18 Repeat Problem 1-17 for air at 10 C, a tube with diameter 25 mm, a steam pressure of 101 kPa, and a tube length of 4 m, and find the heat transfer coefficient in SI units if the heat transfer rate is 1250 W.

CHAPTER TWO

Air-Conditioning Systems

All air-conditioning systems generally have the same basic elements; however, the physical appearance and arrangement of the various components may vary dramatically. Although the same elements are present, the manner in which the systems are controlled and operated to satisfy a given environmental requirement may be quite different. Therefore, various systems that meet the requirements of different building types and uses, load variations, and economic considerations are discussed in the following sections.

The original basic system of air conditioning was the forced warm air heating and ventilating system with centrally located equipment that distributed tempered air through ducts. It was found that cooling and dehumidification equipment added to these systems produced satisfactory air conditioning in spaces where heat gains were relatively uniform throughout the conditioned area. The use of this system was made complicated by the presence of variable heat sources within the space served. When this occurred, the area had to be divided into sections or zones and the central system supplemented by additional equipment and controls.

As the science of air conditioning progressed, variations in the basic designs were required to meet the functional and economic demands of individual buildings. Now air-conditioning systems are categorized according to the means by which the controllable heating and cooling is accomplished in the conditioned area. They are further classified to accomplish specific purposes by special equipment arrangements.

2-1 THE BASIC CENTRAL SYSTEM

The basic central system is an all-air, single-zone system that is used as part of most systems. It can be designed for low-, medium-, and high-pressure air distribution. Normally the equipment is located outside the conditioned area in a basement, on the roof, or in a service area at the core of a commercial building. It can be installed in any convenient area of a factory, particularly in the roof truss area or on the roof. The equipment can be located adjacent to the heating and refrigerating equipment or at a considerable distance from it by using a circulating refrigerant, chilled water, hot water, or steam for energy transfer sources. It is most important that the temperature within the area conditioned by a central system be uniform if a single-zone constant-air-volume duct system is to be used because air temperature is sensed at only one place in the space for control.

Spaces with Uniform Loads

In spaces with relatively large open areas and small external loads such as theaters, auditoriums, department stores, and the public spaces of most buildings, air-conditioning loads are fairly uniform throughout. Adjustments for minor variations can be made by supplying more or less air, by changing the supply air temperatures in the original design, and by balance of the system. In commercial buildings the interior areas generally meet these criteria when local heat sources such as computers are treated separately. These interior areas usually require year-round cooling, and any isolated spaces of limited occupancy may require special attention.

Spaces Requiring Precision Control

Spaces with stringent requirements for cleanliness, humidity, and temperature control, and/or air distribution are usually isolated within the larger building and require precision control. The components of a central system can be selected and assembled to meet the exact requirements of the area. Locating the equipment outside the conditioned space permits routine inspection and maintenance without interference.

Multiple Systems for Large Areas

In spaces such as large office buildings, factories, and large department stores, practical considerations require multiple installation of central systems. The size of the individual system is usually limited only by the physical limitations of the structure. In multiple systems the equipment is often located in the truss space, against the outside wall, or on the roof where it least interferes with the operations within the conditioned space and outdoor air is readily available.

Primary Source of Conditioned Air for Other Systems

There are various systems for controlling conditions in individual zones in which a constant supply of conditioned outdoor air is used for ventilation and for some of the air-conditioning load. This reduces the amount of conditioned air handled by the central system and, consequently, the space required for ductwork. By utilizing high velocities and designing for the resultant higher pressure and sound levels, the ductwork can be further reduced in size.

Systems for Environmental Control

For applications requiring close aseptic or contamination control of the environment, all-air type systems generally are used to provide the necessary air supply to sustain adequate dilution of the controlled space. These applications are usually combinations of supply systems and scavenging exhaust systems to circulate the diluting air through the controlled environment space.

2-2   CENTRAL SYSTEM COMPONENTS

When the air-handling requirements of the system have been determined from the use and physical characteristics of the space to be conditioned, and capacities for both air volumes and thermal exchange have been established, the designer can select and arrange the various components. It is important that equipment be adequate, accessible for easy maintenance, and no more complex in arrangement and control than is necessary to produce the conditions required to meet the design criteria.

Figure 2-1 shows a general arrangement of the component parts of an all-air, central system for year-round conditioning. Although Fig. 2-1 indicates a built-up system, most of the components are available in subassembled sections ready for bolting together in the field or completely assembled by the manufacturer. Figure 2-2 shows a single zone central air handler that contains the fan, heating/cooling coil, and the filter section. The fan is located downstream of the coil and is referred to as a draw-through configuration. A multizone air handler is shown in Fig. 2-3, which has a hot deck and a cold deck similar to

Figure 2-1.   A central air system.

Figure 2-2. Single-zone, draw-through air handler. (Reproduced by permission of the Trane Company, LaCrosse, Wisconsin.)

Figure 2-3.   Multi-zone, blow-through air handler with hot deck and cold deck. (Reproduced by permission of the Trane Company, LaCrosse, Wisconsin.)

that shown in Fig. 2-1. This is called a blow-through configuration. The discharge area of the air handler of Fig. 2-3 may be divided so that several zones may be served with separate temperature control in each zone. Further, this air handler may be used without the dampers in a dual-duct system.

Other components of the central system are the cooling, heating, and preheat coils, which are usually the finned-tube type shown in Fig. 2-4. Dampers are required to control the air temperature leaving the air handler (Fig. 2-3) and to facilitate mixing of return air and outdoor air (Fig. 2-4). The humidifier is usually a type utilizing steam in a commercial system (Fig. 2-5). The fans may be either centrifugal or vaneaxial types as shown in Fig. 2-6. A typical filter is shown in Fig. 2-7.

2-3   CENTRAL MECHANICAL EQUIPMENT

The selection of central mechanical equipment for air-conditioning systems in large buildings chiefly depends on economic factors once the total required capacity has been determined. The decision as to the choice of components depends on the type of fuel available, environmental protection required, structural support, and available space, among other considerations.

In recent years the cost of energy has led to many designs that recover the internal heat from lights people, and equipment to reduce the size of the heating plant.

Among the largest installations of central mechanical equipment are the central cooling and heating plants serving groups of large buildings. These

Figure 2-4. A multi-zone, blow-through air handler showing the coils, fan, filters, and mixing box. (Reproduced by permission of the Trane Company, LaCrosse, Wisconsin.)

Figure 2-5. A commercial-type steam humidifier. (Courtesy of Spirax Sarco, Inc.)

Figure 2-6a. A centrifugal fan. (Reproduced by permission of the Trane Company, LaCrosse, Wisconsin.)

Figure 2-6b. A vane-axial fan. (Reproduced by permission of Hartzell Fan, Inc., Piqua, Ohio.)

Figure 2-7. A unit-type air filter. (Courtesy of the Farr Company, Los Angeles, California.)

provide higher diversity, greater efficiency, less maintenance cost, and lower labor costs compared to individual plants.

Most large buildings, however, have their own central equipment plant and the choice of equipment depends mainly on owning and operating costs.

Heating Equipment

Steam and hot water boilers for heating are manufactured for high or low pressure, using coal, oil, electricity, gas, or waste material for fuel.

Low pressure boilers are rated for a working pressure of 15 psig or 103 kPa for steam, 160 psig or 1.1 MPa for water, with a maximum temperature limitation of 250 F or 121 C. High-rise buildings usually use high-pressure boilers. However, a low-pressure steam boiler may be used with a heat exchanger, called a converter, to heat water, which is then pumped to the various floors in a high-rise building. This eliminates the need for an expensive high-pressure boiler. Packaged boilers with all components and controls assembled as a unit are available. Figure 2-8 shows such a unit.

Heating of the space may be accomplished by heating and circulating air with the air handler as shown in Fig. 2-1 or a radiant system may be used where a single enclosed finned tube is installed along the outer wall near the floor. This will be discussed later.

Figure 2-8.   Packaged fire-tube hot water boiler. (Courtesy of Federal Corporation, Oklahoma City, Oklahoma.)

Refrigeration Equipment

Three major types of compressors are used in systems:

1. Reciprocating: $\frac{1}{16}$ to 150 hp or 50 W to 112 kW.
2. Helical rotary: 100 to 1000 tons or 350 to 3500 kW.
3. Centrifugal: 100 tons or 350 kW to an upper limit of capacity determined only by physical size.

Compressors come with many types of drives including electric, gas, and diesel engines, and gas and steam turbines. Many compressors are purchased as part of a condensing unit; they consist of compressor, drive, condenser, and all necessary safety and operating controls. Reciprocating condensing units are available with air-cooled condenser as part of the unit or arranged for remote installation. Remote air-cooled condensers are available up to 500 tons or 1.8 MW. A commercial size, air-cooled condensing unit is shown in Fig. 2-9. Compressors and refrigeration equipment are discussed fully in Chapter 15.

Chillers

As the name implies, chillers cool water or other fluid that is circulated to a remote location where it is used to cool air with a cooling coil in an air handler. Chillers operating on the vapor compression cycle take many forms ranging in size from about 3 tons to more than a thousand tons. Smaller units

Figure 2-9. A large air-cooled condensing unit. (Courtesy of Carrier Corporation, Syracuse, New York.)

usually use reciprocating compressors with air-cooled condensers, whereas large units usually reject heat to water from cooling towers. Figure 2-10 shows a large centrifugal chiller.

Absorption chillers are also used. The most common type is the indirect fired lithium bromide–water cycle unit. These are available in large units from 50 to 1500 tons or 176 kW to 5 MW capacity. Their generator sections are heated with low-pressure steam, hot water, or other hot liquids. In large installations the absorption chiller is frequently combined with centrifugal compressors driven by steam turbines. Steam from the noncondensing turbine is taken to the generator of the absorption machine. When the centrifugal unit is driven by a gas turbine or an engine, the absorption machine generator is heated with the exhaust-jacket water. Because of rising energy costs many absorption chillers that use primary energy are being replaced with more energy-efficient reciprocating or centrifugal equipment.

Cooling Towers

To remove the heat from the water-cooled condensers of air-conditioning systems the water is usually cooled by contact with the atmosphere. This is

Figure 2-10.   A large centrifugal chiller. (Reproduced by permission of the Trane Company, LaCrosse, Wisconsin.)

accomplished by natural draft of mechanical draft cooling towers or by spray ponds. Of these the mechanical draft tower can be designed for more exacting conditions because it is independent of the wind.

Air-conditioning systems use towers ranging from small package towers of 5 to 500 tons or 17.5 to 1760 kW or intermediate-size towers of 2000 to 4000 tons or 7 to 14 MW. A mechanical draft cooling tower is shown in Fig. 2-11.

Water treatment is a definite requirement for satisfactory operation. Units that operate during the entire year must be protected against freezing.

Pumps

The pumps used in air-conditioning systems are usually single inlet centrifugal pumps. The pumps for the larger system and for heavy duty have a horizontal split case with double-suction impeller for easier maintenance and high efficiency. End-suction pumps, either close coupled or flexible connected, are used for smaller tasks. Figure 2-12*a* shows a medium-sized direct-coupled centrifugal pump and Fig. 2-12*b* shows an inline centrifugal pump. The major applications for pumps in the equipment room are primary and secondary chilled water, hot water, condenser water, steam condensate return, boiler feed water, and fuel oil.

Figure 2-11.   A mechanical-draft cooling tower. (Courtesy of The Marley Company, Mission, Kansas.)

Figure 2-12a.   A single-inlet direct-coupled centrifugal pump. (Courtesy of Pacific Pump Company, Oakland, California.)

Fuels and Energy

Consideration of the type of fuel or energy must be made at the same time as the selection of the energy-consuming equipment to assure maximum efficiency. Chapter 15 of the *ASHRAE Handbook* (2) gives the types and properties of fuels and energy and guidance in their proper use.

The type of fuel or other energy to be used is an important consideration. Obviously, the selection should result in the least use of our natural resources. This will usually also result in least cost.

Piping

Air-conditioning piping systems can be divided into two parts, the piping in the main equipment room and the piping required for the air-handling systems

Figure 2-12b.   A small in-line centrifugal pump. (Courtesy of Taco, Incorporated, Cranston, Rhode Island.)

throughout the building. The air-handling system piping follows procedures developed in detail in Chapter 9. The major piping in the main equipment room consists of fuel lines, refrigerant piping, steam, and water connections.

Ductwork

The air ducts should deliver conditioned air to an area as quietly and economically as possible. Design of the duct system must be coordinated with the building design to avoid last minute changes. Chapter 11 explains this part of the system design.

Instrumentation

At the time of installation all equipment should be provided with adequate gages, thermometers, flow meters, and balancing devices so that system performance is properly established. In addition, capped thermometer wells, gage cocks, capped duct openings, and volume dampers should be provided at strategic points for system balancing. A central control center to monitor and control a large number of control points should be considered for any large and complex air-conditioning system.

Figure 2-13.   Elementary air-temperature control system.

Control

Three elements are necessary for a control system: sensor, controller, and controlled device. These need only a process to give them meaning. In this text the definition of the process is limited to heating, ventilating, and air-conditioning systems.

Figure 2-13 illustrates an elementary control system. Air is shown flowing through a heating coil in a duct. The sensor measures the temperature of the air downstream of the coil and passes the information to the controller. The controller compares the air temperature with a set point and sends a signal to open or close the hot-water valve (the controlled device) as required to maintain a correspondence between the air temperature and the set point. This is a

Figure   2-14.   Elementary   water-temperature control system.

"closed-loop" system, in which the change in temperature caused by a change in valve position will be sensed and additional adjustments will be made as necessary. The air temperature is known as the "controlled variable."

A variation on this control system is shown in Fig. 2-14. In this case it is desired to maintain the water temperature leaving the coil at a fixed temperature. When the load on the coil decreases, the temperature of the water leaving the coil tends to decrease. The sensor transmits this information to the controller, which directs the actuator to adjust the valve (close it slightly). The opposite action occurs with an increase in water temperature. The flow of air is often controlled by dampers as shown in Fig. 2-15 to maintain a given air temperature downstream. The similarity of these examples is obvious and in a complete system these controls are repeated many times. Other controls may start or stop equipment such as chillers, boilers, or fans. Large systems usually require control of the pressure in the duct system by varying the fan speed. These control systems are similar to the foregoing examples. The common thermostat is another example of a simple control system. Most control systems fall in the "closed-loop" classification but "open-loop" systems are sometimes used. In an open-loop system the sensor is not directly affected by the action of the controlled device.

The HVAC system operates most economically when equipment capacity is closely matched to load, and this can be better accomplished by an automatic control system than manually. A completely automatic system with changeover controls, interlocks, and internal monitoring and compensating controls minimizes human intervention and, therefore, the chance of human error.

To satisfy the need for various kinds of control response, there are available a number of types of control actions. These may be broadly classified as follows.

Two-position or On–Off Action   This is the simplest and most obvious control action. An example of on–off action is a thermostat starting and stopping a ventilating fan motor or furnace. Any two-position controller needs a differential to prevent rapid cycling. This differential is the difference between the setting at which the controller operates to one position and the setting at which it changes to the other. In a thermostat this is expressed in degrees of temperature.

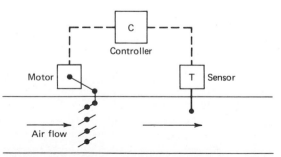

Figure 2-15.   Elementary airflow control system.

Proportional Action Proportional action simply utilizes feedback. Feedback means that the controlled device actuator moves only enough to satisfy the change in the controlled variable. It is not necessary to wait for the controlled variable to respond. The term "modulating control" usually refers to proportional control.

With proportional control we meet some new terms:

Throttling range is the amount of change in the controlled variable required to run the actuator of the controlled device from one end of its stroke to the other.

Set point is the controller setting and is the desired value of the controlled variable.

Control point is the actual value of the controlled variable. If the control point lies within the throttling range of the controller, it is said to be in control. When it exceeds the throttling range, it is said to be out of control.

Offset or control point shift is the difference between the set point and the control point. This is sometimes called drift, droop, or deviation. The amount of offset theoretically possible is determined by the throttling range.

Other aspects of control and some examples will be introduced at appropriate times.

2-4 ALL-AIR SYSTEMS

An all-air system provides complete sensible heating and sensible and latent cooling by supplying only air to the conditioned space. In such systems there may be piping connecting the refrigerating- and heat-producing devices to the air-handling device. No additional cooling is required at the zone. The term *zone* implies a provision or the need for separate thermostatic control, whereas the term *room* implies a partitioned area that may or may not require separate control.

All-air systems may be briefly classified in two basic categories: (1) single-path systems and (2) dual-path systems. Single-path systems contain the main heating and cooling coils in a series air flow path using a common duct distribution system at a common air temperature to feed all terminal apparatus. Dual-path systems contain the main heating and cooling coils in a parallel flow or series-parallel air flow path using either (1) a separate cold and warm air duct distribution system that is blended at the terminal apparatus (dual-duct system) or (2) a single supply duct to each zone with a blending of warm and cold air at the main supply fan.

The all-air system may be adapted to all types of air-conditioning systems for comfort or process work. It is applied in buildings requiring individual control of conditions and having a multiplicity of zones such as office buildings, schools and universities, laboratories, hospitals, stores, hotels, and ships. Air systems are also used for many special applications where a need exists for close control of temperature and humidity, including clean rooms, computer rooms, hospital operating rooms, textile and tobacco factories.

Heating Considerations

Heating may be accomplished by the same duct system used for cooling, by a separate perimeter air system, or by a separate perimeter radiation system using hot water or steam. The radiation system is also a natural convection unit, as shown in Fig. 2-16. It is the function of the perimeter heating system to offset the heat losses through the exterior envelope of the building. A perimeter heating system is usually used in conjunction with the all-air systems. However, its greatest application is with variable-air-volume systems used for cooling only. During the times when heat is required, the cooling system is used for ventilation only.

Single-zone System

The simplest all-air system is a supply unit serving a single zone. The unit can be installed either within a zone or remote from the space it serves and may operate with or without ductwork.

Figure 2-16.   A perimeter radiation and convection heating system. (Courtesy Sterling Radiator, Division of Reed National Corporation.)

Figure 2-17. A single-zone, constant-volume, all-air system.

A single-zone system responds to only one set of space conditions. Thus, it is limited in application to where variations in load are uniform throughout the zone. Figure 2-17 shows a schematic of a single-zone constant-volume all-air system.

Reheat Systems

The reheat system is a modification of the single-zone constant-volume system. Its purpose is to permit zone or space control for areas of unequal loading, or to provide heating or cooling of perimeter areas with different exposures, or for process or comfort applications where close control of space conditions is desired. As the word *reheat* implies, the application of heat is a secondary process, being applied to either preconditioned primary air or re-circulated room air. A single low-pressure reheat system is produced when a heating coil is inserted in the duct system. The more sophisticated systems utilize higher pressure duct designs and pressure-reduction devices to permit system balancing at the reheat zone. The medium for heating may be hot water, steam, or electricity.

Conditioned air is supplied from a central unit at a fixed cold air temperature designed to offset the maximum cooling load in the space. The control thermostat activates the reheat unit when the temperature falls below the upper limit of the controlling instrument's setting. A schematic arrangement of the components for a typical reheat system is shown in Fig. 2-18. To conserve energy reheat should not be used unless absolutely necessary. At the very least, reset control should be provided to maintain the cold air at the highest possible temperature to satisfy the space cooling requirement.

Variable-volume System

The variable-volume system compensates for varying load by regulating the volume of air supplied through a single duct. Special zoning is not required

Figure 2-18.   Arrangement of components for a reheat system.

because each space supplied by a controlled outlet is a separate zone. Figure 2-19 is a schematic of a true variable-air-volume (VAV) system.

Significant advantages of the variable-volume system are low initial cost and low operating costs. The first cost of the system is far lower in comparison with other systems that provide individual space control because it requires only single runs of duct and a simple control at the air terminal. Where diversity of loading occurs, smaller equipment can be used and operating costs are generally the lowest among all the air systems. Because the volume of air is reduced with a reduction in load, the refrigeration and fan horsepower follow closely the actual air-conditioning load of the building. During intermediate and cold seasons outdoor air can be used for economy in cooling. In addition the system is virtually self-balancing.

Until recently there were two reasons why variable-volume systems were not recommended for applications with loads varying more than 20 percent. First, throttling of conventional outlets down to 50 or 60 percent of their maximum design volume flow might result in the loss of control of room air motion with noticeable drafts resulting. Second, the use of mechanical throttling dampers produces noise, which increases proportionally with the amount of throttling.

Figure 2-19.   Variable air-volume system.

Improvements in volume-throttling devices and aerodynamically designed outlets have helped overcome these problems and extended the potential application of variable-volume systems. This system can now handle interior areas as well as building perimeter areas where load variations are greatest, and where throttling to 10 percent of design volume flow is often necessary.

Although some heating may be done with a variable volume system, it is primarily a cooling system and should be applied only where cooling is required the major part of the year. Buildings with internal spaces with large internal loads are the best candidates. A secondary heating system should be provided for boundary surfaces during the heating season. Baseboard perimeter heat is often used. During the heating season, the VAV system simply provides tempered ventilation air to the exterior spaces.

Reheat may be used in conjunction with the VAV system. In this case the air is throttled to some predetermined ratio and then reheat takes over to temper the air.

An important aspect of VAV system design is fan control. There are significant fan power savings where fan speed is reduced in relation to the volume of air being circulated. This topic is discussed in detail in Chapter 11.

Single-duct variable-volume systems should be considered in applications where full advantage can be taken of their low cost of installation and operation. Applications exist for office buildings, hotels, hospitals, apartments, and schools. Further details of the use of VAV systems may be obtained from reference 1.

Dual- or Double-duct System

In the dual-duct system the central station equipment supplies warm air through one duct run and cold air through the other. The temperature in an individual space is controlled by mixing the warm and cool air in proper proportions. Variations of the dual-duct system are possible, with one form shown in Fig. 2-20.

For best performance some form of constant volume regulation should be incorporated into the system to maintain a constant flow of air. Without this the system is difficult to control because of the wide variations in system static pressure that occur from the normal demand from loading changes.

Figure 2-20. Dual-duct system.

Many double-duct systems are installed in office buildings, hotels, hospitals, schools, and large laboratories. A common characteristic of these multiroom buildings is their highly variable sensible heat load. This system provides great flexibility in satisfying multiple loads and in providing prompt and opposite temperature response as required.

Space or zone thermostats may be set once to control year-round temperature conditions. All outdoor air can be used when the outdoor temperature is low enough to handle the cooling load. A dual-duct system should be provided with control that will automatically reset the cold air supply to the highest temperature acceptable and the hot air supply to the lowest temperature acceptable.

Variable air volume may be incorporated into the dual-duct system, with various arrangements possible. Two supply fans are usually used in this case, one for the hot deck and one for the cold deck, with each controlled by the static pressure downstream in each duct.

From the energy conservation viewpoint the dual-duct system has the same disadvantage as reheat. Although many of these systems are in operation, few are now being designed and installed.

Multizone System

The multizone central station units provide a single supply duct for each zone and obtain zone control by mixing hot and cold air at the central unit in response to room or zone thermostats. For a comparable number of zones this system provides greater flexibility than the single duct and involves lower cost than the dual-duct system, but it is physically limited by the number of zones that may be provided at each central unit.

Typical multizone equipment is similar in some respects to the dual-duct system, but the two air streams are proportioned within the equipment instead of being mixed at each space served, and the proper temperature air is provided as it leaves the equipment. Figure 2-21 shows a sketch of a multizone system.

Figure 2-21. Multizone system.

The system conditions groups of rooms or zones by means of a blow-through central apparatus having heating and cooling coils in parallel downstream from the fan.

The multizone, blow-through system is applicable to locations and areas having high sensible heat loads and limited ventilation requirements. The use of many duct runs and control systems can make initial costs of this system high compared to other all-air systems. Also to obtain very fine control this system might require larger refrigeration and air-handling equipment, which should be considered in estimating both initial and operating costs.

The use of these systems with simultaneous heating and cooling is now discouraged for energy conservation. However, through the use of outdoor air and either heating or cooling, satisfactory control may be attained in many applications.

2-5 AIR-AND-WATER SYSTEMS

In the all-air systems discussed in the previous section the spaces within the building are cooled solely by air supplied to them from the central air-conditioning equipment. In contrast, in an air-and-water system both air and water are distributed to each space to perform the cooling function. In virtually all air–water systems both cooling and heating functions are carried out by changing the air or water temperatures (or both) to permit control of space temperature during all seasons of the year.

There are several basic reasons for the use of this type of system. Because of the greater specific heat and much greater density of water compared to air, the cross-sectional area required for the distribution pipes is markedly less than that required for ductwork to accomplish the same cooling task. Consequently, the quantity of air supplied can be low compared to an all-air system, and less building space need be allocated for the cooling distribution system.

The reduced quantity of air is usually combined with a high-velocity method of air distribution to minimize the space required. If the system is designed so that the air supply is equal to the air needed to meet outside air requirements or that required to balance exhaust (including exfiltration) or both, the return air system can be eliminated for the areas conditioned in this manner.

The pumping horsepower necessary to circulate the water throughout the building is usually significantly less than the fan horsepower to deliver and return the air. Thus not only space but also operating cost savings can be realized.

Systems of this type have been commonly applied to office buildings, hospitals, hotels, schools, better appartment houses, research laboratories, and other buildings. Space saving has made these systems beneficial in high-rise structures.

The air side of air-and-water systems is comprised of central air-conditioning equipment, a duct distribution system, and a room terminal. The air is supplied at constant volume and is often referred to as primary air to distinguish it from room air that is recirculated over the room coil.

The water side in its basic form consists of a pump and piping to convey water to the heat transfer surface within each conditioned space. The heat exchange surface may be a coil that is an integral part of the air terminal (as with induction units), a completely separate component within the conditioned space (radiant panel), or either (as is true of fan–coil units).

Individual room temperature control is obtained by varying the capacity of the coil (or coils) within the room by regulation of either the water flow through it or the air flow over it. The coil may be converted to heating service during the winter, or a second coil or a heating device within the space may provide heating capacity depending on the type of system.

Air-and-water systems are categorized as two-pipe, three-pipe, and four-pipe systems. They are basically similar in function and all incorporate both cooling and heating capabilities for all-season air conditioning. However, arrangements of the secondary water circuits and control systems differ greatly.

Air–Water Induction System

The basic arrangement for air–water induction units is shown in Fig. 2-22. Centrally conditioned primary air is supplied to the unit plenum at high

Figure 2-22.   Air–water induction unit.

pressure. The plenum is acoustically treated to attenuate part of the noise generated in the duct system and in the unit. A balancing damper is used to adjust the primary air quantity within limits.

The high-pressure air flows through the induction nozzles and induces secondary air from the room and over the secondary coil. This secondary air is either heated or cooled at the coil depending on the season, the room requirement, or both. Ordinarily no latent cooling is accomplished at the room coil, but a drain pan is provided to collect condensed moisture resulting from unusual latent loads of short duration. The primary and secondary air are mixed and discharged to the room.

Induction units are usually installed at a perimeter wall under a window, but units designed for overhead installation are available. During the heating season the floor-mounted induction unit can function as a convector during off hours with hot water to the coil and without a primary air supply.

Fan–Coil Conditioner System

The fan–coil conditioner unit is a versatile room terminal that is applied to both air–water and water-only systems.

The basic elements of fan–coil units are a finned-tube coil and a fan section, Fig. 2-23. The fan section recirculates air continuously from within the perimeter space through the coil, which is supplied with either hot or chilled water. In addition, the unit may contain an auxiliary heating coil, which is usually of the electric resistance type but which can be of the steam or hot water type. Thus the recirculated room air is either heated or cooled. Primary air made up of outdoor air sufficient to maintain air quality is supplied by a separate central system usually discharged at ceiling level. The primary air is normally tempered

1. Finned tube coil
2. Fan scrolls
3. Filter
4. Fan motor
5. Auxiliary condensate pan
6. Coil connections
7. Return air opening
8. Discharge air opening
9. Water control valve

Figure 2-23.   Typical fan–coil unit.

to room temperature during the heating season, but is cooled and dehumidified in the cooling season. The primary air may be shut down during unoccupied periods to conserve energy.

2-6   ALL-WATER SYSTEMS

All-water systems are those with fan–coil, unit ventilator, or valance-type room terminals, with unconditionined ventilation air supplied by an opening through the wall or by infiltration. Cooling and humidification are provided by circulating chilled water or brine through a finned coil in the unit. Heating is provided by supplying hot water through the same or a separate coil using two-, three-, or four-pipe water distribution from central equipment. Electric heating or a separate steam coil may also be used. Humidification is not practical in all-water systems unless a separate package humidifier is provided in each room.

The greatest advantage of the all-water system is its flexibility for adaptation to many building module requirements.

A fan–coil system applied without provision for positive ventilation or one taking ventilation air through an aperture is one of the lowest first-cost central station type perimeter systems in use today. It requires no ventilation air ducts, is comparatively easy to install in existing structures, and as with any central station perimeter system utilizing water in pipes instead of air ducts, its use results in considerable space savings throughout the building.

All-water systems have individual room control with quick response to thermostat settings and freedom from recirculation of air from other conditioned space. These systems have remote chilling and heating equipment. When fan–coil units are used with the three- or four-pipe water arrangements, each is its own zone with a choice of heating or cooling at all times and no seasonal changeover is required. All-water systems can be installed in existing buildings with a minimum of interference in the use of occupied space.

There is no positive ventilation unless openings to the outside are used, and then ventilation can be affected by wind pressures and stack action on the building. Special precautions are required at each unit with an outside air opening to prevent freezing of coil and potential water damage from rain. Because of these problems and energy conservation considerations, it is becoming standard practice to eliminate the outdoor air feature of these systems and rely on other means to provide outdoor air. This type of system is not recommended for applications requiring high indoor air quality.

Seasonal changeover is required in most climates with a two-pipe system, and zoning and piping are required to reduce operating difficulties during intermediate seasons when a sun-exposed zone may need cooling while other zones need heat. When heating, units cannot be used as convectors in unoccupied rooms and fans must be kept running.

If a two-pipe system has only one pump, the same quantity of hot and cold water is circulated even though the requirements for each may be different.

With three- and four-pipe systems hot and cold water may be furnished throughout the year.

Maintenance and service work has to be done in the occupied areas; as the units become older, the fan noise can become objectionable. Each unit requires a condensate drain line. It is difficult to limit bacterial growth in the unit. In extremely cold weather it is often necessary to close the outside air dampers to prevent freezing of coils, reducing ventilation air to that obtained by infiltration. Filters are small and inefficient and require frequent changing to maintain air volume.

Figure 2-24 illustrates a typical unit ventilator, used in all water systems, with two separate coils, one used for heating and the other for cooling with a four-pipe system. In some cases the unit ventilator may have only one coil, such as the fan coil of Fig. 2-23.

The heating coil may use hot water, steam, or electricity. The cooling coil can be either a chilled water coil or a direct expansion refrigerant coil. Heating and cooling coils are sometimes combined in a single coil by providing separate

Figure 2-24. Typical air-conditioning unit ventilator with separate coils.

tube circuits for each function. In such cases the effect is the same as having two separate coils.

Unit ventilator capacity control is essentially the same as described for fan coils in the previous section. In most cases, however, the control system will be slightly more sophisticated.

2-7 UNITARY AIR CONDITIONERS

Unitary air-conditioning equipment consists of factory-matched components for inclusion in air-conditioning systems that are field designed to meet the needs of the user.

The following list of variations is indicative of the vast number of types of unitary air conditioners available.

1. Arrangement: single or split (evaporator connected in the field).
2. Heat rejection: air cooled, evaporative condenser, water cooled.
3. Unit exterior: decorative for in-space application, functional for equipment room and ducts, weatherproofed for outdoors.
4. Placement: floor standing, wall mounted, ceiling suspended, roof mounted.
5. Indoor air: vertical upflow, counterflow, horizontal, 90 and 180 degree turns, with fan, or for use with forced air furnace.
6. Locations: *indoor*—exposed with plenums or furred in ductwork, concealed in closets, attics, crawl spaces, basements, garages, utility rooms, or equipment rooms; *wall*—built in, window, transom; *outdoor*—rooftop, wall mounted, or on ground.
7. Heat: intended for use with upflow, horizontal, or counterflow forced-air furnace, combined with furnace, combined with electrical heat, combined with hot water or steam coil.

Unitary air conditioners are designed with fan capability for duct work.

Heat pumps (Sec. 2-8) are also offered in many of the same types and capacities as unitary air conditioners.

Packaged reciprocating and centrifugal water chillers can be considered as unitary air conditioners, particularly when applied with unitary-type chilled water blower coil units.

Figure 2-25 depicts a small single package air conditioner while Fig. 2-26 shows a large commercial packaged system.

The nearly infinite combination of coil configurations, evaporator temperatures, air-handling arrangements, refrigerating capacities, and other variations that are available in central systems are not possible with unitary systems. Consequently, in many respects a higher level of design ingenuity and performance is required to develop superior system performance using unitary equipment than for central systems.

The need for zoning is widely recognized because it is desirable to give the occupant some control over his or her individual environment. Unitary

Figure 2-25. Rooftop installation of air-cooled single-package or single-package year-round unit.

equipment tends to be a single zone system; however, for large single spaces the use of multiple units can achieve the same objectives as a large multiple-zone system.

Room Air Conditioners

A room air conditioner is an encased assembly designed as a unit primarily for mounting in a window, through a wall, or as a console. The basic function of a room air conditioner is to provide comfort by cooling, dehumidifying, filtering or cleaning, and circulating the air in one room. It may also provide ventilation by introducing outdoor air into the room, and by exhausting the room air to the outside. The conditioner may be designed to provide heating by reverse cycle (heat pump) operation or by electric resistance elements. Room air conditioners do not generally provide the best all-around environment and are considered to be appliances. Figure 2-27 shows a schematic view of a typical room air conditioner.

Through-the-wall Conditioner Systems

A through-the-wall system uses an air-cooled room air conditioner designed for mounting through the wall and in many cases is capable of providing both the heating and cooling function. Design and manufactor parameters vary widely. Specification grades range from appliance grade through heavy-duty commercial grade. The latter is called a *packaged terminal air conditioner*.

The through-the-wall concept incorporates a complete air-cooled refrigeration and air-handling system in an individual package; it utilizes space normally occupied by the building wall for equipment with the remainder projecting inside the room.

Each packaged terminal air conditioner has a self-contained, direct-expansion cooling system, heating coil (electric, hot water, or steam), and pack-

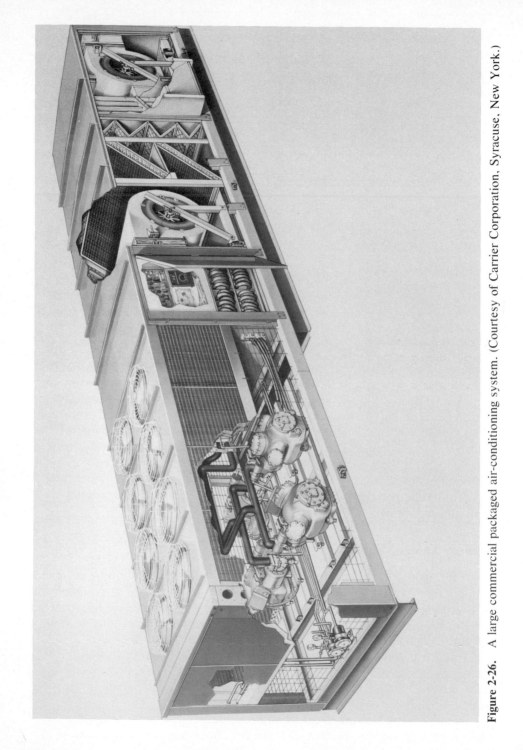

Figure 2-26. A large commercial packaged air-conditioning system. (Courtesy of Carrier Corporation, Syracuse, New York.)

Figure 2-27. Schematic view of room air conditioner.

aged controls. Figure 2-28 shows a typical unit; it consists of a combination wall sleeve and room cabinet, combination heating and cooling chassis, and outdoor louver.

Through-the-wall systems are best applied in multizone applications. The initial cost of the through-the-wall system of air conditioning in multiroom applications is considerably less than central systems adapated to provide simultaneous functions of either heating or cooling in each room under control of the room occupants. With a through-the-wall system of heating and air

Figure 2-28. Packaged terminal air conditioner with combination heating and cooling chassis.

conditioning, considerable space is saved by eliminating both ductwork and equipment rooms.

The through-the-wall system is limited to multizone systems and generally cannot be used economically in large spaces. One of the most popular applications has been in hotels and motels. This system is also applicable in renovation of existing buildings because there is less disruption and construction-forced sacrifice of rentable space than with alternative systems.

2-8 HEAT PUMP SYSTEMS

The term *heat pump* as applied to HVAC systems is a system in which refrigeration equipment is used such that heat is taken from a heat source and given up to the conditioned space when heating service is wanted and is removed from the space and discharged to a heat sink when cooling and dehumidification are desired. The thermal cycle is identical with that of ordinary refrigeration, but the application is equally concerned with the cooling effect produced at the evaporator and the heating effect produced at the condenser. In some applications both the heating and cooling effects obtained in the cycle are utilized.

Unitary heat pumps (as opposed to applied heat pumps) are shipped from the factory as a complete preassembled unit including internal wiring, controls, and piping. Only the ductwork, external power wiring, and condensate piping are required to complete the installation. For the split unit it is also necessary to connect the refrigerant piping between the indoor and outdoor sections. In appearance and dimensions, casings of unitary heat pumps closely resemble those of conventional air-conditioning units having equal capacity.

Capacities of unitary heat pumps range from about $1\frac{1}{2}$ to 25 tons or 5 to 90 kW, although there is no specific limitation. This equipment is almost universally used in residential and the smaller commercial and industrial installations. The multiunit type of installation with a number of individual units of 2 to 20 tons or 7 to 70 kW of cooling capacity is particularly advantageous to obtain zoning and to provide simultaneous heating and cooling. It may also be used for heat reclaiming to conserve energy by connecting the units to a common water circuit.

Large central heat pumps of modern design within the capacity range of about 30 to 1000 horsepower or 20 to 750 kW of compressor-motor rating are now operating in a substantial number of buildings. A single or central system is generally used throughout the building but in some instances the total capacity is divided among several separate heat pump systems to facilitate zoning.

Heat Pump Types

The air-to-air heat pump is the most common type. It is particularly suitable for factory-built unitary heat pumps and has been widely used for residential and commercial applications. Outdoor air offers a universal heat-source, heat-sink medium for the heat pump. Extended-surface, forced-convection heat transfer coils are normally employed to transfer the heat between the air and the refrigerant.

The capacity of an air-to-air heat pump is highly dependent on the outdoor temperature. Therefore, it is usually necessary to provide supplemental heat. The most common type of supplemental heat for heat pumps in the United States is electrical-resistance heat. This is usually installed in the air-handler unit and is designed to turn on automatically, sometimes in stages, as the indoor temperature drops. Heat pumps that have fossil fuel-fired supplemental heat are referred to as hybrid or bivalent heat pumps. These are more common in Europe than in the United States.

Air-to-water heat pumps are sometimes used in large buildings where zone control is necessary and are also sometimes employed for the production of hot or cold water in industrial applications as well as heat reclaiming. Heat pumps for water heating are commercially available in residential sizes.

A water-to-water heat pump uses water as the heat source and sink for both cooling and heating operation. Heating–cooling changeover may be accomplished in the refrigerant circuit, but in many cases it is more convenient to perform the switching in the water circuits.

A water-to-air heat pump uses water as a heat source and sink and uses air to transmit heat to or from the conditioned space. Water is a satisfactory and in many cases an ideal heat source. Well water is particularly attractive because of its relatively high and nearly constant temperature, generally about 50 F or 10 C in northern areas and 60 F or 16 C and higher in the south. However, abundant sources of suitable water are not always available and the application of this type of system is limited. Frequently sufficient water may be available from wells, but the condition of the water may cause corrosion in heat exchangers or it may induce scale formation. Other considerations to be made are the costs of drilling, piping, and pumping and the means for disposing of used water.

Surface or stream water may be utilized, but under reduced winter temperatures the cooling spread between inlet and outlet must be limited to prevent freeze-up in the water chiller, which is absorbing the heat.

Under certain industrial circumstances waste process water such as spent warm water in laundries and warm condenser water may be a source for specialized heat pump operations.

Water-refrigerant heat exchangers generally take the form of direct expansion water coolers either of the shell-and-coil type or of the shell-and-tube type. They are circuited so that they can be used as a refrigerant condenser during the heating cycle and as a refrigerant evaporator during the cooling cycle.

Closed-loop Systems

In many cases a building may require cooling in interior zones while needing heat in exterior zones. The needs of the north zones of a building may also be different from those of the south. In some cases a closed-loop heat pump

Figure 2-29.   Schematic of a closed-loop heat pump system.

system is a good choice. Closed-loop systems may be solar assisted. A closed-loop system is shown in Fig. 2-29.

Individual water-to-air heat pumps in each room or zone accept energy from or reject energy to a common water loop, depending on whether that area has a call for heating or for cooling. In the ideal case the loads will balance and there will be no surplus or deficiency of energy in the loop. If cooling demand is such that more energy is rejected to the loop than is required for heating, the surplus is rejected to the atmosphere by a cooling tower. In the other case an auxiliary boiler furnishes any deficiency.

The ground has been used successfully as a source-sink for heat pumps, using either a vertical or a horizontal run of plastic pipe as the ground-coupling device. Water from the heat pump exchanges heat with the surrounding earth before being returned back to the heat pump. Tests and analysis have shown rapid recovery in earth temperature around the well after the heat pump cycles off. Proper sizing of the well depends on the nature of the soil surrounding the well, the water table level, and the efficiency of the heat pump. Reference 4 deals with the design of ground-coupled systems.

2-9 HEAT RECOVERY SYSTEMS

In large commercial applications considerable heat energy is generated internally and may require removal even during the coldest weather. This condition usually occurs within the central spaces, which do not have exterior walls. It is necessary to exhaust considerable quantities of air from large commercial structures because of the introduction of outdoor ventilation air. Considerable savings in energy can be realized if the heat energy from the interior spaces and the exhaust air can be recovered and used in heating the exterior parts of the structure. Heat energy may also be recovered from waste water.

Figure 2-30.   Air-to-water type of heat-recovery system.

Redistribution of heat energy within a structure can be accomplished through the use of heat pumps of the air-to-air or water-to-water type. Another approach is the use of the dual-path systems described earlier.

Recovery of heat energy from exhaust air is accomplished through the use of air-to-air heat exchangers, rotating (periodic type) heat exchangers, and air-to-water heat exchangers connected by a circulating water loop. Sometimes spray systems are used; they may contain desiccants to enhance latent heat transfer. Figures 2-30 and 2-31 illustrate the air-to-water and rotating heat recovery systems while Fig. 2-32 shows how an air-to-air system might be arranged. The air-to-air and rotating type systems are the most effective in recovering energy but require that the intake and exhaust to the building be at the same location, whereas the air-to-water system may have the exhaust and intake at widely separated locations; however, the air-to-water system is not

Figure 2-31.   Rotating-type heat exchanger used for heat
recovery.

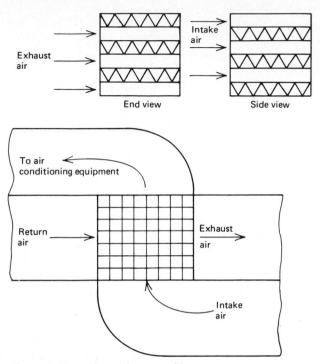

Figure 2-32.   Air-to-air type of heat-recovery system.

as effective because of the added thermal resistance of the water. To prevent freezing, glycol must also be introduced, which further reduces the effectiveness of the air-to-water system.

All of the previously described systems may also be effective during the cooling season when they function to cool and dehumidify outdoor ventilation air.

2-10 ECONOMIZER SYSTEMS

It is often possible to cool a space either totally or partially using outdoor air. This will be true during the spring and fall of the year and at night in northern regions and at high altitudes. Outdoor air can also be used to cool interior spaces, which may otherwise require operation of the cooling equipment. Because this type of cooling results in economies of energy and money, the general term *economizer* has been adopted to describe these systems.

The design of economizers are quite varied but generally have provisions for introducing outdoor air and exhausting indoor air (Fig. 2-33). There is a great variety of control arrangements, and the economizer may be combined with the heat recovery system. It is important to sense both outdoor temperature and humidity so that hot humid air is not introduced to the space. This would defeat the purpose of the economizer. Adequate precautions must be taken to prevent contamination of the space by dirty or odorous outdoor air.

Figure 2-33.   Arrangement of components in an economizer system.

2-11 SUMMARY

It is apparent from the preceding sections that the entire subject of HVAC systems is complex and that the types and combinations of equipment are numerous. The design engineer must constantly keep acquainted with the literature from the various manufacturers to stay abreast with new innovations. It is the business of the sales or application engineer to be particularly well versed on available equipment and to give the best advice possible to the design engineer.

REFERENCES

1. *ASHRAE Handbook, Systems Volume,* American Society of Heating, Refrigerating and Air-Conditioning Engineers, New York, 1984.

2. *ASHRAE Handbook, Fundamentals Volume,* American Society of Heating, Refrigerating and Air-Conditioning Engineers, New York, 1985.

3. "ASHRAE Standard 90A-80, Energy Conservation in New Building Design," American Society of Heating, Refrigerating and Air-Conditioning Engineers, New York, 1980.

4. "Design/Data Manual for Closed-Loop Ground Coupled Heat Pump Systems," American Society of Heating, Refrigerating and Air-Conditioning Engineers, Inc., Atlanta, Ga., 1985.

PROBLEMS

2-1 Consider the small single-story office building shown in Fig. 2-34. Lay out an all-air central system using an air handler with two zones. There is space between the ceiling and roof for ducts. The air handler is equipped with a direct expansion cooling coil and a hot water heating coil. Show all associated equipment schematically.

Figure 2-34.   Floor plan of small office building.

2-2 Suppose the building of Problem 2-1 is to use a combination air–water system where fan coil units in each room are used for heating. Schematically lay out this part of the system with related equipment. Discuss the general method of control for (a) the supplied air and (b) the fan–coil units.

2-3 Lay out a year-round all-water system for the building of Problem 2-1. Show all equipment schematically. Discuss the control and operation of the system in the summer, winter, and between seasons.

2-4 Apply single-package year-round rooftop type unit(s) to the single story building in Figure 2-34.

2-5 Suppose a variable air volume (VAV) all-air system is to be used to condition the space shown in Figure 2-35. Assume the space is the ground floor of a multistory office building. Describe the system using a schematic diagram. The lighting and occupant load are variable. Discuss the general operation of the system during the (a) colder months and (b) the warmer months.

2-6 Devise a central equipment arrangement for the system of Problem 2-5 that would save energy during the winter months. Sketch the system schematically.

2-7 Suppose an air-to-water heat pump is used to condition each space of Figure 2-35 where the water side of each heat pump is connected to a common water circuit. Sketch this system schematically showing all necessary additional equip-

Figure 2-35.   Schematic floor plan of one floor of a large building.

ment. Discuss the operation of this system during the (a) colder months, (b) warmer months, and (c) intermediate months.

2-8   A building such as that shown in Figure 2-35 requires some outdoor air. Explain and show schematically how this may be done with the system of Problem 2-5. Incorporate some sort of heat recovery device in the system. What controls would be necessary?

2-9   How can an economizer be used to advantage during the (a) winter months, (b) the summer months, and (c) the intermediate seasons?

2-10   The system proposed in Problem 2-7 requires the distribution of outdoor ventilation air to each space. Sketch a central air-handler system for this purpose that has energy recovery equipment and an economizer. Do not sketch the air distribution system. Discuss the control of this system assuming that the air will always be delivered at 72 F.

2-11   Make a single-line, block diagram of an all-water cooling system. The system has unit ventilators in each room with a packaged water chiller, and pumps. Explain how the system will be controlled.

2-12   Sketch a diagram of an air–water system that uses fan coils around the perimeter and an overhead air distribution system from a central air handler. Show a hot water boiler, chiller, and water distribution pumps. Explain the operation of the system in the summer and in the winter. What kind of controls does the system need?

2-13   Make a sketch of a variable-volume system with a secondary perimeter heating system for a perimeter zone. Discuss the operation and control of the system for the different seasons of the year.

2-14   Diagram a combination air-to-air heat recovery and economizer system. Describe the operation and control of the system for various times of the year.

2-15   A large manufacturing facility requires hot and chilled water and electricity in its operation. Describe how internal combustion engines operating on natural gas could provide part or all of these needs using heat recovery and generating electricity. The objective would be to save energy.

2-16   Thermal storage is often used to smooth the demand for cooling in large buildings. Imagine that the chiller can also make ice during the nighttime hours for use later when the peak cooling demand is high. Make a sketch of such a central plant and describe its operation for a typical daily cycle. How would this system benefit the building owner?

2-17   Make a sketch of a single-zone system for a small building that uses a ground-coupled heat pump. Show all the major parts of the system including the ground heat exchanger. Discuss operation of the system in summer and winter.

2-18   Sketch a variable air volume reheat system that has four zones. Discuss the operation of a typical zone.

2-19   Sketch a dual-duct VAV system. Show the fans and a typical zone. Describe a way to control the speed of the fans as the terminal devices reduce the air flow to the various zones.

CHAPTER THREE

Moist Air Properties and Conditioning Processes

A thorough understanding of the properties of moist air and the ability to analyze the various processes involving air is basic to the HVAC engineer. Atmospheric air makes up the environment in almost every design situation.

In 1911 Willis H. Carrier made a significant contribution to the air-conditioning field when he published relations for moist air properties together with a psychrometric chart. These formulas became fundamental to the industry.

In about 1945 Goff and Gratch (1) published thermodynamic properties for moist air that were for many years the most accurate available. New formulations have recently been developed at the National Bureau of Standards, based on the thermodynamic temperature scale (2, 3). The properties based on these formulations are the basis for the thermodynamic properties of moist air given in the 1985 *ASHRAE Handbook, Fundamentals Volume* (5) and Table A-2 of this text. Thelkeld (4) has shown that errors in calculation of the major properties will be less than 0.7 percent when perfect gas relations are used. This chapter emphasizes the use of the perfect gas relations.

Material in this chapter emphasizes the thermodynamic analysis. That is, only the states at the beginning and end of a process are considered. In the final analysis the nature of the physical process, the path of the process, must also be included. This important part of the problem is covered in Chapters 12, 13, and 14, which deal with heat and mass transfer processes.

3-1 MOIST AIR AND THE STANDARD ATMOSPHERE

Atmospheric air is a mixture of many gases plus water vapor and countless pollutants. Aside from the pollutants, which may vary considerably from place to place, the composition of the dry air alone is relatively constant, varying slightly with time, location, and altitude. The *ASHRAE Handbook, Funda-*

Table 3-1   COMPOSITION OF DRY AIR

Constituent	Molecular Weight	Volume Fraction
Oxygen	32.000	0.2095
Nitrogen	28.016	0.7809
Argon	39.944	0.0093
Carbon dioxide	44.010	0.0003

mentals Volume gives the following approximate composition of dry air by volume fraction:

Nitrogen	0.78084
Oxygen	0.20948
Argon	0.00934
Carbon dioxide	0.00031
Neon, helium, methane, sulfur dioxide, hydrogen, and other minor gases	0.00003

In 1949 a standard composition of dry air was fixed by the International Joint Committee on Psychrometric Data (4) as shown in Table 3-1.

Based on the composition of air in Table 3-1, the molecular weight M_a of dry air is 28.965, and the gas constant R_a is

$$R_a = \frac{\overline{R}}{M_a} = \frac{1545.32}{28.965} = 53.352 \text{ (ft-lbf)/(lbm-R)} \quad \text{or} \quad 287 \text{ J/(kg-K)} \quad (3\text{-}1)$$

where $\overline{R}$ is the universal gas constant; $\overline{R} = 1545.32$ (ft-lbf)/(lb mole-R) or 8314 J/(kg mole-K).

The basic medium in air-conditioning practice is a mixture of dry air and water vapor. The amount of water vapor may vary from zero to a maximum determined by the temperature and pressure of the mixture. The latter case is called saturated air, a state of neutral equilibrium between the moist air and the liquid or solid phases of water. The molecular weight of water is 18.015 and the gas constant for water vapor is

$$R_v - \frac{1545.32}{18.015} = 85.78 \text{ (ft-lbf)/(lbm-R)} \quad \text{or} \quad 462 \text{ J/(kg-K)} \quad (3\text{-}2)$$

The *ASHRAE Handbook* (5) gives the following definition of the U.S. Standard Atmosphere:

1. Acceleration due to gravity is constant at 32.174 ft/sec² or 9.807 m/s².
2. Temperature at sea level is 59.0 F, 15 C or 288.1 K.
3. Pressure at sea level is 29.921 in. of mercury or 101.039 kPa.*
4. The atmosphere consists of dry air, which behaves as a perfect gas.

*Standard atmospheric pressure is also commonly taken to be 14.696 lbf/in.² or 101.325 kPa, which corresponds to 30.0 in. of mercury.

Table 3-2   CONSTANTS FOR EQ. 3-4

Constant	$H \leq 4000$ ft or 1220 m		$H > 4000$ ft or 1200 m	
	IP	SI	IP	SI
a	29.92	101.325	29.42	99.436
b	-0.001025	-0.01153	-0.0009	-0.010

$$Pv = \frac{P}{\rho} = R_a T \tag{3-3}$$

Standard sea level density computed using Eq. 3-3 with the standard temperature and pressure is 0.0765 lbm/ft³ or 1.225 kg/m³. The *ASHRAE Handbook* (5) summarizes standard atmospheric data for altitudes up to 60,000 ft or 18,291 m. Atmospheric pressure may be computed as a function of elevation by the following relation:

$$P = a + bH \tag{3-4}$$

where the constants a and b are given in Table 3-2 and H is the elevation above sea level in feet or meters. The pressure P is in in. Hg or kPa.

Elevation above sea level is given in Table C-1 for many locations in the United States.

3-2   FUNDAMENTAL PARAMETERS

Moist air up to about three atmospheres pressure obeys the perfect gas law with sufficient accuracy for engineering calculations. The Gibbs Dalton law for a mixture of perfect gases states that the mixture pressure is equal to the sum of the partial pressures of the constituents.

$$P = p_1 + p_2 + p_3 \tag{3-5}$$

For moist air

$$P = p_{N_2} + p_{O_2} + p_{CO_2} + p_A + p_v \tag{3-6}$$

Because the various constituents of the dry air may be considered to be one gas, it follows that the total pressure of moist air is the sum of the partial pressures of the dry air and the water vapor.

$$P = p_a + p_v \tag{3-7}$$

Each constituent in a mixture of perfect gases behaves as if the others were not present. To compare ideal gas behavior with actual table values, consider a saturated mixture of dry air and water vapor at 80 F. Table A-1 gives the saturation pressure of water as 0.5073 lbf/in.^{2}. For saturated air this is the partial pressure of the vapor. The mass density is 1/632.8 or 0.001580 lbm/ft^3. By using Eq. 3-3 we get

$$\frac{1}{v} = \rho = \frac{p_v}{R_v T} = \frac{0.5073(144)}{85.78(459.67 + 80)}$$

or

$$\rho = 0.001578 \, \frac{\text{lbm}}{\text{ft}^3}$$

This result is accurate within about 0.1 percent. For superheated conditions the agreement is generally better.

Humidity ratio W (sometimes called the specific humidity) is the ratio of the mass of the water vapor m_v to the mass of the dry air m_a in the mixture.

$$W = \frac{m_v}{m_a} \tag{3-8}$$

Relative humidity ϕ is the ratio of the mole fraction of the water vapor x_v in a mixture to the mole fraction x_s of the water vapor in a saturated mixture at the same temperature and pressure

$$\phi = \left| \frac{x_v}{x_s} \right|_{t,P} \tag{3-9}$$

For a mixture of perfect gases the mole fraction is equal to the partial pressure ratio of each constituent. The mole fraction of the water vapor is

$$x_v = \frac{p_v}{P} \tag{3-10}$$

Using Eq. 3-9 and letting p_s stand for the partial pressure of the water vapor in a saturated mixture, we may express the relative humidity as

$$\phi = \frac{P_v/P}{P_s/P} = \frac{p_v}{p_s} \tag{3-11}$$

Since the temperature of the dry air and the water vapor are assumed to be the same in the mixture

$$\phi = \frac{p_v/R_v T}{p_s/R_v T} = \left[\frac{\rho_v}{\rho_s} \right]_{t,P} \tag{3-12}$$

where the densities ρ_v and ρ_s are referred to as the absolute humidities of the water vapor (mass of water per unit volume of mixture).

By using the perfect gas law we can derive a relation between the relative humidity ϕ and the humidity ratio W:

$$m_v = \frac{p_v V}{R_v T} = \frac{p_v V M_v}{\bar{R} T} \tag{3-13}$$

and

$$m_a = \frac{p_a V}{R_a T} = \frac{p_a V M_v}{\bar{R} T} \tag{3-14a}$$

and

$$W = \frac{M_v p_v}{M_a p_a} \tag{3-14}$$

For the air–water vapor mixture, Eq. 3-14 reduces to

$$W = 0.6129 \frac{p_v}{p_a} \tag{3-14a}$$

Combining Eqs. 3-11 and 3-14a gives

$$\phi = \frac{W p_a}{0.6219 p_s} \tag{3-15}$$

Degree of saturation μ is the ratio of the humidity ratio W to the humidity ratio W_s of a saturated mixture at the same temperature and pressure.

$$\mu = \left[\frac{W}{W_s} \right]_{t,P} \tag{3-16}$$

This parameter is most useful when the data of Table A-2 are used.

Dew point temperature t_d is the temperature of saturated moist air at the same pressure and humidity ratio as the given mixture. It may also be stated that as a mixture is cooled at constant pressure, the temperature at which condensation first begins is the dew point.

The *enthalpy* of a mixture of perfect gases is equal to the sum of the enthalpies of each constituent and is usually referenced to a unit mass of one constituent. For the air–water vapor mixture the mass of dry air used as the reference because the amount of water vapor may vary during some processes.

$$i = i_a + W i_v \tag{3-17}$$

Each term has the units of energy per unit mass of dry air. With the assumption of perfect gas behavior the enthalpy is a function of temperature only. If zero Fahrenheit or Celsius is selected as the reference state where the enthalpy of dry air is zero, and if the specific heat c_{pa} and c_{pv} are assumed to be constant, simple relations result

$$i_a = c_{pa} t \tag{3-18}$$
$$i_v = i_g + c_{pv} t \tag{3-19}$$

where the enthalpy of saturated water vapor i_g at 0 F is 1061.2 Btu/lbm and 2501.3 kJ/kg at 0 C.

Using Eqs. 3-17, 3-18, and 3-19 with c_{pa} and c_{pv} taken as 0.240 and 0.444 Btu/(lbm-F), respectively.

$$i = 0.240t + W(1061.2 + 0.444t) \text{ Btu/lbma} \qquad (3\text{-}20)$$

In SI units Eq. 3-20 becomes

$$i = 1.0t + W(2501.3 + 1.86t) \text{ kJ/kga} \qquad (3\text{-}20a)$$

where c_{pa} and c_{pv} are 1.0 and 1.86 kJ/(kg-C), respectively.

Example 3-1.   Compute the enthalpy of saturated air at 60 F or 15.56 C and standard atmospheric pressure in both English and SI units.

Solution.   Equations 3-20 and 3-20a will be used to compute enthalpy; however, the humidity ratio W_s must first be determined from Eq. 3-14a.

$$W_s = 0.6219 \frac{p_s}{p_a} = 0.6129 \frac{p_s}{P - p_s}$$

From Table A-1, $p_s = 0.2563$ psia or 1767.9 Pa. Then

$$W_s = 0.6219 \left[\frac{0.2563}{14.696 - 0.2563} \right] = 0.01104 \text{ lbmv/lbma}$$

or

$$W_s = 0.6219 \left| \frac{1767.9}{101325 - 1767.9} \right| = 0.01104 \text{ kgv/kga}$$

$$i_s = (0.24)60 + 0.01104[1061.2 + (0.444)60] = 26.41 \text{ Btu/lbma}$$

and Eq. 3-20a gives

$$i_s = 15.56 + 0.01104[2501.3 + (1.86)15.56] = 43.49 \text{ kJ/kga}$$

The results of Example 3-1 may be compared with the precise data of Table A-2. The enthalpy calculated using ideal gas relations is about 0.25 percent low but quite satisfactory for engineering calculations.

3-3   ADIABATIC SATURATION

Examination of the equations discussed in the previous section indicates that at a given pressure and temperature of an air–water vapor mixture one additional property is required to completely specify the state, except at saturation. Theoretically any of the parameters discussed (ϕ, W, or i) would be acceptable; however, there is no practical way to measure any of them in a real situation. The concept of adiabatic saturation provides a convenient solution to this problem.

Consider the device shown in Fig. 3-1. The apparatus is assumed to operate such that the moist air leaving at point 2 is saturated. The temperature t_2 is

Figure 3-1. Schematic of adiabatic saturation device.

then defined as the *adiabatic saturation temperature* t^*, and the relative humidity ϕ_2 is 100 percent. The term *thermodynamic wet bulb temperature* is also used to described t^*. If we assume that the device operates in a steady-flow–steady-state manner, an energy balance on the control volume yields

$$i_{a1} + W_1 i_{v1} + (W^*_{s2} - W_1) i^*_w = W^*_{s2} i^*_{v2} + i^*_{a2} \tag{3-21}$$

or

$$W_1(i_{v1} - i^*_w) = c_{pa}(t^* - t_1) + W^*_{s2}(i^*_{v2} - i^*_w) \tag{3-21a}$$

and

$$W_1(i_{v1} - i^*_w) = c_{pa}(t^* - t_1) + W^*_{s2} i^*_{fg2} \tag{3-21b}$$

Solving for W_1, yields

$$W_1 = \frac{c_{pa}(t^* - t_1) + W^*_{s2} i^*_{fg2}}{(i_{v1} - i^*_w)} \tag{3-21c}$$

where the * superscript refers to the adiabatic saturation temperature t^*. It is clear from Eq. 3-21c that W_1 is a function of t_1, t^*, P_1, P_2.

$$W^*_{s2} = 0.6219 \frac{P_{v2}}{(P_2 - P_{v2})} \tag{3-14a}$$

and $P_{v2} = P_{s2}$ at t^*. The enthalpy of vaporization i^*_{fg2} depends only on t^*; the enthalpy of the vapor i_{v1} is a function of t_1 and i^*_w is a function of t^*. Therefore, the humidity ratio of an air–water vapor mixture can theoretically be determined by measuring the entering and leaving temperatures and pressures for the adiabatic saturator.

Because the measurement of pressure and temperature is relatively easy, the adiabatic saturation process provides a convenient method of determining the state of moist air. Consider the following example.

Example 3-2. The pressure entering and leaving an adiabatic saturator is 14.696 lbf/in.², the entering temperature is 80 F, and the leaving temperature is 64 F. Compute the humidity ratio W_1 and the relative humidity ϕ_1.

Solution. Because the mixture leaving the device is saturated, $p_{v2} = p_{s2}$, and W_2 can be calculated using Eq. 3-14a.

$$W_{s2}^* = 0.6219 \left| \frac{0.2952}{14.696 - 0.2952} \right| = 0.0127 \text{ lbmv/lbma}$$

Now using Eq. 3-21c and data from Table A-1, we get

$$W_1 = \frac{c_{pa}(t^* - t_1) + W_{s2}^* i_{g2}^*}{(i_{v1} - i_w^*)}$$

$$W_1 = \frac{0.24(64 - 80) + (0.0127 \times 1057.3)}{(1096.4 - 32.09)} = 0.0090 \text{ lbmv/lbma}$$

Then using Eq. 3-14a

$$W_1 = 0.6219 \frac{p_{v1}}{14.696 - p_{v1}} = 0.0090 \text{ lbmv/lbma}$$

$$p_{v1} = 0.2098 \text{ psia}$$

Finally, from Eq. 3-11

$$\phi_1 = \frac{p_{v1}}{p_{s1}} = \frac{0.2098}{0.5073} = 0.414$$

It is therefore evident that the state of moist air can be completely determined from pressure and temperature measurements. However, the adiabatic saturator is not a practical device because it would have to be infinitely long in the flow direction. A practical method of determining the approximate adiabatic saturation temperature is discussed in the next section.

3-4 WET BULB TEMPERATURE AND THE PSYCHROMETRIC CHART

A practical device used in place of the adiabatic saturator is the *psychrometer*. This apparatus consists of two thermometers, or other temperature-sensing elements, one of which has a wetted cotton wick covering the bulb (Fig. 3-2). The temperatures indicated by the psychrometer are called the *wet bulb* and the *dry bulb* temperatures. The dry bulb temperature corresponds to t_1 in Fig. 3-1 and the wet bulb temperature is an approximation to t^* in Fig. 3-1, whereas P_1 and P_2 are equal to barometric pressure. The combination heat-and-mass-transfer process from the wet bulb thermometer is not the same as the adiabatic saturation process; however, the error is relatively small when the wet bulb thermometer is used under suitable conditions. The difference between the wet bulb and thermodynamic wet bulb temperature should be less than about 0.5 F or 0.25 C for atmospheric pressure and temperatures above 32 F or 0 C, where the difference between the dry bulb and wet bulb temperatures is less than about 20 F or 11 C, and where no unusual radiation conditions

Figure 3-2. Typical hand-held psychrometer. (Courtesy of Central Scientific Company, Chicago, Illinois.)

exist; an unshielded wet bulb thermometer is assumed to be in air that has a velocity greater than 100 ft/min or 0.5 m/s. If thermocouples are used, the velocity may be somewhat lower with similar accuracy. Figure 3-3 shows a psychrometer properly installed in a duct to meet the foregoing conditions.

To facilitate engineering computations, a graphical representation of the properties of moist air has been developed and is known as a *psychrometric chart.* Richard Mollier was the first to use such a chart with enthalpy as a coordinate. Modern-day charts are somewhat different but still retain the enthalpy coordinate feature. ASHRAE has developed five of the Mollier-type charts to cover the necessary range of variables. Figure 3-4 is an abridgment of ASHRAE Chart 1 that covers the normal range of variables at standard atmospheric pressure. The charts are based on the precise data of Table A-2; within the readability of the charts, however, agreement with the perfect gas relations is very good. Details of the actual construction of the charts may be found in references 5 and 6. Charts for both English and SI units are provided in the packet in the back of the book.

Figure 3-3. A psychrometer installed in a duct.

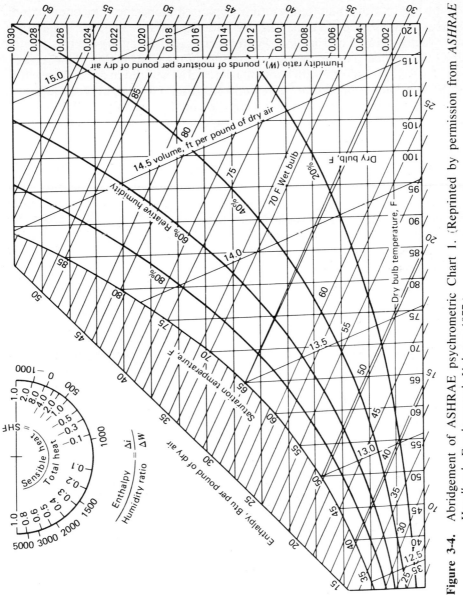

Figure 3-4. Abridgement of ASHRAE psychrometric Chart 1. (Reprinted by permission from *ASHRAE Handbook, Fundamentals Volume*, 1977.)

In Fig. 3-4 dry bulb temperature is plotted along the horizontal axis in degrees Fahrenheit or Celsius. The dry bulb temperature lines are straight but not exactly parallel and incline slightly to the left. Humidity ratio is plotted along the vertical axis on the right-hand side of the chart in lbmv/lbma or kgv/kga. The scale is uniform with horizontal lines. The saturation curve with values of the wet bulb temperature curves upward from left to right. Dry bulb, wet bulb, and dew point temperatures all coincide on the saturation curve. Relative humidity lines with a shape similar to the saturation curve appear at regular intervals. The enthalpy scale is drawn obliquely on the left of the chart with parallel enthalpy lines inclined downward to the right. Although the wet bulb temperature lines appear to coincide with the enthalpy lines, they diverge gradually in the body of the chart and are not parallel to one another. The spacing of the wet bulb lines is not uniform. Specific volume lines appear inclined from the upper left to the lower right and are not parallel. A protractor with two scales appears at the upper left of ASHRAE Chart 1. One scale gives the sensible heat ratio and the other the ratio of enthalpy difference to humidity ratio difference. Notice that the enthalpy, specific volume, and humidity ratio scales are all based on a unit mass of dry air and not a unit mass of the moist air.

Example 2-3.   Read the properties of moist air at 75 F db, 60 F wb, and standard sea level pressure from ASHRAE Psychrometric Chart 1.

Solution.   The intersection of the 75 F db and 60 F wb lines defines the given state. This point on the chart is the reference from which all the other properties are determined.

   Humidity Ratio, W. Move horizontally to the right and read $W = 0.0077$ lbmv/lbma on the vertical scale. It is not uncommon for W to be given in "grains." One pound mass is equal to 7000 grains. Then $W = 7000 \times 0.0077 = 53.9$ grains/lbma.

   Relative Humidity, ϕ. Interpolate between the 40 and 50 percent relative humidity lines and read $\phi = 41$ percent.

   Enthalpy, i. Follow a line of constant enthalpy upward to the left and read $i = 26.4$ Btu/lbma on the oblique scale.

   Specific Volume, v. Interpolate between the 13.5 and 14.0 specific volume lines and read $v = 13.65$ ft^3/lbma.

   Dew Point Temperature, t_d. Move horizontally to the left from the reference point and read $t_d = 50$ F on the saturation curve.

   Enthalpy, i (alternate method). The nomograph in the upper left-hand corner of Chart 1 gives the difference D between the enthalpy of unsaturated moist air and enthalpy of saturated air at the same wet bulb temperature. Then $i = i_s + D$. For this example $i_s = 26.5$ Btu/lbma, $D = -0.1$ Btu/lbma, and $i = 26.5 - 0.1 = 26.4$ Btu/lbma.

3-5 CLASSIC MOIST AIR PROCESSES

The most powerful analytical tools of the air-conditioning design engineer are the *first law of thermodynamics* or energy balance, and the *conservation of mass* or mass balance. These conservation laws are the basis for the analysis of moist air processes. It is customary to analyze these processes by using the bulk average properties at the inlet and outlet of the device being studied. In actual practice the properties may not be uniform across the flow area especially at the outlet, and a considerable length may be necessary for complete mixing.

In this section we will consider the basic processes that are a part of the analysis of most systems.

Heating or Cooling of Moist Air

When air is heated or cooled without the loss or gain of moisture, the process yields a straight horizontal line on the psychrometric chart because the humidity ratio is constant. Such processes can occur when moist air flows through a heat exchanger. In cooling, if the surface temperature is below the dew point temperature of the moist air, dehumidification will occur. This process will be considered later. Figure 3-5 shows a schematic of a device used to heat or cool air. Under steady-flow–steady-state conditions the energy balance becomes

$$\dot{m}_a i_2 + \dot{q} = \dot{m}_a i_1 \tag{3-22}$$

The energy balance technique yields a positive number for $\dot{q}$ for both cooling and heating, and the direction of the heat transfer is implied by the terms heating and cooling. It is to be emphasized that

$$i_1 = i_{a1} + W_1 i_{v1} \tag{3-23}$$

and

$$i_2 = i_{a2} + W_2 i_{v2} \tag{3-24}$$

where i_1 and i_2 may be obtained from the psychrometric chart. The convenience of the chart is evident. Figure 3-6 shows heating and cooling processes. Because

Figure 3-5.   Schematic of a heating or cooling device.

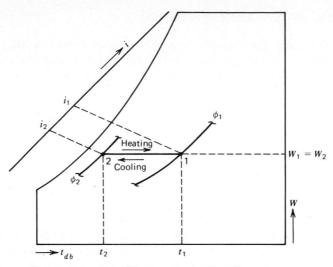

Figure 3-6.   Sensible heating and cooling process.

the moist air has been assumed to be a perfect gas, Eq. (3-22) may be rearranged
and written

$$\dot{q} = \dot{m}_a c_p (t_1 - t_2) \quad \text{(cooling or heating)} \tag{3-25}$$

where

$$c_p = c_{pa} + W c_{pv} \tag{3-26}$$

In the temperature range of interest, c_{pa} = 0.24 Btu/(lbma-F) or 1.0 kJ/
(kga-C), c_{pv} = 0.45 Btu/(lbmv-F) or 1.86 kJ/(kga-C), and W is the order of 0.01.
Then c_p is about 0.245 Btu/(lbma-F) or 1.02 kJ/(kga-C).

Example 3-4.   Find the heat transfer rate required to warm 1500 cfm (ft³/
min) of air at 60 F and 90 percent relative humidity to 120 F without the
addition of moisture.

Solution.   Equations (3-22) or (3-25) may be used to find the required heat
transfer rate. First it is necessary to find the mass flow rate of the dry air.

$$\dot{m}_a = \frac{\overline{V}_1 A_1}{v_1} = \frac{\dot{Q}}{v_1} \tag{3-27}$$

The specific volume is read from Chart 1 at t_1 = 60 F and ϕ = 90 percent
as 13.31 ft³/lbma:

$$\dot{m}_a = \frac{1500(60)}{13.31} = 6762 \text{ lbma/hr}$$

Also from Chart 1, i_1 = 25.3 Btu/lbma and i_2 = 40 Btu/lbma. Then by
using Eq. (3-22), we get

Figure 3-7. Schematic of a cooling and dehumidifying device.

$$\dot{q} = 6762(40.0 - 25.3) = 99,400 \text{ Btu/hr}$$

or from Eq. (3-25)

$$\dot{q} = (6762)(0.245)(120 - 60) = 99,400 \text{ Btu/hr}$$

Example 3-4 shows that the relative humidity decreases when the moist air is heated. The reverse process of cooling results in an increase in relative humidity.

Cooling and Dehumidifying of Moist Air

When moist air is cooled to a temperature below its dew point, some of the water vapor will condense and leave the air stream. Figure 3-7 shows a schematic of a cooling and dehumidifying device, and Fig. 3-8 shows the process

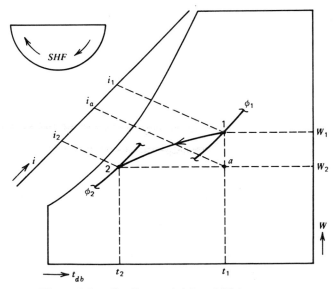

Figure 3-8. Cooling and dehumidifying process.

on the psychrometric chart. Although the actual process path will vary considerably depending on the type of surface, surface temperature, and flow conditions, the heat and mass transfer can be expressed in terms of the initial and final states. By referring to Fig. 3-7, we see that the energy balance becomes

$$\dot{m}_a i_1 = \dot{q} + \dot{m}_a i_2 + \dot{m}_w i_w \qquad (3\text{-}28)$$

and the steady flow mass balance for the water in the air is

$$\dot{m}_a W_1 = \dot{m}_w + \dot{m}_a W_2 \qquad (3\text{-}29)$$

Combining Eqs. (3-28) and (3-29) we get

$$\dot{q} = \dot{m}_a(i_1 - i_2) - \dot{m}_a(W_1 - W_2)i_w \qquad (3\text{-}30)$$

Equation (3-30) represents the total amount of heat transfer from the moist air. The last term on the right-hand side of Eq. (3-30) is usually small compared to the others and is often neglected. The following example illustrates this point.

Example 3-5.   Moist air at 80 F db and 67 F wb is cooled to 58 F db and 80 percent relative humidity. The volume flow rate is 2000 cfm and the condensate leaves at 60 F. Find the heat transfer rate.

Solution.   Equation (3-30) applies to this process, which is similar to Fig. 3-8. The following properties are read from Chart 1: $v_1 = 13.85$ ft^3/lbma, $i_1 = 31.6$ Btu/lbma, $W_1 = 0.0112$ lbmv/lbma, $i_2 = 22.9$ Btu/lbma, $W_2 = 0.0082$ lbmv/lbma. The enthalpy of the condensate is obtained from Table A-1, $i_w = 28.08$ Btu/lbmw. The mass flow rate $\dot{m}_a$ is obtained from Eq. (3-27).

$$\dot{m}_a = \frac{2000(60)}{13.85} = 8664 \text{ lbma/hr}$$

Then

$$\dot{q} = 8664[(31.6 - 22.9) - (0.0112 - 0.0082)28.08]$$
$$\dot{q} = 8664[(8.7) - (0.084)]$$

The last term, which represents the energy of the condensate, is quite insignificant in this case. For most cooling and dehumidifying processes this will be true. Finally, neglecting the condensate term, $\dot{q} = 75,000$ Btu/hr. A ton of refrigeration is 12,000 Btu/hr. Then $\dot{q} = 6.25$ tons.

The cooling and dehumidifying process involves both sensible and latent heat transfer, where sensible heat transfer is associated with the decrease in dry bulb temperature and the latent heat transfer is associated with the decrease in humidity ratio. These quantities may be expressed as

$$\dot{q}_s = \dot{m}_a c_p(t_1 - t_2) \qquad (3\text{-}31)$$

and

$$\dot{q}_l = \dot{m}_a(W_1 - W_2)i_{fg} \qquad (3\text{-}32)$$

By referring to Fig. 3-8 we may also express the latent heat transfer as

$$\dot{q}_l = \dot{m}_a(i_1 - i_a) \tag{3-33}$$

and the sensible heat transfer is given by

$$\dot{q}_s = \dot{m}_a(i_a - i_2) \tag{3-34}$$

The energy of the condensate has been neglected. Obviously

$$\dot{q} = \dot{q}_s + \dot{q}_l \tag{3-35}$$

The sensible heat factor SHF is defined as $\dot{q}_s/\dot{q}$. This parameter is shown on the semicircular scale of Fig. 3-8. The use of this feature of the chart is shown later.

Heating and Humidifying Moist Air

A device to heat and humidify moist air is shown schematically in Fig. 3-9. This process is generally required during the cold months of the year. An energy balance on the device yields

$$\dot{m}_a i_1 + \dot{q} + \dot{m}_w i_w = \dot{m}_a i_2 \tag{3-36}$$

and a mass balance on the water gives

$$\dot{m}_a W_1 + \dot{m}_w = \dot{m}_a W_2 \tag{3-37}$$

Equations (3-36) and (3-37) may be combined to obtain

$$\frac{i_2 - i_1}{W_2 - W_1} = \frac{\dot{q}}{\dot{m}_a(W_2 - W_1)} + i_w \tag{3-38}$$

or

$$\frac{i_2 - i_1}{W_2 - W_1} = \frac{\dot{q}}{\dot{m}_w} + i_w \tag{3-38a}$$

Equation (3-38) or (3-38a) gives the direction of a straight line that connects the initial and final states on the psychrometric chart. Figure 3-10 shows a typical combined heating and humidifying process.

Figure 3-9. Schematic of a heating and humidifying device.

Figure 3-10.   Typical heating and humidifying process.

A graphical procedure makes use of the semicircular scale on Chart 1 to locate the process line. The ratio of the change in enthalpy to the change in humidity ratio is

$$\frac{\Delta i}{\Delta W} = \frac{i_2 - i_1}{W_2 - W_1} = \frac{\dot{q}}{\dot{m}_w} + i_w \tag{3-39}$$

Figure 3-10 shows the procedure where a straight line is laid out parallel to the line on the protractor through state 1. Although the process may be represented by one line from state 1 to state 2, it is not practical to do this in practice. The heating and humidification processes are usually carried out separately, as shown in Fig. 3-9 and Fig. 3-10 as processes 1-a and a-2. Let us then consider the humidification of air without the addition of heat.

Adiabatic Humidification of Moist Air

When moisture is added to moist air without the addition of heat, Eq. 3-38a becomes

$$\frac{i_2 - i_1}{W_2 - W_1} = i_w = \frac{\Delta i}{\Delta W} \tag{3-40}$$

The direction of the process on the psychrometric chart can vary considerably. If the injected water is saturated vapor at the dry bulb temperature, the process will proceed at a constant dry bulb temperature. If the water enthalpy is greater than the enthalpy of saturated vapor at the dry bulb temperature, the air will be heated and humidified. If the water enthalpy is less than the enthalpy of saturated vapor at the dry bulb temperature, the air will be cooled and humidified. Figure 3-11 shows these processes. One other situation is worthy of

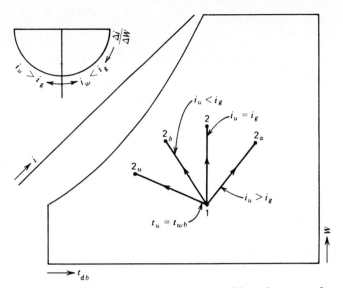

Figure 3-11.   Humidification processes without heat transfer.

mention. When liquid water at the wet bulb temperature is injected, the process follows a line of constant wet bulb temperature, as shown by Fig. 3-11.

Example 3-6   Moist air at 60 F db and 20 percent relative humidity enters a heater and humidifier at the rate of 1600 cfm. It is necessary to heat the air followed by adiabatic humidification so that it leaves at a temperature of 115 F and a relative humidity of 30 percent. Saturated water vapor at 212 F is injected. Determine the required heat transfer rate and mass flow rate of the steam.

Solution.   Figure 3-9 is a schematic of the apparatus. It is first necessary to locate the states as shown in Fig. 3-10 from the given information and Eq. 3-40 using the protractor feature of the psychrometric chart. Process 1-a is sensible heating; therefore, a horizontal line to the right of state 1 is constructed. Process a-2 is determined from Eq. 3-40 and the protractor:

$$\frac{\Delta i}{\Delta W} = i_w = 1151 \text{ Btu/lbm}$$

where i_w is read from Table A-1. A parallel line is drawn from state 2 as shown in Fig. 3-10. State a is determined by the intersection on lines 1-a and a-2. The heat transfer rate is then given by

$$\dot{q} = \dot{m}_a(i_a - i_1)$$

where

$$\dot{m}_a = \frac{\dot{Q}(60)}{v_1} = \frac{(1600)(60)}{13.15} = 7300 \text{ lbma/hr}$$

and i_1 and i_a are read from Chart 1 as 16.8 and 28.6 Btu/lbma, respectively. Then

$$\dot{q} = 7300(28.6 - 16.8) = 86,140 \text{ Btu/hr}$$

The mass flow rate of the steam is given by

$$\dot{m}_w = \dot{m}_a(W_2 - W_1)$$

where W_2 and W_1 are read from Chart 1 as 0.01275 and 0.0022 lbmv/lbma, respectively. Then

$$\dot{m}_w = 7300(0.01275 - 0.0022) = 77.0 \text{ lbmv/hr}$$

Adiabatic Mixing of Two Streams of Moist Air

The mixing of air streams is quite common in air-conditioning systems. The mixing process usually occurs under adiabatic conditions and with steady flow. Figure 3-12 illustrates the mixing of two air steams. An energy balance gives

$$\dot{m}_{a1}i_1 + \dot{m}_{a2}i_2 = \dot{m}_{a3}i_3 \qquad (3\text{-}41)$$

The mass balance on the dry air is

$$\dot{m}_{a1} + \dot{m}_{a2} = \dot{m}_{a3} \qquad (3\text{-}42)$$

and the mass balance on the water vapor is

$$\dot{m}_{a1}W_1 + \dot{m}_{a2}W_2 = \dot{m}_{a3}W_3 \qquad (3\text{-}43)$$

By combining Eqs. 3-41, 3-42, and 3-43 and eliminating $\dot{m}_{a3}$, we obtain the following result:

$$\frac{i_2 - i_3}{i_3 - i_1} = \frac{W_2 - W_3}{W_3 - W_1} = \frac{\dot{m}_{a1}}{\dot{m}_{a2}} \qquad (3\text{-}44)$$

The state of the mixed streams must lie on a straight line between states 1 and 2. This is shown on Fig. 3-13. It may be further inferred from Eq. 3-44 that the length of the various line segments are proportional to the masses of dry air mixed.

Figure 3-12.   Schematic of the adiabatic mixing of two air streams.

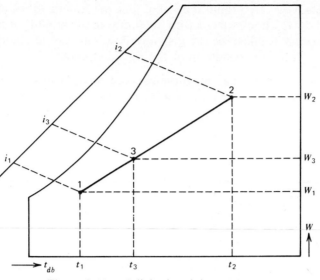

Figure 3-13.   Adiabatic mixing process.

$$\frac{\dot{m}_{a1}}{\dot{m}_{a2}} = \frac{\overline{32}}{\overline{13}}; \quad \frac{\dot{m}_{a1}}{\dot{m}_{a3}} = \frac{\overline{32}}{\overline{12}}; \quad \frac{\dot{m}_{a2}}{\dot{m}_{a3}} = \frac{\overline{13}}{\overline{12}} \qquad (3\text{-}45)$$

The truth of these statements is most easily shown by solving Eq. 3-44 for i_3 and W_3.

$$i_3 = \frac{\dfrac{\dot{m}_{a1}}{\dot{m}_{a2}} i_1 + i_2}{1 + \dfrac{\dot{m}_{a1}}{\dot{m}_{a2}}} \qquad (3\text{-}44a)$$

$$W_3 = \frac{\dfrac{\dot{m}_{a1}}{\dot{m}_{a2}} W_1 + W_2}{1 + \dfrac{\dot{m}_{a1}}{\dot{m}_{a2}}} \qquad (3\text{-}44b)$$

Clearly for given states 1 and 2, a straight line will be generated when various values of $\dot{m}_{a1}/\dot{m}_{a2}$ are used and the result plotted on the psychometric chart. It is also clear that the location of state 3 on the line is proportional to $\dot{m}_{a1}/\dot{m}_{a3}$. Consider the case when $\dot{m}_{a1}/\dot{m}_{a2} = 1$, for example. This fact provides a very convenient graphical procedure for solving mixing problems in contrast to the use of Eqs. 3-41, 3-42, and 3-43.

It should be noted that the mass flow rate is used when the graphical procedure is employed; however, the volume flow rates may be used to obtain approximate results.

Example 3-7. Two thousand cubic feet per minute (cfm) of air at 100 F db and 75 F wb are mixed with 1000 cfm of air at 60 F db and 50 F wb. The process is adiabatic, at a steady flow rate and at standard sea level pressure. Find the condition of the mixed streams.

Solution. A combination graphical and analytical solution is first obtained. The initial states are first located on Chart 1 as illustrated on Fig. 3-13 and connected with a straight line. Equations 3-42 and 3-43 are combined to obtain

$$W_3 = W_1 + \frac{\dot{m}_{a2}}{\dot{m}_{a3}}(W_2 - W_1) \tag{3-46}$$

By using the property values from Chart 1, we obtain

$$\dot{m}_{a1} = \frac{1000(60)}{13.21} = 4542 \text{ lbma/hr}$$

$$\dot{m}_{a2} = \frac{2000(60)}{14.4} = 8332 \text{ lbma/hr}$$

$$W_3 = 0.0053 + \frac{8332}{(4542 + 8332)}(0.013 - 0.0053)$$

$$W_3 = 0.0103 \text{ lbmv/lbma}$$

The intersection of W_3 with the line connecting states 1 and 2 gives the mixture state 3. The resulting dry bulb temperature is 86 F and the wet bulb temperature is 68 F.

The complete graphical procedure could also be used where

$$\frac{\overline{13}}{\overline{12}} = \frac{\dot{m}_{a2}}{\dot{m}_{a3}} = \frac{8332}{(8332 + 4542)} = 0.65$$

and

$$\overline{13} = 0.65(\overline{12})$$

The length of line segments $\overline{12}$ and $\overline{13}$ depends on the scale of the psychrometric chart used. However, when the length $\overline{13}$ is laid out along $\overline{12}$ from state 1, state 3 is accurately determined.

3-6 SPACE AIR CONDITIONING—DESIGN CONDITIONS

The complete air-conditioning system may involve two or more of the processes just considered. For example, in the air conditioning of a space during the summer, the air supplied must have a sufficiently low temperature and moisture content to absorb the total heat gain of the space. Therefore, as the air flows through the space it is heated and humidified. If the system is a closed loop, the air is then returned to the conditioning equipment, where it is cooled and dehumidified and supplied to the space again. If fresh air is required in the space, outdoor air may be mixed with the return air before it goes to the cooling and dehumidifying equipment. During the winter months the same general processes occur but in reverse. Various systems were described in Chapter 2 that carry out these conditioning processes with some variations.

Sensible Heat Factor

The *sensible heat factor* (SHF) was defined in Section 3-5 as the ratio of the sensible heat transfer to the total heat transfer for a process:

$$\text{SHF} = \frac{\dot{q}_s}{\dot{q}_s + \dot{q}_l} = \frac{\dot{q}_s}{\dot{q}} \tag{3-47}$$

If we recall Eqs. 3-33 and 3-34 and refer to Fig. 3-8, it is evident that the SHF is related to the parameter $\Delta i/\Delta W$. The SHF is plotted on the inside scale of the protractor on Chart 1. The following examples will demonstrate the usefulness of the SHF.

Example 3-8.   Conditioned air is supplied to a space at 15 C db and 14 C wb at the rate of 0.5 m³/s. The sensible heat factor for the space is 0.70 and the space is to be maintained at 24 C db. Determine the sensible and latent cooling loads for the space.

Solution.   Chart 1a can be used to solve this problem conveniently. A line is drawn on the protractor through a value of 0.7 on the SHF scale. A parallel line is then drawn from the initial state (15 C db and 14 C wb) to the intersection of the 24 C db line, which defines the final state. Figure 3-14 illustrates the procedure. The total heat transfer rate for the process is given by

$$\dot{q} = \dot{m}_a(i_2 - i_1)$$

and the sensible heat transfer is given by

$$\dot{q}_s = (\text{SHF})\dot{q}$$

and

$$\dot{m}_a = \dot{Q}/v_1 = \frac{0.5}{0.827} = 0.605 \text{ kg/s}$$

where v_1 is read from Chart 1a. Also from Chart 1a, $i_1 = 39.3$ kJ/kg dry air and $i_2 = 52.6$ kJ/kg dry air. Then

$$\dot{q} = 0.605(52.6 - 39.3) = 8.04 \text{ kJ/s} = 8.04 \text{ kW}$$
$$\dot{q}_s = \dot{q}(\text{SHF}) = 8.04(0.7) = 5.63 \text{ kW}$$

and

$$\dot{q}_l = \dot{q} - \dot{q}_s = 2.41 \text{ kW}$$

Figure 3-14. Psychrometric process for Example 3-8.

The process 1-2 of Example 3-8 and its extension to the left is called the *condition line* for the space. Assuming that state 2, the space condition, is fixed, air supplied at any state on the condition line will satisfy the load requirements. However, as the point is moved, different quantities of air must be supplied to the space. In general, the closer point 1 is to point 2, the more air is required, and the converse is also true.

We will now consider several examples of single-path, constant-flow systems. Heat losses and gains to the ducts and fan power will be neglected for the time being.

Example 3-9. A given space is to be maintained at 78 F db and 65 F wb. The total heat gain to the space has been determined to be 60,000 Btu/hr of which 42,000 Btu/hr is sensible heat transfer. The outdoor air requirement of the occupants is 500 cfm. The outdoor air has a temperature and relative humidity of 90 F and 55 percent, respectively. Determine the quantity and state of the air supplied to the space and the required capacity of the cooling and dehumidifying equipment.

Solution. A simplified schematic is shown in Fig. 3-15. The given quantities are shown and stations are numbered for reference. Losses in connecting ducts will be neglected. Let us first consider the steady flow process for the conditioned space. By Eq. 3-47 the sensible heat factor is

$$\text{SHF} = \frac{42,000}{60,000} = 0.7$$

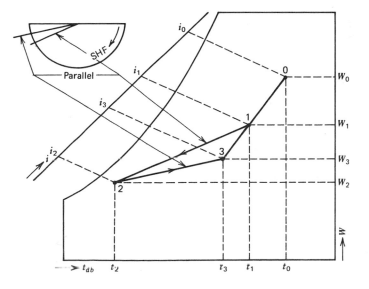

Figure 3-15. Single-line sketch of cooling and dehumidifying system for Example 3-9.

The state of the air entering the space lies on the line defined by the SHF on psychrometric Chart 1. Therefore, state 3 is located as shown on Fig. 3-16 and a line drawn through the point parallel to the SHF = 0.7 line on the protractor. State 2 may be any point on the line and is determined by the operating characteristics of the equipment, desired indoor air quality, and by what will be comfortable for the occupants. These aspects of the problem will be developed later. For now assume that the dry bulb temperature t_2 cannot be more than 20 F less than t_3. Then t_2 = 58 F and state

Figure 3-16. Psychrometric processes for Example 3-9.

Figure 3-16a. Psychrometric processes for Example 3-9,
showing the effect of fans and heat gain.

2 is determined. The air quantity required may now be found from an
energy balance on the space

$$\dot{m}_{a2}i_2 + \dot{q} = \dot{m}_{a3}i_3$$

or

$$\dot{q} = \dot{m}_{a2}(i_3 - i_2)$$

and

$$\dot{m}_{a2} = \frac{\dot{q}}{(i_3 - i_2)}$$

From Chart 1, $i_3 = 30$ Btu/lbma, $i_2 = 23$ Btu/lbma, and

$$\dot{m}_{a2} = \dot{m}_{a3} = \frac{60{,}000}{(30 - 23)} = 8570 \text{ lbma/hr}$$

Also from Chart 1, $v_2 = 13.21$ ft^3/lbma and

$$\dot{Q}_2 = \dot{m}_{a2}v_2 = \frac{8570(13.21)}{60} = 1890 \text{ cfm}$$

Attention is now directed to the cooling and dehumidifying unit. However,
state 1 must be determined before continuing. A mass balance on the mixing
section yields

$$\dot{m}_{a0} + \dot{m}_{a4} = \dot{m}_{a1} = \dot{m}_{a2}$$

$$\dot{m}_{a0} = \frac{\dot{Q}_0}{v_0}; \qquad v_0 = 14.23 \text{ ft}^3/\text{lbma}$$

$$\dot{m}_{a0} = \frac{(500 \times 60)}{14.23} = 2110 \text{ lbma/hr}$$

Then

$$\dot{m}_{a4} = \dot{m}_{a2} - \dot{m}_{a0} = 8570 - 2110 = 6460 \text{ lbma/hr}$$

By using the graphical technique discussed in Example 3-7 and referring to Fig. 3-16, we see that

$$\frac{\overline{31}}{\overline{30}} = \frac{\dot{m}_{a0}}{\dot{m}_{a1}} = \frac{2110}{8570} = 0.246$$

$$\overline{31} = 0.246(\overline{30})$$

State 1 is located at 81 F db and 68 F wb. A line constructed from state 1 to state 2 on Chart 1 then represents the process taking place in the conditioning equipment. An energy balance gives

$$\dot{m}_{a1} i_1 = \dot{q}_c + \dot{m}_{a2} i_2$$

or

$$\dot{q}_c = \dot{m}_{a1}(i_1 - i_2)$$

From Chart 1, $i_1 = 32.4$ Btu/lbma

$$\dot{q}_c = 8570(32.4 - 23) = 80,560 \text{ Btu/hr} = 6.7 \text{ tons}$$

The sensible heat factor (SHF) for the cooling unit is found to be 0.6 using the protractor of Chart 1 (Fig. 3-16). Then

$$\dot{q}_{cs} = 0.6(80,560) = 48,340 \text{ Btu/hr}$$

and

$$\dot{q}_{cl} = 80,560 - 48,340 = 32,220 \text{ Btu/hr}$$

The sum of $\dot{q}_{cs}$ and $\dot{q}_{cl}$ is known as the coil refrigeration load in contrast to the space cooling load.

In an actual system fans are required to move the air and some energy may be gained from this. In addition some heat may be lost or gained in the duct system. Referring to Fig. 3-15, the supply fan is located just downstream of the cooling unit and the return fan is just upstream of the mixing box. All of the power input to the fans is manifested as a sensible energy input to the air, just as if heat were transferred. Heat may also be gained in the supply and return ducts. The power input to the supply air fan and the heat gain to the supply air duct may be summed as shown on Chart 1, Fig. 3-16a, as process 1'-2. It is assumed that all of the supply fan power input is transformed to internal energy by the time the air reaches the space, state 2. Likewise, heat is gained from point 3 to point 4 and the return fan power input occurs between 4 and 4' as shown in Fig. 3-16a. The condition line for the space, 2-3, is the same as it was before, when the fans and heat gain were neglected. However, the requirements of the cooling unit have changed. Process 1-1' now shows that the capacity of the coil must be greater to offset the fan power input and duct heat gain.

An alternate approach to the analysis of the cooling coil in Example 3-9 uses the so-called coil *bypass factor*. Note that when line 1-2 of Fig. 3-16 is extended, it intersects the saturation curve at point d. This point represents the apparatus dewpoint temperature of the cooling coil. The coil cannot cool all of the air passing through it to the coil surface temperature. This fact makes the coil perform in a manner similar to what would happen if a portion of the air were brought to the coil temperature and the remainder bypassed the coil unchanged. A dehumidifying coil thus produces unsaturated air at a higher temperature than the coil temperature. Again referring to Fig. 3-16, notice that in terms of the length of the line d-1, the length d-2 is proportional to the mass of air bypassed, and the length 1-2 is proportional to the mass of air not by-passed. Because the line d-1 is inclined, it is approximately true that

$$b = \frac{t_2 - t_d}{t_1 - t_d} \tag{3-48}$$

and

$$1 - b = \frac{t_1 - t_2}{t_1 - t_d} \tag{3-49}$$

where b is the fraction of air bypassed, or coil bypass factor, expressed as a decimal, and where the temperatures are dry bulb values. Now

$$\dot{q}_{cs} = \dot{m}_{a1} c_p (t_1 - t_2) \tag{3-50}$$

or

$$\dot{q}_{cs} = \dot{m}_{a1} c_p (t_1 - t_d)(1 - b) \tag{3-50a}$$

Example 3-10.   Find the bypass factor for the coil of Example 3-9 and compute the sensible and latent heat transfer rates.

Solution.   The apparatus dewpoint temperature obtained from Chart 1 as indicated in Fig. 3-16 is 46 F. Then from Eq. 3-48 the bypass factor is

$$b = \frac{58 - 46}{81 - 46} = 0.343 \quad \text{and} \quad 1 - b = 0.657$$

Equation 3-50a expresses the sensible heat transfer rate as

$$\dot{q}_{cs} = 8570(0.245)(81 - 46)(0.657) = 48,280 \text{ Btu/hr}$$

The coil sensible heat factor is used to compute the latent heat transfer rate. From Example 3-9, the SHF is 0.6, then the total heat transfer rate is

$$\dot{q}_t = \dot{q}_{cs}/\text{SHF} = 48,280/0.6 = 80,470 \text{ Btu/hr}$$

and

$$\dot{q}_{cl} = \dot{q}_t - \dot{q}_{cs} = 80,470 - 48,280 = 32,190 \text{ Btu/hr}$$

It should be noted that the bypass factor approach results in values about 12 percent high in this case.

In Example 3-10 the outdoor air was hot and humid. This is not always the case and state 0 can be almost anywhere on Chart 1. For example, the southwestern part of the United States is hot and dry during the summer and evaporative cooling can often be used to advantage under these conditons. A simple system of this type is shown in Fig. 3-17. The dry outdoor air flows through an adiabatic spray chamber and is cooled and humidified. An energy balance on the spray chamber will show that the enthalpies i_o and i_1 are equal; therefore, the process is as shown in Fig. 3-18. Ideally the cooling process terminates at the space condition line. The air then flows through the space and is exhausted. Large quantities of air are required, and this system is not satisfactory where the outdoor relative humidity is high. If W_o is too high the process 0-1 cannot intersect the condition line.

Evaporative cooling can be combined with a conventional system as shown in Fig. 3-19 when outdoor conditions are suitable. There are a number of possibilities. First, if the outdoor air is just mixed with return air without evaporative cooling, the ideal result would be state 1 in Fig. 3-20. The air would require only sensible cooling to state 2 on the condition line. Second, outdoor air could be cooled by evaporation to state 0', Fig. 3-20, and then mixed with return air resulting in state 1'. Sensible cooling would then be required from state 1' to state 2. Finally, the outdoor air could ideally be evaporatively cooled

Figure 3-17. A simple evaporative cooling system.

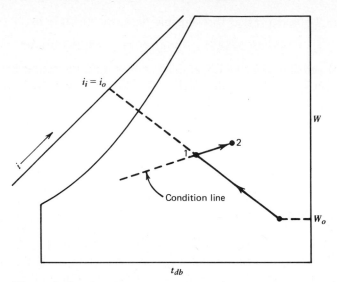

Figure 3-18. Psychrometric diagram for the evaporative cooling system of Figure 3-17.

all the way to state 1″. This would require the least power for sensible cooling, and the air supplied to the space would be 100 percent outdoor air.

Examples of other single-path systems such as VAV or multizone could be presented; however, under the full-flow design condition these systems operate the same as the simple system of Figs. 3-15 and 3-16. They will be discussed further in the section on part-load operation.

The contrasting problem of space air conditioning during the winter months may be solved in a manner similar to Example 3-9.

Example 3-11. A space is to be maintained at 75 F and 50 percent relative humidity. Heat losses from the space are 225,000 Btu/hr sensible and 56,250

Figure 3-19. Combination evaporative and regular cooling system.

Figure 3-20. Psychrometric diagram for Figure 3-19.

Btu/hr latent. The latent heat transfer is due to the infiltration of cold dry air. The outdoor air required is 1000 cfm and is at 35 F and 80 percent relative humidity. Determine the quantity of air supplied at 120 F, the state of the supply air, the size of the furnace or heating coil, and the humidifier characteristics.

Solution. Figure 3-21 is a schematic for the problem; it contains the given information and reference points. First consider the conditioned space

$$\text{SHF} = \frac{225,000}{(225,000 + 56,250)} = 0.80$$

Figure 3-21. The heating and humidifying system for Example 3-11.

The state of the supply air lies on a line drawn through state point 3 parallel to the SHF = 0.8 line on the protractor of Chart 1. Figure 3-22 shows this construction. State 2 is located at 120 F dry bulb and the intersection of this line. An energy balance on the space gives

$$\dot{m}_{a2} i_2 = \dot{q} + \dot{m}_{a3} i_3$$

or

$$\dot{q} = \dot{m}_{a2}(i_2 - i_3)$$

From Chart 1, i_2 = 42 Btu/lbma, i_3 = 28.2 Btu/lbma, and

$$\dot{m}_{a2} = \frac{\dot{q}}{(i_2 - i_3)} = \frac{281{,}250}{(42 - 28.2)}$$

$$\dot{m}_{a2} = 20{,}400 \text{ lbma/hr}$$

From Chart 1, v_2 = 14.89 ft³/lbma and

$$\dot{Q}_2 = \frac{20{,}400}{60} \times 14.89 = 5060 \text{ cfm}$$

To find the conditions at state 1 the mixing process must be considered. A mass balance on the mixing section yields

$$\dot{m}_{a0} + \dot{m}_{a4} = \dot{m}_{a1} = \dot{m}_{a2}$$

or

$$\dot{m}_{a4} = \dot{m}_{a2} - \dot{m}_{a0}$$

$$\dot{m}_{a0} = \frac{\dot{Q}_0}{v_0} \quad \text{and} \quad v_0 = 12.53 \text{ ft}^3/\text{lbma}$$

$$\dot{m}_{a0} = \frac{1000 \times 60}{12.53} = 4790 \text{ lbma/hr}$$

and

$$\dot{m}_{a4} = 20{,}400 - 4790 = 15{,}600 \text{ lbma/hr}$$

Using the graphical technique and referring to Fig. 3-22, we obtain

$$\frac{\overline{31}}{\overline{30}} = \frac{\dot{m}_{a0}}{\dot{m}_{a1}} = \frac{4790}{20{,}400} = 0.235$$

$$\overline{31} = 0.235(\overline{30})$$

State 1 is then located at 66 F db and 57 F wb. The line $\overline{12}$ constructed on Chart 1, Fig. 3-22, represents the heating and humidifying process that must take place in the heating and humidifying unit. An energy balance gives

$$\dot{m}_{a1} i_1 + \dot{q}_h + \dot{m}_w i_w = \dot{m}_{a2} i_2$$

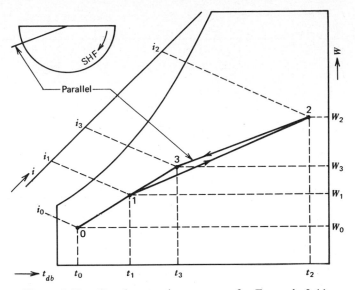

Figure 3-22. Psychrometric processes for Example 3-11.

and a mass balance on the water gives

$$\dot{m}_{a1}W_1 + \dot{m}_w = \dot{m}_{a2}W_2$$

or

$$\dot{m}_w = \dot{m}_{a1}(W_2 - W_1)$$

Then

$$\dot{q}_h = \dot{m}_{a1}(i_2 - i_1) - \dot{m}_{a1}(W_2 - W_1)i_w$$

Assume that ordinary tap water at 55 F is used in the humidifier. Then $i_w = 23.08$ Btu/lbm from Table A-1. From Chart 1, $i_1 = 24.6$ Btu/lbma, $W_1 = 0.008$ lbmw/lbma, and $W_2 = 0.0119$ lbmw/lbma.

$$\dot{q}_h = 20{,}400[(42 - 24.6) - (0.0119 - 0.008)23.08]$$
$$\dot{q}_h = 353{,}000 \text{ Btu/hr}$$

The amount of water supplied to the humidifier is

$$\dot{m}_w = \dot{m}_{a1}(W_2 - W_1) = 20{,}400(0.0119 - 0.008)$$
$$\dot{m}_w = 79.6 \text{ lbm/hr} = 1.33 \text{ lbm/min}$$

It is usually necessary to use a preheat coil to heat the outdoor air to a temperature above the dew point of the air in the equipment room so that condensation will not form on the air ducts upstream of the regular heating coil. Figure 3-23 shows this arrangement. The outdoor air is heated to state 0' where it is mixed with return air, resulting in state 1. The mixed air is then

Figure 3-23.   Heating system with preheat of outdoor air.

heated to state 1', where it is humidified to state 2 on the condition line for supply to the space. Figure 3-24 shows the states on Chart 1.

3-7   SPACE AIR CONDITIONING—OFF-DESIGN CONDITIONS

The previous section treated the common space air conditioning problem assuming that the system was operating steadily at the design condition. Actually the space requires only a part of the designed capacity of the conditioning equipment most of the time. A control system functions to match the required cooling or heating of the space to the conditioning equipment by varying one

Figure 3-24.   Psychrometric diagram for Figure 3-23.

or more system parameters. For example, the quantity of air circulated through the coil and to the space may be varied in proportion to the space load, as in the *variable-air-volume* (VAV) system. Another approach is to circulate a constant amount of air to the space, but some of the return air is diverted around the coil and mixed with air coming off the coil to obtain a supply air temperature that is proportional to the space load. This is known as face and bypass control, because face and bypass dampers are used to divert the flow. Another possibility is to vary the coil surface temperature with respect to the required load by changing the temperature or the amount of heating or cooling fluid entering the coil. This technique is usually used in conjunction with VAV and face and bypass systems. However, control of the coolant temperature or quantity may be the only variable in some small systems.

These and other control methods will be considered separately later. Figure 3-25*a* illustrates what might occur when the load on a variable air volume system decreases. The solid lines represent the full load design condition, whereas the broken lines illustrate a part-load condition where the amount of air circulated to the space and across the coil has decreased. Note that the state of the outdoor air 0′ has changed and could be almost anyplace on the chart under part-load conditions. Due to the lower air flow rate through the coil, the air is cooled to a lower temperature and humidity. The thermostat maintains the space temperature, but the humidity in the space will decrease. This explains why control of the water temperature or flow rate is desirable. Decreasing the water flow rate will cause point 2′ to move upward and to the right to a position where the room process curve will terminate at point 3.

The behavior of a constant air volume, face and bypass system is shown

(a)

Figure 3-25.   Psychrometric processes for off-design conditions. (*a*) Variable air-volume control.

Figure 3-25. Psychrometric processes for off-design conditions. (*b*) Face and bypass control. (*c*) Water-temperature control.

in Fig. 3-25*b*. The total design air flow rate is flowing at state 2, 3, and 3′ but a lower flow rate occurs at state 2′, leaving the coil. Air at states 2′ and 1′ is mixed downstream of the coil to obtain state 4. The total design flow rate and the enthalpy difference, $h_{3'} - h_4$, then match the space load. Note that the humidity at state 4 is lower than necessary, which lowers state 3′ below the design value. At very small space loads, point 4 may be located very near state 1′ on the condition line. In this case the humidity in the space may become high. This is a disadvantage of a multizone face and bypass system. Control of the coil water temperature can help to correct this problem.

A constant air volume system with either water temperature or flow rate control is shown in Fig. 3-25c. In this case both the temperature and humidity of the air leaving the coil increase and the room process curve 2'-3' may not terminate at state 3. In fact, there will be cases when state 3' will lie above state 3, causing an uncomfortable condition in the space. For this reason, water control alone is not usually used in commercial applications, but is used in conjunction with VAV and face and bypass systems as discussed before. In fact, all water coils should have control of the water flow rate. This is also important to the operation of the water chiller and piping system. The following example illustrates the analysis of a VAV system with variable water temperature.

Example 3-12.   A variable air volume system operates as shown in Fig. 3-26. The solid lines show the full-load design condition of 100 tons with a room SHF of 0.75. At the estimated minimum load of 15 tons with SHF of 0.9, the air flow rate is decreased to 20 percent of the design value and all outdoor air is shut off. Estimate the supply air and apparatus dew point temperatures of the cooling coil for minimum load, assuming that state 3 does not change.

Solution.   The solution is best carried out using Chart 1, as shown in Fig. 3-26. Because the outdoor air is off during the minimum load condition, the space condition and coil process lines will coincide as shown by line 3-2'-d'. This line is constructed by using the protractor of Chart 1 with a SHF of 0.9. The apparatus dew point is seen to be 55 F, as compared with 50 F for the design condition. The air flow rate for the design condition is given by

Figure 3-26.   Schematic psychrometric processes for Example 3-12.

$$\dot{m}_2 = \dot{q}/(i_3 - i_2)$$
$$\dot{m}_2 = 100(12,000)/(29.4 - 23.2) = 193,550 \text{ lbma/hr}$$

or

$$\dot{Q}_2 = \dot{m}_2 v_2/60 = 193,550(13.25)/60 = 42,740 \text{ cfm}$$

Then, the minimum volume flow rate is

$$\dot{Q}_m = 0.2(42,740) = 8,550 \text{ cfm}$$

and the minimum mass flow rate may be estimated by assuming a value for $v_{2'}$.

$$\dot{m}_m = 8550(60)/13.28 = 38,630 \text{ lbma/hr}$$

State point 2′ may then be determined by computing $i_{2'}$.

$$i_{2'} = i_3 - \frac{\dot{q}_m}{\dot{m}_m} = 29.4 - 15(12,000)/38,630 = 24.7 \text{ Btu/lbma}$$

Then from Chart 1, the air condition leaving the coil is 60.5 F db and 57.5 F wb. Calculation of the coil water temperature is beyond the scope of this analysis; however, the water temperature would be increased about 7 degrees from the design to the minimum load condition.

Reheat was mentioned as a variation on the simple constant flow and VAV systems to obtain control under part load conditions. Figure 3-27 shows how this affects the psychrometric analysis for a typical zone. After the air leaves the cooling coil at state 2, it is then heated to state 2′ and enters the zone at

Figure 3-27. A simple constant flow system with reheat.

Figure 3-28. Schematic of a dual-duct system.

a higher temperature to accommodate the part-load condition. A VAV reheat system operates similarly. The purpose of a dual-duct system is to adjust to highly variable conditions from zone to zone and to give accurate control. Figure 3-28 is a schematic of such a system showing one typical zone and Fig. 3-29 shows the processes on Chart 1. Part of the air is cooled to state 2 and part of the air is heated to state 3. A thermostat controls dampers in the terminal unit to mix the air at states 2 and 3 in the correct proportion, state 4, for supply

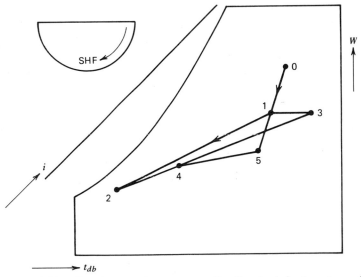

Figure 3-29. Psychrometric processes for the dual-duct system of Figure 3-28.

Figure 3-30. Psychrometric processes for an economizer
cycle.

to the zone. Process 4-5 is on the condition line for the zone. The flow rates
of the two air streams may also vary as with regular VAV.

The economizer cycle is a system used during part-load conditions when
outdoor temperature and humidity are favorable to saving operating energy.
One must be cautious in the application of such a system, however, if the
desired space conditions are to be maintained. Once the cooling equipment and
especially the coil have been selected, there are limitations on the quantity and
state of the outdoor air. The coil apparatus dew point can be used as a guide
to avoid impossible situations. For example, a system is designed to operate
as shown by the solid process lines in Fig. 3-30. Assume that the condition line
2-3, does not change, but state 0 changes to state 0′. Theoretically a mixed
state 1′ located anyplace on the line 0′-3 could occur, but the air must be cooled
and humidified to state 2. To do this the coil apparatus dew point must be
reasonable. Values below about 48 F are not economical to attain. Therefore,
state 1′ must be controlled to accommodate the coil. It can be seen in Fig.
3-30 that moving of state 1′ closer to state 0′ lowers the coil apparatus dew
point rapidly and soon reaches the condition where the coil process line will
not intersect the saturation curve, indicating an impossible condition. It is
obvious in Fig. 3-30 that less energy is required to cool the air from state 1′ to
2 than from state 1 to 2. There are many other possibilities, which must be
analyzed on their own merits. Some may require more or less outdoor air,
humidification, or reheat to reach state 2 in Fig. 3-30.

REFERENCES

1. J. A. Goff and S. Gratch, "Thermodynamics of Moist Air," *ASHVE Transactions,* Vol. 51, 1945.

2. R. W. Hyland and A. Wexler, "Formulations for the Thermodynamic Properties of the Saturated Phases of H_2O from 173.15 K to 473.15 K," *ASHRAE Transactions,* Vol. 89, Part 2A, 1983.

3. R. W. Hyland and A. Wexler, "Formulations for the Thermodynamic Properties of Dry Air from 173.15 K to 473.15 K, and of Saturated Moist Air from 173.15 K to 372.15 K, at Pressures to 5 MPa," *ASHRAE Transactions,* Vol. 89, Part 2, 1983.

4. James L. Threlkeld, *Thermal Environmental Engineering,* 2nd ed., Prentice-Hall, Englewood Cliffs, N.J., 1970.

5. *ASHRAE Handbook, Fundamentals Volume,* American Society of Heating, Refrigerating and Air-Conditioning Engineers, Atlanta, Ga., 1985.

6. R. B. Stewart, R. J. Jacobsen, and J. H. Becker, "Formulations for Thermodynamic Properties of Moist Air at Low Pressures as Used for Construction of New ASHRAE SI Unit Psychrometric Charts," *ASHRAE Transactions,* Vol. 89, Part 2, 1983.

PROBLEMS

3-1 Compute the local atmospheric pressure at elevations ranging from sea level to 6000 ft (1830 m) in (a) inches of mercury and (b) kPa.

3-2 Calculate values of humidity ratio, enthalpy, and specific volume for saturated air at 14.696 psia pressure using perfect gas relations and Table A-1 for temperatures of (a) 70 F, and (b) 20 F. (Note: All properties are per lbma.)

3-3 Rework Problem 3-2 for an atmospheric pressure corresponding to an elevation of (a) 5280 ft and (b) 1600 m.

3-4 A duct has moist air flowing at a rate of 5000 ft³/min (2.36 m³/s). What is the mass flow rate of the dry air, where the pressure inside the duct corresponds to (a) sea level, and (b) 6000 ft (1830 m)?

3-5 Compute the enthalpy of moist air at 60 F (16 C) and 80 percent relative humidity for an elevation of (a) sea level and (b) 5000 ft (1525 m).

3-6 The condition within a room is 70 F db, 50 percent relative humidity, and 14.696 psia pressure. The inside surface temperature of the windows is 40 F. Will moisture condense on the window glass?

3-7 Assume that the dimensions of the room of Problem 3-6 are 30 by 15 by 8 ft. Calculate the mass of water vapor in the room.

3-8 Moist air exists at a dew point temperature of 18 C, a relative humidity of 60 percent, and a pressure of 96.5 kPa. Determine (a) the humidity ratio and (b) the volume in m³/kga.

3-9 Compute the dew point temperature for moist air at 80 F (27 C) and 50 percent relative humidity for pressures corresponding to (a) sea level and (b) 5000 ft (1225 m).

3-10   The temperature of a certain room is 22 C and the relative humidity is 50 percent. The barometric pressure is 100 kPa. Find (a) the partial pressures of the air and water vapor, (b) the vapor density, and (c) the humidity ratio of the mixture.

3-11   Air with a dry bulb temperature of 70 F and a wet bulb temperature of 65 F is at a barometric pressure of 29.92 in. Hg. Without making use of the psychrometric chart find (a) the relative humidity of the air, (b) the vapor density, (c) the dew point temperature, (d) the humidity ratio, and (e) the volume occupied by the mixture associated with a pound mass of dry air.

3-12   Air is supplied to a room at 72 F db and 68 F wb from outside air at 40 F db and 37 F wb. The barometric pressure is 29.92 in. Hg. Find (a) the dew point temperatures of the inside and outside air, (b) the moisture added to each pound of dry air, (c) the enthalpy of the outside air, and (d) the enthalpy of the inside air.

3-13   Air is cooled from 75 F db and 70 F wb until it is saturated at 55 F. Find (a) the moisture moved per pound of dry air, (b) the heat removed to condense the moisture, (c) the sensible heat removed, and (d) the total amount of heat removed.

3-14   The dry bulb and thermodynamic wet bulb temperature are measured to be 75 F and 62 F, respectively, in a room. Compute the humidity ratio and relative humidity for the air at (a) sea level and (b) 5000 ft (1225 m).

3-15   To save energy, the environmental conditions in a room are to be regulated so that the dry bulb temperature will be greater than or equal to 78 F and the dew point temperature will be less than or equal to 64 F. Compute the maximum relative humidity that can occur for standard barometric pressure.

3-16   A chilled water cooling coil receives 2.5 m³/s of air at 25 C db, 20 C wb. It is necessary for the air to leave the coil at 13 C db, 12 C wb. Assume sea level pressure.
        (a) Determine the SHF and the apparatus dew point temperature.
        (b) Compute the bypass factor.
        (c) Compute total and sensible heat transfer rates from the air using enthalpy difference and the SHF.
        (d) Compute the total and sensible heat transfer rate from the air using the bypass factor and the SHF.
        (e) Compare the results of (c) and (d).

3-17   The dry bulb and wet bulb temperatures are measured to be 78 F and 65 F, respectively, in an air duct. Make use of psychrometric Charts 1 and 1a to find the enthalpy, specific volume, humidity ratio, and relative humidity in (a) English units and (b) SI units.

3-18   The air in Problem 3-17 is heated to a temperature of 110 F. Make use of Charts 1 and 1a and compute the heat transfer rate if 4000 ft³/min (1.9 m³/s) is flowing at state 1 in (a) English units and (b) SI units.

3-19   The air in Problem 3-17 is cooled to 50 F (10 C) dry bulb and 90 percent relative humidity. Make use of Charts 1 and 1a to compute the total heat transfer rate, the sensible heat transfer rate, the latent heat transfer rate, and the sensible heat factor (SHF) if 4000 ft³/min (1.9 m³/s) is flowing at state 1 in (a) English units and (b) SI units.

3-20 Air at 100 F db and 65 F wb is humidified adiabatically with steam. The steam supplied contains 20 percent moisture (quality of 0.80) at 14.7 psia. If the air is humidified to 60 percent relative humidity, what is the dry bulb temperature of the humidified air? The barometer is at 29.92 in. Hg.

3-21 Air at 84 F db and 60 F wb and at 29.92 in. Hg is humidified with the dry bulb temperature remaining constant. Wet steam is supplied for humidification at 20 psia. What quality must the steam have (a) to provide saturated air and (b) to provide air at 70 percent relative humidity?

3-22 Rework Problem 3-20 for air at 38 C and 18 C dry bulb and wet bulb temperatures, respectively, with steam at 101.3 kPa using Chart 1*a*.

3-23 Rework Problem 3-21 for standard sea level pressure and dry bulb and wet bulb temperatures of 30 C and 16 C, respectively, with steam at 140 kPa pressure.

3-24 Air at 38 C db and 20 C wb and 101.325 kPa is mixed adiabatically with liquid water supplied at 60 C, in such proportions that the mixture has a relative humidity of 80 percent. Find the dry bulb temperature of the mixture.

3-25 It is desired to heat and humidify 2000 cfm (1.0 m³/s) of air from an initial state defined by a temperature of 60 F (16 C) dry bulb and relative humidity of 30 percent to a final state of 110 F (43 C) dry bulb and 30 percent relative humidity. The air will first be heated by a hot water coil followed by adiabatic humidification using saturated vapor at 5 psig (34.5 kPa). Using the psychrometric chart, find the heat transfer for the heating coil and the mass flow rate of the water vapor, and sketch the processes on a skeleton chart showing pertinent data. Use (a) English units and (b) SI units.

3-26 Air at 40 F (5 C) db and 35 F (2C) wb is mixed with warm air at 100 F (38 C) db and 77 F (25 C) wb in the ratio of 2 lbm(kga) cool air to 1 lbm(kga) of warm air. (a) Compute the humidity ratio and enthalpy of the mixed air and (b) find the humidity ratio and enthalpy using the psychrometric chart.

3-27 Air at 10 C db and 5 C wb is mixed with air at 25 C db and 18 C wb in a steady-flow process at standard atmospheric pressure. The volume flow rates are 10 m³/s and 6 m³/s, respectively. (a) Compute the mixture conditions. (b) Find the mixture conditions using Chart 1*a*.

3-28 Rework Problem 3-27, using Chart 1*a*, assuming that the mixture condition can be computed on the basis of the volume flow rates rather than mass flow rate. What is the percent error in the mixture enthalpy and humidity ratios?

3-29 A zone in a building has a design cooling load of 136,000 Btu/hr (40 kW) of which 110,000 Btu/hr (32 kW) is sensible cooling load. The space is to be maintained at 76 F (24 C) dry bulb temperature and 50 percent relative humidity. Locate the space condition line on Charts 1 and 1*a*.

3-30 Refer to Problem 3-29 and assume that the air can be supplied to the space at 55 F (13 C). Compute the volume flow rate of the air required in (a) English units and (b) SI units.

3-31 Rework Problem 3-29 using Charts 1H and 1H*a* for (a) 5000 ft and (b) 1500 m elevation, respectively.

3-32 Rework Problem 3-30 using Charts 1H and 1Ha for (a) 5000 ft and (b) 1500 m elevation, respectively.

3-33 A meeting hall is to be maintained at 25 C db and 18 C wb. The barometric pressure is 101.3 kPa. The space has a load of 58.6 kW sensible and 58.6 kW latent. The temperature of the supply air to the space cannot be lower than 18 C db. (a) How many kg/s of air must be supplied? (b) What is the required wet bulb temperature of the supply air? (c) What is the sensible heat ratio?

3-34 A space is to be maintained at 72 F (22 C) and 30 percent relative humidity during the winter months. The sensible heat loss from the space is 500,000 Btu/hr (146 kW) and the latent heat loss due to infiltration is 50,000 Btu/hr (14.6 kW). Construct the condition line on (a) Chart 1 and (b) Chart 1a.

3-35 What is the maximum relative humidity to prevent condensation on pipes carrying water at 50 F through a room that has an air temperature of 70 F?

3-36 A space is to be maintained at 21 C dry bulb. It is estimated that the inside wall surface temperature could be as low as 7 C. What maximum relative and specific humidities can be maintained without condensation on the walls?

3-37 Outdoor air with a temperature of 40 F db and 35 F wb and with a barometric pressure of 29 in. Hg is treated and humidified under steady-flow conditions to a final temperature of 70 F db and 40 percent relative humidity. (a) Find the mass of water vapor added to each pound mass of dry air. (b) If the water is supplied at 50 F, how much heat is added per pound mass of dry air?

3-38 Outdoor air at 95 F db and 79 F wb and at a barometric pressure of 29.92 in. Hg is cooled and dehumidified under steady conditions until it becomes saturated at 60 F. (a) Find the mass of water condensed per pound of dry air. (b) If the condensate is removed at 60 F, what quantity of heat is removed per pound of dry air?

3-39 Outdoor air at 38 C db and 26 C wb is cooled to 15 C db and 14 C wb. The process occurs at 101.3 kPa barometric pressure. Determine (a) the mass of water condensed per kilogram of dry air and (b) the sensible and latent heat transfer per kilogram of dry air and the sensible heat factor.

3-40 Moist air enters a refrigeration coil at 89 F db and 75 F wb at a rate of 1400 cfm. The apparatus dew point temperature of the coil is 55 F. (a) If 3.5 tons of refrigeration are available, find the dry bulb temperature of the air leaving the coil. (b) Compute the bypass factor. Assume sea level pressure.

3-41 Saturated steam at a pressure of 25 psia is sprayed into a stream of moist air. The initial condition of the air is 55 F db and 45 F wb temperature. The mass rate of air flow is 2000 lbma/min. Barometric pressure is 14.696 psia. Determine (a) how much steam must be added in lbm/min to produce a saturated air condition and (b) the resulting temperature of the saturated air.

3-42 Saturated water vapor at 100 C is used to humidify a stream of moist air. The air enters the humidifier at 13 C db and 2 C wb at a flow rate of 2.5 m³/s. The pressure is 101.35 kPa. Determine (a) the mass flow rate of the steam required to saturate the air and (b) the temperature of the saturated air.

3-43 Moist air at 70 F db and 45 percent relative humidity is recirculated from a room and mixed with outdoor air at 97 F db and 83 F wb. Determine the mixture dry bulb and wet bulb temperatures if the volume of recirculated air is three times the volume of outdoor air. Assume sea level pressure.

3-44 A structure has a calculated cooling load of 10 tons of which 2.5 tons is latent load. The space is to be maintained at 76 F db and 50 percent relative humidity. Ten percent by volume of the air supplied to the space is outdoor air at 100 F db and 50 percent relative humidity. The air supplied to the space cannot be less than 56 F db. Assume sea level pressure and find (a) the minimum amount of air supplied to the space in cfm, (b) the amounts of return air and outdoor air in cfm, (c) the conditions and volume flow rate of the air entering the cooling coil, and (d) the capacity and SHF for the cooling coil.

3-45 Rework Problem 3-44 for an elevation of 5000 feet.

3-46 A building has a calculated cooling load of 410 kW. The latent portion of the load is 100 kW. The space is to be maintained at 25 C db and 50 percent relative humidity. Outdoor air is at 38 C and 50 percent relative humidity, and 10 percent by mass of the air supplied to the space is outdoor air. Air is to be supplied to the space at not less than 18 C. Assume sea level pressure and find (a) the minimum amount of air supplied to the space in m³/s, (b) the volume flow rates of the return air, exhaust air, and outdoor air, (c) the condition and volume flow rate of the air entering the cooling coil, and (d) the capacity, apparatus dew point temperature, bypass factor, and SHF of the cooling coil.

3-47 Rework Problem 3-46 for an elevation of 1500 m.

3-48 A building has a total heating load of 200,000 Btu/hr. The sensible heat factor for the space is 0.8. The space is to be maintained at 72 F db and 40 percent relative humidity. Outdoor air at 40 F db and 20 percent relative humidity in the amount of 1000 cfm is required. Air is supplied to the space at 120 F db. Find (a) the conditions and amount of air supplied to the space, (b) the temperature rise of the air through the furnace, (c) the amount of water at 50 F required by the humidifier, and (d) the capacity of the furnace. Assume sea level pressure.

3-49 Rework Problem 3-48 for an elevation of 5000 feet.

3-50 The system of Problem 3-44 has a supply air fan located just downstream of the cooling coil. The total power input to the fan is 2.0 hp. It is also estimated that heat gain to the supply duct system is 500 Btu/hr. Rework Problem 3-44 taking the fan and duct heat gain into account. Make a sketch of the processes.

3-51 The system of Problem 3-46 has a supply air fan located just downstream of the coil and a return air fan just upstream of the mixing box. The power input to the supply fan is 18 kW and the power input to the return fan is 12 kW. Rework Problem 3-46 taking the fan power into account. Make a sketch of the system processes on a skeleton psychrometric chart.

3-52 A large warehouse located in Denver, Colorado (elevation = 5000 feet or 1500 m), is to be conditioned using an evaporative cooling system. Assume that the space is to be maintained at 80 F (27 C) and 50 percent relative humidity by a 100 percent outdoor air system. Outdoor design conditions are 91 F (33 C) db and 59 F (15C) wb. The cooling load is estimated to be 100 tons (352 kW) with a sensible heat factor of 0.9. The supply air fan is located just downstream of the spray chamber and is estimated to require a power input of 20 hp (15 kw). Determine the volume flow rate of air to the space and sketch the processes on a skeleton psychrometric chart in (a) English units and (b) SI units.

3-53 Consider a conventional cooling system designed for use at high elevation (5000 ft or 1500 m). The space is to be maintained at 75 F (24 C) db and 40 percent relative humidity, and outdoor design conditions are 100 F (38 C) and 10 percent relative humidity. Outdoor air is to be mixed with return air such that it can be cooled sensibly to 50 F (10 C) where it crosses the condition line. The air is then supplied to the space. Sketch the processes on Chart 1H or 1H*a* and compute the volume flow rate of the supply air and the percent outdoor air by mass per ton of cooling load for a SHF of 0.7 in (a) English units and (b) SI units.

3-54 A space heating system is designed as shown in Fig. 3-23 for a large zone in a building. Under design conditions for Kansas City, Missouri, air enters the preheat coil at 6 F (-14 C) and essentially 0 percent relative humidity. The outdoor air is heated to 60 F (16 C) and mixed with return air. It is then heated and humidified in separate processes to 105 F (40 C) and 30 percent relative humidity (RH) for supply to the space. Saturated vapor at 2.0 psig is used for humidification. Twenty-five percent of the supply air is outdoor air by mass. The total space heating load is 500,000 Btu/hr (145 kW) and the space design conditions are 70 F (21 C) and 30 percent RH. Sketch the psychrometric processes and compute the supply air volume flow rate, the heat transfer rates in both coils, and the steam flow rate in (a) English units and (b) SI units.

3-55 A space is to be maintained at 78 F (26 C) db and 68 F (20 C) wb. The cooling system is a variable air volume (VAV) type where the quantity of air supplied and the supply air temperature are controlled. Under design conditions, the total cooling load is 150 tons (530 kW) with a sensible heat factor of 0.6 and the supply air temperature is 60 F (16 C) db. At minimum load, about 18 tons (63 kW) with SHF of 0.8, the air quantity may be reduced no more than 80 percent by volume of the full load design value. Determine the supply air conditions for minimum load. Show all of the conditions on a psychrometric chart for (a) English units and (b) SI units. Assume sea level pressure.

3-56 Rework Problem 3-55 for an elevation of 5000 feet (1500 m).

3-57 Investigate the feasibility of conditioning a space with 100 percent outdoor air and a direct expansion cooling coil that requires a fixed volume flow rate of air per ton of cooling capacity. Explain. Assuming this is not feasible with a fixed flow rate, describe a different kind of coil and cooling medium to accomplish the process.

3-58 A 50-ton constant-volume space air-conditioning system uses face and bypass and water temperature control. At the design condition the space is to be maintained at 77 F (25 C) db and 50 percent relative humidity with 55 F (13 C) db supply air at 90 percent relative humidity. Outdoor air is supplied at 95 F (35 C) db, 60 percent relative humidity with a ratio of 1 lbm (kgm) to 5 lbm (kgm) return air. A part-load condition exists where total space load decreases by 50 percent

and the SHF increases to 90 percent. The outdoor air condition changes to 85 F (29 C) db and 70 percent relative humidity. Assume sea level pressure.
(a) At what temperature must the air be supplied to the space under the part-load condition?
(b) If the air leaving the coil has a dry bulb temperature of 60 F (15 C), what is the ratio of the air bypassed to that flowing through the coil?
(c) What is the apparatus dew point temperature for both the design and part-load conditions?
(d) Show all the processes on a psychrometric chart.

3-59 Rework Problem 3-58 for an elevation of 5000 feet (1500 m).

3-60 A condition exists where it is necessary to cool and dehumidify air from 80 F db, 67 F wb to 60 F db and 54 F wb.
(a) Discuss the feasibility of doing this in one process with a cooling coil. (*Hint:* Determine the apparatus dew point temperature for the process.)
(b) Describe a practical method of achieving the required process and sketch on a psychrometric chart.

3-61 Consider one zone of a dual-duct conditioning system. Conditions in the zone are to be maintained at 75 F (24 C) and 50 percent relative humidity (RH). Cold deck air is at 51 F (11 C) and 90 percent RH while the hot deck air is outdoor air at 90 F (32 C) and 30 percent RH. The sensible heat factor for the zone is 0.65. Assume sea level pressure. In what proportion must the warm and cold air be mixed to satisfy the space condition? If the total space load is 60 tons (210 kW), what is the toal volume flow rate of air supplied to the zone? Sketch the states and processes on a psychrometric chart. (a) Use English units and (b) SI units.

3-62 Rework Problem 3-61 for an elevation of 5000 ft (1500 m).

3-63 In Problem 3-61 return air is cooled by a water coil to the cold deck condition. Sketch the processes for the entire system on a psychrometric chart and determine the coil load (for the one zone) and volume flow rate in (a) English units and (b) SI units.

3-64 Calculate the humidity ratio of moist air at 78 F db with degree of saturation equal to 0.4 using Table A-2. Barometric pressure is 29.92 in. Hg.

3-65 During the winter months it is possible to cool and dehumidify a space using outdoor air. Suppose an interior zone of a large building is designed to have a supply air flow rate of 5000 cfm, which can be all outdoor air. The cooling load is constant at 10 tons with a SHF of 0.8 the year round. Indoor conditions are 78 F db and 67 F wb.
(a) What is the maximum outdoor air dry bulb temperature and humidity ratio that would satisfy the load condition?
(b) Consider a different time when the outdoor air has a temperature of 40 F db and 20 percent relative humidity. Return air and outdoor air may be mixed to cool the space, but humidification will be required. Assume that saturated water vapor at 14.7 psia is used to humidify the mixed air and compute the amounts of outdoor and return air in cfm.

(c) At another time, outdoor air is at 70 F db with a relative humidity of 90 percent. The cooling coil is estimated to have a minimum apparatus dew point of 50 F. What amount of outdoor and return air should be mixed before entering the coil to satisfy the given load condition?

(d) What is the refrigeration load for the coil of part (c) above?

3-66 An economizer mixes outdoor air with room return air to reduce the refrigeration load on the cooling coil.

(a) For a space condition of 25 C db and 20 C wb, describe the maximum wb and db temperatures that will reduce the coil load.

(b) Suppose a system is designed to supply 5 m³/s at 18 C db and 17 C wb to a space maintained at the conditions given in (a) above. What amount of outdoor air at 20 C db and 90 percent relative humidity can be mixed with the return air if the coil SHF is 0.6?

(c) What is the apparatus dew point and the bypass factor in (b) above?

(d) Compare the coil refrigeration load in (b) above with the outdoor air to that without outdoor air.

CHAPTER FOUR

Comfort and Health

Air conditioning was defined in Chapter 1 as the simultaneous control of temperature, humidity, cleanliness, odor, and air circulation as required by the occupants of the space. In this chapter, we shall be concerned with the conditions that actually provide a comfortable and healthful environment. Experience has shown that not everyone within a given space can be made completely comfortable by one set of conditions. This is due to a number of factors, many of which cannot be completely explained. However, clothing, age, sex, and the level of activity of each person are considerations. The factors that influence comfort are temperature, radiation, humidity, air motion, and the quality of the air with regard to odor, dust, and bacteria. With a complete air conditioning system all of these factors may be controlled simultaneously. It is found that in most cases a reasonably comfortable environment can be maintained when two or three of these factors are controlled. The *ASHRAE Handbook, Fundamentals Volume* (1) is probably the most up-to-date and complete source of information relating to the physiological aspects of thermal comfort. ASHRAE comfort standard 55 (11) defines acceptable thermal comfort as an environment that at least 80 percent of the occupants will find thermally acceptable. This definition gives insight into the complex problem of making as many people as possible comfortable with one set of conditions.

4-1 PHYSIOLOGICAL CONSIDERATIONS

A complex regulating system in the body acts to maintain the deep body temperature at approximately 98.6 F or 36.9 C. If the environment is maintained at suitable conditions so that the body can easily maintain an energy balance, a feeling of comfort will result. When the environmental conditions transfer energy away from the body too rapidly or if the environment is so warm that there is a problem in transferring energy from the body, discomfort will result. Obviously, if these conditions are carried to an extreme and the regulating system cannot maintain the normal body temperature, unhealthy conditions result.

Two basic mechanisms within the body seem to control the body temperature. The first is a decrease in the internal energy production as the body temperature rises or falls. This internal process is generally called metabolism. The second is the control of the rate of heat dissipation by changing the rate of cutaneous blood circulation (the blood circulation near the surface of the skin). In this way heat transfer from the body can be increased or decreased.

All forms of heat transfer as well as mass transfer are important in maintaining body temperature. Heat transfer to or from the body is principally by convection and conduction and, therefore, the air motion in the immediate vicinity of the body is a very important factor. Radiation exchange between the body and surrounding surfaces, however, can be a deciding factor in achieving comfort if the surfaces surrounding the body are at a different temperature than the air.

Another very important regulatory function of the body is sweating. Under very warm conditions great quantities of moisture can be released by the body to help cool itself.

The metabolic rate depends on the level of activity such as rest, work, or exercise. Naturally the higher the metabolic rate, the higher the heat transfer rate from the body to the environment to maintain an energy balance. A general knowledge of the amounts of energy expended is useful to engineering. Table 4-1, which is taken from the *ASHRAE Handbook* lists approximate metabolic rates for various types of activities. Notice that a new unit is introduced called the "met." One met is equal to 18.4 Btu/(hr-ft^2) or 58.2 W/m^2 of surface area,

Table 4-1   METABOLIC RATE AT DIFFERENT
TYPICAL ACTIVITIES

Activity	Metabolic Rate in Met Units
Reclining	0.8
Seated, quietly	1.0
Sedentary activity	1.2
(office, dwelling, lab, school)	
Standing, relaxed	1.2
Light activity, standing	1.6
(shopping, lab, light industry)	
Medium activity, standing	2.0
(shop assistant, domestic work, machine work)	
High activity	3.0
(heavy machine work, garage work)	

Source: Abridged by permission from the *ASHRAE Handbook of Fundamentals,* 1977.

and is based on the approximate energy generated by an average sedentary man. This corresponds to approximately 350 Btu/hr or 105 W. A normal healthy man 20 years of age has a maximum energy capacity of approximately 12 mets, and he can maintain about 50 percent of his maximum capacity on a continuous basis. The maximum levels for women tend to be about 30 percent lower than those for men. Well-trained athletes may have a maximum energy capacity as high as 20 mets.

Tolerance to high temperature and humidity depends on ability to transfer heat to such an environment. This takes place to a large extent through sweating. There is a limit to the length of time that sweating can continue because this process means that water in the body will be depleted. This dehydration leads to reduced sweating, and a rise in body temperature; the result is a heat stroke. A practical cooling limit is about 110 Btu/(ft²-hr) or 350 W/m², which corresponds to approximately 0.037 lbm/min or 0.28 g/s of sweating for the average man.

Survival in extreme cold is determined by the ability to maintain a thermal energy balance. The problem is to reduce the rate of heat loss from the body, or to generate energy at a rate equal to that at which it is being lost. A person usually maintains this heat balance by wearing adequate clothing. From a practical viewpoint the lower limit for useful outdoor activity while wearing the best clothing is about − 30 F or − 34 C.

4-2   ENVIRONMENTAL INDICES

There are many parameters to describe the environment in terms of comfort. These indices are generally subdivided into three groups entitled "the direct," "the rationally derived," and "the empirical." Not all of these indices are directly useful by the design engineer in his or her day-to-day work and many of these are defined mainly for research purposes. However, we shall mention all of them to provide some background for the discussion to follow.

Direct Indices

The direct indices are most useful to the design engineer, with *dry bulb temperature* the single most important index of comfort. This is especially true when the relative humidity is between 40 and 60 percent. The dry bulb temperature is especially important for comfort in the colder regions. When humidity is high, the significance of the dry bulb temperature is less.

The *dew point temperature* is a good single measure of the humidity of the environment, and is directly related to the water vapor pressure in saturated air. The usefulness of the dew point temperature in specifying comfort conditions is, however, limited.

The *wet bulb temperature* is useful in describing comfort conditions in the regions of high temperature and high humidity where dry bulb temperature has less significance. For example, the upper limit for tolerance of the average individual with normal clothing is a wet bulb of about 86 F or 30 C when the air movement is in the neighborhood of 50 to 75 ft/min or 0.25 to 0.38 m/s.

Relative humidity, although considered to be a direct index, has no real meaning in terms of comfort unless the accompanying dry bulb temperature is also known. Very high or very low relative humidity is generally associated with discomfort, however.

The last direct index, *air movement,* is the most difficult of the direct indices to describe. The convective heat transfer from the body depends on the velocity of the air moving over it. Evidence shows that one is more comfortable in a warm humid environment if the air movement is high. Yet if the temperature is low, one becomes uncomfortable if the air movement is too high. Generally, when air motion in the vicinity of an individual is in the neighborhood of 50 ft/min or 0.25 m/s, the average person will be comfortable.

Rationally Derived Indices

The rationally derived indices have less direct use in design. Nevertheless they form a basis from which we can draw various conclusions about comfort conditions. The *mean radiant temperature,* for example, is the uniform surface temperature of an imaginary black enclosure with which a person, also assumed to be a black body, exchanges the same heat by radiation as in the actual environment. This parameter is important when surrounding surfaces are at a temperature different from the body.

The *operative temperature* is the uniform temperature of an imaginary enclosure with which an individual exchanges the same heat by radiation and convection as in the actual environment. This index attempts to include the effect of convection as well as radiation. For usual practical applications the operative temperature is the mean of the dry bulb and mean radiant temperatures at a given location in the space and is referred to as the *adjusted dry bulb* temperature (11).

The *humid operative temperature* is the uniform temperature of an environment at 100 percent relative humidity with which a person will exchange the same heat from the skin surface by radiation, convection, conductance through clothing, and evaporation as in the actual environment.

Note that the three temperatures just defined progressively bring in the effect of radiation, convection, and mass transfer in the cooling or heating of the human body, as the case may be. The humid operative temperature is particularly important because it takes into account all three of the heat transfer mechanisms employed by the body to maintain a constant temperature. The humid operative temperature also involves *skin wettedness,* defined later, which is related to sweating and is a physiological rather than an environmental variable.

The *heat stress index* is the ratio of the total evaporative heat loss required for thermal equilibrium of the body to the maximum possible evaporative heat lost to the environment. Closely related to this is the index of *skin wettedness,* which is the ratio of the observed skin sweating to the maximum sweating

possible for the environment as defined by the skin temperature, air temperature, humidity, air movement, and clothing. Skin wettedness has been shown
to be a more reliable parameter for describing a sense of discomfort or un-
pleasantness than the usual temperature sensation.

Figure 4-1 shows a comparison of the humid operative temperature, lines
of constant wettedness, comfort, and tolerance plotted on a psychrometric
chart (4). This plot is an initial step toward obtaining a more directly useful
comfort chart. At the lower dry bulb temperatures in the neighborhood of the
comfortable region, the humid operative temperatures and the lines of constant
wettedness have essentially the same trends. The divergence of these two
indices may be noted in the upper right of the chart. This indicates the increased
importance of humidity in the higher temperature regions.

Empirical Indices

The empirical indices are attempts to use one particular parameter to
include all the effects in the environment that influence comfort. The first of

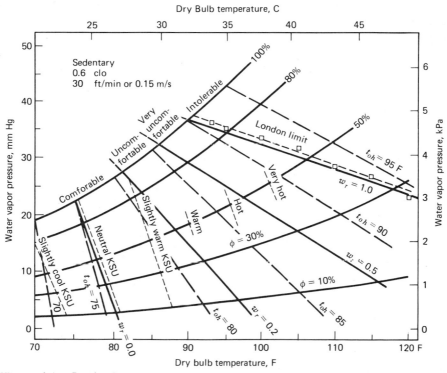

Figure 4-1.   Loci of constant humid operative temperature t_{oh} and of constant
wettedness l_r compared with KSU measurement of temperature sensation, Pierce Laboratory observations of warm discomfort, and the
London Limit for Heat Tolerance. (Adapted by permission from
ASHRAE Handbook, Fundamentals Volume, 1977.)

these and probably the oldest is the *effective temperature, ET*. It attempts to combine the effect of dry bulb, wet bulb, and air movement to yield equal sensations of warmth or cold. In other words, as humidity becomes high, the dry bulb temperature needs to be lower and as humidity becomes lower, dry bulb temperature needs to be higher to give the same sensation of comfort. The original effective temperature concept over-emphasizes the effect of humidity in the cooler regions, underemphasizes the effect of humidity under warm conditions, and does not fully account for air motion under hot humid conditions. Finally its use is limited to sedentary conditions.

The *black globe temperature* (5) is the equilibrium temperature of a 6-in. (150-mm) diameter black globe and is used as a single temperature index describing the combined physical effects of dry bulb temperature, air movement, and the radiant energy received from various surrounding areas.

The *corrected effective temperature* (5) is the black globe temperature used in place of the dry bulb temperature of the original effective temperature scale to include the effects of any intense radiant heat source in the surrounding environment.

The *wet bulb globe temperature* (6) is a weighted average of the dry bulb, a naturally convected wet bulb, and the globe temperature according to the relationship

$$WBGT = 0.7t_{wb} + 0.2t_g + 0.1t_{db} \qquad (4\text{-}1)$$

The index includes the combined effect of low-temperature radiant heat, solar radiation, and air movement.

The last empirical index is the *wind chill index* (7). This parameter attempts to include the effects of air motion and dry bulb temperature on comfort.

4-3   COMFORT CONDITIONS

Intensive work in the field of comfort has been conducted at Kansas State University (3, 8), and most of the following is based on that work. A book on thermal comfort has been written by Fanger (2) and is useful as a background on the subject.

The perception of comfort, temperature, and thermal acceptability is related to one's rate of metabolic heat production, its rate of transfer to the environment, and the resulting physiological adjustments and body temperatures. The heat transfer rate is influenced by the environmental factors of air temperature, thermal radiation, air movement, and humidity, and by the personal factors of activity and clothing. Thermal sensations can be described by feelings of hot, warm, slightly warm, neutral, slightly cool, cool, and cold. Judgment as to whether the environment is thermally acceptable is related to environmental parameters and to thermal sensation. In a uniform thermal environment the 80 percent thermal acceptability limits occur at conditions that produce thermal sensations near slightly cool and slightly warm. Clothing, through its insulation properties, is an important modifier of body heat loss and comfort. Clothing insulation can be described in terms of its clo value [1 clo = 0.88 (ft²-hr-F)/Btu = 0.155 (m²-C)/W]. A heavy two-piece business suit and accessories has an insulation value of about 1 clo, whereas a pair of shorts is about 0.05 clo. The operative or adjusted db temperatures and clo values corresponding to the optimum sensation of neutral, and the 80 percent thermal acceptability limits of Standard 55 (11) are given in Fig. 4-2.

The insulation value of clothing worn indoors is influenced by the season and outside weather conditions. During the summer months, typical clothing in commercial establishments consists of lightweight dresses, lightweight slacks, short-sleeved shirts or blouses, and accessories. These ensembles have insulation values ranging from 0.35 to 0.6 clo. In the winter heating season thicker, heavier clothing is worn. A typical winter ensemble may include heavy slacks or skirt, long-sleeved shirt or blouse, warm sweater or jacket, and appropriately warm accessories such as stockings, shoes, and underwear. These ensembles have insulation values ranging from 0.8 to 1.2 clo. At other times, between seasons, the clothing may have an insulation value in the range of 0.6 to 0.8 clo. Because of the seasonal clothing habits of building occupants, the temperature range for comfort in summer is higher than for winter. The acceptable range of operative temperatures and humidities for the winter and summer is defined on the psychrometric chart of Fig. 4-3. The zones overlap in the

Figure 4-2.   Clothing insulation necessary for various levels of comfort at a given temperature. (Adapted by permission from ASHRAE Standard 55-81, 1981.)

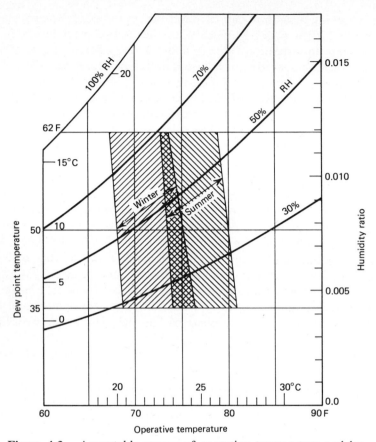

Figure 4-3. Acceptable ranges of operative temperature and humidity for persons clothed in typical summer and winter clothing, at light, mainly sedentary, activity. (Adapted by permission from ASHRAE Standard 55-81, 1981.)

73–75 F (23–24 C) range. In this region people in summer dress would tend to be slightly cool, whereas those in winter clothing would be near the slightly warm sensation. Due to individual, clothing, and activity differences, the boundaries of each comfort zone are not actually as sharp as shown in Fig. 4-3.

The air temperature in a room generally increases from floor to ceiling. If this increment is sufficiently large, local discomfort can occur. To prevent this, the vertical air temperature difference between head and ankles should not exceed 5 F (3 C).

To minimize foot discomfort, the surface temperature of the floor for people wearing appropriate indoor footwear should be between about 65 F (18.3 C) and 84 F (29 C).

Comfort conditions for clothing levels different from those just described can be determined approximately by lowering the temperature ranges of Fig. 4-3 by 1 F (0.6 C) for each 0.1 clo of increased clothing.

In general, children below 12 years of age require a reduction in temperature of 1 F or 0.6 C, whereas adults over 60 may require an increase of 1 F or 0.6 C.

Humidity is described in terms of dew point temperature. In the zone occupied by sedentary or near sedentary people the dew point temperature should not be less than 35 F (1.7 C) or greater than 62 F (16.7 C). The thermal effect of humidity on the comfort of sedentary persons is small (Fig. 4-3). The upper and lower humidity limits are based on considerations of comfort, respiratory health, mold growth, and other moisture-related phenomena. It should be noted that humidification in winter may need to be limited to prevent condensation on windows and other building surfaces. To minimize respiratory distress in winter, relative humidity should be as high as the structure will allow without condensation, but not more than 50 percent.

Within the thermally acceptable temperature ranges of Fig. 4-3, there is no minimum air movement that is necessary for thermal comfort. The maximum average air movement allowed in the occupied zone is lower in winter than in summer. In winter, the average air movement in the occupied zone should not exceed 30 fpm (0.15 m/s). If the temperature is less than the optimum, the maintenance of low air movement is important to prevent local draft discomfort.

In summer, the average air movement in the occupied zone should not exceed 50 fpm (0.25 m/s). The comfort zone can be extended above 79 F (26 C), however, if the average air movement is increased 30 fpm (0.152 m/s) for each degree F (0.6 C) of increased temperature to a maximum temperature of 82.5 F (28 C).

The mean radiant temperature can be as important as air temperature in effecting heat loss and comfort. The mean radiant temperature is given by

$$t_{mr} = t_{bg} + C\overline{V}^{1/2}(t_{bg} - t_a) \tag{4-2}$$

where

t_{bg} = black globe temperature, F or C
C = constant, 0.157 English units and 2.20 SI units
$\overline{V}$ = air velocity, ft/min or m/s
t_a = air dry bulb temperature, F or C

For an indoor environment, where air movement is low, the operative temperature is approximately the average of air temperature and mean radiant temperature. When the mean radiant temperature in an occupied zone differs from the air temperature, adjustments should be made to keep the operative temperature within the appropriate comfort zone. With the increasing need to conserve energy, adjustments may be mandated in the allowable winter and summer temperatures. The extent to which comfort can be achieved by clothing adjustments is indicated in Fig. 4-2.

It is often necessary for an engineer to determine the acceptability of comfort conditions by making measurements. ASHRAE Standard 55 describes procedures for making these measurements and analyzing the data.

Example 4-1.   Determine the operative temperature for a work station in a room near a large window where the dry bulb and black globe temperatures are measured to be 75 F and 81 F, respectively. The air velocity is estimated to be 30 ft/min at the station.

Solution.   The operative temperature depends on the mean radiant temperature, which is given by Eq. 4-2:

$$t_{mr} = 81 + 0.157(30)^{1/2}(81 - 75) = 86 \text{ F}$$

Then the operative temperature is

$$t_o = \frac{t_a + t_{mr}}{2} = \frac{75 + 86}{2} = 81 \text{ F}$$

A person working at this location would probably be uncomfortable.

High Activity

The comfort conditions discussed so far relate to sedentary activity, which accounts for most applications. However, in research laboratories, machine shops, assembly lines, and general manufacturing areas, for example, the occupants may be engaged in rather active work. The comfort zone temperatures of Fig. 4-3 should be decreased when the average steady-state activity level of the occupants is higher than sedentary or slightly active (1.2 met). The acceptable operative temperature depends on both the time average activity level and the clothing insulation. The temperature can be calculated with the following equation, which is appropriate only between 1.1 and 3 mets:

$$t_{o_{\text{active}}} = t_{o_{\text{sedentary}}} - 5.4(1 + \text{clo})(\text{met} - 1.1); \quad \text{deg F} \qquad (4\text{-}3)$$

The minimum allowable operative temperature for these activities is 59 F (15 C). Table 4-1 gives the met levels of some common activities. One might expect people to remove some of their normal clothing when working hard.

4-4   CONTROL OF INDOOR AIR QUALITY

Smell plays an important role in human survival. Through it a person is able to detect danger, food, and other things relating to safety and health. However, even pleasant odors when confined to a given space will become objectionable, when the intensity reaches an excessive level. Until recently the control of air quality came about rather naturally. Most structures had considerable infiltration, and excessive amounts of outdoor air were often used because of poor understanding of the needs for contaminant control.

The indoor air quality question has become increasingly important as a result of several factors. Modern construction has greatly reduced infiltration, and there is very little diffusion of air through the building materials themselves. Also outdoor air requirements have been reduced considerably because the quality of outdoor air is now lower and because excessive outdoor air may add considerably to the cooling and heating load, a very important consideration in times of energy shortages.

To eliminate or reduce an odor or gas so that it is undetectable, it is necessary either to remove the offending material or to dilute the air until the concentration is too low for perception. The sources of odors are many. Outdoor air as a source of odors has increased in importance because it contains a high percentage of pollutants from automotive exhausts, furnace effluents, and other industrial sources. In recent years problems have arisen from contaminants generated by the building and its contents. Examples include asbestos dust, radon gas, and vapors of formaldehyde and mercury. In industrial spaces odors may arise from printing ink, dyes, synthetics, plastics, and rubber products. In offices, arenas, theaters, and other enclosed spaces, obnoxious odors consist mainly of body odors and tobacco smoke. Tobacco smoke contains many irritants, some of which may even impair visibility. Other sources are paint, upholstery, drapes, rugs, food, food preparation, and pets. The air-conditioning equipment may also contribute odors. Cooling coils collect dirt and lint, both of which are moistened by the condensate; this helps mildew form and leads to consequent objectionable odors.

Humidity has an effect on odor perception, and research has shown that increasing humidity at constant dry bulb temperature has the same effect as lowering the odor level of cigarette smoke as well as that of pure vapors. This effect is more pronounced for some odors than others. Dry bulb temperature has some effect on odor perception. For example, an increase in dry bulb temperature at constant specific humidity seems to lower the odor level of cigarette smoke slightly. It has been shown that if the conditioned space is held at a relative humidity of about 45 to 60 percent, odor perception and irritation will be at a minimum.

Contaminants may be removed by physical or chemical means. The principal methods involve ventilation, air washing or scrubbing, condensation, adsorption, chemical reaction, and odor modification. Other methods are chemical absorption, oxidation, vapor neutralization, and combustion. Ventilation, adsorption, chemical reaction, and odor modification are most effectively used in air-conditioning applications. If the contaminant is a gas or vapor, like ammonia, water sprays are most practical for removal. But if the gases are insoluble in water, these methods are fairly ineffective. It should also be kept in mind that when various odorants are collected in water, they must ultimately be disposed of and cannot generally be dumped in the environment.

A method closely related to ventilation is isolation or exhaust. This method is generally used in kitchen and restroom areas where air is exhausted directly from the space to outdoors. This requires that makeup air be introduced, which takes on the same function as outdoor air. The various methods of air quality control will now be discussed.

Ventilation

The dominating function of outdoor air is to control air quality, and spaces that are more or less continuously occupied require some outdoor air. Research shows that the required outdoor air is dependent on the rate of contaminant generation and the maximum acceptable contaminant level. Misunderstanding of these factors has led to considerable confusion in the past concerning the quantity of outdoor air required. It can generally be stated that in most cases more outdoor air than necessary is supplied. However, some overzealous attempts to save energy through reduction of outdoor air have caused poor-quality indoor air. ASHRAE Standard 62 (10) defines acceptable air quality as ambient air in which there are no known contaminants at harmful concentrations and with which a substantial majority of the people exposed do not express dissatisfaction. Standard 62 defines ventilation requirements and specifies allowable contaminant concentrations for ventilation air. Table 4-2, reprinted from Standard 62, prescribes the basic requirements for acceptable air quality and Table 4-3, also from Standard 62, gives the minimum outdoor air requirements. Standard 62 should be consulted for additional details and definitions. However, some critical definitions will be considered before proceeding. *Ventilation air* is the combination of outdoor air, of acceptable quality, and recirculated air from the conditioned space that after passing through the air-conditioning unit becomes *supply air*. The ventilation air may be 100 percent

Table 4-2   AMBIENT AIR QUALITY STANDARDS

| Contaminant | Long Term | | Short Term | |
	Level	Time	Level	Time
Carbon monoxide			40 mg/m³	1 hr
			10 mg/m³	8 hr
Lead	1.5 μg/m³	3 mo.		
Nitrogen dioxide	100 μg/m³	Year		
Oxidants (ozone)			235 μg/m³	1 hr
Particulates	75 μg/m³	Year	260 μg/m³	24 hr
Sulfur dioxide	80 μg/m³	Year	365 μg/m³	24 hr

[a]U.S. Environmental Protection Agency, National Primary and Secondary Ambient Air Quality Standards, Code of Federal Regulations, Title 40 Part 50 (40C.F.R. 50). Pertinent local regulations should also be checked. Some regulations may be more restrictive than those given here, and additional substances may be regulated.

Source: Reprinted by permission from ASHRAE Standard 62-81, 1981.

Table 4-3 OUTDOOR AIR REQUIREMENTS FOR VENTILATION

	Estimated occupancy, persons per 1000 ft² or 100 m²/floor area. Use only when design occupancy is not known.	Outdoor Air Requirements				Comments
		cfm/person	(L/s)/person	cfm/ft²	(L/s)/m²	
COMMERCIAL FACILITIES						
Retail Stores						
Sales floors and showrooms						
Basement and street floors	30			0.3	1.5	
Upper floors	20			0.2	1.0	
Storage areas (serving sales and storerooms)	15			0.15	0.75	
Dressing rooms	—			0.2	1.0	
Malls and arcades	20			0.2	1.0	
Shipping and receiving areas	10			0.15	0.75	
Warehouses	5			0.05	0.25	
Elevators	—			1.0	5.0	
Smoking rooms	70	60	30			
Public Spaces						
Corridors and utility rooms				0.05	0.25	
Public restrooms		50	25	37.5	—	

Table 4-3   OUTDOOR AIR REQUIREMENTS FOR VENTILATION, continued

	Estimated occupancy, persons per 1000 ft² or 100 m²/floor area. Use only when design occupancy is not known.	Outdoor Air Requirements		Comments
		cfm/locker	(L/s)/locker	
COMMERCIAL FACILITIES				
Locker and dressing rooms		.05	.25	
Sports and Amusement Facilities				
Ballrooms and discos	100	25	13	
Bowling alleys (seating area)	70	25	13	
Playing floors (gymnasium)	30	20	10	
Spectator areas	150	15	8	
Game rooms (e.g., card and billiard rooms)	70	25	13	
Swimming pools	—	0.5	2.5	
Hotels, Motels, Resorts, Dormitories, and Correctional Facilities				
Bedrooms (single, double), cfm/room		30	15	
Living rooms (suites), cfm/room		30	15	
Baths, toilets (attached to bedrooms), cfm/room		35	18	

Lobbies	30	15	8			
Conference rooms (small)	50	20	10			
Large assembly rooms	120	15	8			
Offices						
Office space	7	20	10			
Waiting spaces	60	15	8			
INSTITUTIONAL FACILITIES (*for areas not listed, refer to Commercial Facilities*)						
Hospitals, Nursing, and Convalescent Homes						
Patient rooms	10	25	13			
Medical procedure areas	20	15	8			
Operating rooms, delivery rooms	20	30	15			Procedures generating contaminants may require higher rates
Recovery and intensive care rooms	20	15	8			Air shall not be recirculated
Autopsy rooms				0.5	2.5	
Physical therapy areas	20	15	8			
Educational Facilities						
Classrooms	50	15	8			
Laboratories	30	20	10			
Training shops	30	20	10			
Music rooms	50	15	8			
Libraries	20	15	8			
Playing floor (gym.)	30	20	10			—

Source: Adapted by permission from ASHRAE Standard 62-81, 1981.

Table 4-3a   OUTDOOR AIR FOR VENTILATION
             OF RESIDENTIAL BUILDINGS

Application	Outdoor Air Requirements	Comments
Living areas	0.35 air changes per hour but not less than 15 cfm (7.5 L/s) per person	For calculating the air changes per hour, the volume of the dwelling shall include all areas within the conditioned space. The ventilation is normally satisfied by infiltration and natural ventilation. Dwellings with tight enclosures may require supplementary ventilation supply for fuel-burning appliances, including fireplaces; and mechanically exhausted appliances. Occupant loading shall be based on the number of bedrooms as follows: first bedroom, two persons; each additional bedroom one person. Where higher densities are known, they shall be used.
Kitchens	100 cfm (50 L/s) 50 cfm (25 L/s)	Intermittent mechanical exhaust, natural ventilation (opened windows) for control of moisture, odors, gases, vapors, and particles.
Garages separate for each dwelling unit	100 cfm (50 L/s) per car	Normally satisfied by infiltration or natural ventilation.
Common for several units	1.5 cfm/ft² [7.5 L/(sm²)]	

outdoor air. The term *makeup* air may be used synonymously with *outdoor* air and the terms *return* and *recirculated* air are often used interchangeably.

The discussion here pertains to the required ventilation air to maintain indoor air quality. A situation could exist where the supply air required to match the heating or cooling load is greater than the ventilation air. In that case an increased amount of air would be recirculated to meet this condition.

Standard 62 describes two procedures by which indoor air quality may be controlled. The first, known as the *Ventilation Rate Procedure,* achieves indoor air quality indirectly by prescribing the minimum amount of ventilation air and various means to condition that air. The second approach, known as the *Indoor Air Quality Procedure,* specifies maximum permissible concentrations of certain contaminants in indoor air but does not prescribe ventilation rates or air treatment methods. The ventilation rate procedure is the most commonly used approach.

The Ventilation Rate Procedure has five elements.

1. Acceptable outdoor air quality, Table 4-2.
2. Outdoor air treatment.
3. Minimum ventilation rates, Table 4-3.
4. Criteria for reduction of outdoor air when mixed with recirculated air.
5. Criteria for noncontinuous ventilation.

Acceptable outdoor air quality is generally established in Table 4-2 although some additional contaminants are also specified in Standard 62. Several steps are outlined in Standard 62 to determine acceptability of local outdoor air.

When the outdoor air contaminant levels exceed the values given in Table 4-2, the air must be treated to control the offending contaminants. Filtering equipment necessay to do this is described in Sec. 4-5.

Indoor air quality is considered acceptable when the required rates of acceptable outdoor air of Table 4-3 are provided. The table lists the required volume flow rates per person for a variety of indoor spaces.

It is presumed that the contamination produced is generally in proportion to the number of persons in a space. The table lists estimated density of people for design purposes where such values may not be known. In residential applications, volume flow rate is specified in air changes per hour.

A minimum supply of outdoor air is necessary to dilute the carbon dioxide produced by metabolism and expired from the lungs. This value, 15 cfm or 7.5 L/s per person, allows an adequate factor of safety to account for health variations and some increased activity levels. Therefore, outdoor air requirements should never be less than 15 cfm or 7.5 L/s per person regardless of the treatment of the recirculated air. At this time no practical equipment exists to remove carbon dioxide.

In some cases exhaust air from one space can be used as supply air to another space where different contaminants are generated (corridors and office spaces exhausted through toilet rooms, or dining areas exhausted through kitchens); this air is considered equivalent to acceptable outdoor air.

The number of persons cannot always be estimated accurately or may vary considerably. In other cases, a space may require ventilation to remove contamination generated within the space, but unrelated to human occupancy (e.g., radon or formaldehyde outgassing from building materials or furniture). For these cases, Table 4-3 lists quantities in cfm/ft^2 or L/(s-m^2).

When spaces are unoccupied, ventilation is not generally required. Ventilation may be required, however, to prevent accumulation of contaminants injurious to people, contents, or structure. The requirements for ventilation quantities given in Table 4-3 are for 100 percent outdoor air when the outdoor air meets the specifications for air quality given in Table 4-2. Although these quantitites are for 100 percent outdoor air, they also set the amount of air required to dilute contaminants to acceptable levels. Therefore, it is necessary that at least this amount of air be circulated at all times that the building is in use. If proper air filters and adequate temperature control are provided, part of this air may be recirculated, but the outdoor air portion must never be less than 15 cfm or 7.5 L/s per person. The filtering system for the recirculated air may be located in the recirculated air stream or in the plenum, which mixes the outdoor air and return air.

The filtering system used to reclaim recirculated air should be designed to remove both particulate and gaseous contaminants. The required air cleaner efficiency and recirculation rate for the system should be sufficient to provide an indoor air quality equivalent to that obtained using outdoor air as specified in Table 4-2 and at a rate specified in Table 4-3. The filter efficiency and recirculation rate will be discussed in Section 4-5, "The Cleaning of Air."

Ventilating systems for spaces not continuously occupied may be shut off under certain conditions. Start-up of the system should lag or lead occupancy depending on the source of contaminants. When contaminants are associated only with occupants and are dissipated by natural means during unoccupied periods, start-up may lag occupancy. When contaminants generated in the space are independent of occupants or their activities do not present health hazards during short-term exposure, start-up should lead occupancy so that acceptable conditions will exist at the start of occupancy. In some situations local codes may override these limits.

Washing

The air washer is used to control temperature and humidity; it is also used to remove particulate matter. As these functions are carried out, removal of odorous vapors or particles is also accomplished.

Generally air washers work in three ways. Probably the most important method is absorption. When absorption occurs, the concentration of a vapor in the air is high in relation to the concentration of that chemical in the liquid, and the contaminant is absorbed in the liquid. The second method is by condensation of the vapor, much the same as water vapor is condensed from air. The third method referred to before is removal of particles that contain odorous materials from the air by a direct washing action. Although water, with or without additives, is a common liquid used in scrubbers and washers, other liquids may be used. Obviously the liquids used in an air washer or scrubber must be maintained in an odor-free condition or odors will be transferred to the air. Therefore, new water must be continuously added or the liquid must be regenerated. In air-conditioning applications it is generally required that large quantities of air be moved through the washer or scrubber without an excessive drop in pressure.

Adsorption

Adsorption is the adhesion of molecules to the surface of a solid called the adsorbent; this contrasts to absorption, which is basically the dissolving of molecules into a substance. Because the process of adsorption is greatly dependent on the available surface area, the materials used must contain large surface areas usually taking the form of minute pores in a solid substance. There are wide varities of adsorbents available, including charcoal, zeolite, silica gel, activated alumina, mica, and others. Activated charcoal can adsorb a variety of gases, vapors, and a broader spectrum of organic substances than most sorbents and is therefore the most nearly universal adsorbent in general use. The basic raw material from which the activated charcoal is manufactured has little significance to the user; however, the basic material may have some influence on the density, pore structure, and hardness. Charcoal has its greatest selectivity toward molecules of high molecular weight and is less effective for lighter gases such as ammonia or ethylene. Chemically active gases may be better accommodated if the charcoal is impregnated with some other substance.

Because there must be intimate contact between the gaseous odorants in the air and the adsorbent, the air to be cleaned is usually passed through a bed of the adsorbent material. These beds may range from one-half to four inches in depth. The charcoal or other adsorbent has a limited life and must be removed or replaced as its capacity decreases. Further information relating to the use of adsorbents may be found in the Systems volume of the *ASHRAE Handbook* series.

Chemical Reactions

A great many chemical reactions are possible to neutralize airborne odorants or other contaminants. In general these may be categorized as specific reagents for specific contaminants, or broad spectrum reagents for a broad range of contaminants. These reactions are usually accomplished in a manner similar to the air washer except that solutions of reagent chemicals are used instead of water. It is also possible to use solid reagents or activated solids containing reagents in the same way in which the activated charcoal was described before. Another method is to inject a reagent gas into the contaminated air. This is usually done only when air is being exhausted because the concentration of reagents can also cause a problem. For additional information in the use of this method, the readers should refer to the Systems volume of the *ASHRAE Handbook* series.

Combustion

Two types of combustion methods are used for odor control. The first method utilizes high-temperature oxidation in a combustion chamber fired directly with a flame; the second employs oxidation at lower but still elevated temperatures on catalytic surfaces. Combustion methods have merit when the recovered odor would not have significant economic value, the concentration of odorant is sufficient to contribute appreciable heat of combustion, the effluent is already well above ambient temperature, and the wasted heat can be used economically. The combustion methods of controlling odors in air-conditioning and heating applications are not very numerous, and again the reader is referred to the Systems volume of the *ASHRAE Handbook* series.

Odor Masking and Counteraction

Odor masking is the technique of introducing a pleasant odor to cover an unpleasant odor. Obviously this can be a hazardous undertaking because what might be pleasant to one person may be unpleasant to another. However, many odorants for this purpose are available along with equipment for introducing them into the conditioned air. In some cases there is no alternative but to use this type of control.

Odor counteraction is the art of mixing two odorous vapors together such that both odors tend to be diminished. This type of control tends to be inexact, and it is not an area for amateur experimentation. Where other methods fail or cannot be used, equipment can be obtained to use this type of control. Odor masking and counteraction should be considered as methods of last resort.

4-5 THE CLEANING OF AIR

Atmospheric dust is a complex mixture of smoke, mists, fumes, dry granular particles, and fibers suspended in air. In addition air may contain living organisms such as mold spores, bacteria, and plant pollens that may cause diseases or allergic responses. Particles in the atmosphere range in size from less than 10^{-6} m up to the order of magnitude of the dimension of leaves and insects. The particles exist in all sizes and shapes and generally it is impossible to design one air cleaner that is best for all applications.

Different degrees of air cleanliness are required for various applications. The minimum standards for air quality are defined by ASHRAE Standard 62 as discussed in Section 4-4. Another consideration is the fact that dust particles in the air may damage the conditioned spaces or its furnishings. For example, the walls may be discolored if even the smallest components of the atmospheric dust are not removed. Unfortunately the smallest particles cause the most discoloring. Therefore, in this case, electronic air cleaners or high performance dry filters are required. In clean-room applications or when radioactive particles are present, extremely high-performance filters should be employed. One of the most effective ways of controlling allergic disorders is to remove the allergens from the air by using high-performance filters. Of all the characteristics affecting the performance of an air filter, particle size is the most important; however, the cleaning efficiency of a filter is affected to some extent by the velocity of the air stream. The major factor influencing filter design and selection is the degree of air cleanliness required. In general the cost of the filter or filter system will increase as the size of the particles to be removed decreases. Maintenance will also increase in the same way.

There are three operating characteristics that distinguish the various types of air cleaners. These are the *efficiency*, the *air flow resistance*, and the *dust-holding capacity*. Efficiency measures the ability of the air cleaner to remove particulate matter from an air stream, and the average efficiency over the life of the filter is the most meaningful for most applications. The air flow resistance is the loss in total pressure across the filter at a given air flow rate. This of course is a very important consideration. The dust-holding capacity defines the amount of a particular type of dust an air cleaner can hold while operating at a specific air flow rate and maximum air flow resistance. Sometimes the dust-holding capacity refers to the value attained before the filter efficiency is seriously reduced. Therefore, to have a complete rating of an air cleaner, data should be made available on efficiency, resistance, and dust-holding capacity. The effect of the dust loading on the efficiency and resistance is also important. Details of the standardized methods by which air cleaners are tested and rated are given in ASHRAE Standard 52 (11). In general, however, the gravimetric efficiency of a filter gives the ratio of the mass of the dust arrested by the filter to the total mass of the dust fed to the filter. The second type of efficiency is called the dust-spot efficiency. This is designed to rate filters for their ability to prevent discoloration. In this case white filter paper targets are used and the efficiency is determined through light transmission measurements and expressed in terms of efficiency. The dust-holding capacity is actually given as an absolute measure of the mass of dust held by the filter. This rating gives an indication of how often the filter may have to be renewed. Of course this will vary considerably depending on the application. Air flow resistance or pressure drop is probably the simplest of all the tests made and is usually given in terms of the clean filter. Naturally, as the filter becomes loaded, pressure drop will increase. The pressure drop or resistance depends on the square of the velocity; therefore, if a filter is rated for a certain velocity, it is a simple matter to convert to some other velocity. The efficiency of a filter depends on the particle sizes, and manufacturers make different kinds of filter media for different applications. Figure 4-4 is an example of gravimetric efficiency as a function of particle size. Here it can be seen that with the use of various filter media the efficiency is different. With the larger particle sizes all of the media are fairly efficient, but when the particle sizes become very small, efficiency varies considerably with the media selected.

Figure 4-4. Gravimetric efficiency of high-performance dry-media filters.

Mechanisms of Particle Collection

As air passes through a filter the airstream is made to turn. The particles, being more dense than the air, tend to keep going in a straight line. When the particles collide with the fibers they stick or are embedded. This mechanism is generally called viscous impingement. Interception is a special case where the particles collide head on with the fibers and the effects of inertia are unimportant. This method is independent of velocity. Another important mechanism is referred to as "straining." Here the fibers are so close together that a particle trying to go between them is arrested. This mechanism usually occurs in the collection of large particles and lint. Diffusion can be an important effect in the case of very fine particles. The very small particles are bombarded by the random motion of the air molecules and are driven into the filter fibers, where they are arrested. Turbulence in the airstream helps in this regard. Electrostatic effects are used in one type of filter. Charged particles are forced toward a plate or grid having an opposite charge. This mechanism is employed by the electronic air cleaner and is usually most effective with very small particles.

There are three broad categories of air cleaners. The first, the *fibrous media unit filter,* is characterized by a gradual increase in accumulated dust load that increases pressure drop up to some maximum permissible value. During this time the efficiency also increases. If allowed to load too heavily with dust, however, the efficiency will again fall off.

The second category, the *renewable media filter,* uses a media usually of a fibrous type. In this case new media is continually introduced into the air-stream, and loaded media is removed so that efficiency, resistance, and loading remain essentially constant.

The third classification is referred to as *electronic air cleaners.* These units have essentially constant pressure drop and efficiency unless their elements become severely dust-loaded. Various combinations of the foregoing types may be used. For example, a fibrous media filter may be placed both upstream and downstream of an electronic filter, the upstream unit to remove large particles, and the downstream mechanical unit to catch any particles that may blow off the plates of the electronic precipitator. Another attractive combination is to use a lower efficiency filter upstream of a very high-efficiency filter to extend the life of the more expensive high-efficiency filter.

Types of Filter Media

Viscous impingement filters are usually flat panels made up of coarse fibers and have a high porosity. Usually the filter media is coated with a viscous substance that acts as an adhesive for particles impinging on the fibers. Design air velocity through the media is usually in the range of 250 to 700 ft/min or 1.3 to 3.6 m/s. These filters are characterized by low pressure drop, low cost, and high efficiency on lint, but low efficiency on normal atmospheric dust. The most common media are glass fibers, metallic wools, expanded metals, foils, crimped screens, and random matted wire. Table 4-4 gives some typical per-formance data for viscous impingement filters. This type of filter varies so much in material, quality, and construction that precise performance data cannot be given.

Table 4-4   PERFORMANCE OF UNIT VISCOUS-IMPINGEMENT FILTERS

Filter Thickness, Inches	ASHRAE Weight Arrestance, Percent	ASHRAE Atmospheric Dust-spot Efficiency, Percent	ASHRAE Dust-holding Capacity Pounds per 1000 cfm Cell	Pressure Loss, Inches of Water at 350 fpm
1	20–50	5–10	0.16–0.32	0.08
$1\frac{3}{4}$	50–75	5–15	0.26–0.08	0.14
$2\frac{1}{2}$	60–80	5–20	0.4–1.2	0.20
4	70–85	10–25	0.53–1.7	0.32

Source: Adapted by permission from *ASHRAE Handbook, Equipment Volume,* 1983.

Another type is called the dry media air filter. The absence of an adhesive on the filter medium does not necessarily make it a dry-type air filter. The most distinguishing characteristic of this type of filter is the smaller and much more closely spaced fibers that form a dense mat or pad. The efficiency of dry-type air filters is usually much higher than viscous impingement filters, and there is usually a much larger variety of media available. An accordionlike arrangement of the media is usually employed. Table 4-5 contains some typical performance data for dry media filters. Table 4-6 gives engineering data related to the efficiency data of Fig. 4-4. The Equipment volume of the *ASHRAE Handbook* series may be consulted for additional information concerning various types of media and application.

Table 4-5   PERFORMANCE OF DRY MEDIA FILTERS

Filter Media Type	ASHRAE Weight Arrestance, Percent	ASHRAE Atmospheric Dust-spot Efficiency, Percent	ASHRAE Dust-holding Capacity, kg per m³/s Cell
Finer open cell foams and textile denier nonwovens	70–80	15–30	0.380–0.900
Thin, paperlike mats of glass fibers, cellulose	80–90	20–35	0.190–0.380
Mats of glass fiber multiply cellulose, wool felt	85–90	25–40	0.190–0.380
Mats of 5 to 10 μm fibers of 6 to 12 mm thickness	90–95	40–60	0.570–1.150
Mats of 3 to 5 μm fibers, 6 to 12 mm thickness	>95	60–80	0.380–0.950
Mats of 1 to 4 μm fibers, mixture of various fibers and asbestos	>95	80–90	0.380–0.760
Mats of $\frac{1}{2}$ to 2 μm fibers (usually glass fibers)	NA[a]	90–98	0.190–0.570
Wet laid papers of mostly submicron glass and asbestos fibers (HEPA filters)	NA	NA	1.060–2.120
Membrane filters (membranes of cellulose acetate, nylon, etc., having micron-size holes)	NA	NA	NA

[a]NA or not applicable means that test method cannot be applied to this level of filter.

Source: Adapted by permission from *ASHRAE Handbook, Equipment Volume,* 1983.

Table 4-6   ENGINEERING DATA—HIGH-PERFORMANCE DRY-MEDIA FILTERS (CORRESPONDS TO EFFICIENCY DATA OF FIGURE 4-4)

Standard Size	Meter Inch	0.3 × 0.6 × 0.2 12 × 24 × 8		0.3 × 0.6 × 0.3 12 × 24 × 12		0.6 × 0.6 × 0.2 24 × 24 × 8		0.6 × 0.6 × 0.3 24 × 24 × 12		Pressure Loss Inches of Water	Pa
Rated Capacity[a]		ft³/min	m³/s	ft³/min	m³/s	ft³/min	m³/s	ft³/min	m³/s		
	M-2A[b]	900	0.42	1025	0.48	1725	0.81	2000	0.94	0.15	37.4
Media	M-15	900	0.42	1025	0.48	1725	0.81	2000	0.94	0.35	87.2
Type	M-100	650	0.30	875	0.41	1325	0.62	1700	0.80	0.40	100.0
	M-200	450	0.21	630	0.29	920	0.43	1200	0.56	0.40	100.0
Effective filtering area		ft²	m²	ft²	m²	ft²	m²	ft²	m²		
All media types		14.5	1.35	20.8	1.93	29.0	2.69	41.7	3.87		

[a]Filters may be operated from 50 to 120 percent of the rated capacities with corresponding changes in pressure drop.

[b]The M-2A is available in 2 in. thickness and standard sizes with a nominal rating of 0.28 in. H_2O at 500 fpm face velocity

121

Electronic Air Cleaners

The term *electronic air cleaner* usually refers to the types of cleaners used in air-conditioning applications. However, by using much higher potentials, the same general-type apparatus is used for cleaning of stack gases. Electronic air cleaners are of two general types. The first of these is the "ionizing plate electronic air cleaner" shown schematically in Fig. 4-5. The positive ions, generated at the high-potential ionizer wire, flow across the airstream, striking and adhering to any dust particles carried by the airstream. These particles then pass into the system of charged and grounded plates, where they are driven to the plates by the force exerted by the electric field on the charges they carry. As the dust particles reach the plates, they are removed from the airstream. A typical electronic air cleaner will use a dc potential of 12,000 volts for the ionizing field and will have about 6000 volts maintained between the plates. This type of air cleaner is an efficient, low-pressure-drop device for removing fine dust and smoke particles. Sometimes the collector plates are coated with a special adhesive to improve the dust-holding capacity. Cleaning is generally accomplished by washing the plates in hot water with or without a detergent. Some units are supplied with servicing equipment from the factory.

The second type of electronic air cleaner is called the "charged media nonionizing electronic air cleaner." This type combines certain characteristics of both the dry filter and electronic air cleaner. It consists of a dielectric filtering medium usually arranged in pleats as in a typical dry filter. The filtering medium is in contact with a grid consisting of alternately grounded and charged members, usually at a potential of about 12,000 volts. An intense but nonuniform electrostatic field is thus created through the dielectric media. Airborne particles approaching the field are polarized and drawn to the fibers of the medium. In

Figure 4-5.   Diagrammatic cross section of ionizing-type electronic air cleaner. (Reprinted by permission from *ASHRAE Handbook, Equipment Volume*, 1983.)

this type of filter the medium is replaced as with other media filters. It has a higher pressure drop than the ionized plate type of filter. The charged medium is ineffective when the relative humidity becomes greater than about 70 percent.

A drawback to the electronic-type air cleaner is that charged particles can sometimes be carried through the cleaner into the space. If this happens on a large scale, a considerable buildup of space charge will result, which drives the charged dust particles to the wall and causes blackening of the surface.

Selection and Maintenance

In selecting a filter system for a particular application, three factors should be carefully weighed: one, the degree of air cleanliness required; two, disposal of the dirt after it is removed from the air; and three, the amount and type of dust in the air to be filtered. These factors determine the initial cost, the operating cost, and the extent of maintenance required. Viscous impingement filters do not have efficiencies as high as can be expected from dry types of filters but their initial cost and upkeep are generally low. This unit type of filter requires more attention than the moving type of filter, which replenishes itself. However, the moving-type filter has a higher initial cost. Developments have improved the efficiencies of dry-type filters so that they approximate the efficiency of electronic types and cost less. However, they are more expensive than the viscous-impingement type and pressure loss is the greatest for this type of filter. Electronic air cleaners have a high initial cost. But they have higher efficiencies, especially when the particles are very small. They also have very low pressure drop.

Filtering and Recirculation

It was mentioned in Section 4-4 that the amount of outdoor air can be reduced and supplemented with recirculated air when proper filtering is done. Under normal conditions when the concentrations of contaminants in the outdoor air are much less than those specified in Table 4-2 and the concentrations indoors are greater than those outdoors, the recirculation rate is given by

$$\dot{Q}_r = \frac{\dot{Q}_O - \dot{Q}_m}{\eta} \tag{4-4}$$

where

$\dot{Q}_O$ = outdoor air rate from Table 4-3, ft^3/min-person or L/s-person
$\dot{Q}_m$ = minimum outdoor air rate to provide acceptable indoor air quality but not less than 15 ft^3/min-person or 7.5 L/s-person
η = efficiency of filter or gas removal device

Equation 4-4 results from a mass balance on a particular contaminant, such as cigarette smoke, for a system like Fig. 4-6. The recirculating system will produce the same indoor air quality as a system using the specified outdoor air given

Figure 4-6.   Schematic of recirculating ventilation system.

in Table 4-3. Note that it has been assumed that the outdoor air is clean. The total air supply rate then becomes

$$\dot{Q}_s = \dot{Q}_r + \dot{Q}_m \qquad (4\text{-}5)$$

The following examples demonstrate the calculation procedure.

Example 4-1.   Consider an office space. A ventilation rate of 20 cfm per person is required from Table 4-3. However, only 15 cfm per person is required as a minimum. Assume the M-15 filter media of Fig. 4-4 is 90 percent efficient in removing contaminants like tobacco smoke and compute the recirculation and total air supply rates if the outdoor air rate is reduced to 15 cfm per person. The outdoor air concentration of total suspended particulates is negligible.

Solution.   Equations 4-3 and 4-4 apply to this problem where $\dot{Q}_o$ is 20 cfm per person, $\dot{Q}_m$ is 15 cfm per person, and η is 90 percent. Using Eq. 4-3

$$\dot{Q}_r = \frac{20 - 15}{0.9} = 6 \text{ cfm/person}$$

then from Eq. 4-4

$$\dot{Q}_s = 6 + 15 = 21 \text{ cfm/person}$$

Note that the total supply rate has been increased from 20 to 21 cfm/person; however, the outdoor air rate is only 15 cfm/person.

Example 4-2.   According to Table 4-3, dormitory bedrooms require 15 L/s-room. To save energy it is desirable to reduce the outdoor air rate to a minimum of 7.5 L/s-room. If the outdoor air has negligible contaminants, compute the recirculation and total air supply rates, assuming the M-2A media of Fig. 4-4 is used and the efficiency is for the $0\text{--}5 \times 10^{-3}$ mm particle size range.

Solution.   The required air flow rates may be computed using Eqs. 4-3 and 4-4. The filter efficiency is 79 percent as read from Fig. 4-4 and the flow rates $\dot{Q}_o$ and $\dot{Q}_m$ are 15 and 7.5 L/s-room, respectively. Then from Eq. 4-3

$$\dot{Q}_r = \frac{15 - 7.5}{0.79} = 10 \text{ L/s-room}$$

and

$$\dot{Q}_s = 10 + 7.5 = 17.5 \text{ L/s-room}$$

This approach would result in significant energy savings because the outdoor air rate has been reduced by 50 percent.

Filter System Design

After the filter type and media have been selected as discussed previously, the filter system must be sized to accommodate the required air flow rate and flow resistance. Filters may be placed at different locations in the air-conditioning system. For example, outdoor air and return air may be filtered separately and then mixed, or they may be mixed and then filtered. The air is usually filtered before entering the air-conditioning equipment rather than afterward. The following examples are typical.

Example 4-3.   Suppose that the office space of Example 4-1 involves 95 people and the outdoor air and recirculated air are mixed before filtering. Design a filter system that has a pressure loss of no more than about 0.30 in. water in the clean condition.

Solution.   Table 4-6 gives application data for the M-15 filter media cited in Example 4-1. Several sizes are shown, none of which match the requirements of this problem. Therefore, two or more modules must be combined and the capacity adjusted to obtain the required pressure loss performance. Let us try the 12 × 24 × 12 size, which is rated at 1025 cfm with 0.35 in. water pressure loss. To a close approximation the pressure loss is proportional to the square of the air velocity or flow rate. Then the capacity with a pressure loss of 0.3 in. water is

$$\dot{Q} = \dot{Q}_c \left[\frac{\Delta P}{\Delta P_c} \right]^{1/2} = 1025(0.3/0.35)^{1/2} = 949 \text{ cfm}$$

where subscript c refers to the rated conditions. The total air flow rate is

$$\dot{Q} = (\dot{Q}_s)(95) = (21)(95) = 1995 \text{ cfm}$$

and the number of filter modules is

$$N = 1995/949 = 2.1$$

Because N must be an integer, two modules may be used. A somewhat higher pressure loss will result.

Example 4-4.   An air handler has been selected that is designed for 5400 cfm with provisions for six 16 × 20 × 2 in. M-2A filters (see Table 4-6).

Assuming that the unobstructed face area of the filters is 90 percent of the actual area, estimate the pressure loss for the clean filters.

Solution.   Footnote *b* of Table 4-6 indicates that these filters have a pressure loss of 0.28 in. water when the face velocity is 500 ft/min. The pressure loss is proportional to the face velocity squared. The face velocity is given by

$$\overline{V}_f = \frac{\dot{Q}}{A_f} = (5400)(144)/(6)(16)(20)(0.9) = 450 \text{ ft/min}$$

Then

$$\Delta P = 0.28\left(\frac{450}{500}\right)^2 = 0.23 \text{ in. water}$$

REFERENCES

1. *ASHRAE Handbook, Fundamentals Volume,* American Society of Heating, Refrigerating and Air-Conditioning Engineers, Atlanta, Ga., 1985.

2. F. O. Fanger, *Thermal Comfort, Analysis and Applications in Environmental Engineering,* McGraw-Hill, New York, 1970.

3. F. H. Rohles, J. E. Woods, and R. G. Nevins, "The Effect of Air Movement and Temperature on the Thermal Sensations of Sedentary Man," *ASHRAE Transactions,* Vol. 80, Part I, 1974.

4. R. G. Nevins et al., "Temperature-Humidity Chart for Thermal Comfort of Seated Persons," *ASHRAE Transactions,* Vol. 72, Part I, 1966.

5. T. Bedford and C. G. Warner, "The Globe Thermometer in Studies of Heating and Ventilating," *J. Hygiene,* Vol. 35, 1935.

6. C. P. Yalou and D. Minard, "Control of Heat Casualties at Military Training Centers," *AMA Archives of Industrial Health,* Vol. 16, 1957.

7. P. A. Siple and C. F. Passcl, "Measurements of Dry Atmospheric Cooling in Subfreezing Temperatures," *Proceedings of the American Philosophical Society,* Vol. 89, 1945.

8. F. H. Rohles and R. G. Nevins, "The Nature of Thermal Comfort for Sedentary Man," *ASHRAE Transactions,* Vol. 77, Part I, 1971.

9. ASHRAE Standard 62-87, "Ventilation for Acceptable Indoor Air Quality," American Society of Heating, Refrigerating and Air-Conditioning Engineers, Atlanta, Ga., 1987.

10. ASHRAE Standard 55-88, "Thermal Environmental Conditions for Human Occupancy," American Society of Heating, Refrigerating and Air-Conditioning Engineers, Atlanta, Ga., 1988.

11. ASHRAE Standard 52-76, "Air Cleaning Devices Used in General Ventilation for Removing Particulate Matter, Methods of Testing," American Society of Heating, Refrigerating and Air-Conditioning Engineers, New York, 1976.

12. A. P. Gagge, J. A. J. Stolwijk, and Y. Nishi, "An Effective Temperature Scale Based on a Simple Model of Human Physiological Regulatory Response," *ASHRAE Transactions,* Vol. 77, Part I, 1971.

13. A. P. Gagge, G. M. Rapp, and J. D. Hardy, "The Effective Radiant Field and Operative Temperature Necessary for Comfort," *ASHRAE Transactions,* Vol. 73, Part I, 1967.

14. *ASHRAE Handbook, Equipment Volume,* American Society of Heating, Refrigerating and Air-Conditioning Engineers, Atlanta, Ga., 1983.

PROBLEMS

4-1 A mixed group of men and women occupy a space maintained at 76 F db and 62 F wb. All are lightly clothed, with sedentary activity. (a) By using Fig. 4-3, draw a conclusion about the general comfort of the group. The MRT is 78 F. (b) Suppose the group is moving around (light activity) instead of sitting. What is your conclusion about their comfort? (c) Suppose the group is composed of retired people, 65 years of age or more, playing cards. How would you change the room conditions to ensure their comfort?

4-2 An air-conditioning system is to be designed for a combination shop and laboratory facility. The space will be occupied by men engaged in active work such as hammering, sawing, walking, climbing, and so on. The men will be dressed in short sleeves. (a) Select comfort conditions for the space, assuming the occupants are to be comfortable. (b) It is expected that the toolroom keeper will work in a fenced-in area in the space with light activity. What is your recommendation about how he or she should dress to be comfortable?

4-3 Suppose that a space is used for both athletic events and classroom space at different times. The relative humidity can be maintained at 40 percent. At what temperature should the thermostat be set when the space is used for (a) classroom? (b) basketball practice?

4-4 A space used for sedentary activity by occupants in light clothing has a dry bulb temperature of 25 C and the air motion is about 0.1 m/s. The mean radiant temperature is about 29 C. (a) Can the occupants be expected to be comfortable? Why? (b) If not, what changes to the room air can be made to improve the comfort?

4-5 For sedentary activity, light clothing, and an air temperature equal to the mean radiant temperature, what is the approximate relationship between air motion (m/s) and the air temperature to maintain comfort? Consider a velocity range of 0.2 to 0.8 m/s.

4-6 Compute the operative temperature in a space where the mean air velocity is 30 ft/min (0.15 m/s) and the dry bulb and black globe temperatures are 74 F (23 C) and 78 F (26 C), respectively, in (a) English units and (b) SI units.

4-7 You are asked to investigate a complaint that an office space being maintained at 76 F (24 C) and 40 percent relative humidity is uncomfortably warm. You measure the black globe temperature to be 80 F (27 C) and estimate the mean air velocity to be 40 ft/min (0.2 m/s). What is your recommendation to correct the situation? Assume that you cannot change the building or fixtures.

4-8 Consider the work environment of a model shop where people are standing, walking, lifting, and so on, around various types of shop equipment. Air motion is about 20 ft/min (0.2 m/s) and a typical black globe temperature measurement is 70 F (21 C). The shop technicians are dressed in shirts and trousers (clo = 0.5 (ft^2-hr-F)/Btu or 0.078 (m^2-C)/W. Estimate the air dry bulb temperature for comfortable conditions in (a) F and (b) C. Assume the relative humidity to be normal (40 to 50 percent).

4-9 To save energy, an office building cooling system is controlled so that the minimum dry bulb temperature in the space is 78 F with a dew point temperature of 63 F. (a) Comment on the comfort expectations for the space. (b) What general type of clothing would you recommend? (c) Assuming the dry bulb temperature is fixed, what could be done to improve comfort conditions?

4-10 During an emergency energy shortage thermostats were ordered set at 65 F to conserve heating fuel. Using Figs. 4-2 and 4-3 discuss the comfort expectations of the following: (a) an executive dressed in a vested business suit, (b) a secretary sitting and typing, (c) a factory worker, (d) a customer in a grocery store, and (e) a patient in a doctor's examining or waiting room.

4-11 Air motion is a parameter that may be adjusted to improve comfort when space temperatures are fixed. Discuss the general effect of increased or decreased air motion when the space temperature is low in winter and high in the summer.

4-12 A method of saving energy in large, chilled water cooling systems is to increase the temperature of the water circulating in the system. (a) How will this affect the space comfort conditions? (b) Would there be certain periods of time when this would be feasible? Discuss.

4-13 The fan power required to circulate air to and from the conditioned space is often minimized by reducing air flow to conserve energy. (a) Is this consistent with high space temperatures in the summer months? Explain. (b) Would the same be true for low space temperatures in the winter? Explain.

4-14 The productivity of people is related to their comfort. Give an example where an overzealous attempt to save energy by raising or lowering the space temperature might result in an increase in overall energy use and greater overall cost.

4-15 A classroom in a school is designed for 100 people. (a) What is the minimum amount of clean outdoor air required? (b) The floor area is 1500 ft^2. What is the outdoor air ventilation requirement on the basis of floor area, Table 4-3?

4-16 A specification calls for viscous impingement filters with an ASHRAE weight arrestance efficiency of 75 percent. Pressure loss must not exceed 0.10 in. of water. Select a filter (thickness and face area) to handle 1000 cfm using Table 4-4.

4-17 Select a filter system that will have a gravimetric efficiency of at least 95 percent in the particle size range of $0-5 \times 10^{-3}$ mm. The system must handle 2200 cfm and the maximum pressure drop across the filters must be less than or equal to 0.25 in. of water. Space limits the depth of the filters to 8 in.

4-18 An M-100, $24 \times 24 \times 12$ filter described in Table 4-6 is suggested for use with 1200 cfm of air. What pressure loss can be expected?

4-19 A filter system to handle 5600 cfm of air is to be designed using the M-2A media (Table 4-6) and the $12 \times 24 \times 8$ modules. The lost pressure must be 0.10 in. of water or less when the filters are clean. How many modules are required?

4-20 The M-200, $0.6 \times 0.6 \times 0.2$ filters of Table 4-6 are to be used with a volume flow rate of 0.38 m³/s. What minimum pressure loss can be expected?

4-21 Design a filter system to handle 20 m³/s of air using the M-15, $0.6 \times 0.6 \times 0.3$ filters of Table 4-6. The pressure loss in the clean condition must be 70 Pa or less.

4-22 It is necessary to design a relatively high-efficiency filter system for a residence to handle 1200 cfm of air with a maximum pressure loss of 0.085 in. water when the filters are clean. (a) Consider the M-2A media of Table 4-6 in the 2 in. thickness and determine the face area required. (b) The physical layout of the equipment would permit the use of either $12 \times 24 \times 2$ in. or $16 \times 20 \times 2$ in. sizes. Select the size and number of modules to best match the requirements.

4-23 A laboratory involving animal research must use either 100 percent outdoor air or 25 percent outdoor air with high-performance filters for the return air. Gravimetric efficiency must be at least 99 percent in the size $0-5 \times 10^{-6}$ m. The computed heat gain of the laboratory is 3 tons with a sensible heat factor (SHF) of 0.7. The cooling and dehumidifying unit requires a fixed air flow rate of 350 cfm per ton. Inside conditions are 78 F db and 40 percent relative humidity, whereas outdoor conditions are 95 F db and 50 percent relative humidity. (a) Investigate the feasibility, and find the required amount of air and the size of the cooling unit when 100 percent outdoor air is used. (b) Find the required amount of air and the size of the cooling unit when 25 percent outdoor air is used and the remainder is recirculated through the high-performance filters. (c) Design the filter system using the data of Fig. 4-4 and Table 4-6 so that the maximum pressure loss is 0.125 in. water with clean filters.

4-24 A smoking area in a factory is designed for 75 people. Design a filter and air-circulation system whereby the outdoor air rate can be reduced to 15 cfm/person. Outdoor and recirculated air should be mixed before filtering. Outdoor air contaminants are negligible. The filter media must have a gravimetric efficiency of about 80 percent in the $0-5 \times 10^{-3}$ mm particle range and the pressure loss must not exceed 0.10 in. water.

4-25 Consider a game room where smoking is permitted. Suppose a filter system is available with an efficiency of 80 percent in removing tobacco smoke. Compute the air supply rates assuming the outdoor air rate is set at the minimum level and contaminants in the outdoor air are negligible.

4-26 Suppose the game room of Problem 4-25 is designed for 40 people. The cited filter media is rated at 0.2 in. water pressure loss with a face velocity of 600 ft/min. What net filter face area is required to obtain a pressure loss of 0.15 in. water when clean?

4-27 For a 3000 ft² combination gymnasium, recreation, and exercise operation it is desirable to reduce the outdoor air rate to a minimum by use of filtering and air recirculation. (a) Design a system using filters selected from Table 4-4. Use the minimum values of the given weight arrestance efficiency. Pressure loss should not exceed about 0.20 in. water. Outdoor air contaminants are negligible. (b) Suppose the cooling load dictates a larger supply air rate than the ventilation rate. Discuss how this would influence your choice of a filter.

4-28 A zone consisting of a large classroom isolated from the outdoor environment except for ventilation air has a capacity of 225 people. The cooling load is 125,000 Btu/hr (37 kW) with a sensible heat factor of 0.7. The outdoor air required for this case is the minimum, 15 cfm (7.5 L/s) per person. (a) Compute the required amount of ventilation air (supply air) on the basis of the cooling load assuming that the space dry bulb is 75 F (24 C) and 50 percent relative humidity (RH) and that the air is supplied at 90 percent RH. (b) What is the minimum amount of supply air based on air quality? (c) Compare (a) and (b) and resolve any inconsistency between the two.

CHAPTER FIVE

Heat Transmission in Building Structures

The design of an acceptable air-conditioning system is dependent on a good estimate of the heat gain or loss in the space to be conditioned. In the usual structure the walls and roofs are rather complex assemblies of materials. Windows are often made of two or more layers of glass with air spaces between them and usually have drapes or curtains. In basements, floors and walls are in contact with the ground. Because of these conditions precise calculation of heat transfer rates is difficult, but experience and experimental data make reliable estimates possible. Because most of the calculations require a great deal of repetitive work, tables that list coefficients and other data for typical situations are used. Thermal resistance is a very useful concept and will be used extensively in the following sections. Thermal capacitance is an important concept in all transient analysis computations; however, discussion of specific heat and density will be included in another chapter.

Generally all three modes of heat transfer—conduction, convection, and radiation—are important in building heat gain and loss. Solar radiation will be treated in a separate section, in Chapter 6, because it creates some unusual complexities in geometry and consequently in computation. Long wavelength radiation, such as occurs in air gaps, will be treated in this chapter.

5-1 BASIC HEAT-TRANSFER MODES

In the usual situation all three modes of heat transfer occur simultaneously. In this section, however, they will be considered separately for clarity and ease of presentation.

Thermal conduction is the mechanism of heat transfer between parts of a continuum because of the transfer of energy between particles or groups of particles at the atomic level. The Fourier equation expresses steady-state conduction in one dimension as follows:

$$\dot{q} = -kA \frac{dt}{dx} \tag{5-1}$$

where

$\dot{q}$ = heat transfer rate, Btu/hr or W
k = thermal conductivity, Btu/(hr-ft-F) or W/(m-C)
A = area normal to heat flow, ft² or m²
$\dfrac{dt}{dx}$ = temperature gradient, F/ft or C/m

Equation 5-1 incorporates a negative sign because $\dot{q}$ flows in the positive direction of x when dt/dx is negative.

Consider the flat wall of Fig. 5-1a where uniform temperatures t_1 and t_2 are assumed to exist on each surface. If the thermal conductivity, the heat transfer rate, and the area are constant, Eq. 5-1 may be integrated to obtain

$$\dot{q} = \frac{-kA(t_2 - t_1)}{(x_2 - x_1)} \tag{5-2}$$

Another very useful form of Eq. 5-2 is

$$\dot{q} = \frac{-(t_2 - t_1)}{R'} \tag{5-2a}$$

where R' is the thermal resistance defined by

$$R' = \frac{x_2 - x_1}{kA} = \frac{\Delta x}{kA} \tag{5-3}$$

The thermal resistance for a unit area of material is very commonly used in handbooks and in the HVAC literature. In this book this quantity, sometimes called the "R-factor," is referred to as the *unit thermal resistance,* or simply the *unit resistance, R.* For a plane wall the unit resistance is

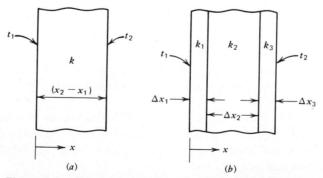

Figure 5-1. Nomenclature for conduction in plane walls.

$$R = \frac{\Delta x}{k} \tag{5-3a}$$

Thermal resistance R' is analogous to electrical resistance, and $\dot{q}$ and $(t_2 - t_1)$ are analogous to current and potential difference in Ohm's law. This analogy provides a very convenient method of analyzing a wall or slab made up of two or more layers of dissimilar material. Figure 5-1b shows a wall constructed of three different materials. The heat transferred by conduction is given by Eq. 5-2a, where

$$R' = R'_1 + R'_2 + R'_3 = \frac{\Delta x_1}{k_1 A} + \frac{\Delta x_2}{k_2 A} + \frac{\Delta x_3}{k_3 A} \tag{5-4}$$

Although the foregoing discussion is limited to a plane wall where the cross-sectional area is a constant, a similar procedure applies to a curved wall. Consider the long, hollow cylinder shown in cross section in Fig. 5-2. The surface temperatures t_i and t_o are assumed to be uniform and steady over each surface. The material is assumed to be homogeneous with a constant value of thermal conductivity. Integration of Eq. 5-1 with k and $\dot{q}$ constant but A a function of r yields

$$\dot{q} = \frac{2\pi k L}{\ln\left(\dfrac{r_o}{r_i}\right)} (t_i - t_o) \tag{5-5}$$

where L is the length of the cylinder. Here the thermal resistance is

$$R' = \frac{\ln\left(\dfrac{r_o}{r_i}\right)}{2\pi k L} \tag{5-6}$$

Cylinders made up of several layers may be analyzed in a manner similar to the plane wall where resistances in series are summed as shown in Eq. 5-4 except that the individual resistances are given by Eq. 5-6.

Table 5-1 summarizes the thermal resistances for several common situations, and Tables 5-2 and 5-2a give the thermal conductivity k for a wide variety

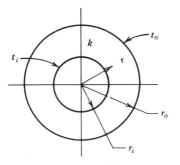

Figure 5-2. Radial heat flow in a hollow cylinder.

Table 5-1 THERMAL RESISTANCES FOR SOME STEADY-STATE CONDUCTION PROBLEMS

Number	System	Expressions for the Resistance R' $\dot{q} = \Delta t/R'$ (Btu/hr) or W
1.	Cylinder buried in a semi-infinite medium having a temperature at a great distance t_∞. The ground surface is assumed adiabatic.	$R' = \dfrac{\left(\ln\dfrac{2L}{D}\right)\left(1 + \dfrac{\ln(L/2z)}{\ln(2L/D)}\right)}{2\pi kL}$
2.	A vertical cylinder placed in a semi-infinite medium having an adiabatic surface and temperature t_∞ at a great distance.	$R' = \dfrac{\ln(4L/D)}{2\pi kL}$
3.	Conduction between inside and outside surfaces of a rectangular box having uniform inside and outside temperatures. Wall thickness Δx is less than any inside dimension.	$R' = \dfrac{1}{k\left[\dfrac{A}{\Delta x} + 0.54\Sigma L + 1.2\Delta x\right]}$ $\Sigma L = $ sum of all 12 inside lengths $\Delta x = $ thickness of wall $A = $ inside surface area

For system 1 figure: t_s, z, D, L, $D \ll z \ll L$, t_∞

For system 2 figure: t_s, L, D, t_∞

For system 3 figure: Δx, t_i, t_o

Table 5-2 THERMAL PROPERTIES OF BUILDING AND INSULATING MATERIALS AT A MEAN TEMPERATURE OF 75 F (ENGLISH UNITS)

Material	Description	Density ρ $\dfrac{\text{lbm}}{\text{ft}^3}$	Thermal Conductivity k $\dfrac{\text{Btu-in.}}{\text{ft}^2\text{-hr-F}}$	Unit Conductance C $\dfrac{\text{Btu}}{\text{hr-ft}^2\text{-F}}$	Unit Resistance — Per Inch Thickness $1/k$ $\dfrac{\text{ft}^2\text{-hr-F}}{\text{Btu-in.}}$	Unit Resistance — For Thickness Listed $1/C$ $\dfrac{\text{hr-ft}^2\text{-F}}{\text{Btu}}$	Specific Heat $\dfrac{\text{Btu}}{\text{lbm-F}}$
Building Board Boards, panels, subflooring, wood-based sheathing, panel products	Asbestos-cement board						
	$\frac{1}{4}$ in. or 6 mm	120	—	16.5	—	0.06	0.24
	Gypsum or plasterboard						
	$\frac{3}{8}$ in. or 10 mm	50	—	3.10	—	0.32	0.26
	$\frac{1}{2}$ in. or 13 mm	50	—	2.22	—	0.45	—
	Plywood	34	0.80	—	1.25	—	0.29
	$\frac{1}{4}$ in. or 6 mm	34	—	3.20	—	0.31	0.29
	$\frac{3}{8}$ in. or 10 mm	34	—	2.13	—	0.47	0.29
	$\frac{1}{2}$ in. or 13 mm	34	—	1.60	—	0.62	0.29
	$\frac{3}{4}$ in. or 20 mm	34	—	1.07	—	0.93	0.29
	Insulating board and sheathing, regular density						
	$\frac{1}{2}$ in. or 13 mm	18	—	0.76	—	1.32	0.31
	$\frac{25}{32}$ in. or 20 mm	18	—	0.49	—	2.06	0.31
	Hardboard, high density, standard tempered	63	1.00	—	1.00	—	0.32
	Particle board						
	Medium density	50	0.94	—	1.06	—	0.31
	Underlayment						
	$\frac{5}{8}$ in. or 16 mm	40	—	1.22	—	0.82	0.29
	Wood subfloor						
	$\frac{3}{4}$ in. or 20 mm	—	—	1.06	—	0.94	0.33

Table 5-2 THERMAL PROPERTIES OF BUILDING AND INSULATING MATERIALS AT A MEAN TEMPERATURE OF 75 F (ENGLISH UNITS), continued

Material	Description	Density ρ $\frac{lbm}{ft^3}$	Thermal Conductivity k $\frac{Btu\text{-}in.}{ft^2\text{-}hr\text{-}F}$	Unit Conductance C $\frac{Btu}{hr\text{-}ft^2\text{-}F}$	Unit Resistance Per Inch Thickness $1/k$ $\frac{ft^2\text{-}hr\text{-}F}{Btu\text{-}in.}$	Unit Resistance For Thickness Listed $1/C$ $\frac{hr\text{-}ft^2\text{-}F}{Btu}$	Specific Heat $\frac{Btu}{lbm\text{-}F}$
Building Paper	Vapor—permeable felt	—	—	16.7	—	0.06	—
	Vapor—seal, two layers of mopped 15-lb felt	—	—	8.35	—	0.12	—
Finish Flooring Materials	Carpet and fibrous pad	—	—	0.48	—	2.08	0.34
	Carpet and rubber pad	—	—	0.81	—	1.23	0.33
	Tile—asphalt, linoleum, vinyl, or rubber	—	—	20.0	—	0.05	0.30
Insulating Materials Blanket and batt	Mineral fiber—fibrous form processed from rock, slag, or glass						
	Approximately 2–2¾ in. or 50–70 mm	0.3–2.0	—	0.143	—	7	0.17–0.23
	Approximately 3–3½ in. or 75–90 mm	0.3–2.0	—	0.091	—	11	0.17–0.23
	Approximately 5¼–6½ in. or 135–165 mm	0.3–2.0	—	0.053	—	19	0.17–0.23
Board and slabs	Cellular glass	8.5	0.38	—	2.63	—	0.24
	Glass fiber, organic bonded	4–9	0.25	—	4.00	—	0.23
	Expanded polystyrene—molded beads	1.0	0.28	—	3.57	—	0.29
	Expanded polyurethane—R-11 expanded	1.5	0.16	—	6.25	—	0.38
	Mineral fiber with resin binder	15	0.29	—	3.45	—	0.17

Loose Fill						
Mineral fiber—rock, slag, or glass						
Approximately 3.75–5 in. or 75–125 mm	0.6–2.0	—	—	—	11	0.17
Approximately 6.5–8.75 in. or 165–222 mm	0.6–2.0	—	—	—	19	0.17
Approximately 7.5–10 in. or 191–254 mm	—	—	—	—	22	0.17
Approximately 7¼ in. or 185 mm	—	—	—	—	30	0.17
Silica aerogel	7.6	0.17	—	5.88	—	—
Vermiculite (expanded)	7–8	0.47	—	2.13	—	—
Roof Insulation						
Preformed, for use above deck						
Approximately ½ in. or 13 mm	—	—	0.72	—	1.39	—
Approximately 1 in. or 25 mm	—	—	0.36	—	2.78	—
Approximately 2 in. or 50 mm	—	—	0.19	—	5.56	—
Cellular glass	9	0.4	—	2.5	—	0.24
Masonry Materials						
Concretes						
Lightweight aggregates including expanded shale, clay, or slate;	200	5.2	—	0.19	—	—
expanded slags; cinders;	100	3.6	—	0.28	—	—
pumice; vermiculite; also	80	2.5	—	0.40	—	—
cellular concretes	40	1.15	—	0.86	—	—
	20	0.70	—	1.43	—	—
Sand and gravel or stone aggregate (not dried)	140	12.0	—	0.08	—	—
Masonry Units						
Brick, common	120	5.0	—	0.20	—	—
Brick, face	130	9.0	—	0.11	—	—
Concrete blocks, three-oval core—sand and gravel aggregate						
4 in. or 100 mm	—	—	1.4	—	0.71	—
8 in. or 200 mm	—	—	0.9	—	1.11	—
12 in. or 300 mm	—	—	0.78	—	1.28	—

Table 5-2 THERMAL PROPERTIES OF BUILDING AND INSULATING MATERIALS AT A MEAN TEMPERATURE OF 75 F (ENGLISH UNITS), continued

Material	Description	Density ρ $\dfrac{\text{lbm}}{\text{ft}^3}$	Thermal Conductivity k $\dfrac{\text{Btu-in.}}{\text{ft}^2\text{-hr-F}}$	Unit Conductance C $\dfrac{\text{Btu}}{\text{hr-ft}^2\text{-F}}$	Unit Resistance Per Inch Thickness $1/k$ $\dfrac{\text{ft}^2\text{-hr-F}}{\text{Btu-in.}}$	For Thickness Listed $1/C$ $\dfrac{\text{hr-ft}^2\text{-F}}{\text{Btu}}$	Specific Heat $\dfrac{\text{Btu}}{\text{lbm-F}}$
Masonry Units (continued)	lightweight aggregate (expanded shale, clay slate or slag; pumice)						
	3 in. or 75 mm	—	—	0.79	—	1.27	—
	4 in. or 100 mm	—	—	0.67	—	1.50	—
	8 in. or 200 mm	—	—	0.50	—	2.00	—
	12 in. or 300 mm	—	—	0.44	—	2.27	—
Plastering Materials	Cement plaster, sand, aggregate	116	5.0	—	0.20	—	—
	Gypsum plaster:						
	Lightweight aggregate						
	$\frac{1}{2}$ in. or 13 mm	45	—	3.12	—	0.32	—
	$\frac{5}{8}$ in. or 16 mm	45	—	2.67	—	0.39	—
	Lightweight aggregate on metal lath						
	$\frac{3}{4}$ in. or 20 mm	—	—	2.13	—	0.47	—

Roofing						
Asbestos-cement shingles	120	—	4.76	—	0.21	—
Asphalt roll roofing	70	—	6.50	—	0.15	—
Asphalt shingles	70	—	2.27	—	0.44	—
Built-in roofing						
$\frac{3}{8}$ in. or 10 mm	70	—	3.00	—	0.33	0.35
Slate, $\frac{1}{2}$ in. or 13 mm	—	—	20.00	—	0.05	—
Wood shingles—plain or plastic film faced	—	—	1.06	—	0.94	0.31
Siding Materials (on flat surface)						
Shingles						
Asbestos-cement	120	—	4.76	—	0.21	—
Siding						
Wood, drop, 1 in. or 25 mm	—	—	1.27	—	0.79	0.31
Wood, plywood, $\frac{3}{8}$ in. or 10 mm, lapped	—	—	1.59	—	0.59	0.29
Aluminum or steel, over sheathing, hollowbacked	—	—	1.61	—	0.61	—
Insulating board—backed nominal, $\frac{3}{8}$ in. or 10 mm	—	—	0.55	—	1.82	—
Insulating board—backed nominal, $\frac{3}{8}$ in. or 10 mm, foil-backed	—	—	0.34	—	2.96	—
Architectural glass	—	—	10.00	—	0.10	—
Woods						
Maple, oak, and similar hardwoods	45	1.10	—	0.91	—	0.30
Fir, pine, and similar softwoods	32	0.80	—	1.25	—	0.33
Metals						
Aluminum (1100)	171	1536	—	0.00065	—	0.214
Steel, mild	489	314	—	0.00318	—	0.120
Steel, stainless	494	108	—	0.00926	—	0.109

Source: Abstracted by permission from *ASHRAE Handbook, Fundamentals Volume*, 1985.

Table 5-2a   THERMAL PROPERTIES OF BUILDING AND INSULATING MATERIALS AT A MEAN TEMPERATURE OF 24 C (SI UNITS)

Material	Description	Density ρ $\dfrac{kg}{m^3}$	Thermal Conductivity k $\dfrac{W}{m\text{-}C}$	Unit Conductance C $\dfrac{W}{m^2\text{-}C}$	Unit Resistance		Specific Heat $\dfrac{kJ}{kg\text{-}C}$
					Per Meter Thickness $1/k$ $\dfrac{m\text{-}C}{W}$	For Thickness Listed $1/C$ $\dfrac{m^2\text{-}C}{W}$	
Building Board							
Boards, panels, subflooring, sheathing, woodbased panel products	Asbestos-cement board						
	$\frac{1}{4}$ in. or 6 mm	1922	—	93.7	—	0.011	1.00
	Gypsum or plasterboard						
	$\frac{3}{8}$ in. or 10 mm	800	—	17.6	—	0.057	1.09
	$\frac{1}{2}$ in. or 13 mm	800	—	12.6	—	0.078	—
	Plywood	545	0.12	—	8.70	—	1.21
	$\frac{1}{4}$ in. or 6 mm	545	—	18.2	—	0.055	1.21
	$\frac{3}{8}$ in. or 10 mm	545	—	12.1	—	0.083	1.21
	$\frac{1}{2}$ in. or 13 mm	545	—	9.09	—	0.110	1.21
	$\frac{3}{4}$ in. or 20 mm	545	—	6.08	—	0.165	1.21
	Insulating board and sheathing, regular density						
	$\frac{1}{2}$ in. or 13 mm	288	—	4.32	—	0.232	1.30
	$\frac{25}{32}$ in. or 20 mm	288	—	2.78	—	0.359	1.30
	Hardboard, high density, standard tempered	1010	0.14	—	6.94	—	1.34
	Particleboard						
	Medium density	800	0.14	—	7.35	—	1.30
	Underlayment						
	$\frac{5}{8}$ in. or 16 mm	640	—	6.93	—	0.144	1.21
	Wood subfloor						
	$\frac{3}{4}$ in. or 20 mm	—	—	6.02	—	0.166	1.38

Building Paper						
Vapor—permeable felt	—	—	94.8	—	0.011	—
Vapor—seal, two layers of mopped 15-lb felt	—	—	47.4	—	0.021	—
Finish Flooring Materials						
Carpet and fibrous pad	—	—	2.73	—	0.367	1.42
Carpet and rubber pad	—	—	4.60	—	0.217	1.38
Tile—asphalt, linoleum, vinyl, or rubber	—	—	113.0	—	0.009	1.26
Insulating Materials Blanket and batt						
Mineral fiber—fibrous form processed from rock, slag, or glass. Approximately 2–2¾ in. or 50–70 mm	4.8–32	—	0.812	—	1.23	0.71–0.96
Approximately 3–3½ in. or 75–90 mm	4.8–32	—	0.517	—	1.94	0.71–0.96
Approximately 5¼–6½ in. or 135–165 mm	4.8–32	—	0.301	—	3.32	0.71–0.96
Board and slabs						
Cellular glass	136	0.0548	—	18.2	—	1.0
Glass fiber, organic bonded	64–144	0.036	—	27.8	—	0.96
Expanded polystyrene—molded beads	16	0.040	—	25.0	—	1.2
Expanded polyurethane—R-11 expanded	24	0.023	—	43.5	—	1.6
Mineral fiber with resin binder	240	0.042	—	23.9	—	0.71
Loose Fill						
Mineral fiber—rock, slag, or glass						
Approximately 3.75–5 in. or 75–125 mm	9.6–32	—	—	—	1.94	0.71
Approximately 6.5–8.75 in. or 165–22 mm	9.6–32	—	0.45	—	3.35	0.71
Approximately 7.5–10 in. or 191–254 mm	—	—	0.28	—	3.87	0.71
Approximately 7¼ in. or 185 mm	—	—	0.23	—	5.28	0.71
Silica aerogel	122	0.025	—	40.8	—	—
Vermiculite (expanded)	122	0.068	—	14.8	—	—

Table 5-2a THERMAL PROPERTIES OF BUILDING AND INSULATING MATERIALS AT A MEAN TEMPERATURE OF 24 C (SI UNITS), continued

Material	Description	Density ρ $\frac{kg}{m^3}$	Thermal Conductivity k $\frac{W}{m\text{-}C}$	Unit Conductance C $\frac{W}{m^2\text{-}C}$	Unit Resistance Per Meter Thickness $1/k$ $\frac{m\text{-}C}{W}$	Unit Resistance For Thickness Listed $1/C$ $\frac{m^2\text{-}C}{W}$	Specific Heat $\frac{kJ}{kg\text{-}C}$
Roof Insulation	Preformed, for use above deck						
	Approximately ½ in. or 13 mm	—	—	4.1	—	0.24	1.0
	Approximately 1 in. or 25 mm	—	—	2.0	—	0.49	2.1
	Approximately 2 in. or 50 mm					0.93	
	Cellular glass	144	0.058	1.1	17.3	—	3.9
							1.0
Masonry Materials Concretes	Lightweight aggregates including expanded shale, clay, or slate; expanded slags; cinders; pumice; vermiculite; also cellular concretes	3200	0.75	—	1.32	—	—
		1600	0.52	—	1.94	—	—
		1280	0.36	—	2.77	—	—
		640	0.17	—	6.03	—	—
		320	0.10	—	10.0	—	—
	Sand and gravel or stone aggregate (not dried)	2242	1.73	—	0.58	—	—

Masonry Units

Brick, common	1922	0.72	—	1.39	—	—
Brick, face	2082	1.30	—	0.77	—	—
Concrete blocks, three-oval core—sand and gravel aggregate						
4 in. or 100 mm	—	—	8.0	—	0.13	—
8 in. or 200 mm	—	—	5.1	—	0.20	—
12 in. or 300 mm	—	—	4.4	—	0.23	—
lightweight aggregate (expanded shale, clay slate or slag; pumice)						
3 in. or 75 mm	—	—	4.5	—	0.22	—
4 in. or 100 mm	—	—	3.8	—	0.26	—
8 in. or 200 mm	—	—	2.8	—	0.35	—
12 in. or 300 mm	—	—	2.5	—	0.40	—

Plastering Materials

Cement plaster, sand aggregate,	1858	0.72	—	1.39	—	—
Gypsum plaster:						
Lightweight aggregate						
$\frac{1}{2}$ in. or 13 mm	721	—	17.7	—	0.056	—
$\frac{5}{8}$ in. or 16 mm	721	—	15.2	—	0.066	—
Lightweight aggregate on metal lath						
$\frac{3}{4}$ in. or 20 mm	—	—	12.1	—	0.83	—

Roofing

Asbestos-cement shingles	1922	—	27.0	—	0.037	—
Asphalt roll roofing	1121	—	36.9	—	0.027	—
Asphalt shingles	1121	—	12.9	—	0.078	—
Built-up roofing						
$\frac{3}{8}$ in. or 10 mm	1121	—	17.0	—	0.059	—
Slate, $\frac{1}{2}$ in. or 13 mm	—	—	113.6	—	0.009	—
Wood shingles—plain or plastic film faced	—	—	6.02	—	0.166	—

Table 5-2a THERMAL PROPERTIES OF BUILDING AND INSULATING MATERIALS AT A MEAN TEMPERATURE OF 24 C (SI UNITS), continued

Material	Description	Density ρ $\dfrac{kg}{m^3}$	Thermal Conductivity k $\dfrac{W}{m\text{-}C}$	Unit Conductance C $\dfrac{W}{m^2\text{-}C}$	Unit Resistance Per Meter Thickness $1/k$ $\dfrac{m\text{-}C}{W}$	Unit Resistance For Thickness Listed $1/C$ $\dfrac{m^2\text{-}C}{W}$	Specific Heat $\dfrac{kJ}{kg\text{-}C}$
Siding Materials (on flat surface)	Shingles						
	Asbestos-cement	1922	—	27.0	—	3.70	—
	Siding						
	Wood, drop, 1 in. or 25 mm	—	—	7.21	—	0.139	1.30
	Wood, plywood, $\frac{3}{8}$ in. or 10 mm, lapped	—	—	9.03	—	0.111	1.21
	Aluminum or steel, over sheathing, hollowbacked	—	—	9.14	—	0.109	—
	Insulating board—backed nominal, $\frac{3}{8}$ in. or 10 mm	—	—	3.12	—	0.320	—
	Insulating board backed nominal, $\frac{3}{8}$ in. or 10 mm, foil-backed	—	—	1.93	—	0.518	—
	Architectural glass	—	—	56.8	—	0.018	—
Woods	Maple, oak, and similar hardwoods	721	0.159	—	6.3	—	1.26
	Fir, pine, and similar softwoods	513	0.115	—	8.67	—	1.38
Metals	Aluminum (1100)	2739	221.5	—	0.0045	—	0.896
	Steel, mild	7833	45.3	—	0.022	—	0.502
	Steel, stainless	7913	15.6	—	0.064	—	0.456

Source: Adapted by permission from *ASHRAE Handbook, Fundamentals Volume,* 1985.

of building and insulating materials. Other useful data are also given in Tables 5-2 and 5-2a; for example, the reciprocal of the unit thermal resistance and the *unit thermal conductance, C*. Note that k has the units of Btu-in./(ft²-hr-F) or W/(m-C). With Δx given in inches or meters, respectively, the unit thermal conductance C is given by

$$C = \frac{1}{R} = \frac{k}{\Delta x} \text{ Btu/(hr-ft}^2\text{-F)} \quad \text{or} \quad \text{W/(m}^2\text{-C)} \tag{5-7}$$

Thermal convection is the transport of energy by mixing in addition to conduction. Convection is associated with fluids in motion generally through a pipe or duct or along a surface. In the very thin layer of fluid next to the surface the transfer of energy is by conduction. In the main body of the fluid mixing is the dominant energy transfer mechanism. A combination of conduction and mixing exists between these two regions. The transfer mechanism is complex and highly dependent on whether the flow is laminar or turbulent.

The usual, simplified approach in convection is to express the heat transfer rate as

$$\dot{q} = hA(t - t_w) \tag{5-8}$$

where

$\dot{q}$ = heat transfer rate from fluid to wall, Btu/hr or W
h = film coefficient, Btu/(hr-ft²-F) or W/(m²-s)
t = bulk temperature of the fluid, F or C
t_w = wall temperature, F or C

The film coefficient h is sometimes called the *unit surface conductance* or alternatively the *convective heat transfer coefficient*. Equation 5-8 may also be expressed in terms of thermal resistance:

$$\dot{q} = \frac{t - t_w}{R'} \tag{5-8a}$$

where

$$R' = \frac{1}{hA} \text{ (hr-F)/Btu} \quad \text{or} \quad \text{C/W} \tag{5-9}$$

or

$$R = \frac{1}{h} = \frac{1^*}{C} \text{ (hr-ft}^2\text{-F)/Btu} \quad \text{or} \quad \text{(m}^2\text{-C)/W} \tag{5-9a}$$

The thermal resistance given by Eq. 5-9 may be summed with the thermal resistances arising from pure conduction given by Eq. 5-3 or 5-6.

---

*Note that the symbol for conductance is C, in contrast to the symbol for the temperature in Celsius degrees, C.

The film coefficient h appearing in Eqs. 5-8 and 5-9 depends on the fluid, the fluid velocity, the flow channel, and the degree of development of the flow field. (That is, whether the fluid has just entered the channel or is relatively far from the entrance.) Many correlations exist for predicting the film coefficient under various conditions. Correlations for forced convection are given in Chapter 3 of the *ASHRAE Handbook* (1).

In convection it is important to recognize the mechanism that is causing the fluid motion to occur. When the bulk of the fluid is moving relative to the heat transfer surface, the mechanism is called *forced convection*, because such motion is usually caused by a blower, fan, or pump that is forcing the flow. In forced convection buoyancy forces are negligible. In *free convection,* on the other hand, the motion of the fluid is due entirely to buoyancy forces, usually confined to a layer near the heated or cooled surface. The surrounding bulk of the fluid is stationary and exerts a viscous drag on the layer of moving fluid. As a result inertia forces in free convection are usually small. Free convection is often referred to as *natural convection.*

Natural or free convection is an important part of HVAC applications. However, it is one of the most difficult physical situations to describe, and the predicted film coefficients have a greater uncertainty than those of forced convection. Various empirical relations for natural convection film coefficients can be found in reference 1.

Most building structures have forced convection along outer walls or roofs, and natural convection occurs inside narrow air spaces and on the inner walls. There is considerable variation in surface conditions, and both the direction and magnitude of the air motion on outdoor surfaces are very unpredictable. The film coefficient for these situations usually ranges from about 1.0 Btu/(hr-ft^2-F) or 6 W/(m^2-C) for free convection up to about 6 Btu/(hr-ft^2-F) or 35 W/(m^2-C) for forced convection with an air velocity of about 15 miles per hour, 20 ft/sec or 6 m/s. Because of the low film coefficients, especially with free convection, the amount of heat transferred by thermal radiation may be equal to or larger than that transferred by convection.

Thermal radiation is the transfer of thermal energy by electromagnetic waves and is an entirely different phenomenon from conduction and convection. In fact thermal radiation can occur in a perfect vacuum and is actually impeded by an intervening medium. The direct net transfer of energy by radiation between two surfaces that see only each other and that are separated by a nonabsorbing medium is given by

$$\dot{q}_{12} = \frac{\sigma(T_1^4 - T_2^4)}{\dfrac{1 - \epsilon_1}{A_1\epsilon_1} + \dfrac{1}{A_1F_{12}} + \dfrac{1 - \epsilon_2}{A_2\epsilon_2}} \tag{5-10}$$

where

σ = Boltzmann constant, 0.1713×10^{-8} Btu/(hr-ft^2-R^4) or 5.673×10^{-8} W/(m^2-K^4)

T = absolute temperature, R or K
ϵ = emittance
A = surface area, ft^2 or m^2
F = configuration factor, a function of geometry only (Chapter 6)

In Eq. 5-10 we assume that both surfaces are "gray" (where the emittance ϵ equals the absorptance α). This assumption can often be justified. The student is referred to Parker et al. (2) or other textbooks on heat transfer for a more complete discussion of thermal radiation. Figure 5-3 shows situations where radiation may be a significant factor. For the wall

$$\dot{q}_i = \dot{q}_w = \dot{q}_r + \dot{q}_o$$

and for the air space

$$\dot{q}_i = \dot{q}_r + \dot{q}_c = \dot{q}_o$$

The resistances can be combined to obtain an equivalent overall resistance R' with which the heat transfer rate can be computed using Eq. (5-2a):

$$\dot{q} = \frac{-(t_o - t_i)}{R'} \tag{5-2a}$$

The thermal resistance for radiation is not easily computed, however, because of the fourth power temperature relationship of Eq. (5-10). For this reason and because of the inherent uncertainty in describing the physical situation, theory and experiment have been combined to develop combined or effective unit thermal resistances and unit thermal conductances for many typical surfaces and air spaces. Table 5-3 gives effective film coefficients and unit thermal resistances as a function of wall position, direction of heat flow, air velocity, and surface emittance for exposed surfaces such as outside walls. Table 5-3a gives representative values of emittance ϵ for some building and insulating

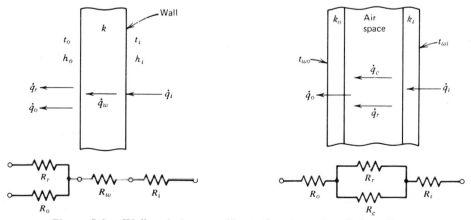

Figure 5-3. Wall and air space illustrating thermal radiation effects.

Table 5-3 SURFACE UNIT CONDUCTANCES AND UNIT RESISTANCES FOR AIR[a]

		Surface Emittances											
		ε = 0.9				ε = 0.2				ε = 0.05			
		h		R		h		R		h		R	
	Direction of Heat Flow	Btu hr-ft²-F	W m²-C	hr-ft²-F Btu	m²-C W	Btu hr-ft²-F	W m²-C	hr-ft²-F Btu	m²-C W	Btu hr-ft²-F	W m²-C	hr-ft²-F Btu	m²-C W
Position of Surface													
Still Air													
Horizontal	Upward	1.63	9.26	0.61	0.11	0.91	5.2	1.10	0.194	0.76	4.3	1.32	0.232
Sloping—45 degrees	Upward	1.60	9.09	0.62	0.11	0.88	5.0	1.14	0.200	0.73	4.1	1.37	0.241
Vertical	Horizontal	1.46	8.29	0.68	0.12	0.74	4.2	1.35	0.238	0.59	3.4	1.70	0.298
Sloping—45 degrees	Downward	1.32	7.50	0.76	0.13	0.60	3.4	1.67	0.294	0.45	2.6	2.22	0.391
Horizontal	Downward	1.08	6.13	0.92	0.16	0.37	2.1	2.70	0.476	0.22	1.3	4.55	0.800
Moving Air													
(any position) wind is 15 mph or 6.7 m/s (for winter)	Any	6.0	34.0	0.17	0.029								
Wind is 7½ mph or 3.4 m/s (for summer)	Any	4.0	22.7	0.25	0.044								

[a]Conductances are for surfaces of the stated emittance facing virtual blackbody surroundings at the same temperature as the ambient air. Values are based on a surface-air temperature difference of 10 deg F and for surface temperature of 70 F.

Source: Adapted by permission from *ASHRAE Handbook, Fundamentals Volume,* 1985.

Table 5-3a   REFLECTANCE AND EMITTANCE OF VARIOUS SURFACES AND
EFFECTIVE EMITTANCES OF AIR SPACE

Surface	Reflectance in Percent	Average Emittance ϵ	Effective Emittance E of Air Space	
			With One Surface Having Emittance ϵ and Other 0.90	With Both Surfaces of Emittance ϵ
Aluminum foil, bright	92–97	0.05	0.05	0.03
Aluminum sheet	80–95	0.12	0.12	0.06
Aluminum coated paper, polished	75–84	0.20	0.20	0.11
Steel, galvanized, bright	70–80	0.25	0.24	0.15
Aluminum paint	30–70	0.50	0.47	0.35
Building materials—wood, paper, glass, masonry, nonmetallic paints	5–15	0.90	0.82	0.82

Source: Adapted by permission from ASHRAE Handbook, Fundamentals Volume, 1985.

materials. For example, a typical vertical brick wall exposed to the outdoors has an effective emittance E of about 0.8 to 0.9. In still air the average film coefficient, from Table 5-3, is about 1.46 Btu/(hr-ft²-F) or 8.29 W/(m²-C) and the unit thermal resistance is 0.68 (hr-ft²-F)/Btu or 0.12 (m²-C)/W.

If the surface were highly reflective, $E = 0.05$, the film coefficient would be 0.59 Btu/(hr-ft²-F) [3.4 W/(m²-C)] and the unit thermal resistance would be 1.7 (hr-ft²-F)/Btu or 0.298 (m²-C)/W. It is evident that thermal radiation is a large factor when natural convection occurs. If the wind velocity were to increase to 15 mph or about 7 m/s, the average film coefficient would increase to about 6 Btu/(hr-ft²-F) or 34 W/(m²-C). With higher air velocities the relative effect of radiation diminishes.

Tables 5-4, 5-5, and 5-6 give conductances and resistances for air spaces as a function of position, direction of heat flow, air temperature, and the effective emittance of the space. The effective emittance E is given by

$$\frac{1}{E} = \frac{1}{\epsilon_1} + \frac{1}{\epsilon_2} - 1 \qquad (5\text{-}11)$$

where ϵ_1 and ϵ_2 are for each surface of the air space. The effect of radiation is quite apparent in Tables 5-4, 5-5, and 5-6, where the thermal resistance may be observed to decrease by a factor of two or three as E varies from 0.05 to 0.82.

The preceding paragraphs cover thermal resistances arising from conduction, convection, and radiation. Equation 5-4 may be generalized to give the equivalent resistance of n resistors in series.

$$R'_e = R'_1 + R'_2 + R'_3 + \cdots + R'_n \qquad (5\text{-}12)$$

Figure 5-4 is an example of a wall being heated or cooled by a combination of convection and radiation on each surface and having five different resistances through which the heat must be conducted. The equivalent thermal resistance R'_e for the wall is given by Eq. 5-12 as

$$R'_e = R'_i + R'_1 + R'_2 + R'_3 + R'_o \qquad (5\text{-}13)$$

Each of the resistances may be expressed in terms of fundamental variables using Eqs. 5-3 and 5-9.

Figure 5-4. Wall with thermal resistances in series.

Table 5-4 UNIT THERMAL RESISTANCE OF A PLANE $\frac{3}{4}$ INCH (20 mm) AIR SPACE[a]

Position of Air Space	Direction of Heat Flow	Mean Air Temperature F	C	Temperature Difference F	C	E = 0.05 ft²-hr-F/Btu	m²-C/W	E = 0.2 ft²-hr-F/Btu	m²-C/W	E = 0.82 ft²-hr-F/Btu	m²-C/W
Horizontal	Up	90	32	10	6	2.22	0.391	1.61	0.284	0.75	0.132
		50	10	30	17	1.66	0.292	1.35	0.238	0.77	0.136
		50	10	10	6	2.21	0.389	1.70	0.299	0.87	0.153
		0	−18	20	11	1.79	0.315	1.52	0.268	0.93	0.164
		0	−18	10	6	2.16	0.380	1.78	0.313	1.02	0.180
45° Slope	Up	90	32	10	6	2.78	0.490	1.88	0.331	0.81	0.143
		50	10	30	17	1.92	0.338	1.52	0.268	0.82	0.144
		50	10	10	6	2.75	0.484	2.00	0.352	0.94	0.166
		0	−18	20	11	2.07	0.364	1.72	0.303	1.00	0.176
		0	−18	10	6	2.62	0.461	2.08	0.366	1.12	0.197
Vertical	Horizontal	90	32	10	6	3.24	0.571	2.08	0.366	0.84	0.148
		50	10	30	17	2.77	0.488	2.01	0.354	0.94	0.166
		50	10	10	6	3.46	0.609	2.35	0.414	1.01	0.178
		0	−18	20	11	3.02	0.532	2.32	0.408	1.18	0.210
		0	−18	10	6	3.59	0.632	2.64	0.465	1.26	0.222
45° Slope	Down	90	32	10	6	3.27	0.576	2.10	0.370	0.84	0.148
		50	10	30	17	3.23	0.569	2.24	0.394	0.99	0.174
		50	10	10	6	3.57	0.629	2.40	0.423	1.02	0.180
		0	−18	20	11	3.57	0.629	2.63	0.463	1.26	0.222
		0	−18	10	6	3.91	0.689	2.81	0.495	1.30	0.229

[a]Effective emittance of the space E is given by Eq. 5-11. Credit for an air space resistance value cannot be taken more than once and only for the boundary conditions established. Resistances of horizontal spaces with heat flow downward are substantially independent of temperature difference.

Source: Adapted by permission from ASHRAE Handbook, Fundamentals Volume, 1985.

Table 5-5 UNIT THERMAL RESISTANCES OF A PLANE 3.5 INCH (89 mm) AIR SPACE[a]

Position of Air Space	Direction of Heat Flow	Mean Air Temperature F	C	Temperature Difference F	C	E = 0.05 ft²-hr-F/Btu	m²-C/W	E = 0.2 ft²-hr-F/Btu	m²-C/W	E = 0.82 ft²-hr-F/Btu	m²-C/W
Horizontal	Up	90	32	10	6	2.66	0.468	1.83	0.322	0.80	0.141
		50	10	30	17	2.01	0.354	1.58	0.278	0.84	0.148
		50	10	10	6	2.66	0.468	1.95	0.343	0.93	0.164
		0	−18	20	11	2.18	0.384	1.79	0.315	1.03	0.181
		0	−18	10	6	2.62	0.461	2.07	0.365	1.12	0.197
45° Slope	Up	90	32	10	6	2.96	0.521	1.97	0.347	0.82	0.144
		50	10	30	17	2.17	0.382	1.67	0.294	0.86	0.151
		50	10	10	6	2.95	0.520	2.10	0.370	0.96	0.169
		0	−18	20	11	2.35	0.414	1.90	0.335	1.06	0.187
		0	−18	10	6	2.87	0.505	2.23	0.393	1.16	0.204
Vertical	Horizontal	90	32	10	6	3.40	0.598	2.15	0.479	0.85	0.150
		50	10	30	17	2.55	0.449	1.89	0.333	0.91	0.160
		50	10	10	6	3.40	0.598	2.32	0.409	1.01	0.178
		0	−18	20	11	2.78	0.490	2.17	0.382	1.14	0.201
		0	−18	10	6	3.33	0.586	2.50	0.440	1.23	0.217
45° Slope	Down	90	32	10	6	4.33	0.763	2.49	0.438	0.90	0.158
		50	10	30	17	3.30	0.581	2.28	0.402	1.00	0.176
		50	10	10	6	4.36	0.768	2.73	0.481	1.08	0.190
		0	−18	20	11	3.63	0.639	2.66	0.468	1.27	0.224
		0	−18	10	6	4.32	0.761	3.02	0.532	1.34	0.236

[a]Effective emittance of the space E is given by Eq. 5-11. Credit for an air space resistance value cannot be taken more than once and only for the boundary conditions established. Resistances of horizontal spaces with heat flow downward are substantially independent of temperature differences.

Source: Adapted by permission from ASHRAE Handbook, Fundamentals Volume, 1985.

Table 5-6 UNIT THERMAL RESISTANCES OF PLANE HORIZONTAL AIR SPACES WITH HEAT FLOW DOWNWARD, TEMPERATURE DIFFERENCE 10 DEG F (6 DEG C)

| Thickness | | Mean Temperature | | $E = 0.05$ | | $E = 0.2$ | | $E = 0.82$ | |
Inch	mm	F	C	$\dfrac{\text{ft}^2\text{-hr-F}}{\text{Btu}}$	$\dfrac{\text{m}^2\text{-C}}{\text{W}}$	$\dfrac{\text{ft}^2\text{-hr-F}}{\text{Btu}}$	$\dfrac{\text{m}^2\text{-C}}{\text{W}}$	$\dfrac{\text{ft}^2\text{-hr-F}}{\text{Btu}}$	$\dfrac{\text{m}^2\text{-C}}{\text{W}}$
$\frac{3}{4}$	20	90	32	3.29	0.579	2.10	0.370	0.85	0.150
		50	10	3.59	0.632	2.41	0.424	1.02	0.180
		0	−18	4.02	0.708	2.87	0.505	1.31	0.231
$1\frac{1}{2}$	38	90	32	5.35	0.942	2.79	0.491	0.94	0.166
		50	10	5.90	1.04	3.27	0.576	1.15	0.203
		0	−18	6.66	1.17	4.00	0.707	1.51	0.266
$3\frac{1}{2}$	89	90	32	8.19	1.44	3.41	0.601	1.00	0.176
		50	10	9.27	1.63	4.09	0.720	1.24	0.218
		0	−18	10.32	1.82	5.08	0.895	1.64	0.289

Source: Adapted by permission from *ASHRAE Handbook, Fundamentals Volume*, 1985.

$$R'_e = \frac{1}{h_i A_i} + \frac{\Delta x_1}{k_1 A_1} + \frac{\Delta x_2}{k_2 A_2} + \frac{\Delta x_3}{k_3 A_3} + \frac{1}{h_o A_o} \tag{5-14}$$

The film coefficients may be read from Table 5-3 and the thermal conductivities may be obtained from Table 5-2. For this case, a plane wall, the areas in Eq. 5-14 are all equal and cancel.

In the general case the area that is properly a part of the resistance may vary and unit thermal resistances may have to be adjusted. Consider the insulated pipe shown in Fig. 5-5. Convection occurs on the inside and outside surfaces while heat is conducted through the pipe wall and insulation. The overall thermal resistance for the pipe of Fig. 5-5 is

$$R'_e = R'_o + R'_2 + R'_1 + R'_i \tag{5-15}$$

or using Eqs. 5-6 and 5-9

$$R'_e = \frac{1}{h_o A_o} + \frac{\ln\left(\dfrac{r_3}{r_2}\right)}{2\pi k_2 L} + \frac{\ln\left(\dfrac{r_2}{r_1}\right)}{2\pi k_1 L} + \frac{1}{h_i A_i} \tag{5-16}$$

Equation 5-16 may also be written as

$$R'_e = \frac{1}{h_o A_o} + \frac{(r_3 - r_2)}{k_2 A_{m2}} + \frac{(r_2 - r_1)}{k_1 A_{m1}} + \frac{1}{h_i A_i} \tag{5-16a}$$

where

$$A_m = \frac{2\pi L(r_o - r_i)}{\ln\left(\dfrac{r_o}{r_i}\right)} \tag{5-17}$$

Equation 5-16a has a form quite similar to Eq. 5-14; however, the areas are all unequal. The thermal resistance on the outside surface is reduced considerably by the relatively large area. Where area changes occur in the direction of heat

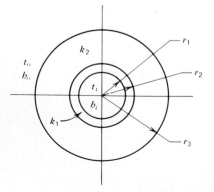

Figure 5-5.   Insulated pipe in convective environment.

flow, unit resistances or conductances can only be used with appropriate area weighting factors.

Thermal resistances may also occur in parallel. In theory the parallel resistances can be combined into an equivalent thermal resistance in the same way as electrical resistances.

$$\frac{1}{R'_e} = \frac{1}{R'_1} + \frac{1}{R'_2} + \frac{1}{R'_3} + \cdots \frac{1}{R'_n} \tag{5-18}$$

In most heat transfer situations with parallel heat flow paths, however, lateral flow occurs, which may invalidate Eq. 5-18. The effect of lateral heat transfer between two thermal conductors is to lower the equivalent resistance of the system. However, when the ratio of the larger to the smaller of the thermal resistances is less than about 5, Eq. 5-18 gives a reasonable approximation of the equivalent thermal resistance. For large variations in the thermal resistance of parallel conduction paths, numerical techniques using the computer are suggested.

The concept of thermal resistance is very useful and convenient in the analysis of complex arrangements of building materials. After the equivalent thermal resistance has been determined for a configuration, however, the overall unit thermal conductance, usually called the *overall heat transfer coefficient* U, is frequently used.

$$U = \frac{1}{R'A} = \frac{1}{R} \text{ Btu/(hr-ft}^2\text{-F)} \quad \text{or} \quad \text{W/(m}^2\text{-C)} \tag{5-19}$$

The heat transfer rate is then given by

$$\dot{q} = UA\Delta t \tag{5-20}$$

where

UA = conductance, Btu/(hr-F) or W/C
A = surface area, ft^2 or m^2
Δt = overall temperature difference, F or C

For a plane wall the area A is the same throughout the wall. In dealing with a curved wall, we select the area for convenience of calculation. For example, in the problem of heat transfer through the ceiling-attic-roof combination, it is usually most convenient to use the ceiling area. The area selected is then used to determine the appropriate value of U from Eq. 5-19.

5-2 TABULATED OVERALL HEAT-TRANSFER COEFFICIENTS

For convenience of the designer, tables have been constructed that give overall coefficients for many common building sections including walls and floors, doors, windows, and skylights. The tables used in the *ASHRAE Handbook* have a great deal of flexibility and are summarized in the following pages.

Walls and Roofs

Walls and roofs vary considerably in the materials from which they are constructed. Therefore, the thermal resistance or the overall heat transfer coefficient is usually computed on an individual basis using Eqs. 5-14 and 5-19. This procedure is demonstrated for a wall and a roof in Tables 5-7a and 5-7b. Note that in each case an element has been changed. The tabular presentation makes it simple to recalculate the thermal resistance. In each case the unit thermal resistance and the overall heat-transfer coefficient have been computed for one set of conditions.

Example 5-1.   A frame wall is modified to have $3\frac{1}{2}$ in. of mineral fiber insulation between the studs. Compute the overall heat-transfer coefficient U if the unit thermal resistance without the insulation is 4.44 (hr-ft²-F)/Btu.

Solution.

Total unit resistance given	4.44
Deduct the air space unit resistance, Table 5-5	− 1.01
Add insulation unit resistance given in Table 5-2	11.00
Total R in (hr-ft²-F)/Btu	14.43

Then based on one square foot, we see that

$$U = \frac{1}{R} = \frac{1}{14.43} = 0.07 \text{ Btu/(hr-ft}^2\text{-F)}$$

Equation 5-18 may be used to correct for framing (2 × 4 studs on 16 in. centers)

$$\frac{1}{R'_c} = \frac{1}{R'} + \frac{1}{R'_f} \qquad \text{or} \qquad U_c A_t = UA_b + U_f A_f$$

where

A_t = total area
A_b = area between studs
A_f = area occupied by the studs

The unit thermal resistance of a section through the 2 × 4 stud is equal to the total resistance less the resistance of the air gap plus the resistance of the stud from Table 5-2. A 2 × 4 stud is only $3\frac{1}{2}$ in. deep.

$$R_f = \frac{1}{U_f} = 4.4 - 0.97 + (1.25)(3.5) = 7.81$$

or

$$U_f = 0.128 \text{ Btu/(hr-ft}^2\text{-F)}$$

Table 5-7a COEFFICIENTS OF TRANSMISSION U OF MASONRY CAVITY WALLS, Btu/(hr-ft²-F)[a]

Replace Cinder Aggregate Block with 6-in. Lightweight Aggregate Block with Cores Filled (New Item 4 in col. 2)

Construction	Resistance (R) 1 Between Furring	1 At Furring	2 Between Furring	2 At Furring
1. Outside surface (15 mph wind)	0.17	0.17	0.17	0.17
2. Face brick, 4 in.	0.44	0.44	0.44	0.44
3. Cement mortar, 0.5 in.	0.10	0.10	0.10	0.10
4. Concrete block, cinder aggregate, 8 in.	1.72	1.72	2.99	2.99
5. Reflective air space, 0.75 in. (50 F mean; 30 deg F temperature difference)	2.77	—	2.77	—
6. Nominal 1-in. × 3-in. vertical furring	—	0.94	—	0.94
7. Gypsum wallboard, 0.5 in., foil backed	0.45	0.45	0.45	0.45
8. Inside surface (still air)	0.68	0.68	0.68	0.68
Total Thermal Resistance (R)	$R_i = 6.33$	$R_s = 4.50$	$R_i = 7.60$	$R_s = 5.77$

Construction No. 1: $U_i = 1/6.33 = 0.158$; $U_s = 1/4.50 = 0.222$. With 20% framing (typical of 1-in. × 3-in. vertical furring on masonry @ 16-in. o.c.), $U_{av} = 0.8 (0.158) + 0.2 (0.222) = 0.171$

Construction No. 2: $U_i = 1/7.60 = 0.132_4 U_s = 1/5.77 = 0.173$.
With framing unchanged, $U_{av} = 0.8(0.132) + 0.2(0.173) = 1.40$

[a] U factor may be converted to W/(m²·C) by multiplying by 5.68.

Source: Adapted by permission from *ASHRAE Handbook, Fundamentals Volume,* 1977.

Table 5-7b   COEFFICIENTS OF TRANSMISSION U OF FLAT BUILT-UP ROOFS,[a] Btu/(hr-ft²-F)[b]

Add Rigid Roof Deck Insulation, $C = 0.24$ ($R = 1/C$) (New Item 7 in col. 2)

Construction (Heat Flow Up)	1	2
	Resistance (R)	
1. Outside surface (15 mph wind)	0.17	0.17
2. Built-up roofing, 0.375 in.	0.33	0.33
3. Rigid roof deck insulation (none)	—	4.17
4. Concrete slab, lightweight aggregate, 2 in.	2.22	2.22
5. Corrugated metal deck	0	0
6. Metal ceiling suspension system with metal hanger rods	0[c]	0[c]
7. Nonreflective air space, greater than 3.5 in. (50 F mean; 10 deg F temperature difference)	0.93[a]	0.93[a]
8. Metal lath and lightweight aggregate plaster, 0.75 in.	0.47	0.47
9. Inside surface (still air)	0.61	0.61
Total Thermal Resistance (R)	4.73	8.90

Construction No. 1: $U_{avg} = 1/4.73 = 0.211$
Construction No. 2: $U_{avg} = 1/8.90 = 0.112$

[a] Use largest air space (3.5 in.) value shown in Table 5-5.
[b] U factor may be converted to W/m²-C by multiplying by 5.68.
[c] Area of hanger rods is negligible in relation to ceiling area.

Source: Adapted by permission from *ASHRAE Handbook, Fundamentals Volume,* 1977.

Then using Eq. 5-18 we get

$$U_c = \frac{(0.07)(14.5) + (0.128)(1.5)}{16} = 0.075 \text{ Btu/(hr-ft}^2\text{-F)}$$

Example 5-2.   Compute the overall average coefficient for the roof-ceiling combination shown in Table 5-7b with 3.5 in. of mineral wool insulation in the ceiling space rather than the rigid roof deck insulation.

Solution.   Total unit resistance of ceiling floor combination in Table 5-7b with no insulation is 4.73 (hr-ft²-F)/Btu. Assume air space greater than 3.5 in.

Total resistance without insulation	4.73
Add mineral fiber insulation, 3.5 in.	11.00
Total R [(hr-ft²-F)/Btu]	15.73
Total U [Btu/(ft²-hr-F)]	0.064

It should be noted that the data given in Tables 5-7 and Examples 1 and 2 are based on

1.   Steady-state heat transfer.
2.   Ideal construction methods.
3.   Surrounding surfaces at ambient air temperature.
4.   Variation of thermal conductivity with temperature negligible.

Some caution should be exercised in applying calculated overall heat transfer coefficients such as those of Tables 5-7 because the effects of poor workmanship and materials are not included. Although a safety factor is not usually applied, a moderate increase in U may be justified in some cases.

The overall heat-transfer coefficients obtained for walls and roofs should always be adjusted for framing, as shown in Tables 5-7 using Eq. 5-18. This adjustment will normally be 5 to 12 percent of the unadjusted coefficient.

The coefficients of Tables 5-7 have all been computed for a 15 mph wind velocity on outside surfaces and should be adjusted for other velocities. The data of Table 5-3 may be used for this purpose. The *ASHRAE Handbook* (1) has conversion tables for this purpose also.

The following example illustrates the calculation of an overall heat transfer coefficient for an unvented roof-ceiling system.

Example 5-3.   Compute the overall heat-transfer coefficient for the roof-ceiling combination shown in Fig. 5-6. The wall assembly is similar to Table 5-7a with an overall heat-transfer coefficient of 0.16 Btu/(hr-ft²-F). The roof assembly is similar to Table 5-7b without the ceiling and has a conductance of 0.13 Btu/(hr-ft²-F) between the air space and the outdoor air. The ceiling has a conductance of 0.2 Btu/(hr-ft²-F) between the conditioned space and the ceiling air space. The air space is 2.0 ft in the

Figure 5-6. Section of a roof-ceiling combination.

vertical direction. The ceiling has an area of 15,000 ft² and a perimeter of 500 ft.

Solution. It is customary to base the overall heat-transfer coefficient on the ceiling area horizontal to the floor. Note that heat can enter or leave the air space through the roof or the wall that closes the space around the perimeter. The thermal resistances of the roof and the wall are parallel and in series with the resistance of the ceiling. Then for roof and wall

$$C_{rw} = C_w A_w + C_r A_r$$

and

$$R'_{rw} = \frac{1}{C_{rw}} = \frac{1}{C_w A_w + C_r A_r}$$

Further

$$R'_o = R'_{rw} + R'_c$$

and

$$R'_o = \frac{1}{C_w A_w + C_r A_r} + \frac{1}{C_c A_c}$$

Substitution yields

$$R'_o = \frac{1}{(0.16)(2)(500) + (0.13)(15,000)} + \frac{1}{(0.2)(15,000)}$$

$$R'_o = 0.000807 = \frac{1}{U_o A_c}$$

then

$$U_o = \frac{1}{(0.000807)(15,000)} = 0.083 \text{ Btu/(hr-ft}^2\text{-F)}$$

Ceiling spaces should be vented to remove moisture from the insulation, but only moderate ventilation rates are required. The effect of ventilation on the transfer of heat through the air space is not significant provided the ceiling is insulated with a unit thermal resistance of about 11 or more. This is true for both winter and summer conditions. It once was thought that increased ventilation during the summer would dramatically reduce the heat gain to the inside space; however, this is apparently incorrect (3). It is generally not economically feasible to use power ventilation. The main reason for the ineffectiveness of ventilation is the fact that most of the heat transfer through the attic is by thermal radiation between the roof and the ceiling insulation. The use of reflective surfaces is therefore much more useful in reducing heat transfer. It is recommended that calculation of the overall transmission coefficient for ceiling spaces be computed using the approach of Example 5-3 with appropriate unit resistances and assuming no ventilation.

Windows and Light Transmitting Panels

Table 5-8 contains overall heat transfer coefficients for windows, skylights, and other light-transmitting panels. Notice that there are values for summer and winter coefficients. This difference arises because coefficients for use in the winter assume a different velocity than in the summer. Table 5-8 applies only for air-to-air heat transfer and does not account for solar radiation, which will be discussed in Chapter 6.

Doors

Table 5-9 gives overall heat transfer coefficients for common doors. Again note the difference between summer and winter. Solar radiation has not been included.

Concrete Floors and Walls Below Grade

The heat transfer through basement walls and floors depends on the temperature difference between the inside air and the ground, the wall or floor material, and the conductivity of the ground. All of these factors involve considerable uncertainty. Mitalas (4) has studied the below-grade heat-transfer problem and developed methods that accurately predict heat losses for basement walls and floors 3 ft (0.9 m) or more below grade. When the design conditions across the United States are considered, the average coefficients in Tables 5-10 and 5-11, derived from reference 4, give satisfactory results for design load calculations. The heat loss is given by

$$\dot{q} = UA(t_i - t_g) \tag{5-21}$$

Table 5-8 OVERALL COEFFICIENTS OF HEAT TRANSMISSION (*U*-FACTOR) OF WINDOWS AND SKYLIGHTS, Btu/(hr-ft²-F)[a]

Description	Summer[j]		Winter[i]		Exterior Horizontal Panels (Skylights)	
	No Indoor Shade	Indoor Shade[k]	No Indoor Shade	Indoor Shade[k]	Summer[l]	Winter[m]
Flat Glass[b]						
Single glass	1.04	0.81	1.10	0.83	0.83	1.23
Insulating glass, double[c]						
$\frac{3}{16}$-in. air space[d]	0.65	0.58	0.62	0.52	0.57	0.70
$\frac{1}{4}$-in. air space[d]	0.61	0.55	0.58	0.48	0.54	0.65
$\frac{1}{2}$-in. air space[e]	0.56	0.52	0.49	0.42	0.49	0.58
$\frac{1}{2}$-in. air space, low emittance coating[f]						
$e = 0.20$	0.38	0.37	0.32	0.30	0.36	0.48
$e = 0.40$	0.45	0.44	0.38	0.35	0.42	0.52
$e = 0.60$	0.51	0.48	0.43	0.38	0.46	0.56

Insulating glass, triple[c]					
$\frac{1}{4}$-in. air space[d]	0.44	0.40	0.39	0.31	
$\frac{1}{2}$-in. air space[g]	0.39	0.36	0.31	0.26	
Storm Windows					
1 in. to 4 in. air spaces[d]	0.50	0.48	0.50	0.42	
Plastic bubbles[n]					
Single walled				0.80	1.15
Double walled				0.46	0.70

[a] See Table 5-8a for adjustments for various windows and sliding patio doors.
[b] Emittance of uncoated glass surface = 0.84.
[c] Double and triple refer to number of lights of glass.
[d] 0.125-in. glass.
[e] 0.25-in. glass.
[f] Coating on either glass surface facing air space; all other glass surfaces uncoated.
[g] Window design: 0.25-in. glass, 0.125-in. glass, 0.125-in. glass, 0.25-in. glass.
[i] 15 mph outdoor air velocity; 0 F outdoor air; 70 F inside air temperature; natural convection.
[j] 7.5 mph outdoor air velocity; 89 F outdoor air; 75 F inside air natural convection; solar radiation 248.3 Btu/(hr·ft^2).
[k] Values apply to tightly closed venetian and vertical blinds, draperies, and roller shades.
[l] For heat flow down.
[m] For heat flow up.
[n] Based on area of opening, not total surface area.

Source: Adapted by permission from *ASHRAE Handbook, Fundamentals Volume*, 1985.

Table 5-8a   ADJUSTMENT FACTORS FOR COEFFICIENTS
             U OF TABLE 5-8

Description	Single Glass	Double or Triple Glass	Storm Windows
Windows			
All glass	1.0	1.0	1.0
Wood sash—80% glass	0.9	0.95	0.90
Metal sash—80% glass	1.0	1.20	1.20
Sliding glass doors			
Metal frame	1.0	1.10	
Wood frame	0.95	1.0	

Source: Adapted by permission from *ASHRAE Handbook, Fundamentals Volume,*
1985.

where

U = overall heat-transfer coefficient from Table 5-10 or 5-11, Btu/(hr-ft²-F) or W/(m²-C)
A = wall or floor surface area below 3 ft (0.9 m), ft or m
t_i = inside air temperature, F or C

and

$$t_g = t_{avg} - A \tag{5-22}$$

where

t_{avg} = average winter temperature, F or C (Table C-2)
A = amplitude of ground temperature variation about t_{avg}, F or C (Fig. 5-7)

Studies have shown that the heat losses from below-grade walls and floors are far more dependent on the ground temperature near the surface than the deep ground temperature. Ground surface temperature is known to vary about a mean value by an amplitude A that varies with geographic location (Fig. 5-7). The mean ground surface temperature is assumed to be the average winter temperature.

When below-grade spaces are conditioned as living space, the walls should be furred and finished with a vapor barrier, insulating board, and some type of finish layer such as paneling. This will add thermal resistance to the wall. The basement floor should also be finished by laying down an insulating barrier and floor tile or carpet. The overall coefficients for the finished wall or floor may be computed as

$$R'_a = R' + R'_f = \frac{1}{UA} + R'_f = \frac{1}{U_a A} \tag{5-23}$$

Table 5-9 COEFFICIENTS OF TRANSMISSION U FOR DOORS

Nominal Thickness	Description	Wood No Storm Door Btu hr-ft²-F	W m²-C	Storm Door[a] Wood Btu hr-ft²-F	W m²-C	Storm Door[a] Metal Btu hr-ft²-F	W m²-C	Summer, No Storm Door Btu hr-ft²-F	W m²-C
$1\frac{3}{8}$ in. or 35 mm	Solid core	0.39	2.22	0.26	1.48	0.28	1.59	0.38	2.16
$1\frac{3}{8}$ in. or 35 mm	Panel door	0.57	3.24	0.33	1.87	0.37	2.10	0.54	3.07
$1\frac{3}{4}$ in. or 45 mm	Solid core	0.33	1.87	0.28	1.59	0.25	1.42	0.32	1.82
$2\frac{1}{4}$ in. or 57 mm	Solid core	0.27	1.53	0.20	1.14	0.21	1.19	0.26	1.48
	Steel door Heavy type								
$1\frac{3}{4}$ in. or 45 mm									
	Solid urethane foam core	0.40	2.27					0.39	2.21
	Solid urethane core with thermal break	0.19	1.08	0.16	0.91	0.17	0.97	0.18	1.02

[a]Approximately 50 percent glass for wood doors; values for metal doors are independent of glass percentage.

Source: Adapted by permission from ASHRAE Handbook, Fundamentals Volume, 1985.

Table 5-10 AVERAGE OVERALL HEAT-TRANSFER COEFFICIENTS FOR BASEMENT WALLS (REF. 4)

| Depth Below Grade | | Wall Uninsulated | | Wall Insulated to a Depth of 2 Ft Below Grade | | | | Wall Insulated Over Full Inside Surface | | | |
Feet	Meter	Btu/hr-ft²-F	W/m²-C	R-5 Btu/hr-ft²-F	R-0.88 W/m²-C	R-10 Btu/hr-ft²-F	R-1.76 W/m²-C	R-5 Btu/hr-ft²-F	R-0.88 W/m²-C	R-10 Btu/hr-ft²-F	R-1.76 W/m²-C
4	1.22	0.200	0.035	0.140	0.025	0.130	0.023	0.088	0.015	0.054	0.010
5	1.52	0.180	0.032	0.140	0.025	0.130	0.023	0.082	0.014	0.050	0.009
6	1.83	0.170	0.030	0.140	0.025	0.130	0.023	0.079	0.014	0.048	0.008
7	2.13	0.160	0.028	0.140	0.025	0.130	0.023	0.076	0.013	0.047	0.008

Figure 5-7. Lines of constant amplitude of ground surface temperature. (Reprinted by permission from *ASHRAE Handbook, Fundamentals Volume,* 1985.)

Floor Slabs at Grade Level

When considering the heat losses for floor slabs at or near grade level, we must take into account two situations. The first is the unheated slab where heat is supplied to the space from above. The second situation results when the air duct system is installed beneath the slab with air discharged at the perimeter of the structure. In both cases most of the heat loss is from the edge of the slab. When compared with the total heat losses of the structure, this loss may not be significant; however, from the viewpoint of comfort the loss is important. Proper insulation around the perimenter of the slab is essential in severe climates to ensure a warm floor.

Figure 5-8 shows typical placement of edge insulation for a floor slab. Insulation may also be placed on the outside of the foundation wall extending

Table 5-11 AVERAGE OVERALL HEAT-TRANSFER COEFFICIENTS FOR UNINSULATED BASEMENT FLOORS, 3 FT OR MORE BELOW GRADE (REF. 4)

Wall Uninsulated		Wall Insulated to a Depth of 2 Ft Below Grade		Wall Insulated Over Full Inside Surface	
Btu hr-ft²-F	W m²-C	Btu hr-ft²-F	W m²-C	Btu hr-ft²-F	W m²-C
0.025	0.004	0.036	0.006	0.045	0.008

Figure 5-8. Edge insulation for a floor slab.

down to the footing. Table 5-12 contains heat-loss factors for slabs. Note that the heat-loss factors are expressed as heat-transfer rate per unit length of perimeter per degree temperature difference. For summer conditions the heat transfer to the floor slab is negligible.

The heat loss from the slab is expressed as

$$\dot{q} = U'P(t_i - t_o) \tag{5-24}$$

where

U' = heat loss coefficient, Btu(hr-ft-F) or W/(m-C) (Table 5-12)
P = perimeter of slab, ft or m
t_i = inside air temperature, F or C
t_o = outdoor design temperature, F or C

Crawl Spaces

The most important factor in calculating heat losses from a crawl space is its temperature. A heat balance on the crawl space taking into account the various gains and losses will yield the temperature. Heat is transferred to the crawl space through the floor and lost through the foundation wall and the ground much like a slab on grade. The inside or outside of the foundation wall may be insulated, and insulation may extend inward from the base of the foundation wall. The problem of crawl space temperature and heat loss is treated in Chapter 7.

Horizontal Pipes and Flat Surfaces

It is often necessary to compute the heat transfer from pipes, ducts, or flat surfaces. Typically these surfaces are in a free convection environment where radiation may also be important. This problem is complicated because the free convection heat-transfer coefficient and the thermal radiation both depend on the surface temperature, which may not be accurately known. Other variables are the surface finish, ambient air temperature, and air motion. Therefore it is customary to present overall heat-transfer coefficients as shown in

Table 5-12 HEAT LOSS COEFFICIENTS FOR SLAB FLOOR CONSTRUCTION

Construction	Insulation	Degree Days (Base 65 F)					
		2950		5350		7433	
		Btu/hr-ft-F	W/m-C	Btu/hr-ft-F	W/m-C	Btu/hr-ft-F	W/m-C
8-in. (200-mm) block wall, brick face	Uninsulated	0.62	1.07	0.68	1.18	0.72	1.25
	R-5.4 (R-0.95) slab to footing	0.48	0.83	0.50	0.87	0.56	0.97
4-in. (100-mm) block wall, brick face	Uninsulated	0.80	1.39	0.84	1.45	0.93	1.61
	R-5.4 (R-0.95) slab to footing	0.47	0.81	0.49	0.85	0.54	0.94
Metal stud wall, stucco	Uninsulated	1.15	1.99	1.20	2.08	1.34	2.32
	R-5.4 (R-0.95) slab to footing	0.51	0.88	0.53	0.92	0.58	1.00
Poured concrete wall, with perimeter ducts	Uninsulated	1.84	3.19	2.12	3.67	2.73	4.73
	R-5.4 (R-0.95) slab to footing	0.64	1.11	0.72	1.25	0.90	1.56

Source: Reprinted by permission from ASHRAE Handbook, Fundamentals Volume, 1985.

Table 5-13   TRANSMISSION COEFFICIENTS U FOR HORIZONTAL BARE STEEL PIPES AND FLAT SURFACES,[a] Btu/(hr-ft²-F)[b]

Pipe Size Inches	Temperature Difference F Between Pipe Surface and Surrounding Air, Air at 80 F[c]									
	50	100	150	200	250	300	350	400	450	500
$\frac{1}{2}$	2.12	2.48	2.80	3.10	3.42	3.74	4.07	4.47	4.86	5.28
$\frac{3}{4}$	2.08	2.43	2.74	3.04	3.35	3.67	4.00	4.40	4.79	5.21
1	2.04	2.38	2.69	2.99	3.30	3.61	3.94	4.33	4.72	5.14
$1\frac{1}{4}$	2.00	2.34	2.64	2.93	3.24	3.55	3.88	4.27	4.66	5.07
$1\frac{1}{2}$	1.98	2.31	2.61	2.90	3.20	3.52	3.84	4.23	4.62	5.03
2	1.95	2.27	2.56	2.85	3.15	3.46	3.78	4.17	4.56	4.97
$2\frac{1}{2}$	1.92	2.23	2.52	2.81	3.11	3.42	3.74	4.12	4.51	4.92
3	1.89	2.20	2.49	2.77	3.07	3.37	3.69	4.08	4.46	4.87
$3\frac{1}{2}$	1.87	2.18	2.46	2.74	3.04	3.34	3.66	4.05	4.43	4.84
4	1.85	2.16	2.44	2.72	3.01	3.32	3.64	4.02	4.40	4.81
$4\frac{1}{2}$	1.84	2.14	2.42	2.70	2.99	3.30	3.61	4.00	4.38	4.79
5	1.83	2.13	2.40	2.68	2.97	3.28	3.59	3.97	4.35	4.76
Vertical surface	1.84	2.14	2.42	2.70	3.00	3.30	3.62	4.00	4.38	4.79
Horizontal surface } Facing upward	2.03	2.37	2.67	2.97	3.28	3.59	3.92	4.31	4.70	5.12
Horizontal surface } Facing downward	1.61	1.86	2.11	2.36	2.64	2.93	3.23	3.60	3.97	4.37

[a]Values are for flat surfaces greater than 4 ft² or 0.4 m².

[b]U in W/(m²-C) equals Btu/(hr-ft²-F) times 5.678.

[c]The temperature difference in C equals F divided by 1.8. An air temperature of 27 C corresponds to 80 F.

Source: Adapted by permission from *ASHRAE Handbook, Fundamentals Volume,* 1985.

Table 5-13, which is for bare steel pipe and flat surfaces. The values given are based on experimental data and analytical solutions for combined free convection and thermal radiation. They are reliable when the surface temperature is known and the surrounding air is still. If the average bulk temperature of the fluid flowing in the pipe is known, a good estimate of the surface temperature can be made because of the low thermal resistance of the pipe wall.

Buried Pipe

To make calculations of the heat transfer to or from buried pipes it is necessary to know the thermal properties of the earth. The thermal conductivity of soil varies considerably with the analysis and moisture content. Typically the range is 0.33 to 1.33 Btu/(ft-hr-F) or 0.58 to 2.3 W/(m-C). A reasonable estimate can be made using the equation for thermal resistance given in Table 5-1 with a known value of k. Thermal conductivity data for various soils and moisture contents are given in reference 1.

5-3   MOISTURE TRANSMISSION

The transfer of moisture through building materials and between the building surfaces and moist air follows theory directly analogous to conductive and convective heat transfer. Fick's law, which has the same form as Eq. 5-1

$$\dot{m}_w = -DA \frac{dC}{dx} \tag{5-25}$$

governs the diffusion of moisture in a substance. Convective transport of moisture may be expressed as

$$\dot{m}_w = h_m A (C - C_w) \tag{5-26}$$

which is similar to Eq. 5-4. This subject is discussed in detail in Chapter 12. The important point here is that moisture moves from a location where the concentration is high to one where it is low. This movement and accumulation of moisture can cause severe damage to the structure if not controlled.

During the coldest months, the moisture concentration tends to be greatest in the interior space. It is transferred to the walls and ceilings and, if not retarded, diffuses outward into the insulation. The moisture reduces the thermal resistance of the insulation and in some cases it may freeze. Ceilings have been known to fail structurally due to an accumulation of ice.

During the summer months, the moisture transfer process is reversed. This case is not as severe as that for the winter; however, the moisture is still harmful to the insulation and condensation may occur on some inside surfaces.

The transfer of moisture and the resulting damage is controlled through the use of barriers or retardants such as aluminum foil, thin plastic film, or other such material, and through the use of ventilation. Analysis of the problem shows that the moisture retarder should be near the warmest surface to prevent moisture from entering the insulation. Because the winter months are the most critical time, the barrier is usually installed between the inside finish layer and the insulation. During the summer months, the problem can usually be controlled by natural ventilation. This is the most important reason for ventilating an attic in both summer and winter. About 0.5 cfm/ft^2 or 0.15 m^3/(m^2-min) is required to remove the moisture from a typical attic. This can usually be accomplished through natural effects. Walls sometimes have provisions for a small amount of ventilation, but usually the outer wall covering is adequate to retard moisture.

REFERENCES

1. *ASHRAE Handbook, Fundamentals Volume,* American Society of Heating, Refrigerating and Air-Conditioning Engineers, Atlanta, Ga., 1985.

2. J. D. Parker, J. H. Boggs, and E. F. Blick, *Introduction to Fluid Mechanics and Heat Transfer,* Addison-Wesley, Reading, Mass., 1969.

3. NBS Special Publication 548 "Summer Attics and Whole-House Ventilation," U.S. Department of Commerce/National Bureau of Standards, Washington, D.C., 1978.

4. G. P. Mitalis, "Basement Heat Loss Studies at DBR/NRC," National Research Council of Canada, Division of Building Research, Ottawa, 1982.

PROBLEMS

5-1 Compute the overall thermal resistance for a 2-in. steel pipe with 1 in. of insulation. The inside and outside film coefficients are 500 and 2 Btu/(hr-ft^2-F), respectively, and the insulation has a thermal conductivity of 0.2 Btu-in./(ft^2-hr-F).

5-2 What is the unit thermal resistance for an inside partition made up of $\frac{3}{8}$-in. gypsum board on each side of 2 × 4 in. studs? (Neglect the studing.)

5-3 Compute the overall thermal resistance of a wall made up of 100-mm face brick and 100-mm common brick with a 20-mm air gap between. There is 13 mm of gypsum plaster on the inside. Assume a 7 m/s wind velocity on the outside and still air inside.

5-4 Assuming that the 2 × 4 studs form a parallel heat transfer path, compute the unit thermal resistance for the partition of Problem 5-2. The studs are on 16 in. centers.

5-5 The pipe of Problem 5-1 has water flowing inside with a heat-transfer coefficient of 500 Btu/(hr-ft^2-F) and is exposed to air on the outside with a film coefficient of 2.0 Btu/(hr-ft^2-F). Compute the overall heat-transfer coefficient.

5-6 The partition of Problem 5-2 has still air on each side. Compute the overall heat transfer coefficient.

5-7   Compute the overall heat-transfer coefficient for a frame construction wall made of brick veneer with 3-in. insulation bats between the studs; the wind velocity is $7\frac{1}{2}$ mph.

5-8   In an effort to save energy it is proposed to change the standard frame wall construction from 2×4 studs on 16 in. centers to 2×6 studs on 24 in. centers. Make a table similar to Table 5-7 showing both constructions. Use $5\frac{1}{2}$-in. and $3\frac{1}{2}$-in. fibrous glass insulation. Compare the two different constructions.

5-9   Use the data in Table 5-3 for still air and estimate what fraction of the heat transfer for a vertical wall is pure convection. Explain.

5-10   Consider a ceiling space formed by an infinite flat roof and horizontal ceiling. The inside surface of the roof has a temperature of 135 F and the top side of the ceiling insulation has a temperature of 115 F. Estimate the heat transferred by radiation and convection separately and compare them. Assume both surfaces have an emittance of 0.9.

5-11   A wall is 12 ft wide and 8 ft high and has an overall heat transfer coefficient of 0.1 Btu/(hr-ft²-F). It contains a solid wood door, $80 \times 32 \times 1\frac{1}{2}$ in., and a single glass window, 60×30 in. Assuming parallel heat flow paths for the wall, door, and window, find the overall thermal resistance and overall heat transfer coefficient for the combination. Assume winter conditions. The window is metal sash, 80 percent glass.

5-12   A wall exactly like the one described in Table 5-7a has dimensions of 10×3 m. The wall has a total window area of 5 m² made of double-insulating glass with a 6-mm air space. There is a steel door, 2×1 m, that has a mineral fiber core. Assuming winter conditions, compute the effective overall heat transfer coefficient for the combination.

5-13   Estimate the unit thermal resistance for a vertical $\frac{3}{4}$-in. (20-mm) air space. The air space is near the outside surface of a wall that has a large thermal resistance near the inside surface. The outdoor temperature is 5 F (-15 C). Assume nonreflective surfaces.

5-14   Compute the heat-transfer rate per square foot through a flat, built-up roof-ceiling combination similar to that shown in Table 5-7b. The ceiling is metal lath and plaster with 4-in. fibrous glass batts above. Indoor and outdoor temperatures are 70 and 5 F, respectively.

5-15   Compute the overall heat transfer for a single glass window and compare with the values given in Table 5-8 for summer conditions with and without indoor shade.

5-16   Compute the overall heat-transfer coefficient for a $1\frac{3}{4}$-in. (45-mm) solid core wood door and compare to the value given in Table 5-9.

5-17   Determine the overall heat-transfer coefficient for a single glass window with an aluminum sash. Compare this with double insulating glass ($\frac{1}{4}$-in. air space) in a wood sash (80 percent glass).

5-18   A basement wall extends 6 ft below grade, is fully insulated (R-10), and is 6 in. thick. The inside is finished with $\frac{1}{2}$-in. insulating board, plastic vapor seal, and $\frac{1}{4}$-in. plywood paneling. Compute the overall heat-transfer coefficient for the wall.

5-19   Estimate the overall heat-transfer coefficient for a basement floor that has been covered with carpet and fibrous pad. The wall is fully insulated.

5-20   A basement is 20 × 20 ft (6 × 6 m) and 7 ft (2.13 m) below grade. The walls are insulated to a depth of 2 ft (0.6 m) below grade with R-5 (R-0.88) insulation. (a) Estimate the overall heat-transfer coefficients for the walls and floor. (b) Assuming the basement is located in Chicago, Illinois, compute the heat loss from the basement.

5-21   Rework Problem 5-20 assuming that the walls are finished on the inside with R-11 insulation and ⅜-in. gypsum board and the floor has a carpet and pad.

5-22   Estimate heat loss from the floor slab of a building with a perimeter heating system. The building has dimensions of 15 × 30 m with a block and brick wall. The insulation is made of 40 mm of expanded polystyrene. The building is located in Kansas City.

5-23   Rework Problem 5-10 assuming both surfaces have an emittance of 0.05.

5-24   Compute the overall heat-transfer coefficient for (a) an ordinary vertical single glass window. (b) Assume the window has a roller shade. Compute the overall heat-transfer coefficient assuming a 3½-in. (89-mm) air space between the shade and the glass. (c) Compare the results of (a) and (b) with values from Table 5-8.

5-25   Refer to Table 5-7a and compute the overall transmission coefficient for the same construction with 1 in. sheathing of expanded polystyrene molded beads and wood siding in place of the brick.

5-26   A 24 × 40 ft (7.3 × 12.2 m) building has a full basement with walls extending 8 ft (2.4 m) below grade. The inside of the walls is finished with R-5 (R-0.7) insulation, a thin vapor barrier, and ½-in. (12.7-mm) gypsum board. Estimate an overall heat-transfer coefficient for the walls.

5-27   The floor of the basement described in Problem 5-26 is finished with a thin vapor barrier, ⅝-in. (16-mm) particle board underlayment, and carpet with rubber pad. Estimate an overall heat-transfer coefficient for the floor.

5-28   Assume that the ground temperature t_g is 40 F (10 C) and that the inside temperature is 68 F (20 C) in Problem 5-26 and estimate the temperatures between the wall and insulation and between the gypsum board and insulation.

5-29   Use the temperatures given in Problem 5-28 and compute the temperature between the underlayment and the carpet pad in Problem 5-27.

5-30   A floor slab with heating ducts below has R-5 (R-0.88), insulation from the slab down to the footing. Estimate the heat loss per unit length of perimeter for a location where the degree days are about 5400 F-day.

5-31   A small single-story frame office building in Syracuse, New York is constructed with a concrete slab floor with an overhead type heating system. Estimate the heat loss per unit length of perimeter. Assume (a) R-5 (R-0.88) edge insulation; (b) no edge insulation. (c) Give a good reason to use the edge insulation other than to save energy.

5-32   Two 3-in. steel pipes cross a basement room near the ceiling. One pipe supplies hot water to a heating unit at 150 F, and the other pipe returns water from the unit at 130 F. If the room temperature is about 75 F, estimate the heat transfer from each pipe per linear foot.

5-33 Estimate the heat transfer rate to a 100-ft length of buried steel pipe carrying chilled water at 40 F. The pipe is 30 in. deep and has a 4-in. diameter. Assume the temperature of the deep ground is 50 F. The thermal conductivity of the earth is about 8 Btu-in./(hr-ft²-F).

5-34 Estimate the heat loss from 100 m of buried hot water pipe. The mean water temperature is 49 C. The copper pipe with 10 mm of insulation, $k = 0.05$ W/(m-C), is buried 1 m below the surface and is 100 mm in diameter. Assume a thermal conductivity of the earth of 1.4 W/(m-C) and a deep ground temperature of 10 C.

5-35 Consider a 50-ft length of rectangular duct with inside dimensions of 12 × 24 in. with 1 in. of fibrous glass insulation. Estimate the approximate heat gain per foot of length if the inside air temperature is 60 F and the outside air temperature is 110 F. The average film coefficient is 7 Btu/(hr-ft²-F). Estimate an approximate outside film coefficient from Table 5-3.

5-36 A large beverage cooler resembles a small building and is to be maintained at about 35 F (2 C). The walls and ceiling are well insulated and are finished on the inside with plywood. Assume that the outdoor temperature drops below 35 F (2 C) for only short periods of time during the winter. Where should the vapor retardant be located? Explain what might happen if the retardant is improperly located.

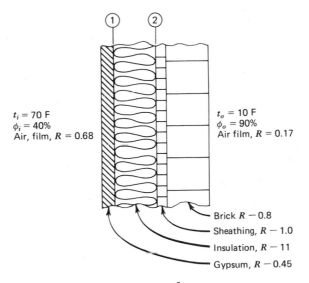

$t_i = 70$ F
$\phi_i = 40\%$
Air, film, $R = 0.68$

$t_o = 10$ F
$\phi_o = 90\%$
Air film, $R = 0.17$

Brick $R - 0.8$
Sheathing, $R - 1.0$
Insulation, $R - 11$
Gypsum, $R - 0.45$

R in units of (hr-ft²-F)/Btu

Figure 5-9. Wall section for Problem 5-37.

5-37 Consider the wall section shown in Fig. 5-9. (a) Compute the temperatures of surfaces 1 and 2. (b) Assuming that the moist air can diffuse through the gypsum and insulation from the inside, would you expect moisture to condense on surface 1? Explain. (c) Would moisture condense on surface 2? Explain. (d) Where should a vapor retardant be placed?

5-38 Refer to the roof-ceiling assembly shown in Fig. 5-7b and compute the temperature of the metal roof deck when the outdoor temperature is 0 F (−18 C) and the indoor temperature is 72 F (22 C) with RH of 40 percent (a) with the rigid insulation and (b) without the insulation. (c) Would you expect any condensation problems on the underside of the metal deck? Explain.

5-39 Consider the wall section shown in Fig. 5-7a and estimate the temperature of the inside surface of the concrete block at the furring. The outdoor temperature is −5 F (−21 C) and the inside temperature is 70 F (21 C) with a relative humidity of 40 percent. Would you recommend a vapor retardant? If so where would you place it? Explain.

CHAPTER SIX

Solar Radiation

The sun is the source of most of the energy used by humans. It drives the winds and ocean currents, it furnishes the energy required for plants to grow, it created our oil and coal resources, and it furnishes warmth to us both directly and indirectly.

Solar radiation has important effects on both the heat gain and heat loss of a building. This effect depends to a great extent on both the location of the sun in the sky and the clearness of the atmosphere as well as on the nature and orientation of the building. It is useful at this point to discuss ways of predicting the variation of the sun's location in the sky during the day and with the seasons for various locations on the earth's surface. It is also useful to know how to predict, for specified weather conditions, the solar irradiation of a surface at any given time and location on the earth, as well as the total radiation striking a surface over a specified period of time.

6-1 THERMAL RADIATION

Thermal radiation is an electromagnetic phenomenon. Its properties and characteristics are similar in many respects to other forms of electromagnetic radiations, such as radio waves, visible light, ultraviolet light, X rays, and gamma rays.

The entire electromagnetic spectrum is continuous, extending from very small wavelengths to those of many thousands of meters in length. All electromagnetic radiation, when absorbed by a system, produces thermal energy, that is, a heating effect. It has been discovered that this heating effect is strongly dependent on wavelength and that practically all of the heating effect is concentrated in one particular wavelength band. To this type of radiation, which originates in the thermal agitation of particles of matter such as atoms, ions, or molecules, the term *thermal radiation* is applied. Thermal radiation is approximately that portion of the electromagnetic spectrum with wavelengths of from 0.1×10^{-6} m up to approximately 100×10^{-6} m. For the sake of convenience, the term *micron* is often used (1 μm = 10^{-6} m); therefore, the approximate range of thermal radiation is from 0.1 to 100 μm. A portion of the shorter wavelengths in this range is visible to the human eye.

The total thermal radiation that impinges on a surface from all directions and from all sources is called the *total or global irradiation (G)*. Its units are Btu/(hr-ft²) or W/m².

The thermal radiation energy that falls on a surface is subject to absorption and reflection as well as transmission through transparent bodies. *Absorption* is the transformation of the radiant energy into thermal energy stored by the molecules. *Reflection* is the return of radiation by a surface without change of frequency. In effect the radiation is "bounced" off of the surface. *Transmission* is the passage of radiation through a medium without change of frequency. Energy falling on a surface must be subject to one of these three actions; therefore:

$$\alpha + \rho + \tau = 1$$

where

α is the absorptance, the fraction of the total incident thermal radiation absorbed

ρ is the reflectance, the fraction of the total incident thermal radiation reflected

τ is the transmittance, the fraction of the total incident radiation transmitted through the body

When the material is optically smooth and of sufficient thickness to show no change of reflectance or absorptance with increasing thickness, the terms *reflectivity and absorptivity* are used to describe the reflectance and absorptance, respectively. In much of the literature there is no distinction between these terms.

Radiant energy may originate at a surface or from the interior of a medium because of the temperature of the material. The rate of emission of energy is stated in terms of the *total emissive power (E)*. Its value depends only on the temperature of the system and the characteristics of the material of the system. Some surfaces emit more energy than others at the same temperature. The units of E may be expressed in Btu/(hr-ft²) or W/m². E is the total energy emitted by the surface into the space and is a multidirectional, total quantity.

It follows that radiant energy leaving an opaque surface ($\tau = 0$) comes from two sources: (1) the emitted energy and (2) the reflected irradiation.

A surface that reflects no radiation ($\rho = 0$) is said to be a *blackbody*, since in the absence of emitted or transmitted radiation it puts forth no radiation visible to the eye and thus appears black. A blackbody is a perfect absorber of radiation and is a useful concept and standard for study of the subject of radiation heat transfer. It can be shown that the perfect absorber of radiant energy is also a perfect emitter, thus the perfect radiant emitter is also given the name *blackbody*. For a given temperature, a black emitter exhibits a maximum monochromatic emissive power at wavelength λ_{max}, given by

$$\lambda_{max} = \frac{5215.6}{T} \mu$$

This equation is known as *Wien's displacement law*. The maximum amount of radiation is emitted in the wavelengths around the value of λ_{max}. According to Wien's displacement law, as the the temperature of a black emitter increases, the major part of the radiation that is being emitted shifts to shorter wavelengths. This is an important concept in engineering, since the concept may be applied to approximate behavior of many nonblack emitters. It implies that higher temperature surfaces are primarily emitters of short wavelength radiation and lower temperature surfaces are primarily emitters of long wavelength radiation. The sun, which has a surface temperature of approximately 10,000 F or 6000 K, emits radiation with a maximum in the visible range. Building surfaces, which are at a much lower temperature, emit radiation primarily at much longer wavelengths.

Most surfaces are not blackbodies, but reflect some incoming radiation and emit less radiation than a blackbody at the same temperature. For such real surfaces we define one additional term, the *emittance* ϵ. The emittance is the fraction of the energy that a blackbody surface would emit at the same temperature, or

$$E = \epsilon E_B$$

The emittance can vary with the temperature of the surface and with its conditions, such as roughness, cleaness, and the like. Although the emittance ϵ and the absorptance α of a given surface are identical for radiation at a given wavelength, the emittance of a building surface is most often quite different than its absorptance for solar radiation. The sun, being at a much higher temperature than a building surface, emits a predominance of radiation having relatively short wavelength compared to that of the building surface. The ratio of absorptance for sunlight to the emittance of a surface, combined with convection effects controls the outer surface temperature of a building in sunlight. Sunlight has an additional important effect in transmitting energy into a building through openings such as windows, doors, and skylights.

6-2   THE EARTH'S MOTION ABOUT THE SUN

The earth moves in a slightly elliptical orbit about the sun. The plane in which the earth rotates around the sun (approximately once every $365\frac{1}{4}$ days) is called the *ecliptic plane* or *orbital plane*. The mean distance from the center of the earth to the center of the sun is approximately 92.9×10^6 miles or 1.5×10^8 km. The *perihelion distance*, when the earth is closest to the sun, is 98.3 percent of the mean distance and occurs on January 4. The *aphelion distance*, when the earth is farthest from the sun, is 101.7 percent of the mean distance and occurs on July 5. Because of this the earth receives about 7 percent more total radiation in January than in July.

Figure 6-1 shows the movement of the earth around the sun. As the earth moves it also spins about its own axis at the rate of one revolution each 24 hours. There is an additional motion because of a slow wobble or gyroscopic precession of the earth. The earth's axis of rotation is tilted 23.5 degrees with respect to the orbital plane. As a result of this dual motion and tilt, the position of the sun in the sky, as seen by an observer on earth, varies with the observer's location on the earth's surface and with the time of day and the time of year. For practical purposes the sun is so small as seen by an observer on earth that it may be treated as a point source of radiation.

Figure 6-1 can be used to explain the effect of the earth's tilt and rotation about the sun. At the time of the vernal equinox (March 21) and of the autumnal equinox (September 22 or 23) the sun appears to be directly overhead at the equator and the earth's poles are equidistant from the sun. Equinox means "equal nights" and during the time of the two equinoxes all points on the earth (except the poles) have exactly 12 hours of darkness and 12 hours of daylight.

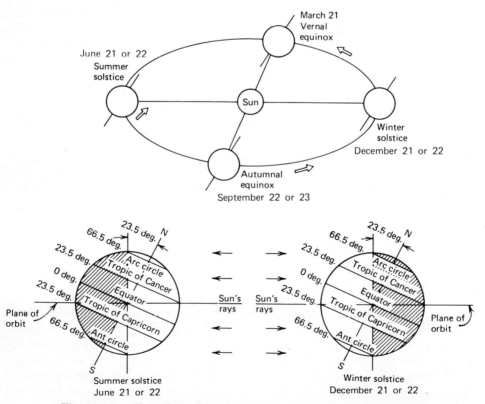

Figure 6-1. The effect of the earth's tilt and rotation about the sun.

During the summer solstice (June 21 or 22) the north pole is inclined 23.5 degrees toward the sun. All points on the earth's surface north of 66.5 degrees north latitude (the Arctic Circle) are in continuous daylight, whereas all points south of 66.5 degrees south latitude (the Antarctic Circle) are in continuous darkness. Relatively warm weather occurs in the northern hemisphere and relatively cold weather occurs in the southern hemisphere. The word solstice means *sun standing still*.

During the summer solstice the sun appears to be directly overhead along the Tropic of Cancer, whereas during the winter solstice it is overhead along the Tropic of Capricorn. The *torrid zone* is the region between, where the sun is at the zenith, directly overhead, at least once during the year. In the *temperate zones* (between 23.5 and 66.5 degrees latitude in each hemisphere) the sun is never directly overhead but always appears above the horizon each day. The *frigid zones* are those zones with latitude greater than 66.5 degrees where the sun is below the horizon for at least one full day (24 hours) each year. In these two zones the sun is also above the horizon for at least one full day each year.

6-3   TIME

Because of the earth's rotation about its own axis, a fixed location on the earth's surface goes through a 24-hour cycle in relation to the sun. The earth is divided into 360 degrees of circular arc by longitudinal lines passing through the poles. Thus, 15 degrees of longitude corresponds to $\frac{1}{24}$ of a day or 1 hour of time. A point on the earth's surface exactly 15 degrees west of another point will see the sun in exactly the same position as the first point after one hour of time has passed. *Universal Time* or *Greenwich Civil Time* (GCT) is the time along the zero longitude line passing through Greenwich, England. *Local Civil Time* (LCT) is determined by the longitude of the observer, the difference being four minutes of time for each degree of longitude, the more advanced time being on meridians further east. Thus when it is 12:00 noon GCT, it would be 7:00 A.M. LCT along the 75th degree west longitude meridian.

Clocks are usually set for the same reading throughout a zone covering approximately 15 degrees of longitude, although the borders of the time zone may be irregular to accommodate local geographical features. The Local Civil Time for a selected meridian near the center of the zone is called the *Standard Time*. The four standard times zones in the lower 48 states and their standard meridians are

Eastern Standard Time, EST 75 degrees
Central Standard Time, CST 90 degrees
Mountain Standard Time, MST 105 degrees
Pacific Standard Time, PST 120 degrees

In much of the United States clocks are advanced one hour during the late spring, summer, and early fall season, leading to *Daylight Savings Time*.

Whereas civil time accounts for days that are precisely 24 hours in length, solar time has slightly variable days because of nonsymmetry of the earth's orbit, irregularities of the earth's rotational speed, and other factors. Time measured by the apparent daily motion of the sun is called *solar time*.

The Local Solar Time (LST) can be calculated from the Local Civil Time (LCT) by a quantity called the *equation of time*, LST = LCT + (equation of time). Values of the equation of time are given in Table 6-1 for each week of the year (2). Values for any day may be obtained by linear interpolation.

Example 6-1. Determine the local solar time (LST) corresponding to 11:00 A.M. CDST on February 8 in the United States at 95 degrees west longitude.

Solution. It is first necessary to convert central daylight savings time to central standard time:

$$CST = CDST - 1 \text{ hour} = 11:00 - 1 = 10:00 \text{ A.M.}$$

Then CST is local civil time at 90 degrees west longitude. Now local civil time (LCT) at 95 degrees west is $5 \times 4 = 20$ minutes less advanced than LCT at 90 degrees west. Then

$$LCT = CST - 20 \text{ min} = 9:40 \text{ A.M.}$$

From Table 6-1 the equation of time is − 14 minutes, 14 second. Then

$$LST = LCT + \text{equation of time}$$
$$LST = 9:40 - 0:14 = 9:26 \text{ A.M.}$$

Table 6-1 THE EQUATION OF TIME

	Day			
Month	**1**	**8**	**15**	**22**
January	− 3:16	− 6:26	− 9:12	− 11:27
February	− 13:34	− 14:14	− 14:15	− 13:41
March	− 12:36	− 11:04	− 9:14	− 7:12
April	− 4:11	2:07	− 0:15	1:19
May	2:50	3:31	3:44	3:30
June	2:25	1:15	0:09	− 1:40
July	− 3:33	− 4:48	− 5:45	− 6:19
August	− 6:17	5:40	− 4:35	− 3:04
September	− 0:15	2:03	4:29	6:58
October	10:02	12:11	13:59	15:20
November	16:20	16:16	15:29	14:02
December	11:14	8:26	5:13	1:47

Source: Reprinted from "The American Ephemeris and Nautical Almanac," U.S. Naval Observatory, Washington, D.C.

6-4 SOLAR ANGLES

Any point on the surface of the earth can be described in relation to the sun's rays at any instant if three fundamental quantities are known. These three quantities are

1. Location on the earth's surface.
2. Time of day.
3. Day of the year.

It is convenient to describe these three quantities by giving the latitude, the hour angle, and the sun's declination, respectively. Figure 6-2 shows a point P located on the surface of the earth in the northern hemisphere. The *latitude* l is the angle between the line OP and the projection of OP on the equatorial plane. This is the same latitude that is commonly used on globes and maps to describe the location of a point with respect to the equator.

The *hour angle h* is the angle between the projection of OP on the equatorial plane and the projection on that plane of a line from the center of the sun to the center of the earth. Fifteen degrees of hour angle corresponds to one hour of time. The hour angle varies from zero at local solar noon to a maximum at sunrise or sunset. Solar noon occurs when the sun is at the highest point in the sky, and hour angles are symmetrical with respect to solar noon. Thus, the hour angles of sunrise and sunset on a given day are identical.

The sun's *declination d* is the angle between a line connecting the center of the sun and earth and the projection of that line on the equatorial plane. Figure 6-3 shows how the sun's declination varies throughout a typical year. On a given day in the year the declination varies slightly from year to year, but for typical HVAC calculations the values from any year are sufficiently

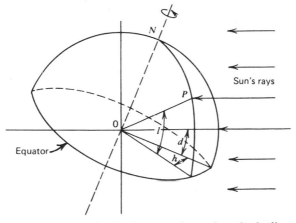

Figure 6-2. Latitude, hour angle, and sun's declination.

Figure 6-3. Variation of sun's declination.

accurate. Table 6-2 shows typical values of the sun's declination for each week. More precise values may be found in reference 2.

It is convenient in HVAC computations to define the sun's position in the sky in terms of the solar altitude β, and the solar azimuth ϕ, which depend on l, h, and d.

The *solar altitude* β (sun's altitude angle) is the angle between the sun's ray and the projection of that ray on a horizontal surface (Fig. 6-4). It is the angle of the sun above the horizon. It can be shown by analytic geometry that the following relationship is true (1):

$$\sin \beta = \cos l \cos h \cos d + \sin l \sin d \tag{6-1}$$

The *sun's zenith* angle Ψ is the angle between the sun's rays and a perpendicular to the horizontal plane at point P, Fig. 6-4. Obviously

$$\beta + \Psi = 90 \text{ degrees} \tag{6-2}$$

Table 6-2   THE SUN'S DECLINATION FOR A TYPICAL YEAR (IN DEGREES AND MINUTES)

Month	Day			
	1	**8**	**15**	**22**
January	− 23:08	− 22:20	− 21:15	− 19:50
February	− 17:18	− 15:13	− 12:55	− 10:27
March	− 7:51	− 5:10	− 2:25	0:21
April	4:16	6:56	9:30	11:57
May	14:51	16:53	18:41	20:14
June	21:57	22:47	23:17	23:27
July	23:10	22:34	21:39	20:25
August	18:12	16:21	14:17	12:02
September	8:33	5:58	3:19	0:36
October	− 2:54	− 5:36	− 8:15	− 10:48
November	− 14:12	− 16:22	− 18:18	− 19:59
December	− 21:41	− 22:38	− 23:14	− 23:27

Source: Reprinted from "The American Ephemeris and Nautical Almanac," U.S. Naval Observatory, Washington, D.C.

Figure 6-4. The solar altitude β, zenith angle ψ, and azimuth angle ϕ.

The daily maximum altitude (solar noon) of the sun at a given location can be shown to be

$$\beta_{\text{noon}} = 90 - |l - d| \qquad (6\text{-}3)$$

where $|(l - d)|$ is the absolute value of $(l - d)$.

The *solar azimuth* angle ϕ is the angle in the horizontal plane measured between south and the projection of the sun's rays on that plane (Fig. 6-4). Again by analytic geometry (1) it can be shown that

$$\cos \phi = (\sin \beta \sin l - \sin d)/(\cos \beta \cos l) \qquad (6\text{-}4)$$

For a vertical surface the angle measured in the horizontal plane between the projection of the sun's rays on that plane and a normal to the vertical surface is called the *wall solar azimuth* γ. Figure 6-5 illustrates this quantity. If ψ is the wall azimuth measured east or west from south, then obviously

$$\gamma = \phi \pm \psi \qquad (6\text{-}5)$$

The *angle of incidence* θ is the angle between the sun's rays and the normal to the surface, as shown in Fig. 6-5. The *angle of tilt* α is the angle between the normal to the surface and the normal to the horizontal surface. It may be shown that

$$\cos \theta = \cos \beta \cos \gamma \sin \alpha + \sin \beta \cos \alpha \qquad (6\text{-}5a)$$

Then for a vertical surface

$$\cos \theta = \cos \beta \cos \gamma \qquad (6\text{-}5b)$$

and for a horizontal surface

$$\cos \theta = \sin \beta \qquad (6\text{-}5c)$$

Values of solar altitude, solar azimuth, and angles of incidence of the sun's rays for various latitudes, times of day, and days of the year are given in tables in reference 12.

Figure 6-5. Wall solar azimuth γ, wall azimuth ψ, and angle of tilt α for an arbitrary tilted surface.

Example 6-2. Find the solar altitude and azimuth at 10:00 A.M. Central Daylight Savings Time on July 22 at 40 degrees north latitude and 85 degrees west longitude.

Solution. The Local Civil Time (LCT) is

$$10:00 - 1:00 + 4(90 - 85) = 9:20 \text{ A.M.}$$

The equation of time is -6 min 19 sec; therefore the Local Solar Time (LST) to the nearest minute is

$$9:20 - 0:06 = 9:14 \text{ A.M.}$$

β is calculated from Eq. 6-1.

$$h = 2 \text{ hr } 46 \text{ min} = 2.767 \text{ hours} = 41.5 \text{ degrees}$$
$$\beta = \sin^{-1}(\cos 40 \cos 41.5 \cos 20.4 + \sin 40 \sin 20.4)$$
$$\beta = 49.6 \text{ degrees}$$

ϕ is calculated from Eq. 6-4

$$\phi = \cos^{-1}\left[\frac{\sin 40 \sin 49.6 - \sin 20.4}{\cos 49.6 \cos 40}\right] = 73.5 \text{ degrees}$$

External Shading

External shading of a window is effective in reducing solar heat gain to a space and may produce reductions of up to 80 percent. A window may be shaded by trees, shrubbery, another building, or by some feature of the structure itself. Figure 6-6 illustrates a window that is set back into the structure where shading may occur on the sides and top depending on the time of day

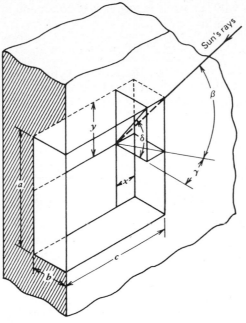

Figure 6-6. Shading of window set back from
the plane of a building surface.

and the direction the window faces. It can be shown that the dimensions x and
y in Fig. 6-6 are given by

$$x = b \tan \gamma \qquad (6\text{-}6)$$

$$y = b \tan \delta \qquad (6\text{-}7)$$

where

$$\tan \delta = \frac{\tan \beta}{\cos \gamma}$$

and

β = sun's altitude angle from Eq. 6-1
γ = wall solar azimuth angle ($\phi \pm \psi$)
ϕ = solar azimuth from Eq. 6-4
ψ = wall azimuth measured east or west from the south

The following rules aid in the computation of the wall solar azimuth angle γ.
    For morning hours with walls facing east of south and afternoon hours
with walls facing west of south:

$$\gamma = |(\phi - \psi)| \qquad (6\text{-}8)$$

For afternoon hours with the walls facing east of south and morning hours with walls facing west of south:

$$\gamma = |(\phi + \psi)| \qquad \qquad (6\text{-}8a)$$

If γ is greater than 90 degrees, the surface is in the shade. Equation 6-7 can be used for an overhang at the top and perpendicular to the window provided that the overhang is wide enough for the shadow to extend completely across the window.

Example 6-3.   A 4×5 ft, $\frac{1}{4}$-in. regular plate glass window faces southwest. The top of the window has a 2-ft overhang that extends a great distance on each side of the window. Compute the shaded area of the window on July 22 at 3:00 P.M. solar time at 40 deg N. latitude.

Solution.   To find the area the dimension y from Eq. 6-7 must be computed. From Eqs. 6-1 and 6-4, β and ϕ are 47.0 and 76.6 degrees, respectively. The wall azimuth for a window facing southwest is 45 degrees. Then for a wall facing west of south and for afternoon hours on July 22 at 3:00 P.M. solar time at 40 deg N. latitude.

$$\gamma = |(\phi - \psi)| = |(76.6 - 45)| = 31.6 \text{ deg}$$

Then

$$y = b \tan \delta = \frac{b \tan \beta}{\cos \gamma}$$

$$y = \frac{2 \tan (47.0)}{\cos (31.6)} = 2.52 \text{ ft}$$

The shaded area is then

$$A_{sh} = 2.52 \times 4 = 10.1 \text{ ft}^2$$

and the sunlit portion has an area of

$$A_{sl} = A - A_{sh} = 20 - 10.1 = 9.9 \text{ ft}^2$$

The shaded portion of a window receives only diffuse radiation.

6-5   SOLAR IRRADIATION

The *mean solar constant* G_{sc} is the rate of irradiation on a surface normal to the sun's rays beyond the earth's atmosphere and at the mean earth–sun distance. According to a fairly recent study (4) the mean solar constant is

$$G_{sc} = 433.4 \text{ Btu/(hr-ft}^2)$$
$$= 1367 \text{ W/m}^2$$

The irradiation from the sun varies about ± 3.5 percent because of the variation in distance between the sun and earth. Because of the large amount of atmospheric absorption of this radiation, and because this absorption is so variable and difficult to predict, precise values of the solar constant do not need to be used in most HVAC calculations.

The radiant energy emitted by the sun closely resembles the energy that would be emitted by a blackbody (an ideal radiator) at about 10,800 F or 5982 C. Figure 6-7 shows the spectral distribution of the radiation from the sun as it arrives at the outer edge of the earth's atmosphere (the upper curve). The peak radiation occurs at a wavelength of about 0.48×10^{-6} m in the green portion of the visible spectrum. Forty percent of the total energy emitted by the sun occurs in the visible portion of the spectrum, between 0.4 and 0.7×10^{-6} m. Fifty-one percent is in the near infrared region between 0.7 and 3.5×10^{-6}. About 9 percent is in the ultraviolet below 0.4×10^{-6} m.

A part of the solar radiation entering the earth's atmosphere is scattered

Figure 6-7.   Spectral distribution of direct solar irradiation at normal incidence during clear days. (Adapted by permission from *ASHRAE Transactions*, Vol. 64, p. 50.)

by gas and water vapor molecules and by cloud and dust particles. The blue color of the sky is a result of the scattering of some of the shorter wavelengths from the visible portion of the spectrum. The familiar red at sunset results from the scattering of longer wavelengths by dust or cloud particles near the earth. Some radiation (particularly ultraviolet) may be absorbed by ozone in the upper atmosphere and other radiation is absorbed by water vapor near the earth's surface. That part of the radiation that is not scattered or absorbed and reaches the earth's surface is called *direct radiation*. It is accompanied by radiation that has been scattered or reemitted, called *diffuse radiation*. Radiation may also be reflected onto a surface from nearby surfaces. The total irradiation G_t on a surface normal to the sun's rays is thus made up of normal direct irradiation G_{ND}, diffuse irradiation G_d, and reflected irradiation G_R.

$$G_t = G_{ND} + G_d + G_R \tag{6-9}$$

The depletion of the sun's rays by the earth's atmosphere depends on the composition of the atmosphere (cloudiness, dust and pollutants present, atmospheric pressure, and humidity). With a given composition on a clear day the depletion is also strongly dependent on the length of the path of the rays through the atmosphere. In the morning or evening, for example, the sun's rays must travel along a much longer path through the atmosphere than they would at noontime. Likewise the sun's rays that hit the polar regions at midday have passed through a longer atmospheric path than those that hit the tropical regions at midday. This length is described in terms of the air mass m, the ratio of the mass of atmosphere in the actual sun–earth path to the mass that would exist if the sun were directly overhead at sea level. The air mass is for practical purposes equal to the cosecant of the solar altitude β multiplied by the ratio of actual atmospheric pressure to standard atmospheric pressure.

Figure 6-7 shows the spectral distribution of direct solar radiation normally incident on a surface at sea level, with air masses equal to 1(β = 90 degrees) and to 5 (β = 11.5 degrees), for a specified value of water vapor and dust in the air denoted by w and d. The area under each of the curves is proportional to the total irradiation that would strike a surface under that particular condition. It can be easily seen that the total radiation is significantly depleted and the spectral distribution is altered by the atmosphere.

ASHRAE Clear Sky Model

It is important to recognize that the value of the solar constant is for a surface outside the earth's atmosphere and does not take into account the absorption and scattering of the earth's atmosphere, which can be significant even for clear days. The value of the solar irradiation at the surface of the earth on a clear day is given by the ASHRAE Clear Sky Model (11):

$$G_{ND} = \frac{A}{\exp(B/\sin \beta)} \tag{6-10}$$

where

G_{ND} = the normal direct irradiation, Btu/(hr-ft^2) or W/m^2

A = apparent solar irradiation at air mass equal zero, Btu/(hr-ft^2) or W/m^2

B = atmospheric extinction coefficient

β = solar altitude

Values of A and B are given in Table 6-3 from reference 11 for the twenty-first day of each month and for an atmospheric clearness number C_N of unity. The data in Table 6-3, when used in Eq. 6-10, do not give the maximum value of G_{ND} that can occur in any given month, but are representative of conditions on average cloudless days. The values of C_N to use as a multiplying factor in Eq. 6-10 are given in Fig. 6-8 for nonindustrial locations in the United States.

The diffuse radiation on a horizontal surface is given by the use of the factor C from Table 6-3.

$$G_d = (C)(G_{ND}) \qquad (6-11)$$

where C is obviously the ratio of diffuse to direct normal irradiation on a horizontal surface. The parameter C is assumed to be a constant for an average clear day for a particular month. In reality the diffuse radiation varies directionally (Fig. 6-9) and changes during the day in a fairly predictable way (see Fig. 6-11).

Table 6-3   EXTRATERRESTRIAL SOLAR RADIATION AND RELATED DATA FOR TWENTY-FIRST DAY OF EACH MONTH, BASE YEAR 1964[a]

	G_0 Btu/(hr-ft^2)	Equation of Time, min	Declination, deg	A Btu/(hr-ft^2)	B (Dimensionless Ratios)	C
Jan	442.7	−11.2	−20.0	390	0.142	0.058
Feb	439.1	−13.9	−10.8	385	0.144	0.060
Mar	432.5	− 7.5	0.0	376	0.156	0.071
Apr	425.3	+ 1.1	+11.6	360	0.180	0.097
May	418.9	+ 3.3	+20.0	350	0.196	0.121
June	415.5	− 1.4	+23.45	345	0.205	0.134
July	415.9	− 6.2	+20.6	344	0.207	0.136
Aug	420.0	− 2.4	+12.3	351	0.201	0.122
Sep	426.5	+ 7.5	0.0	365	0.177	0.092
Oct	433.6	+15.4	−10.5	378	0.160	0.073
Nov	440.2	+13.8	−19.8	387	0.149	0.063
Dec	443.6	+ 1.6	−23.45	391	0.142	0.057

[a]To convert Btu/(hr-ft^2) to W/m^2, multiply by 3.1525.

Source: Reprinted by permission from ASHRAE Handbook, Fundamentals Volume, 1985.

S—Summer
W—Winter

Figure 6-8. Estimated Atmospheric Clearness Numbers in the U.S. for Nonindustrial Localities. (Reprinted by permission from *ASHRAE Handbook, Fundamentals Volume,* 1985.)

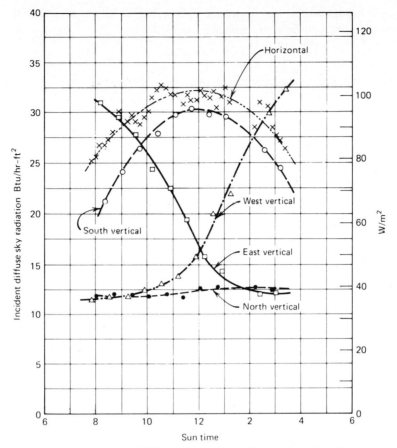

Figure 6-9.   Variation of diffuse solar radiation from a clear sky. (Adapted by permission from *ASHRAE Transactions*, Vol. 69, p. 29.)

An evaluation of the ASHRAE Clear Sky Model was made by Powell (14), who suggested corrections for the effects of refraction at solar zenith angles in excess of 70 degrees and to account for the decrease in atmospheric path length with increased elevation. His corrections make the model applicable to situations where the clearness numbers may not be known.

A critical review of the ASHRAE Solar Radiation Model was given by Galamio and Chatigny (15) in which they show the contradiction of using the concept of a clearness number as a multiplying factor for both the normal direct irradiation and the diffuse radiation. They suggested dividing by the square of the clearness number to give the correct behavior for the diffuse component

$$G_d = \frac{(C)(G_{ND})}{(C_N)^2} \qquad (6\text{-}12)$$

Their model would then combine Eq. 6-12 with an equation for the direct radiation G_D on a surface of arbitrary orientation

$$G_D = G_{ND} \cos \theta \qquad (6\text{-}13)$$

where θ is the angle of incidence of the sun's rays to the surface.

The result is, for a horizontal surface where $\cos \theta = \sin \beta$

$$G_t = G_D + G_d = \left[\cos \theta + \frac{C}{(C_N)^2} \right] G_{ND} \qquad (6\text{-}14)$$

Reference 15 also gives an expression for a cloudy sky model, using Eq. 6-14 as a starting point. This model involves the use of cloud cover information reported in meteorological observations. Also given in Table 6-3 are monthly values of the extraterrestrial solar radiation G_o, equation of time, and solar declination.

To estimate the rate at which diffuse radiation $G_{d\theta}$ strikes a nonhorizontal surface on a clear day, the following equation is used

$$G_{d\theta} = CG_{ND}F_{ws} \qquad (6\text{-}15)$$

In which F_{ws} is the configuration factor or angle factor between the wall and the sky. The *configuration factor* is the fraction of the diffuse radiation leaving one surface that would fall directly on another surface. This factor is sometimes referred to in the literature as the *angle factor*—or the *view, shape, interception, or geometrical factor*. For diffuse radiation this factor is a function only of the geometry of the surface or surfaces to which it is related. It is important to note that the configuration factor is useful for any type of diffuse radiation. For this reason information obtained in illumination, radio, or nuclear engineering studies is often useful to engineers interested in thermal radiation.

The symbol for configuration factor always has two subscripts describing the surface or surfaces that it describes. For example, the configuration factor, F_{12}, applies to the two surfaces numbered 1 and 2. F_{12} would be the fraction of the diffuse radiation leaving surface 1 that would fall directly on surface 2. F_{11} would be the fraction of the diffuse radiation leaving surface 1 that would fall on itself and obviously applies only to nonplane surfaces.

A very important and useful characteristic of configuration factors is the *reciprocity relationship:*

$$A_1 F_{12} = A_2 F_{21} \qquad (6\text{-}16)$$

Its usefulness is in determining configuration factors when the reciprocal factor is known or when the reciprocal factor is more easily obtained than the desired factor.

For example, the fraction of the diffuse radiation in the sky that strikes a given surface would be difficult to determine directly. The fraction of the energy that leaves the surface and "strikes" the sky directly, however, can be easily determined from the geometry

$$F_{ws} = \frac{(1 + \cos \Sigma)}{2} \qquad (6\text{-}17)$$

where Σ is the tilt angle of the surface from horizontal, $\Sigma = (90 - \alpha)$.

The rate at which diffuse radiation from the sky strikes a given surface of area A_w is, per unit area of surface

$$\frac{\dot{q}}{A_w} = \frac{A_s G_d F_{sw}}{A_w}$$

by reciprocity

$$A_s F_{sw} = A_w F_{ws}$$

therefore

$$\frac{\dot{q}}{A_w} = G_d F_{ws}$$

Thus, although the computation involves the irradiation of the sky on the surface or wall, the configuration factor most convenient to use is F_{ws}, the one describing the fraction of the surface radiation that strikes the sky.

The use of the configuration factor assumes that diffuse radiation comes uniformly from the sky in all directions. This is not true for either cloudy or clear skies. In both clear and cloudy skies the location of the sun can distort the uniformity of the diffuse radiation. In the case of cloudiness, the nonuniformity of the clouds in the sky may also cause the diffuse radiation to be nonuniform. Figure 6-9 shows the variation of diffuse solar radiation with sun position on a clear day.

In determining the total rate at which radiation strikes an arbitrarily oriented surface at any time, one must also consider the energy reflected onto the surface. The most common case is reflection of solar energy from the ground to a tilted surface or vertical wall. For such a case the rate at which energy is reflected to the wall is

$$G_R = G_{tH} \rho_g F_{wg} \qquad (6\text{-}18)$$

where

G_R = rate at which energy is reflected onto the wall, Btu/(hr-ft^2) or W/m^2

G_{tH} = rate at which the total radiation (direct plus diffuse) strikes the horizontal surface or ground in front of the wall, Btu/(hr-ft^2) or W/m^2

ρ_g = reflectance of ground or horizontal surface

F_{wg} = configuration or angle factor from wall to ground, defined as the fraction of the radiation leaving the wall of interest that strikes the horizontal surface or ground directly. For a surface or wall at a tilt angle Σ to the horizontal

$$F_{wg} = \frac{(1 - \cos \Sigma)}{2} \qquad (6\text{-}19)$$

Equations 6-10 to 6-19 are useful for design purposes where cooling loads are to be estimated because these equations are for clear days, when solar conditions are most severe.

In some cases it is desirable to be able to estimate the solar radiation for typical conditions, including both clear and cloudy days. In many of these cases one might want to know the total irradiation on a surface over an entire day or series of days or weeks. An example would be the computation of the irradiation on a solar collector panel. Another would be in the determination of energy requirements of a building for a simulation study. In such cases the best information would be based on historical weather data for that particular location. Local weather data are usually best if they are available.

Hourly total (global) horizontal insolation data are available for 26 sites in the SOLMET data base, available from the National Climatic Center, Asheville, North Carolina, 28801. This is rehabilitated data derived from data taken by the National Weather Bureau Service over a time period that generally covers 1953 through 1975.

This data base was expanded to cover 222 additional sites by applying correlations to those locations having historical records of percentage sunshine and/or cloud cover observations. This data base is known as the SOLMET/ERZATZ data base and is also available from the National Climatic Center.

Typical Meteorlogical Year (TMY) Data were developed from the SOLMET and ERZATZ data for a collection of 12 typical months over a typical year. This is also available from NCC.

ASHRAE has made available four weather data tapes, based on the Weather Year for Energy Calculations (WYEC) for 51 cities in the United States and Canada (12).

In the most common situation the typical weather data that are available give the total (or global) solar insolation on a horizontal surface. To use this data for making predictions of insolation on nonhorizontal surfaces, there must be some procedure for determining the relative proportion of the total horizontal radiation that is direct and that proportion that is diffuse. Each part can then be used to determine the rate at which direct and diffuse radiation strikes the surface of interest. In addition, the energy reflected onto the surface must be determined.

Figure 6-10 illustrates the logic involved. The total radiation on a horizontal

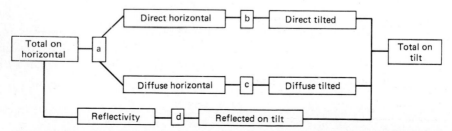

Figure 6-10.   Conversion of horizontal insolation to insolation on tilted surface.

surface is first divided into the direct and diffuse components, step a. One way that this can be accomplished is by using the methods of Page or Liu and Jordan described in references 9 and 16.

If hourly calculations are being made, information relating hourly diffuse to daily diffuse radiation and hourly total to daily total radiation must be used. These curves are given in Fig. 6-11.

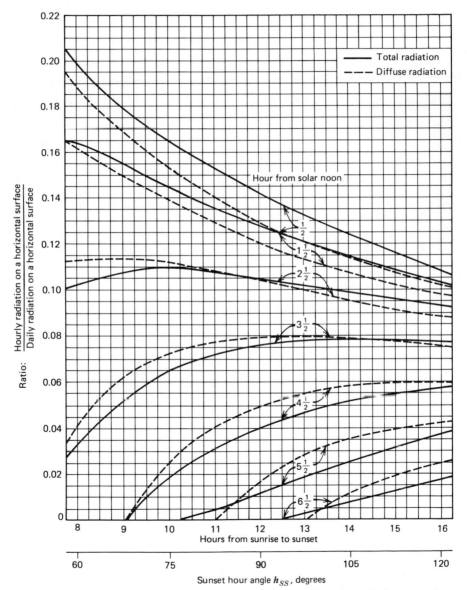

Figure 6-11. Relationships between daily radiation and hourly radiation on a horizontal surface. (Reprinted by permission from *ASHRAE Low Temperature Engineering Application of Solar Energy*, 1967.)

With the total radiation thus divided one can use Eq. 6-13 to determine the direct radiation on the "tilted" surface, step b, and Eq. 6-15 to determine the diffuse radiation on the "tilted" surface, step c. Equation 6-18 is then used with the total horizontal insolation data to determine the radiation reflected, step d. The three components are then added to give the total insolation on the surface of interest.

Example 6-3. Calculate the clear day direct, diffuse, and total solar radiation rate on a horizontal surface at 36 degrees north latitude and 84 degrees west longitude on June 1 at 12:00 noon CST.

Solution. Equations 6-10 and 6-11 are used together with the parameter A, B, and C from Table 6-3. In order to calculate the angles β and θ used in these equations the local solar time must be calculated. The declination is obtained from Table 6-2.

$$\text{LST} = \text{LCT} + (\text{equation of time})$$

$$= 12:00 + \left(\frac{90 - 84}{15}\right)(60) + 0:02:25 = 12:26$$

The hour angle h is given by

$$h = \frac{(26)(15)}{60} = 6.5 \text{ deg and } d = 21 \text{ deg } 57 \text{ min}$$

$$\sin \beta = \cos l \cos d \cos h + \sin l \sin d$$

$$\sin \beta = (0.809)(0.994)(0.928) + (0.588)(0.374)$$

$$\sin \beta = 0.965$$

For a horizontal surface $\cos \theta = \sin \beta$

$$G_{ND} = \frac{A}{\exp\left(\dfrac{B}{\sin \beta}\right)} = \frac{345}{\exp\left(\dfrac{0.205}{0.965}\right)} = 279 \text{ Btu/(hr-ft}^2)$$

The direct radiation is

$$G_D = G_{ND} \cos \theta = (279)(0.965) = 269 \text{ Btu/(hr-ft}^2)$$

The diffuse radiation is

$$G_d = CG_{ND} = (0.134)(279) = 37.4 \text{ Btu/(hr-ft}^2)$$

The total radiation is

$$G_t = G_D + G_d = 269 + 37.4 = 306 \text{ Btu/(hr-ft}^2)$$

A particularly useful curve, Fig. 6-12, gives the ratio of diffuse sky radiation on a vertical surface to that incident on a horizontal surface on a clear day.

Figure 6-12. Ratio of diffuse sky radiation incident upon a vertical surface to that incident upon a horizontal surface during clear days. (Reprinted by permission from *ASHRAE Transactions*, Vol. 69, p. 29.)

Example 6-4. Calculate the total incidence of solar radiation on a window facing south located 6 ft above the ground. In front of the window is a concrete parking area that extends 50 ft south and 50 ft to each side of the window. The window has no setback. The following parameters have been previously computed: $\beta = 69$ degrees 13 minutes, $\phi = 17$ degrees 18 minutes, $G_{ND} = 278$ Btu/(hr-ft^2), $G_{tH} = 293$ Btu/(hr-ft^2), $G_{dH} = 33$ Btu/(hr-ft^2), $\rho_1 = 0.33$, and $F_{wg} = 0.433$.

Solution. The angle of incidence for the window is first computed.

$$\gamma = \phi \pm \psi; \quad \psi = 0$$
$$\gamma = \phi = 17 \text{ degrees } 18 \text{ minutes}$$
$$\cos \theta = \cos \beta \cos \gamma = 0.339$$
$$G_{DV} = G_{ND} \cos \theta = 287(0.339) = 94 \text{ Btu/(hr-ft}^2)$$

From Fig. 6-12

$$\frac{G_{dV}}{G_{dH}} = 0.75$$
$$G_{dV} = 0.75(33) = 25 \text{ Btu/(hr-ft}^2)$$

The reflected component is given by Eq. 6-15 where

$$G_R = 0.33(293)(0.433) = 42 \text{ Btu/(hr-ft}^2)$$

Then

$$G_{tV} = G_{DV} + G_{dV} + G_R = 94 + 25 + 42 = 162 \text{ Btu/(hr-ft}^2)$$

6-6   MEASUREMENT OF DIRECT AND DIFFUSE SOLAR RADIATION

The measurement of direct, diffuse, and total radiation of the sun is essential to research in solar energy and in solar effects on HVAC systems. An instrument that measures the direct solar radiation is called a *pyrheliometer*. The term *pyranometer* describes an instrument that measures the total radiation from both sun and sky. A pyranometer can be used to measure the diffuse (sky) radiation by shielding it from the direct rays of the sun but leaving its view of the sky essentially unobstructed.

The first reliable pyrheliometer was developed by K. Angstrom of Sweden near the end of the nineteenth century (5). This instrument, still in wide use today, measures the electrical power required to heat a shaded metal strip to the same temperature as a blackened metal strip heated by the sun's rays. The response is given in the *angstrom scale of solar radiation*.

In the United States the Smithsonian Institution has done most of the developmental work on pyrheliometers. An early model used a mercury-type instrument and defined the *Smithsonian original scale of solar radiation*. In 1913 an instrument using water to cool a well-insulated, cylindrical blackbody chamber was developed by C. G. Abbot and others (6). This led to a new scale called the *Smithsonian Scale of 1913*. A later improvement in which two chambers, one heated by solar rays and the other shaded and heated electrically, were both cooled by equally divided flows of water. The electric power was adjusted until the exit temperature of both streams was the same. The power required was proportional to the radiation from the sun. This led to the *Smithsonian Scale of 1932*.

In 1956 the International Radiation Conference met in Switzerland and recommended a new scale called the *International Scale of 1956*. Because it was defined in terms of two of the existing scales, no new instrument was required. The International Scale of 1956 is defined as the angstrom scale plus 1.5 percent or the Smithsonian scale of 1913 minus 2.0 percent. This new scale is thought to be within ± 1.0 percent of a true absolute scale. A comparison of the five scales described above is given in Fig. 6-13. Care should be taken to use the appropriate scale for old data or old calibrations.

The Smithsonian water flow type of instruments require too much auxiliary equipment for convenient use in the field. A convenient, secondary-type instrument developed by Abbot at the Smithsonian is widely used and is called the *Smithsonian silver-disk pyrheliometer*. The rate of increase of temperature of a pool of mercury behind a blackened silver disk is measured by a thermopile as the sun's rays are admitted and allowed to fall on the disk. Fig. 6-14 shows this type of device used to measure normally incident direct solar radiation.

A practical device for measuring the total sun and sky radiation (pyranometer) consists of a differential thermopile with the hot junction blackened and the cold junction whitened. The thermopile is covered by a hermetically sealed, precision ground optical glass envelope (Fig. 6-15).

Percent of international scale of 1956

102.4

102.0 — Smithsonian, 1913

101.6

101.2

100.8

100.4

— Smithsonian, original

100.0 — International, 1956

99.6

— Smithsonian, 1932

99.2

99.8

— Angstrom

98.4

Figure 6-13. Comparison of scales of solar radiation.

Figure 6-14. An Eppley normal incidence pyrheliometer. (Courtesy of Eppley Laboratory, Inc., Newport, Rhode Island.)

Figure 6-15.   An Eppley black and white pyranometer. (Courtesy of Eppley Laboratory, Inc., Newport, Rhode Island.)

REFERENCES

1. James L. Threlkeld, *Thermal Environmental Engineering,* Second Edition, Prentice-Hall, Englewood Cliffs, N. J., 1970.

2. U.S. Nautical Almanac Office, "The American Ephemeris and Nautical Almanac," U.S. Naval Observatory, Washington, D.C. (published annually)

3. "Tables of Computed Altitude and Azimuth," Hydrographic Office Bulletin No. 214, U.S. Government Printing Office.

4. C. Frohlich and R. W. Brusa, "Solar Radiation and its Variation in Time," *Solar Physics,* Vol. 74, p. 209, 1981.

5. K. Angstrom, "The Absolute Determination of the Radiation of Heat with the Electric Compensation Pyrheliometer," *Astrophysical Journal,* Vol. 6, (1899).

6. C. G. Abbot, F. E. Fowle, and L. B. Aldrich, "Improvements and Tests of Solar Constant Methods and Apparatus," *Annals of the Astrophysical Observatory,* Vol. 3, (1913).

7. John A. Duffie and William A. Beckman, *Solar Energy Thermal Processes*, Wiley-Interscience, New York, 1974.

8. *Cooling and Heating Load Calculation Manual*, ASHRAE GRP 158, American Society of Heating, Refrigerating, and Air-Conditioning Engineers, New York, 1979.

9. *Applications of Solar Energy for Heating and Cooling of Buildings*, ASHRAE GRP 170, American Society of Heating, Refrigerating, and Air-Conditioning Engineers, New York, 1977.

10. *ASHRAE Handbook, Applications Volume*, American Society of Heating, Refrigerating, and Air-Conditioning Engineers, New York, 1978.

11. *ASHRAE Handbook, Fundamentals Volume*, American Society of Heating, Refrigerating, and Air-Conditioning Engineers, New York, 1977.

12. *ASHRAE Handbook, Fundamentals Volume*, American Society of Heating, Refrigerating, and Air-Conditioning Engineers, Atlanta, Ga., 1985.

13. J. L. Threlkeld and R. C. Jordan. "Direct Solar Radiation Available on Clear Days," *ASHRAE Transactions*, Vol. 64, p. 45, 1958.

14. Gary L. Powell, "The ASHRAE Clear Sky Model—An Evaluation," *ASHRAE Journal*, pp. 32–34, November 1982.

15. N. Galanis and R. Chatigny, "A Critical Review of the ASHRAE Solar Radiation Model," *ASHRAE Transactions*, Vol. 92, Pt. 1, 1986.

16. P. J. Lunde, *Solar Thermal Engineering*, John Wiley & Sons, New York, 1980.

PROBLEMS

6-1 Find the local solar time (LST) on August 22 for the following local times and locations:
(a) 9:00 A.M. EST and 70 degrees west longitude
(b) 1:00 P.M. CST and 95 degrees west longitude
(c) 10:00 A.M. MST and 100 degrees west longitude
(3) 3:00 P.M. PST and 125 degrees west longitude

6-2 What are the hour angles corresponding to the following local solar times: (a) 9:32 A.M., (b) 11:38 A.M., (c) 3:20 P.M., and (d) 12:38 P.M.

6-3 Calculate the sun's altitude and azimuth angles at 9:00 A.M. solar time on September 21 at 40 degrees north latitude. Compare your results with values shown in Table 7, Chapter 26, reference 11.

6-4 Determine the solar time and azimuth angle for sunrise at 50 degrees north latitude on (a) June 15 and (b) December 15.

6-5 On what month, day, and time does the maximum solar altitude angle β occur in (a) Stillwater, Oklahoma, 36 degrees north latitude, (b) Chicago, Illinois, 42 degrees north latitude, and (c) central Florida, 29 degrees north latitude?

6-6 Compute the wall solar azimuth γ for a surface facing 12 degrees west of south located at 40 degrees north latitude and 90 degrees west longitude on October 1 at 3:30 P.M. central daylight saving time (CDST).

6-7 Calculate the angle of incidence for the surface of Problem 6-6 for (a) a vertical orientation, and (b) a 20 degree tilt from the vertical.

6-8 For a location at 85 degrees west longitude and 43 degrees north latitude on July 8, determine (a) the incidence angle of the sun for a horizontal surface at 4:00 P.M. central daylight saving time and (b) the time of sunset in central daylight saving time.

6-9 Calculate the angle of incidence at 10:30 A.M. EST on July 22 for 36 degrees north latitude and 78 degrees west longitude for (a) a horizontal surface, (b) a surface facing southeast, and (c) a surface inclined 40 degrees from the vertical and facing south.

6-10 Calculate the total clear day irradiation of a surface tilted at an angle of 30 degrees from the vertical located at 36 degrees north latitude and 78 degrees west longitude on July 21 at 2:00 P.M. central daylight saving time. The surface faces the southwest. Neglect reflected radiation.

6-11 Compute the reflected irradiation of a window facing southwest over a large lake on a clear day. The location is 36 degrees north latitude and 92 degrees west longitude. The time is June 15 at 3:00 P.M. CST. Assume the water has a diffuse reflectance of 0.06.

6-12 Determine magnitudes of direct, diffuse, and reflected clear day solar radiation incident upon a small vertical surface facing south on March 1 at solar noon for a location at 36 degrees north latitude having a clearness number of 0.95. The reflecting surface is snow-covered ground of infinite extent with a diffuse reflectance of 0.85.

6-13 Estimate the total clear day irradiation of a roof with a one-to-one slope that faces southwest at 36 degrees north latitude. The date is August 22 and the time is 1:00 P.M. LST. Include reflected radiation from the ground with a reflectance of 0.3. The view factor from the roof to the ground is given approximately by $F_{rg} = (\frac{1}{2})(1 - \cos \Sigma)$, where Σ is the slope angle of the roof.

6-14 Compute the time for sunrise and sunset on July 22 in (a) Stillwater, Oklahoma, (b) Chicago, Illinois, (c) Tallahassee, Florida, and (d) San Francisco, California.

6-15 A south-facing window is 4 ft wide by 6 ft tall and is set back into the wall a distance of 8 in. For Atlanta, Georgia, estimate the percent of the window that is shaded for
(a) April 22, 9:00 A.M. solar time
(b) July 22, 12:00 noon solar time
(c) September 22, 5:00 P.M. solar time

6-16 Work Problem 6-15 assuming a long 2-ft overhang located 2 ft above the top of the window.

6-17 Determine the amount of diffuse, direct, and total radiation that would strike a south-facing surface tilted at 45 degrees on a clear December 22 in St. Louis, Missouri:

(a) At 12 noon solar time

(b) At 3 P.M. solar time

Express your answer in Btu/(hr-ft^2). Neglect ground reflectance.

6-18 For the hours of 1 P.M., 2 P.M., 3 P.M., 4 P.M., 5 P.M., 6 P.M., and 7 P.M. CDST, estimate the rate at which solar energy will strike a west-facing window, 3 ft wide by 5 ft high, with no setback. Assume a clear June 21 day at 40 degrees north latitude and 84 degrees west longitude.

6-19 Work Problem 6-18 assuming a 6 in. setback for the window.

6-20 Work Problem 6-18 for a clear day on December 21.

6-21 Work Problem 6-18 assuming a long overhang of 2 ft that is 3 ft above the top of the window.

6-22 Work Problem 6-17 assuming a clear day on July 21.

6-23 Write a computer program to predict the altitude and zenith angles for the sun and the normal direct irradiation. Make a table of these quantities for the twenty-first day of each month for each hour between sunrise and sunset.

6-24 Write a computer program to predict the total irradiation of any building surface by the sun. Include reflected radiation from the ground. *Note:* This problem is an extension of Problem 6-23. Test the program with hand calculations.

6-25 Write a computer program to predict the fraction of sunlit area of a vertical window that may face any arbitrary direction in the northern hemisphere. The window may have setback on an overhang extending to the left and right a great distance. Test the program at some critical conditions, such as a west window at solar noon.

6-26 Modify the program of Problem 6-24 to account for exterior shading of windows using the program of Problem 6-25.

CHAPTER SEVEN

Space Heat Load

Prior to the design of the heating system an estimate must be made of the maximum probable heat loss of each room or space to be heated. There are two kinds of heat losses: (1) the heat transmitted through the walls, ceiling, floor, glass, or other surfaces, and (2) the heat required to warm outdoor air entering the space.

The actual heat loss problem is transient because the outdoor temperature, wind velocity, and sunlight are constantly changing. The transfer function method, discussed in Chapter 8 in connection with the cooling load, may be used under winter conditions to account for changing solar radiation, outdoor temperature, and the energy storage capacity of the structure. During the coldest months, however, sustained periods of very cold, cloudy, and stormy weather with relatively small variation in outdoor temperature may occur. In this situation heat loss from the space will be relatively constant and in the absence of internal heat gains will peak during the early morning hours. Therefore, for design purposes the heat loss is usually estimated for steady-state heat transfer for some reasonable design temperature. Transient analyses are often used to study the actual energy requirements of a structure in simulation studies. In such cases solar effects and internal heat gains are taken into account.

Here is the general procedure for calculation of design heat losses of a structure:

1. Select the outdoor design conditions: temperature, humidity, and wind direction and speed.
2. Select the indoor design conditions to be maintained.
3. Estimate the temperature in any adjacent unheated spaces.
4. Select the transmission coefficients and compute the heat losses for walls, floors, ceilings, windows, doors, and floor slabs.
5. Compute the heat load due to infiltration.

6. Compute the heat load due to outdoor ventilation air. This may be done as part of the air quantity calculation.
7. Sum the losses due to transmission and infiltration.

7-1 OUTDOOR DESIGN CONDITIONS

The ideal heating system would provide enough heat to match the heat loss from the structure. However, weather conditions vary considerably from year to year, and heating systems designed for the worst weather conditions on record would have a great excess of capacity most of the time. The failure of a system to maintain design conditions during brief periods of severe weather is usually not critical. However, close regulation of indoor temperature may be critical for some industrial processes.

Table C-1 contains outdoor temperatures that have been recorded for selected locations in the United States and Canada. The data are based on official weather station records of the U.S. Weather Bureau, U.S. Air Force, U.S. Navy, and the Canadian Department of Transport. At this point we shall consider only data pertaining to the winter months. The $97\frac{1}{2}$ percent value is the temperature equaled or exceeded $97\frac{1}{2}$ percent of the total hours (2160) in December, January, and February. During a normal winter there would be about 54 hours at or below the $97\frac{1}{2}$ percent value. For the Canadian stations the $97\frac{1}{2}$ percent value pertains only to hours in January.

The *ASHRAE Handbook* (1) has a more extensive tabulation of weather data; it also includes temperatures that are equaled or exceeded 99 percent of the total hours in December, January, and February.

The outdoor design temperature should generally be the $97\frac{1}{2}$ percent value as specified by ASHRAE Energy Standards. If the structure is of lightweight construction (low heat capacity), poorly insulated, has considerable glass, or space temperature control is critical, however, the 99 percent values should be considered. The designer must remember that should the outdoor temperature fall below the design value for some extended period, the indoor temperature may do likewise. The performance expected by the owner is a very important factor, and the designer should make clear to the owner the various factors considered in the design.

Unusual wind conditions may result from the location of the structure on a hill or between two other large structures. Abnormally high wind can cause a peak heat load at outdoor temperatures above the design temperature.

7-2 INDOOR DESIGN CONDITIONS

The main purpose of Chapter 4 was to define indoor conditions that make most of the occupants comfortable. Therefore the theories and data presented there should serve as a guide to the selection of the indoor temperature and humidity for heat-loss calculation. It should be kept in mind, however, that the purpose of heat-load calculations is to obtain data on which the heating system components are sized. In most cases, the system will never be set to operate at the design conditions. Indeed, it is practically impossible to control the environment to operate at the design point. Therefore, the use and occupancy of the space is a general consideration from the design temperature point of view. Later, when the energy requirements of the building are computed, the actual conditions in the space and outdoor environment, including internal heat gains, must be considered.

The indoor design temperature should be kept relatively low so that the heating equipment will not be oversized. ASHRAE Energy Standards specify 70 F or 22 C. Even properly sized equipment operates under partial load, at reduced efficiency, most of the time; therefore, any oversizing aggravates this condition and lowers the overall system efficiency. The indoor design value of relative humidity should be compatible with a healthful environment and the thermal and moisture integrity of the building envelope. A maximum relative humidity of 30 percent is recommended.

Frequently, unheated rooms or spaces exist in a structure. These spaces will be at temperatures between the indoor and outdoor design temperatures discussed earlier. The temperature in an unheated space is needed to compute the heat loss and may be estimated by assuming steady-state heat transfer and making an energy balance on the space. This can best be explained by an example.

Example 7-1. Estimate the temperature in the crawl space of Fig. 7-1. The conductance for the floor is 0.20 Btu/(hr-ft²-F) including the air film

Figure 7-1. A crawl space for building.

on each side. The conductance for the foundation wall including the insulation and inside and outside air film resistances is 0.12 Btu/(hr-ft²-F). Assume an indoor temperature of 70 F and an outdoor temperature of 10 F at a location where the degree days are about 3000. The building dimensions are 50 × 75 ft.

Solution. The first step is to make an energy balance on the crawl space. Heat will be gained through the floor and heat will be lost through the foundation wall and through the ground around the perimeter much like a floor slab. Infiltration of outdoor air will also represent a heat loss but that will be neglected in this example.

$$\dot{q}_{fl} = \dot{q}_{fo} + \dot{q}_{ground}$$

or

$$C_{fl}A_{fl}(t_i - t_c) = C_{fo}A_{fo}(t_c - t_o) + U'P(t_c - t_o)$$
$$t_c = \frac{t_o(CA)_{fo} + U'Pt_o + t_i(CA)_{fl}}{(CA)_{fl} + (CA)_{fo} + U'P}$$

Now the area of the floor is 50 × 75 = 3750 ft² and assuming the foundation wall averages a height of 2 ft, the area of the foundation wall is 2(2 × 50) + (2 × 75) = 500 ft². The perimeter of the building is (2 × 50) + (2 × 75) = 250 ft. Referring to Table 5-12, the heat-loss coefficient is estimated to be 0.8 Btu/(hr-ft-F). Then

$$t_c = \frac{10[(0.12 \times 500) + (0.8 \times 250)] + 70(0.20 \times 3750)}{(0.2 \times 3750) + (0.12 \times 500) + (0.8 \times 250)} = 54.6\ F$$

If the infiltration had been considered, the crawl space temperature would be lower.

Example 7-2. Estimate the temperature in the unheated room shown in Fig. 7-2. The structure is built on a slab. The exterior walls of the unheated room have an overall heat transfer coefficient of 0.2 Btu/(hr-ft²-F), the ceiling-attic-roof combination has an overall coefficient of 0.07 Btu/(hr-ft²-F), and the interior walls have an overall coefficient of 0.06 Btu/

Figure 7-2. A structure with an attached unheated room.

(hr-ft²-F). Inside and outdoor design temperatures are 72 F and −5 F, respectively.

Solution. The heat transferred from the heated to the unheated room has a single path, neglecting the door, and may be represented as

$$\dot{q} = U_i A_i (t_i - t_u) \tag{7-1}$$

The heat transferred from the unheated room to the outdoor air has parallel paths through the ceiling and walls (neglecting the door and floor):

$$\dot{q} = U_c A_c (t_u - t_o) + U_o A_o (t_u - t_o) \tag{7-2}$$

The heat loss to the floor can be neglected because of the anticipated low temperature in the room. Equation 7-2 may be written

$$\dot{q} = \frac{(t_u - t_o)}{R'_c} + \frac{(t_u - t_o)}{R'_o} = \frac{(t_u - t_o)}{R'} \tag{7-3}$$

where

$$\frac{1}{R'} = \frac{1}{R'_c} + \frac{1}{R'_o} = U_c A_c + U_o A_o \tag{7-4}$$

By combining Eqs. 7-1 and 7-3, we obtain

$$\frac{(t_i - t_u)}{R'_i} = \frac{(t_u - t_o)}{R'} \tag{7-5}$$

and solving for t_u,

$$t_u = \frac{t_i + (R'_i/R')t_o}{1 + (R'_i/R')} \tag{7-6}$$

$$R'_i = \frac{1}{U_i A_i} = \frac{1}{0.06(8)(22)} = 0.095$$

$$\frac{1}{R'} = U_c A_c + U_o A_o = 0.07(120) + 0.2(22)8 = 43.6$$

$$R' = 0.023 \text{ (hr-F)/Btu}$$

Then from Eq. 7-6

$$t_u = \frac{72 + (0.095/0.023)(-5)}{1 + (0.095/0.023)} = 10 \text{ F}$$

The assumption of negligible heat transfer through the slab is justified by the result.

The temperature of unheated basements is generally between the ground temperature (about 50 F or 10 C) and the inside design temperature unless there are many windows. Therefore a reasonable estimate of the basement temperature is not difficult. However, for a more precise value, the energy balance procedure may be used with data from Chapter 5.

7-3   CALCULATION OF HEAT LOSSES

Transmission

The heat transferred through walls, ceiling, roof, window glass, floors, and doors is all sensible heat transfer, referred to as *transmission heat loss* and computed from

$$\dot{q} = UA(t_i - t_o) \tag{7-7}$$

A separate calculation is made for each different surface in each room of the structure. To ensure a thorough job in estimating the heat losses, a worksheet should be used. A worksheet provides a convenient and orderly way of recording all the coefficients and areas. Summations are conveniently made by room and for the complete structure. The spread sheet computer programs available for all personal computers are quite useful in simple load calculation, or a particular program may be written to carry out the calculations and print the results. Many such programs are available. The overall heat transfer coefficient U of Eq. 7-7 is determined as discussed in Chapter 5, whereas the area A is the net area for the given component for which U was calculated.

Infiltration

All structures have some air leakage or infiltration. This means a heat loss because the cold dry outdoor air must be heated to the inside design temperature and moisture must be added to increase the humidity to the design value. The heat required to increase the temperature is given by

$$\dot{q}_s = \dot{m}_o c_p(t_i - t_o) \tag{7-8}$$

where

$\dot{m}_o$ = mass flow rate of the infiltrating air, lbm/hr or kg/s
c_p = specific heat capacity of the moist air, Btu/(lbm-F) or J/(kg-C)

Infiltration is usually estimated on the basis of volume flow rate at outdoor conditions. Equation 7-8 then becomes

$$\dot{q}_s = \frac{\dot{Q}c_p(t_i - t_o)}{v_o} \tag{7-8a}$$

where

$\dot{Q}$ = volume flow rate, ft³/hr or m³/s
v_o = specific volume, ft³/lbm or m³/kg

The latent heat required to humidify the air is given by

$$\dot{q}_l = \dot{m}_o(W_i - W_o)i_{fg} \tag{7-9}$$

where

$(W_i - W_o)$ = difference in design humidity ratio, lbmv/lbma or kgv/kga

i_{fg} = latent heat of vaporization at indoor conditions, Btu/lbmv or J/kgv

In terms of volume flow rate, Eq. 7-9 becomes

$$\dot{q}_l = \frac{\dot{Q}}{v_o}(W_i - W_o)i_{fg} \tag{7-9a}$$

It is easy to show, using Eqs. 7-8a and 7-9a, that infiltration can account for a large portion of the heating load.

Various methods are used in estimating air infiltration in building structures (1). In this book two approaches to the problem will be discussed. In one method the estimate is based on the characteristics of the windows, walls, and doors and the pressure difference between inside and outside. This is known as the *crack method* because of the cracks around window sash and doors. The other approach is the *air-change method,* which is based on an assumed number of air changes per hour based on experience. The crack method is generally considered to be the most accurate when the window and pressure characteristics can be properly evaluated. However, the accuracy of predicting air infiltration is restricted by the limited information on the air leakage characteristics of the many components that make up a structure (7). The pressure differences are also difficult to predict because of variable wind conditions and stack effect in tall buildings.

Air-change Method.   Experience and judgment are required to obtain satisfactory results with this method. Experienced engineers will often simply make an assumption of the number of air changes per hour (ACH) that a building will experience based on their appraisal of the building type, construction, and use. The range will usually be from 0.5 ACH (very low) to 2.0 ACH (very high). This approach is usually satisfactory for design load calculations but not recommended for the beginner.

Crack Method.   Outdoor air infiltrates the indoor space through cracks around doors, windows, lighting fixtures, joints between walls and floor, and even through the building material itself. The amount depends on the total area of the cracks, the type of crack, and the pressure difference across the crack. The volume flow rate of infiltration may be represented by

$$\dot{Q} = A \, C \, \Delta P^n \tag{7-10}$$

where

> A = effective leakage area of the cracks
> C = flow coefficient, which depends on the type of crack and the nature of the flow in the crack
> ΔP = outside–inside pressure difference, $P_o - P_i$
> n = exponent that depends on the nature of the flow in the crack, $0.4 < n < 1.0$.

Experimental data are required to use Eq. 7-10 directly; however, the relation is useful in understanding the problem. For example, Fig. 7-3 shows the leakage rate for some windows and doors as a function of the pressure difference and the type of crack. The curves clearly exhibit the behavior of Eq. 7-10.

The pressure difference of Eq. 7-10 results from three different effects:

$$\Delta P = \Delta P_w + \Delta P_s + \Delta P_p \tag{7-11}$$

where

> ΔP_w = pressure difference due to the wind
> ΔP_s = pressure difference due to the stack effect
> ΔP_p = pressure difference due to building pressurization

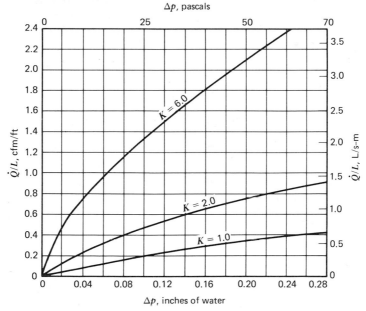

Figure 7-3.   Window and door infiltration characteristics. (Reprinted by permission from ASHRAE GRP 158, *Cooling and Heating Load Calculation Manual*, 1979.)

All of the pressure differences are positive when each causes flow of air to the inside of the building.

The pressure difference due to the wind results from an increase or decrease in air velocity and is described by

$$\Delta P_w = \frac{\rho}{2g_c} (\bar{V}_w^2 - \bar{V}_f^2) \tag{7-12}$$

where ΔP_w has the unit of lbf/ft^2 or Pascals when consistent English or SI units are used. The velocity $\bar{V}_f$ is the final velocity of the wind at the building boundary. Note that ΔP_w is positive when $\bar{V}_w > \bar{V}_f$, which gives an increase in pressure. The velocity $\bar{V}_f$ is not known or easily predictable; therefore, it is assumed equal to zero in this application and a pressure coefficient, defined by

$$C_p = \Delta P_w / \Delta P_{wt} \tag{7-13}$$

is used to account for the fact that $\bar{V}_f$ is not zero. The pressure difference ΔP_{wt} is the computed pressure difference when $\bar{V}_f$ is zero. The pressure coefficient may be positive or negative. Finally, Eq. 7-12 may be written

$$\frac{\Delta P_w}{C_p} = \frac{\rho}{2g_c} \bar{V}_w^2 \tag{7-14}$$

The pressure coefficient depends on the shape and orientation of the building with respect to the wind. To satisfy conditions of flow continuity, the air velocity must increase as it flows around or over a building; therefore, the pressure coefficient will change from a positive to a negative value in going from the windward to the leeward side. The pressure coefficients will also depend on whether the wind approaches normal to the side of the building or at an angle. Table 7-1 gives some approximate values for a rectangular building with normal and quartering wind.

Notice that the pressure coefficients are consistent with the fact that air will generally flow into the building on the windward side and out on the leeward sides.

Table 7-1   PRESSURE COEFFICIENTS FOR A RECTANGULAR BUILDING

Building Wall	Normal Wind	Quartering Wind
Windward	0.95	0.70
Sides	−0.4	—
Leeward	−0.15	−0.50

The stack effect occurs when the air density differs between the inside and outside of a building. On winter days, the lower outdoor temperature causes a higher pressure at ground level on the outside and consequent infiltration. Buoyancy of the warm inside air leads to upward flow, a higher inside pressure at the top of the building, and exfiltration of air. In the summer, the process reverses with infiltration in the upper portion of the building and exfiltration in the lower part.

Considering only the stack effect, there is a level in the building where no pressure difference exists. This is defined as the neutral pressure level. Theoretically, the neutral pressure level will be at the mid-height of the building if the cracks and other openings are distributed uniformly in the vertical direction. When larger openings predominate in the lower portion of the building, the neutral pressure level will be lowered. Similarly, the neutral pressure level will be raised by larger openings in the upper portion of the building. Normally the larger openings will occur in the lower part of the building because of doors. The theoretical pressure difference with no internal separations is given by (6)

$$\Delta P_{st} = \frac{P_o h}{R_a} \frac{g}{g_c} \left(\frac{1}{T_o} - \frac{1}{T_i} \right)$$ (7-15)

where

P_o = outside pressure, psia or Pascals
h = vertical distance from neutral pressure level, ft or m
T_o = outside temperature, R or K
T_i = inside temperature, R or K
R_a = gas constant for air, (ft-lbf)/(lbm-R) or J/(kg-K)

The floors in a conventional building offer resistance to vertical air flow. Furthermore, this resistance varies depending on how stairwells and elevator shafts are sealed. When the resistance can be assumed equal for each floor, a single correction, called the *draft coefficient,* can be used to relate the actual pressure difference ΔP_s to the theoretical value ΔP_{st}.

$$C_d = \frac{\Delta P_s}{\Delta P_{st}}$$ (7-16)

The flow of air from floor to floor causes a decrease in pressure at each floor; therefore, ΔP_s will be less than ΔP_{st} and C_d is less than one. Using the draft coefficient, Eq. 7-15 becomes

$$\Delta P_s = \frac{C_d P_o h g}{R_a g_c} \left(\frac{1}{T_o} - \frac{1}{T_i} \right)$$ (7-17)

Figure 7-4 is a plot of Eq. 7-17 for an inside temperature of 75 F or 24 C, sea level outside pressure, and winter temperatures; however, Fig. 7-4 can be used for summer stack effect with little loss in accuracy.

The draft coefficient depends on the tightness of the doors in the stairwells

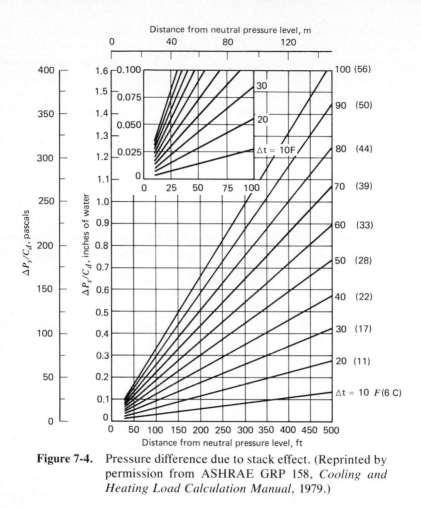

Figure 7-4. Pressure difference due to stack effect. (Reprinted by permission from ASHRAE GRP 158, *Cooling and Heating Load Calculation Manual*, 1979.)

and elevator shafts. Values of C_d range from 1.0 for buildings with no doors in the stairwells to about 0.65 for a modern office building.

Pressurization of the indoor space is accomplished by introducing more makeup air than exhaust air. Prediction of this pressure difference, ΔP_p, is difficult because it depends on the pressure differences due to wind and stack effect. When building pressurization is used, it will vary from zero to about one third of the pressure difference due to wind on the windward side, except ΔP_p is negative.

Figures 7-3, 7-5, and 7-6 and associated Tables 7-2, 7-3, and 7-4 give the infiltration rates, based on experimental evidence, for windows and doors, curtain walls, and commercial swinging doors. Note that the general procedure is the same in all cases, except that curtain wall infiltration is given per unit of wall area rather than crack length. The pressure differences are estimated by

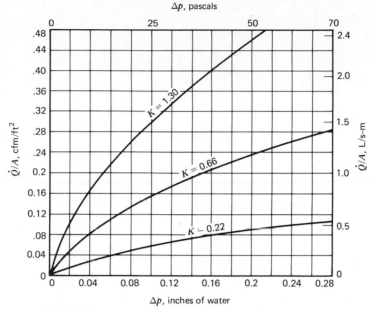

Figure 7-5. Curtain wall infiltration for one room or one floor. (Reprinted by permission from ASHRAE GRP 158, *Cooling and Heating Load Calculation Manual*, 1979.)

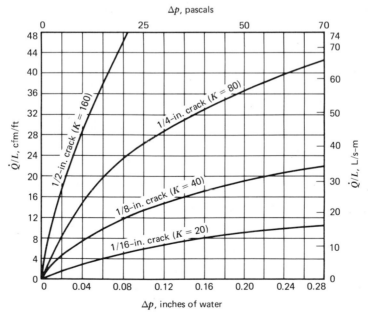

Figure 7-6. Infiltration through closed swinging door cracks. (Reprinted by permission from ASHRAE GRP 158, *Cooling and Heating Load Calculation Manual*, 1979.)

Table 7-2   WINDOW CLASSIFICATION (FOR FIG. 7-3)

	Wood Double-hung (Locked)	Other Types
Tight-fitting Window $K = 1.0$	Weatherstripped average gap ($\frac{1}{64}$-in. crack)	Wood casement and awning windows; weatherstripped Metal casement windows; weatherstripped
Average-fitting Window $K = 2.0$	Non-weatherstripped average gap ($\frac{1}{64}$-in. crack) or Weatherstripped large gap ($\frac{3}{32}$-in. crack)	All types of vertical and horizontal sliding windows; weatherstripped. Note: If average gap ($\frac{1}{64}$-in. crack) this could be tight-fitting window Metal casement windows; Nonweatherstripped Note: If large gap ($\frac{3}{32}$-in. crack) this could be a loose-fitting window
Loose-fitting Window $K = 6.0$	Non-weatherstripped large gap ($\frac{3}{32}$-in. crack)	Vertical and horizontal sliding windows; nonweatherstripped

Source: Reprinted by permission from *ASHRAE GRP 158 Cooling and Heating Load Calculation Manual,* 1979.

Table 7-3   DOOR CLASSIFICATION (FOR FIG. 7-3)

Tight-fitting Door $K = 1.0$	Very small perimeter gap and perfect fit weatherstripping—often characteristic of new doors
Average-fitting Door $K = 2.0$	Small perimeter gap having stop trim fitting properly around door and weatherstripped
Loose-fitting Door $K = 6.0$	Large perimeter gap having poor fitting stop trim and weatherstripped or Small perimeter gap with no weatherstripping

Source: Reprinted by permission from *ASHRAE GRP 158 Cooling and Heating Load Calculation Manual,* 1979.

218

Table 7-4   CURTAIN WALL CLASSIFICATION (FOR FIG. 7-5)

Leakage Coefficient	Description	Curtain Wall Construction
$K = 0.22$	Tight-fitting wall	Constructed under close supervision of workmanship on wall joints. When joints seals appear inadequate, they must be redone
$K = 0.66$	Average-fitting wall	Conventional construction procedures are used
$K = 1.30$	Loose-fitting wall	Poor construction quality control or an older building having separated wall joints

Source: Reprinted by permission from *ASHRAE GRP 158 Cooling and Heating Load Calculation Manual*, 1979.

the methods discussed earlier and the values for the coefficient K are given in Tables 7-2, 7-3, and 7-4. The use of storm sash and storm doors is common. Figure 7-7 gives data for this case. Note that the addition of a storm sash with crack length and a K value equal to the prime window does not reduce the infiltration by one half but by about 35 percent.

Commercial buildings often have a rather large number of people going and coming, which can increase infiltration significantly. Figures 7-8 and 7-9

Figure 7-7.   Infiltration for storm-prime combination windows. (Reprinted by permission from ASHRAE GRP 158, *Cooling and Heating Load Calculation Manual*, 1979.)

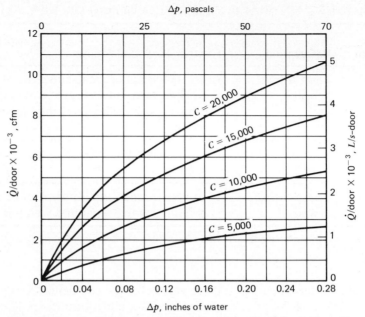

Figure 7-8. Swinging door infiltration characteristics with traffic. (Reprinted by permission from ASHRAE GRP 158, *Cooling and Heating Load Calculation Manual,* 1979.)

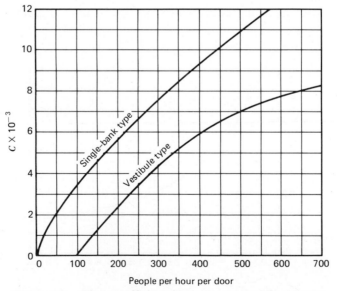

Figure 7-9. Flow coefficient dependence on traffic rate. (Reprinted by permission from ASHRAE GRP 158, *Cooling and Heating Load Calculation Manual,* 1979.)

have been developed to estimate this kind of infiltration for swinging doors. The infiltration rate per door is given in Fig. 7-8 as a function of the pressure difference and a traffic coefficient that depends on the traffic rate and the door arrangement. Figure 7-9 gives the traffic coefficients as a function of the traffic rate and two door types. Single-bank doors open directly into the space; however, there may be two or more doors at one location. Vestibule-type doors are best characterized as two doors in series so as to form an air lock between them. These doors often appear as two pairs of doors in series, which amounts to two vestibule-type doors.

Buildings may be put into two categories for calculation of infiltration. These are low rise, with less than three stories, and high rise, with more than three stories. The stack effect seems to be small in low-rise buildings and wall infiltration is usually very low; therefore, only wind effects and crackage need be considered. In high-rise buildings, the stack effect may be dominant, with a relatively large amount of leakage through the walls and around fixed window panels. All pressure effects as well as window, door, and wall leakage should be considered for high-rise buildings.

Theoretically, it is possible to predict which sides of a building will experience infiltration and which will experience exfiltration by use of the pressure coefficient. However, buildings usually do not have uniformly distributed openings on all sides. This will be particularly true for low-rise buildings. Because flow continuity must be satisfied, it is recommended that the infiltration for low-rise buildings be based on the crack length for the windward side of the structure, but never less than one half the total crack length for the building. There is room for innovation by the designer in making infiltration calculations. For example, a single-story building may have considerable leakage through openings in the ceiling. If so, air may tend to infiltrate on all sides and leave through the ceiling area. Each situation must be evaluated and a rational approach developed. The pressure coefficient approach is more feasible for high-rise buildings because the stack effect tends to cause infiltration at the lower levels and exfiltration at the higher levels in winter and the reverse in summer. Nonuniformity of the cracks and openings tends to be less important for flow continuity here. The following examples demonstrate the use of the data and methods described previously.

Example 7-3.   A 12-story office building is 120 ft tall with plan dimensions of 120 × 80 ft. The structure is of conventional curtain wall construction with all windows fixed in place. There are double vestibule-type doors on all four sides. Under winter design conditions, a wind of 15 mph blows normal to one of the long dimensions. Estimate the pressure differences for all walls for the first and twelfth floors. Consider only wind and stack effects. The indoor–outdoor temperature difference is 60 F.

Solution.   The pressure difference for each effect must first be computed and then combined to find the total. First consider the wind: Equation

7-14 expresses the wind pressure difference where the pressure coefficients may be obtained from Table 7-1 for a normal wind. Then using standard sea level density:

Windward side: $C_p = 0.95$

$$\Delta P_w = \frac{0.95(0.0765)(15 \times 1.47)^2 \ (12)}{2(32.17)62.4} = 0.105 \text{ in. } H_2O$$

Leeward: $C_p = -0.15$

$$\Delta P_w = \frac{0.105}{0.95}(-0.15) = -0.017 \text{ in. } H_2O$$

Sides: $C_p = -0.4$

$$\Delta P_w = \frac{0.105(-0.4)}{0.95} = -0.045 \text{ in. } H_2O$$

The wind effect will be assumed constant with respect to height.

The pressure difference due to stack effect can be computed from Eq. 7-17 or more easily determined from Fig. 7-4. Because there are more openings in the lower part of the building, assume that the neutral pressure level is at the fifth floor instead of the sixth. Also assume that the draft coefficient is 0.8. Then for the first floor, $h = 50$ ft and from Fig. 7-4

$$\frac{\Delta P_s}{C_d} = 0.10$$

and

$$\Delta P_s = 0.10(0.8) = 0.08 \text{ in. } H_2O$$

For the twelfth floor, $h = 70$ ft and

$$\frac{\Delta P_s}{C_d} = -0.12$$

and

$$\Delta P_s = -0.12(0.8) = -0.096 \text{ in. } H_2O$$

The negative sign indicates that the pressure is greater inside the building than on the outside.

The pressure differences may now be summarized for each side where $\Delta P = (\Delta P_w + \Delta P_s)$ in. H_2O.

Orientation	1st Floor	12th Floor
Windward	0.186	0.010
Sides	0.035	-0.141
Leeward	0.063	-0.113

These results show that air will tend to infiltrate on all floors on the wind-ward wall. Infiltration will occur on about the lower four floors on the leeward wall and on about the lower two floors on the side walls. All other surfaces will have exfiltration.

Example 7-4.   Estimate the infiltration rate for the leeward doors of Example 7-3. The doors have $\frac{1}{8}$-in. cracks and the traffic rate is low except at 5:00 P.M., when the traffic rate is 350 people per hour per door for a short time.

Solution.   This problem is solved in two steps to account for crack leakage and infiltration due to traffic. For the design condition, the effect of traffic is negligible; however, it is of interest to compute this component for 5:00 P.M. Figure 7-6 pertains to crack leakage for commercial swinging doors. For a pressure difference of 0.063 in. H_2O and $\frac{1}{8}$-in. cracks, the leakage rate is 10 cfm/ft. The crack length for standard double swinging doors is

$$L = 3(6.75) + 2(6) = 32 \text{ ft}$$

Then

$$\dot{Q} = \left(\frac{\dot{Q}}{L}\right) L = 10(32) = 320 \text{ cfm}$$

Vestibule-type doors will tend to decrease the infiltration rate somewhat like storm sash or a storm door. Assume a 30 percent reduction, then

$$\dot{Q} = (1 - 0.3)\, 320 = 224 \text{ cfm}$$

Figures 7-8 and 7-9 are used to estimate the infiltration due to traffic. The traffic coefficient C is read from Fig. 7-9 for 350 people per hour and for vestibule-type doors as 5000. Then, from Fig. 7-8 at a pressure difference of 0.063 in H_2O,

$$\dot{Q}/\text{door} = 800 \text{ cfm/door}$$

and for two doors

$$\dot{Q} = 1600 \text{ cfm}$$

A part of the crack leakage should be added to this; however, this is somewhat academic. Care should be exercised in including the traffic infiltration in the design heat load. It will usually be a short-term effect.

Example 7-5.   Estimate the net leakage rate for the twelfth floor of the building in Example 7-3.

Solution.   Referring to the pressure differences computed in Example 7-3, it is obvious that the net leakage will be from the inside out on the twelfth floor. Therefore, a great deal of air must be entering the space from the stairwells and elevator shafts. Because the twelfth floor has no movable openings, except to the roof, all leakage is assumed to be through the walls. Figure 7-5 gives data for this case where $K = 0.66$ for conventional construction.

Windward wall: $\Delta P = 0.010$ in. H_2O; $\dot{Q}/A = 0.02$ cfm/ft²
$$\dot{Q} = 0.02(120)10 = 24 \text{ cfm}$$
Side walls: $\Delta P = 0.141$ in. H_2O, $\dot{Q}/A = -0.19$ cfm/ft²
$$\dot{Q} = -0.19(80)(10)2 = -304 \text{ cfm}$$
Leeward wall: $\Delta P = -0.113$ in. H_2O, $\dot{Q}/A = -0.15$ cfm/ft²
$$\dot{Q} = -0.15(120)10 = -180 \text{ cfm}$$

The net leakage rate is then:

$$\dot{Q}_{net} = 24 - 304 - 180 = -460 \text{ cfm}$$

where the negative sign indicates that the flow is from the inside out. To compute the heat loss due to infiltration for this floor, only the 24 cfm on the windward wall should be considered. The net leakage flow of 460 cfm entered the building at other locations where the heat loss should be assigned.

Example 7-6. A single-story building is oriented so that a 15 mph wind approaches at 45 degrees to the windward sides. There are 120 ft of crack for the windows and 20 ft of crack for a door on the windward sides. It is estimated that another 150 ft of miscellaneous cracks exist on the windward as well as the leeward sides. The leeward sides have 130 ft of window cracks and 18 ft of door crack. All windows and doors are average fitting with storm sash and doors that are classified as loose fitting. Estimate the infiltration.

Solution. The major portion of the infiltration for this kind of building will be through the cracks. It is approximately true that air will enter on the windward sides and flow out on the leeward sides with most of the heat loss imposed on the rooms where the air enters. To be conservative, we will use the total crack length and assume that most of the air leaves through the ceiling area with a pressure difference computed for a quartering wind on the windward side. Using Eq. 7-14 and Table 7-1

$$\Delta P_w = \frac{0.70(0.0765)(15 \times 1.47)^2 (12)}{2(32.17)62.4} = 0.078 \text{ in. } H_2O$$

where standard sea level air density has been used. From Tables 7-2 and 7-3, the K factor for the windows and doors is read as 2.0. Then from Fig. 7-3, the leakage per foot of crack is

$$\frac{\dot{Q}}{L} = 0.375 \text{ cfm/ft}$$

and the total infiltration for the prime windows and doors is

$$\dot{Q}_1 = 0.375(250 + 38) = 108 \text{ cfm}$$

The effect of the storm sash and storm door is expressed in Fig. 7-7 where K for the storm sash is 6.0 from Table 7-2 and the crack lengths will be assumed equal for prime and storm sash. Then

$$k_p L_p / k_s L_s = 2.0 \times 288/(6.0 \times 288) = 0.33$$

and

$$\dot{Q}_1 / \dot{Q}_p = 0.88$$
$$\dot{Q}_1 = 0.88(108) = 95 \text{ cfm}$$

In the absence of data for the miscellaneous cracks, assume them to be equivalent to tight-fitting windows (Table 7-2 and Fig. 7-3). Then with a pressure difference of 0.078 in. H_2O, $\dot{Q}/L = 0.2$ cfm/ft and

$$\dot{Q}_2 = 0.2(300) = 60 \text{ cfm}$$

and

$$\dot{Q} = \dot{Q}_1 + \dot{Q}_2 = 95 + 60 = 155 \text{ cfm}$$

Fireplaces, chimneys, and flues can increase infiltration dramatically or necessitate the introduction of outdoor air. In either case the heat loss of the structure is increased. When the fireplace is not in use, large quantities of air will rise through the chimney if the damper is left open; this causes an increased amount of cold air to enter the structure through cracks. If the chimney damper is closed, the effect should be eliminated. However, fireplace dampers usually do not fit tightly and will allow some flow when closed. Therefore, it is recommended that at least 100 cfm or 0.048 m³/s infiltration be allowed for fireplaces.

When the fireplace is in use, the loss of air through the chimney is increased because of greater draft. This increases the infiltration of outdoor air for the complete structure, although the space in the immediate vicinity of the fireplace is warmed and probably will not require heat from the central system. The net effect is to increase the total heat loss from the structure. In the ideal situation, the room containing the fireplace should be provided with independent control of the warm-air heating system. The additional outdoor air required for the chimney should be introduced through the central heating unit and not be allowed to infiltrate. In this way the complete space may be maintained at a comfortable temperature. The outdoor air required for the burning fireplace should be approximately 200 cfm or 0.094 m³/s, some of which may be infiltration air.

Direct-fired warm-air furnaces are sometimes installed within the confines of the conditioned space. If combustion air is not brought in from outdoors, conditioned air from the space will be drawn in and exhausted through the flue.

Infiltration or outdoor air must then enter the structure to make up the loss and contributes to a higher heat loss. Many codes require that combustion air be introduced directly to the furnace from outdoors. Indeed this should always be the rule. For natural gas (methane) the ratio of air to gas on a volume basis is about 10. This is equivalent to 10 ft^3 or 0.28 m^3 of air per 1000 Btu or 1.06×10^6 J input to the furnace.

7-4 HEAT LOSSES FROM AIR DUCTS

The losses of a duct system can be considerable when the ducts are not in the conditioned space. Proper insulation will reduce these losses but cannot completely eliminate them. The loss may be estimated using the following relation:

$$\dot{q} = UA_s \, \Delta t_m \qquad (7\text{-}18)$$

where

$\quad U$ = overall heat transfer coefficient, Btu/(hr-ft^2-F) or W/(m^2-C)
$\quad A_s$ = outside surface area of the duct, ft^2 or m^2
$\quad \Delta t_m$ = mean temperature difference between the air in the duct and the environment, F or C.

When the duct is covered with 1 or 2 in. of fibrous glass insulation with a reflective covering, the heat loss will usually be reduced sufficiently to assume that the mean temperature difference is equal to the difference in temperature between the supply air temperature and the environment temperature. Unusually long ducts should not be treated in this manner and a mean air temperature should be used instead.

Example 7-7. Estimate the heat loss from 1000 cfm of air at 120 F flowing in a 16-in. round duct 25 ft in length. The duct has 1 in. of fibrous glass insulation and the overall heat transfer coefficient is 0.2 Btu/(hr-ft^2-F). The environment temperature is 12 F.

Solution. Equation 7-18 will be used to estimate the heat loss assuming that the mean temperature difference is given approximately by

$$\Delta t_m = t_s - t_a = 12 - 120 = -108 \text{ F}$$

The surface area of the duct is

$$A_s = \frac{\pi(16 + 2)(25)}{12} = 117.8 \text{ ft}^2$$

Then

$$\dot{q} = 0.2(117.8)(-108) = -2540 \text{ Btu/hr}$$

The temperature of the air leaving the duct may be computed from

$$\dot{q} = \dot{m}c_p(t_2 - t_1) = \dot{Q}\rho c_p(t_2 - t_1)$$

or

$$t_2 = t_1 + \frac{\dot{q}}{\rho c_p}$$

$$t_2 = 120 + \frac{-2540}{1000(60)(0.067)(0.24)}$$

$$t_2 = 117.4 \text{ F}$$

Although insulation drastically reduces the heat loss, the magnitude of the temperature difference and surface area must be considered in each case.

Minimum insulation of supply and return ducts is presently specified by ASHRAE Standards as follows:

All duct systems shall be insulated to provide a thermal resistance, excluding film resistance, of

$$R = \frac{\Delta t}{15} \text{ (hr-ft}^2\text{-F)/Btu} \qquad \text{or} \qquad R = \frac{\Delta t}{47.3} \text{ (m}^2\text{-C)/W}$$

where Δt is the design temperature differential between the air in the duct and the surrounding air in F or C. Heat losses from the supply ducts become part of the space heat load and should be summed with transmission and infiltration heat losses. Heat losses from the return air ducts are not a part of the space heat loss but should be added to the heating equipment load.

7-5   AUXILIARY HEAT SOURCES

The heat energy supplied by people, lights, motors, and machinery should always be estimated, but any actual allowance for these heat sources requires careful consideration. People may not occupy certain spaces in the evenings, on weekends, and during other periods and these spaces must generally be heated to a reasonably comfortable temperature prior to occupancy. In industrial plants heat sources, if available during occupancy, should be substituted for part of the heating requirement. In fact there are situations where so much heat energy is available that outdoor air must be used to cool the space. However, sufficient heating equipment must still be provided to prevent freezing of water pipes during periods when a facility is shut down.

7-6   INTERMITTENTLY HEATED STRUCTURES

When a structure is not heated on a continuous basis, the heating equipment capacity may have to be enlarged to assure that the temperature can be raised to a comfortable level within a reasonable period of time. The heat capacity of the building and occupant comfort are important factors when considering the use of intermittent heating. Occupants may feel discomfort if the mean radiant temperature falls below the air temperature. To conserve energy it is becoming a common practice to set back thermostats or to completely shut down equipment during the late evening, early morning, and weekend hours. This is effective and is accompanied by only small sacrifices in comfort when the periods of shutdown are adjusted to suit outdoor conditions and the mass of the structure.

7-7   AIR REQUIRED FOR SPACE HEATING

Computing the air required for heating was discussed in Chapter 3, and took into account sensible and latent effects as well as outdoor air. That procedure is always recommended when sufficient latent heat loss exists or outdoor air is used. There are many cases when the latent heat loss is quite small and may be neglected. The air quantity is then computed from

$$\dot{q} = \dot{m}c_p(t_s - t_r) \tag{7-19}$$

or

$$\dot{q} = \frac{\dot{Q}c_p}{v_s}(t_s - t_r) \tag{7-19a}$$

where

v_s = specific volume of supplied air, ft^3/lbm or m^3/kg
t_s = temperature of supplied air, F or C
t_r = room temperature, F or C

The temperature difference $(t_s - t_r)$ is normally less than 100 F or 55 C. Light commercial equipment operates with a temperature rise of 60 to 80 F or 33 to 44 C, whereas commercial applications will allow higher temperatures. The temperature of the air to be supplied must not be high enough to cause discomfort to occupants before it becomes mixed with room air.

In the unit-type equipment typically used for residences and small commercial buildings, each size is able to circulate a relatively fixed quantity of air. Therefore the air quantity is fixed within a narrow range when the heating equipment is selected. These units have different capacities that change in increments of 10 to 20,000 Btu/hr or about 5 kW according to the model. A slightly oversize unit is usually selected with the capacity to circulate a larger quantity of air than theoretically needed. Another condition that leads to greater quantities of circulated air for heating than needed is the greater air quantity usually required for cooling and dehumidifying. The same fan is used throughout the year and must therefore be large enough for the maximum air quantity required. Some units have different fan speeds for heating and for cooling.

After the total air flow rate required for the complete structure has been determined, the next step is to allocate the correct portion of the air to each room or space. This is necessary for design of the duct system. Obviously the air quantity for each room should be apportioned according to the heating load for that space; therefore,

$$\dot{Q}_{rn} = \dot{Q}(\dot{q}_{rn}/\dot{q}) \tag{7-20}$$

where

$\dot{Q}_{rn}$ = volume flow rate of air supplied to room n, ft^3/min or m^3/s
$\dot{q}_{rn}$ = total heat loss rate of room n, Btu/hr or W

7-8   ESTIMATING FUEL REQUIREMENTS

Following the calculation of the design heat load and selection of the heating system, it is often desirable to estimate the quantity of energy necessary to heat the structure under typical weather conditions and with typical inputs from internal heat sources. This is a distinctly different procedure from design heat load calculations, which are usually made for one set of design conditions neglecting solar effects and internal heat sources. It was mentioned earlier that the building with its air-conditioning equipment and variable indoor and outdoor environmental conditions constitute a complex dynamic system. Simulation of such a system requires a digital computer.

Many computer codes are available that model building systems quite well, and they may be used with all types of structures ranging from residential to heavy commercial. This approach will be discussed in Chapter 8, because the building simulation is usually carried out for a whole year considering heating, cooling, and other energy requirements.

There are cases, however, where computer simulation is not possible or cannot be justified. Residential buildings fall into this category. Reasonable results can be obtained in this case using hand calculation methods such as the degree day or bin method. These methods will be discussed briefly here. Reference 1 discusses energy estimating in detail.

The Degree Day Procedure

The traditional degree day procedure (1) for computing fuel requirements is based on the assumption that on a long-term basis, solar and internal gains for a residential structure will offset heat loss when the mean daily outdoor temperature is 65 F (18 C). It is further assumed that fuel consumption will be proportional to the difference between the mean daily temperature and 65 F or 18 C.

For cities in the United States and Canada, Table C-2 lists the average number of degree days that have occurred over a period of many years; they are listed by month and the yearly totals of these averages are given. Degree days are defined by the relationship

$$DD = \frac{(t - t_a)N}{24} \tag{7-21}$$

where N is the number of hours for which the average temperature t_a is computed and t is 65 F or 18 C. Residential insulation practices have improved dramatically over the last 40 years, however, and internal heat gains have increased. These changes indicate that a temperature less than 65 F should be used for the base; nevertheless, the data now available are based on 65 F. Another factor, which is not accounted for in traditional methods, is the decrease in efficiency of fuel-fired furnaces and heat pumps under partial load. Until methods now under development are completed, a modified degree day procedure that accounts for these factors in an approximate way will be in current use.

The general relation for fuel calculations using this procedure is

$$F = \frac{24DD\dot{q}}{\eta(t_i - t_o) H} C_D \tag{7-22}$$

where

$F =$ the quantity of fuel required for the period desired; the units depend on H

$DD =$ the degree days for period desired, F-day or C-day

$\dot{q} =$ the total calculated heat loss based on design conditions, t_i and t_o. Btu/hr or W

$\eta =$ an efficiency factor that includes the effects of rated full-load efficiency, part-load performance, oversizing, and energy conservation devices

$H =$ the heating value of fuel, Btu or kW-hr per unit volume or mass

$C_D =$ the interim correction factor for degree days based on 65 F or 18 C (Fig. 7-10)

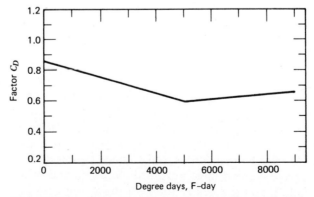

Figure 7-10.   Correction factor for use in Eq. 7-22. (Reprinted by permission from *ASHRAE Handbook, Fundamentals Volume*, 1985.)

Figure 7-10 gives values for the correction factor C_D as a function of yearly degree days. These values were calculated using typical modern single-family construction, and generally agree with electric utility experience (5).

The efficiency factor η of Eq. 7-22 is empirical and will vary from about 0.6 for older heating equipment to about 0.9 for new high efficiency equipment now available. For electric resistance heat, η has a value of 1.0.

It is recommended that more sophisticated methods of energy estimating be considered even for residential structures. The availability and simplicity of personal computers makes more refined methods practical. A serious shortcoming of the degree day method is inability to model heating equipment whose performance depends on outdoor ambient conditions. A heat pump is an example. Degree days are useful in comparing the heating requirements from one location to another. Sometimes degree days are used as a parameter in studying energy data such as utility costs. In Chaper 5 degree days were used in presenting data for floor slab heat losses.

Example 7-8   Estimate the amount of natural gas required to heat a residence in Stillwater, Oklahoma, using the modified degree day method. The heating value of the fuel is 1000 Btu/std ft³. The calculated heat loss from the house is 80,000 Btu/hr with indoor and outdoor design temperatures of 70 F and 0 F, respectively. The furnace efficiency factor is approximately 0.55.

Solution   The average winter temperature and the degree days for Stillwater are estimated to be 48 F and 3725 from Table C-2. Equation 7-22 will give an estimate of the fuel required by the prescribed method.
The correction factor C_D is 0.66 from Fig. 7-10 for 3725 degree days.

$$F = \frac{24(3725)80,000(0.661)}{0.55(70 - 0)1,000} = 122,790 \text{ std ft}^3$$

or $F = 123$ mcf of natural gas.

Bin Method

The energy estimating method discussed previously is based on average conditions and does not take into account actual day-to-day weather variations and the effect of temperature on equipment performance. The *bin method* is a hand calculation procedure where energy requirements are determined at many outdoor temperature conditions. Reference 1 describes this method in detail.

Weather data is required in the form of 5 F bins with the hours of occurrence for each bin. The data should be divided into several shifts and the mean coincident wet bulb temperature for each bin should also be given so that latent load due to infiltration can be computed if desired. Table C-3 is an example of annual bin data for Oklahoma City, Oklahoma taken from reference 6.

The bin method is based on the concept that all the hours during a month, season, or year when a particular temperature (bin) occurs can be grouped together and an energy calculation made for those hours with the equipment operating under those particular conditions. The bin method can be as simplified or complex as the situation may require and applies to both heating and cooling energy calculations. In this chapter a somewhat simplified approach will be used to introduce the method. Refinements are introduced in Chapter 8 where cooling energy calculations are discussed.

The bin method requires a load profile for the building. That is, the heating or cooling required to maintain the conditioned space at the desired conditions as a function of outdoor temperature. Figure 7-11 shows a simplified profile. In some cases more than one profile may be required to accommodate different uses of the building such as occupied and unoccupied periods. The load profiles may be determined in a number of ways. A comprehensive approach will be described in the next chapter; however, more simplified profiles are often satisfactory when only heating is considered and will be used in this chapter. The design heating load represents a good estimate of one point on the unoccupied load profile, since the design load does not include internal loads or solar effects and occurs in the early morning hours when the building is not occupied. This is point d in Fig. 7-11. There is some outdoor temperature where the heating load will be zero, such as point 0 in Fig. 7-11. Solar effects influence the location of point 0. The occupied load profile $d' - 0'$ is influenced by the internal loads due to people and equipment as well as solar effects. After considering the cooling load in the next chapter, a method will be shown to estimate the balance points where neither heating or cooling are required for both profiles. For the present let us rely on experience. For a residence the balance point is approx-

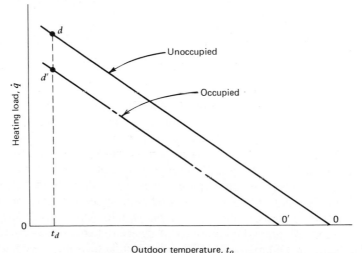

Figure 7-11.   Simplified load profiles.

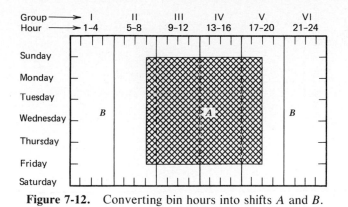

Figure 7-12. Converting bin hours into shifts *A* and *B*.

imately 60 F (16C). The balance point for a commercial building will be lower, depending on occupancy and other internal loads. Assuming that points *d* and 0 on the load profile have been determined, a straight line may be drawn and a linear equation determined to express the load as a function of outdoor temperature.

Assuming that two load profiles are to be used, it is then necessary to reduce the bin data in six time groups (Table C-3) to two time groups or shifts. This is most easily done as shown in Fig. 7-12, where the occupied and unoccupied hours are shown schematically as *A* and *B*, respectively. Table 7-5 shows computation of the fraction of the bin hours in each time group that fall in each shift.

Table 7-6 shows the calculation of the bin hours in each time group for each bin and the summations for each shift. For convenience, Table 7-6a summarizes the annual bin data for Oklahoma City, which was used to develop

Table 7-5   COMPUTATION OF FRACTION OF BIN HOURS
           IN EACH SHIFT

Time Group	Hours in Shift A in Each Group	Days in Shift A in Each Group	Total Occupied Hours in Each Group	Total Hours in Each Group	Shift A Fraction in Each Group	Shift B Fraction in Each Group
I	0	7	0	28	0.0	1.0
II	1	5	5	28	0.18	0.82
III	4	5	20	28	0.71	0.29
IV	4	5	20	28	0.71	0.29
V	2	5	10	28	0.36	0.64
VI	0	7	0	28	0.0	1.0

Table 7-6 CALCULATION OF BIN HOURS FOR EACH SHIFT

Bin Temperature	Shift A Hours Each Time Group						Shift A Hours	Shift B Hours
	I 0.00[a]	II 0.18[a]	III 0.71[a]	IV 0.71[a]	V 0.36[a]	VI 0.00[a]		
102	0	0	0	1	0	0	1	1
97	0	0	4	50	10	0	64	40
92	0	0	39	109	32	0	179	117
87	0	0	82	103	43	0	229	178
82	0	6	105	109	60	0	280	338
77	0	17	94	82	52	0	244	532
72	0	40	98	84	42	0	264	545
67	0	29	70	70	33	0	202	545
62	0	18	67	96	35	0	216	426
57	0	19	75	77	42	0	212	389
52	0	19	97	67	48	0	230	454
47	0	17	71	47	31	0	166	403
42	0	22	70	48	33	0	172	495
37	0	28	63	38	27	0	156	465
32	0	25	53	28	18	0	124	380
27	0	9	19	17	9	0	54	175
22	0	7	16	12	8	0	44	127
17	0	7	12	1	2	0	21	74
12	0	1	2	0	0	0	3	15
						TOTAL	2862	5698

[a]Shift A fraction.

Table 7-6a   ANNUAL BIN HOURS FOR OKLAHOMA
CITY, OKLAHOMA

Bin Temp., F	Time Group						Total Hours
	1–4 I	5–8 II	9–12 III	13–16 IV	17–20 V	21–24 VI	
102	0	0	0	2	0	0	2
97	0	0	5	70	29	0	104
92	0	0	55	153	88	0	296
87	2	0	116	145	120	24	407
82	20	33	148	153	168	96	618
77	121	93	132	115	144	171	776
72	29	221	138	118	117	186	809
67	161	161	98	98	93	136	747
62	120	99	95	135	96	97	642
57	87	104	105	108	116	81	601
52	96	103	137	94	133	121	684
47	98	96	100	66	87	122	569
42	150	121	98	67	91	140	667
37	144	153	89	54	76	105	621
32	107	140	74	40	50	93	504
27	63	51	27	24	24	40	229
22	36	41	23	17	23	31	171
17	19	37	17	1	5	16	95
12	7	7	3	0	0	1	18

Table 7-6. To summarize, the shift A bin hours are used with the occupied load profile and shift B bin hours are used with the unoccupied load profile.

The operating characteristics of the heating equipment as a function of the outdoor temperature are required. This information is supplied by the equipment manufacturer. The efficiency of fossil-fueled equipment such as gas- or oil-fired boilers and furnaces is relatively independent of outdoor temperature; however, the COP* of a heat pump is greatly dependent on outdoor conditions and this must be taken into account. Another factor that should be considered for all equipment is the effect of operating at a partial load. Practically all manufacturers' performance data assume full-load steady-state operation when in fact the equipment operates at partial load most of the time. Figure 7-13 shows the operating characteristics for an air-to-air heat pump with fixed conditions for the heating coil. Table 7-7 is an example of air-to-air heat pump performance data from a manufacturer's catalog from which the curves of Fig.

*Coefficient of performance

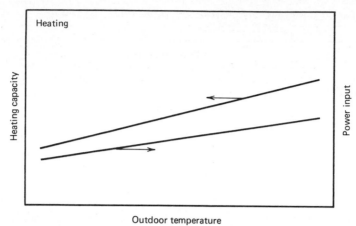

Figure 7-13.   Heat-pump operating characteristics.

7-13 may be plotted. Note that the performance depends on indoor temperature and air flow rate as well as outdoor temperature.

Detailed part-load performance of large equipment is often available from the manufacturer; however, for smaller unit-type equipment a method developed at the National Bureau of Standards (NBS) is normally used. A partial load factor is defined as

$$\text{PLF} = \frac{\text{Theoretical energy required at part load}}{\text{Actual energy required at part load}} \tag{7-23}$$

The theoretical energy required at part load is based on the steady-state operating efficiency, the COP in case of a heat pump. The actual part-load energy required takes into account the loss in efficiency due to start up and shutdown, or other part load operation. To quantify Eq. 7-23

$$\text{PLF} = 1 - D_c \left(1 - \frac{\text{Building load}}{\text{Unit capacity}} \right) \tag{7-24}$$

where D_c is the degradation coefficient, which may be specified by the manufacturer or taken as 0.25 as a default value. For unitary equipment that is controlled by starting and stopping the unit, the part load factor may also be expressed as

$$\text{PLF} = \frac{\text{Theoretical run time}}{\text{Actual run time}} \tag{7-25}$$

The Bin Calculation Method can now be outlined for a typical bin:

1. Determine the building load from the profile shown in Fig. 7-11.
2. Determine the unit capacity from Fig. 7-13.
3. Compute the theoretical run-time fraction, as the ratio of building load to unit capacity.

Table 7-7   HEAT PUMP HEATING CAPACITIES at 6000 CFM

Outdoor Temperature, F	Heating Capacity, Btu/hr × 1000 at Indoor Dry Bulb Temperature, F				Total Power Input, kW at Indoor Dry Bulb Temperature, F			
	60	70	75	80	60	70	75	80
−3	70.5	68.8	68.0	67.2	12.9	13.3	13.5	13.7
2	78.7	76.9	75.9	75.0	13.4	13.8	14.0	14.2
7	87.0	84.9	83.9	82.9	13.8	14.2	14.5	14.7
12	95.2	93.0	91.8	90.7	14.3	14.7	14.9	15.2
17	103.0	101.0	99.8	98.6	14.7	15.2	15.4	15.7
22	111.0	109.0	108.0	106.0	15.0	15.5	15.7	16.0
27	120.0	117.0	115.0	114.0	15.3	15.8	16.0	16.3
32	128.0	125.0	123.0	121.0	15.5	16.0	16.3	16.6
37	140.0	136.0	135.0	133.0	16.0	16.6	16.8	17.1
42	158.0	154.0	152.0	150.0	16.9	17.4	17.7	18.0
47	176.0	172.0	170.0	168.0	17.7	18.3	18.6	18.9
52	188.0	184.0	182.0	179.0	18.2	18.8	19.1	19.4
57	201.0	196.0	193.0	191.0	18.7	19.3	19.7	20.0
62	213.0	208.0	205.0	202.0	19.2	19.9	20.2	20.5
67	225.0	219.0	217.0	214.0	19.7	20.4	20.7	21.0

Note: Correction factor—Value at other air flow = (Value at 6000 CFM × (Cor. Fac.)

Air flow	5250	6750
Heating cap.	0.99	1.01
Power input	1.01	0.99

Table 7-8   BIN ENERGY CALCULATION FOR EXAMPLE 7-9

Bin Temperature	Occupied Hours	Unoccupied Hours	Occupied Load Btu/hr	Unoccupied Load Btu/hr	Equipment Capacity Btu/hr	Occupied PLF
1	2	3	4	5	6	7
Table 7-6	Table 7-6	Table 7-6	Given Eq. 7-26	Given Eq. 7-27	Given Eq. 7-28	Eq. 7-24 $D_c = 0.25$
62	216	426	0	14680	207521	0.75
57	212	389	0	38980	195686	0.75
52	230	454	14280	63280	183851	0.77
47	166	403	38580	87580	172016	0.81
42	172	495	62880	111880	160181	0.85
37	156	465	87180	136180	148346	0.90
32	124	380	111480	160480	136511	0.95
27	54	175	135780	184780	124676	1.00
22	44	127	160080	209080	112841	1.00
17	21	74	184380	233380	101006	1.00
12	3	15	208680	257680	89171	1.00

4. Compute the partial-load fraction from Eq. 7-24.
5. Compute the actual run-time fraction, Eq. 7-25.
6. Compute actual run time as (bin hours) × (actual run-time fraction).
7. Determine the rate of unit input from Fig. 7-13.
8. Compute the energy use, (unit input) × (actual run time).
9. Determine energy cost per unit of energy from local utility rate schedule.
10. Compute energy cost for this bin as ($/kWh) × (energy use).
11. Repeat steps 1 through 10 for all bins.

Certain refinements may be required. For example, a heat pump may not be able to supply enough heat when the outdoor temperature is very low and supplemental electrical resistance heat may be required. Often the fan energy associated with the heat pump may not be accounted for in the performance data and must be added to the calculation. Also when the building load exceeds the unit capacity, the PLF is assumed to be 1.0 because the unit will run continuously. The bin calculation procedure may be carried out by hand; however, a personal computer is especially useful. An example for an air-to-air heat pump is presented next.

Example 7-9. Consider a building in Oklahoma City, Oklahoma, that is operated on two shifts as shown in Fig. 7-12. The load profiles as shown in Fig. 7-11 are given by

Unoccupied PLF	Occupied Run Time Hr	Unoccupied Run Time Hr	Power Input kW	Occupied Electrical Resistance Input kW	Unoccupied Electrical Resistance Input kW	Total Energy kWh
8	9	10	11	12	13	14
Eq. 7-24 $D_c = 0.29$	$4 \times 2/(6 \times 7)$	$5 \times 3/(6 \times 8)$	Given Eq. 7-29	4 − 6	5 − 6	$9 \times (11+12) +$ $10 \times (11+13)$
0.77	0.0	39.3	19.8			776.7
0.80	0.0	96.9	19.3			1867.0
0.84	23.2	186.9	18.8			3941.1
0.88	46.2	233.9	18.2			5108.8
0.92	79.6	373.9	17.7			8039.4
0.98	102.2	435.8	17.2			9259.8
1.00	106.1	446.7	16.7		7.0	12368.6
1.00	58.8	259.4	16.2	3.3	17.6	9908.6
1.00	62.4	235.3	15.7	13.8	28.2	12165.8
1.00	38.3	171.0	15.2	24.4	38.8	10741.5
1.00	7.0	43.3	14.6	35.0	49.4	3123.8
					Total:	77301.1

Shift A, Occupied

$$\dot{q}_o = 267{,}000 - 4860t_o \text{ Btu/hr} \qquad (7\text{-}26)$$

Shift B, Unoccupied

$$\dot{q}_{uo} = 316{,}000 - 4860t_o \text{ Btu/hr} \qquad (7\text{-}27)$$

The heat pump performance is shown in Table 7-7 with a degradation coefficient of 0.25. Compute the energy required to heat the building assuming all the applicable bins in Table 7-6 occur during the heating season and the building is maintained at 70 F during both shifts.

Solution. The load profiles are given in a convenient form for use with the bin method. The balance temperature for each shift may be found by setting $\dot{q}_o$ and $\dot{q}_{uo}$ equal to zero.

Occupied:

$$t_o = 267{,}000/4860 = 55 \text{ F}$$

and

Unoccupied:

$$t_{ou} = 316{,}000/4860 = 65 \text{ F}$$

Therefore, bin temperatures greater than 65 F do not have to be considered.

Equations to express the steady-state heat pump performance can be derived from Table 7-7 as follows. Assuming linear dependence on the outdoor temperature, select two operating capacities and temperatures such as 101,000 Btu/hr at 17 F and 172,000 Btu/hr at 47 F and fit the points to a linear equation of the form $Y = ax + b$. For this case the heating capacity is

$$C = 2367t_o + 60,767 \text{ Btu/hr} \tag{7-28}$$

Using the same approach for the power input

$$P = 0.103t_o + 13.4 \text{ kW} \tag{7-29}$$

Table 7-8 shows the calculation procedure in tabular form. The calculations and source of data are explained for each of the numbered columns. Reading across for a single bin makes the procedure evident. Note that there is a duplicate calculation for each shift. Also note that supplemental heat in the form of electrical resistance is required at about 32 F for each shift.

It should be noted that annual bin data have been used in the preceding example. This was done for brevity and clarity. It would be more accurate to assemble the bin data for all the months during the heating season, say October through April, for the heating-energy calculation and to use data for the months of May through September for a cooling-energy calculation. The reason for this is that a few hours where the bin temperature is below the balance point for heating occur during the summer months when heat will not actually be supplied. The same is true for cooling. A few hours occur in the winter when cooling may be indicated but the air-conditioning system is off. Reference 6 gives bin data on a monthly basis.

REFERENCES

1. *ASHRAE Handbook, Fundamentals Volume,* American Society of Heating, Refrigerating and Air-Conditioning Engineers, Atlanta, Ga., 1985.

2. *ASHRAE Handbook, Systems Volume,* American Society of Heating, Refrigerating and Air-Conditioning Engineers, Atlanta, Ga., 1984.

3. *ASHRAE Cooling and Heating Load Calculation Manual GRP 158,* American Society of Heating, Refrigerating and Air-Conditioning Engineers, New York, 1979.

4. P. E. Janssen, et al., "Calculating Infiltration: An Examination of Handbook Models," *ASHRAE Transactions,* Vol. 86, Part 2, 1980.

5. W. J. Kelnhofer, *Evaluation of the ASHRAE Modified Degree Day Procedure for Predicting Energy Usage by Residential Gas Heating Systems,* American Gas Association, 1979.

6. *Bin and Degree Hour Weather Data for Simplified Energy Calculations*, American Society of Heating, Refrigerating and Air-Conditioning Engineers, Atlanta, Ga., 1986.

PROBLEMS

7-1 Determine the design conditions, both indoor and outdoor, for the following locations: (a) Sioux City, Iowa, (b) Atlanta, Georgia, (c) Syracuse, New York, (d) Denver, Colorado, (e) San Francisco, California, (f) Bismarck, North Dakota.

7-2 Estimate the temperature in the crawl space of a house with floor plan dimensions of 30 × 60 ft. The concrete foundation has an average height of 2 ft and the wall is 6 in. thick. Use winter design conditions for Oklahoma City, Oklahoma.

7-3 Estimate the temperature in the knee space shown in Fig. 7-14. The roof is equivalent to 25 mm of wood on 2 × 6 in. rafters on 0.6 m centers. The walls are 2 × 4 in. studs on 0.4 m centers with 90 mm of insulation. The joist spaces all have 150 mm of insulation. Inside and outside temperatures are 21 and − 12 C.

7-4 Consider the knee space shown in Fig. 7-14. The vertical dimension is 8 ft, the horizontal dimension is 3 ft, and the space is 20 ft long. The walls and roof surrounding the space all have an overall heat-transfer coefficient of about 0.09 Btu/(hr-ft²-F). Assuming an outdoor temperature of 2 F and an indoor temperature of 68 F, make a recommendation concerning the placement of water pipes in the knee space.

7-5 Estimate the temperature in an unheated basement that is completely below ground level with heated space above. The basement is located in Lincoln, Nebraska. Use standard design conditions.

7-6 Rework Example 7-1 assuming that the crawl space experiences an infiltration rate of 0.25 air changes per hour.

7-7 A large single-story business office is fitted with nine loose-fitting, double-hung woodsash windows 3 ft wide by 6 ft high. If the outside wind is 15 mph at a

Figure 7-14.   Sketch of building for Problem 7-3.

temperature of -10 F, what is the reduction in sensible heat loss if the windows are weather stripped? Assume an inside temperature of 70 F. Base your solution on a minimum condition in which half of the crackage undergoes infiltration and half undergoes exfiltration, with a quartering wind.

7-8 Using the crack method compute the infiltration for a tight-fitting swinging door that is used occasionally. The door has dimensions of 0.9×2.0 m and is on the windward side of a house exposed to a 9 m/s wind. Neglect internal pressurization and stack effect.

7-9 A room has two 3×5 ft wood, double-hung windows of average fit that are not weather stripped. The wind is 20 mph and normal to the wall with negligible pressurization of the room. Find the infiltration rate assuming that the entire crack is admitting air.

7-10 Compute the design infiltration rate for the house described in Problem 7-16 assuming an orientation normal to a 15 mph wind. The windows and doors are tight fitting.

7-11 Refer to Example 7-3. (a) Estimate the total pressure difference for each wall for the third and ninth floors. (b) Using design conditions for Denver, Colorado, estimate the heat load due to infiltration for the third floors and ninth floors.

7-12 Refer to Examples 7-3 and 7-4. (a) Estimate the infiltration rates for the windward and side doors for a low traffic rate. (b) Estimate the curtain wall infiltration for the first floor. (c) Compute the heat load due to infiltration for the first floor if the building is located in Cleveland, Ohio.

7-13 A 20-story office building has plan dimensions of 100×60 ft and is oriented at 45 degrees to a 20 mph wind. All windows are fixed in place. There are double vestibule-type swinging doors on the 60-ft walls. The walls are average curtain wall construction and the doors have about $\frac{1}{8}$-in. cracks. (a) Compute the pressure differences for each wall due to wind and stack effect for the first, fifth, fifteenth, and twentieth floors. (b) Plot pressure difference versus height for each wall and estimate which surfaces have infiltration and exfiltration. (c) Compute the total infiltration rate for the first floor assuming a very low traffic rate. (d) Compute the net infiltration rate for the fifteenth floor. (e) Compute the net infiltration rate for the twentieth floor. Neglect any leakage through the roof. Assume winter design conditions for Dallas, Texas.

7-14 Compute the heat gain due to infiltration for the house of Problem 7-10.

7-15 Refer to Problem 7-13. (a) Compute the heat gain due to infiltration for the first floor with the building located in Dallas, Texas. (b) Compute the heat gain due to infiltration for the fifteenth floor. (c) What is the heat gain due to infiltration for the twentieth floor?

7-16 Compute the transmission heat loss for the structure described below. Use design conditions recommended by ASHRAE Standards.

Location:	Kansas City, Missouri
Walls:	Table 5-7a, case 2
Floor:	Concrete slab with 1 in. edge insulation, both horizontal and vertical, with heating ducts under the slab

Windows:	Double-insulating glass, $\frac{1}{4}$-in. air space, 3 × 3 ft, double-hung, three on each side.
Doors:	Wood, $1\frac{3}{4}$-in. with wood storm doors, three each, 3 × 6.75 ft
Roof-ceiling:	Same as Example 5-3
House plan:	Single story, 40 × 55 ft

7-17 Consider a building located in Syracuse, New York. Using ASHRAE Standard recommendations, compute the transmission loss for (a) a wall like Table 5-7a, case 2, which is 3 × 4 m with a window of 1 m² area, (b) a single pane window, (c) a roof-ceiling combination that is like Table 5-7b, case 1 with an area of 16 m².

7-18 Compute the heating load for the structure described on the plans and specifications furnished by the instructor.

7-19 Rework Problem 7-16 for Lincoln, Nebraska. Include infiltration in the analysis.

7-20 A small commercial building has a computed heat load of 200,000 Btu/hr sensible and 25,000 Btu/hr latent. Assuming a 50 F temperature rise for the heating unit, compute the quantity of air to be supplied by the unit using the following methods: (a) Use a psychrometric chart with room conditions of 70 F and 30 percent relative humidity. (b) Calculate the air quantity based on the sensible heat transfer.

7-21 Using the degree day method, estimate the quantity of natural gas required to heat the building of Problem 7-20. The location is Stillwater, Oklahoma. Design conditions are 70 F indoor and 12 F outdoor temperatures. Assume an efficiency factor of 80 percent. The heating value of the fuel is 1000 Btu/std ft³.

7-22 If electric resistance heat were used to heat the building mentioned in Problem 7-21, how much energy would be required in kW-hr assuming a 100 percent efficiency factor? If the electrical energy costs ten cents per kW-hr and natural gas costs $4.5 per mcf, what are the relative heating costs? Assuming a power plant efficiency of 33 percent, compare the total amounts of energy in terms of mcf of gas required to heat the building using a gas furnace and an electric furnace.

7-23 A light commercial building, located in Sioux City, Iowa, has construction and use characteristics much like a residence and a design heat load of 120,000 Btu/hr (35 kW). The structure is heated with a natural gas warm-air furnace and is considered energy efficient. Assuming standard design conditions, estimate the yearly heating fuel requirements.

7-24 Refer to Problem 7-23 and determine the simplified unoccupied load profile assuming a balance point temperature of 60 F (16 C).

7-25 Refer to Problems 7-23 and 7-24. The building has an average internal load of 20,000 Btu/hr (6 kW) due to lights, equipment, and people. Determine the simplified occupied load profile.

7-26 Consider a building that operates on two shifts. The first shift begins at 10:00 A.M. and ends at midnight and the second shift includes all the remaining hours. Compute the bin hours in each shift for Oklahoma City, Oklahoma. Consider bin temperatures of 62 F and less.

7-27 Solve Problem 7-26 for (a) Denver, Colorado, (b) Washington, D.C., (c) Chicago, Illinois.

7-28 Solve Example 7-9 using the shifts of Problem 7-26.

7-29 Solve Example 7-9 for (a) Denver, Colorado, (b) Washington, D.C., (c) Chicago, Illinois.

7-30 Solve Example 7-9 using the shifts of Problem 7-26 for (a) Denver, Colorado, (b) Washington, D.C., (c) Chicago, Illinois.

CHAPTER EIGHT

The Cooling Load

A larger number of variables are considered in making cooling-load calculations than in heating-load calculations. In both situations the actual heat loss or gain is a transient one. As we explained in Chapter 7, however, for design purposes heat loss is usually based on steady-state heat transfer, and the results obtained are usually quite adequate. In design for cooling, however, transient analysis must be used if satisfactory results are to be obtained. This is because the instantaneous heat gain into a conditioned space is quite variable with time, primarily because of the strong transient effect created by the hourly variation in solar radiation. There may be an appreciable difference between the heat gain of the structure and the heat removed by the cooling equipment at a particular time. This difference is caused by the storage and subsequent transfer of energy from the structure and contents to the circulated air. If this is not taken into account, the cooling and dehumidifying equipment will usually be grossly oversized and estimates of energy requirements meaningless.

In determining seasonal energy requirements for either heating or cooling, both the solar inputs and the transient effects due to storage must be considered for accurate results. These thermal simulations are usually made using computer programs with some type of assumed or historical weather data. These computations are often performed on an hourly basis over the period of interest.

8-1 HEAT GAIN, COOLING LOAD, AND HEAT EXTRACTION RATE

It is important to differentiate between *heat gain, cooling load,* and *heat extraction rate. Heat gain* is the rate at which energy is transferred to or generated within a space. It has two components, sensible heat and latent heat, which must be computed and tabulated separately. Heat gains usually occur in the following forms:

1.   Solar radiation through openings.
2.   Heat conduction through boundaries with convection and radiation from the inner surface into the space.
3.   Sensible heat convection and radiation from internal objects.
4.   Ventilation (outside) and infiltration air.
5.   Latent heat gains generated within the space.

The *cooling load* is the rate at which energy must be removed from a space to maintain the temperature and humidity at the design values. The cooling load will generally differ from the heat gain at any instant of time. This is because the radiation from the inside surface of walls and interior objects as well as the solar radiation coming directly into the space through openings does not heat the air within the space directly. This radiant energy is mostly absorbed by floors, interior walls, and furniture, which are then cooled primarily by convection as they attain temperatures higher than that of the room air. Only when the room air receives the energy by convection does this energy became part of the cooling load. The heat storage characteristics of the structure and interior objects determine the thermal lag and therefore the relationship between heat gain and cooling load. For this reason the thermal mass (product of mass and specific heat) of the structure and its contents must be considered in such cases. Figure 8-1 illustrates the phenomenon. The reduction in peak cooling load because of the thermal lag can be quite important in sizing the cooling equipment.

The *heat extraction rate* is the rate at which energy is removed from the space by the cooling and dehumidifying equipment. This rate is equal to the cooling load when the space conditions are constant and the equipment is operating. However, this is rarely true because some fluctuation in room temperature is necessary for the control system to operate. Because the cooling load is also below the peak or design value most of the time, intermittent or variable operation of the cooling equipment is required.

Figure 8-2 shows the relation between heat gain and cooling load and the effect of the mass of the structure. The attenuation and delay of the peak heat

Figure 8-1.   Schematic relation of heat gain to cooling load.

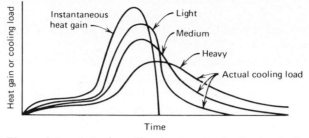

Figure 8-2.   Actual cooling load and solar heat gain for
light, medium, and heavy construction.

gain is very evident especially for heavy construction. Figure 8-3 shows the
cooling load for fluorescent lights that are used only part of the time. The
sensible heat component from people acts in a similar way. The part of the
energy produced by the lights or people that is radiant energy is momentarily
stored in the surroundings. The energy convected directly to the air by the
lights and people, and later by the surrounding objects, goes into cooling load.
The area under the curves of Fig. 8-2 are all approximately equal. This means
that about the same total amount of energy must be removed from the structure
during the day; however, a larger portion is removed during the evening hours
for the heavier constructions.

   To explain the principles of heat gain and cooling load from the computational viewpoint, consider a room enclosed by walls, windows, roof, and floor
with infiltration and internal heat sources. The heat gain for the interior space
at a given time is

$$\dot{q}_\theta = \dot{q}_{i,\theta} + \dot{q}_{r,\theta} + \dot{q}_{s,\theta} + \dot{q}_{L,\theta} \qquad (8\text{-}1)$$

Figure 8-3.   Actual cooling load from fluorescent
lights.

where

$\dot{q}_{i,\theta}$ = convective heat transfer from inside surfaces of boundaries at time θ

$\dot{q}_{r,\theta}$ = radiation heat transfer between the inside surfaces of boundaries and other interior surfaces at time θ

$\dot{q}_{s\theta}$ = rate of solar energy entering through the windows at time θ

$\dot{q}_{L\theta}$ = rate of heat generation by lights, people, and other internal sources at time θ

The sensible cooling load for the space at a given time may be expressed as

$$\dot{q}_{c,\theta} = \dot{q}_{i,\theta} + \dot{q}_{I,\theta} + \dot{q}_{v,\theta} + \dot{q}_{sc,\theta} + \dot{q}_{Lc,\theta} \tag{8-2}$$

where

$\dot{q}_{I,\theta}$ = rate of heat gain due to infiltration at time θ

$\dot{q}_{v,\theta}$ = rate of heat gain due to outdoor ventilation air at time θ

$\dot{q}_{sc,\theta}$ = rate of energy convected from the inner window and other interior surfaces to the room air at time θ

$\dot{q}_{Lc,\theta}$ = rate of heat generated by lights, people, and other internal sources that is convected into the room air at time θ

Equation 8-2 is actually the relation of most interest. In order to solve either Eq. 8-1 or 8-2, the equations describing the conduction heat transfer in the boundary walls, floors, and ceilings must be solved simultaneously because they are coupled through the inside surface temperature. The convective components of the solar and internal heat generation are related to the room air temperature also.

Consider the heat conduction problem. Calculation of heat transfer through a wall or roof section with a variable temperature in the outdoor environment and with a variable solar radiation input on the outside surface requires the heat conduction equation with nonlinear, time-dependent boundary conditions. In addition, the wall is a complex assembly of materials and may be two-dimensional. However, if the wall or roof is a single homogeneous slab, the governing differential equation is

$$\frac{\partial t}{\partial \theta} = \frac{k}{\rho c} \frac{\partial^2 t}{\partial x^2} \tag{8-3}$$

where

t = local temperature at a point in the slab, F or C

θ = time, hour

$k/\rho c$ = thermal diffusivity of the slab, ft^2/hr or m^2/hr

x = length, ft or m

At $x = 0$, the outside surface, the boundary condition is

$$-k_x \left(\frac{\partial t}{\partial x} \right)_{x=0} = h_o(t_o(\theta) - t_{wo}) + \dot{q}_r(\theta) \tag{8-4}$$

where

$\dot{q}_r$ = net solar radiation heat transfer to the exterior at a particular time
t_{wo} = the outer surface temperature
t_o = the outdoor air temperature
h_o = the convective heat-transfer coefficient on the outside

Both t_o and $\dot{q}_r$ are functions of time. At $x = L$, the inside surface, the boundary condition is

$$-k_w \left(\frac{\partial t}{\partial x} \right)_{x=L} = h_i(t_{wi} - t_i) \tag{8-5}$$

The initial condition for Eq. 8-3 is some specified temperature distribution in the wall at a time equal to zero.

The nonlinear, time-dependent boundary condition at the outside surface, Eq. 8-4, is the primary obstacle in obtaining a solution to Eq. 8-3. To circumvent this difficulty the concept of the sol-air temperature is introduced. The *sol-air temperature* t_e is the fictious air temperature that in the absence of all radiation exchanges gives the same rate of heat transfer to the exterior surface as actually occurs by solar radiation and convection. In terms of the sol-air temperature the heat transfer to the outer surface is

$$\dot{q}_o = h_o A(t_e - t_{wo}) \tag{8-6}$$

In terms of the actual outdoor temperature t_o the heat transfer rate is

$$\dot{q}_o = h_o A(t_o - t_{wo}) + \alpha A G_t - \epsilon \Delta R A \tag{8-7}$$

where

h_o = coefficient of heat transfer, Btu/(hr-ft²-F) or W/(m²-C)
ϵ = emittance of the surface
ΔR = difference between the long wavelength radiation incident on the surface from the sky and the radiation emitted from a blackbody at outdoor air temperature, Btu/(hr-ft²) or W/m²
α = absorptance of the wall surface
G_t = total incident solar radiation upon the surface, Btu/(hr-ft²) or W/m²
A = surface area, ft² or m²

Equations 8-6 and 8-7 may be combined to obtain

$$t_e = t_o + \alpha \frac{G_t}{h_o} - \frac{\epsilon \Delta R}{h_o} \tag{8-8}$$

The term $\epsilon \, \Delta R/h_o$ varies from about zero for a vertical surface to about 7 F or 4 C for horizontal surfaces. The ratio α/h_o varies from about 0.15 (hr-ft²-F)/Btu or 0.026 (m²-C)/W for a light-colored surface up to a maximum of about 0.30 (hr-ft²-F)/Btu or 0.053 (m²-C)/W. The total incident solar radiation G_t is computed as shown in Chapter 6.

Note that the sol-air temperature given by Eq. 8-8 is time dependent because t_o and G_t depend on time. However, the variation is approximately harmonic. The boundary condition at the outer surface given by Eq. 8-4 is greatly simplified by the sol-air temperature concept and becomes

$$-k_w \left(\frac{\partial t}{\partial x}\right)_{x=0} = h_o[t_e(\theta) - t_{wo}] \tag{8-9}$$

With this simplification and the assumption that t_e may be represented by a harmonic function, Fourier series solutions of Eq. 8-3 are possible using the digital computer. The solution for a particular wall yields the inside surface temperature as a function of time. The rigorous approach outlined before requires the most refined digital computers. Even so, very few computer programs are in use where space cooling load is computed in this manner. The number of surfaces and separate spaces to be considered adds even more complications.

To simplify and reduce the time required for computations, transform methods have been applied to this problem (1). A digital computer is still necessary, but the speed of calculations is rapid even for small computers. Design cooling loads for one day as well as long-term energy calculations may be done using the transfer function approach. Details of this method are discussed in Sec. 8-11.

Although preferable, it is not always practical to compute the cooling load using the transfer function method; therefore, a hand calculation method has been developed from the transfer function procedure (1, 3) and is referred to as the *Cooling Load Temperature Difference (CLTD)* method. The method involves extensive use of tables and charts and various factors to express the dynamic nature of the problem. The following sections describe the procedure beginning with design conditions that are the same for all calculation methods.

8-2 OUTDOOR AND INDOOR DESIGN CONDITIONS

The problem of selecting outdoor design conditions for calculation of heat gain is similar to that for heat loss. Again it is not reasonable to design for the worst conditions on record because a great excess of capacity will result. The heat storage capacity of the structure also plays an important role in this regard. A massive structure will reduce the effect of overload from short intervals of outdoor temperature above the design value. The *ASHRAE Handbook, Fundamentals Volume* (1) gives extensive outdoor design data. Tabulation of dry bulb and mean coincident wet bulb temperatures that are equaled or exceeded 1, 2½, and 5 percent of the total hours during June through September (2928 hours) are given. For example, a normal summer in Stillwater, Oklahoma, will have about 30 hours at 100 F dry bulb or greater, about 75 hours at 96 F or greater, and about 150 hours at 93 F or greater. Table C-1 gives the $2\frac{1}{2}$ percent values that are recommended for design purposes by ASHRAE Standards. The daily range of temperature given in Table C-1 is the difference between the average maximum and average minimum for the warmest month. The daily range is usually larger for the higher elevations where temperatures may be quite low late at night and during the early morning hours. The daily range has an effect on the energy stored by the structure. The variation in dry bulb temperature for a typical design day may be computed using the peak outdoor dry bulb temperature and the daily range. This is discussed in Sec. 8-10 in connection with the transfer function method.

The local wind velocity for summer conditions is usually taken to be about one half the winter design value but not less than about $7\frac{1}{2}$ mph or 3.4 m/s.

The indoor design conditions are governed by principles outlined in Chapter 4. For the average job in the United States and Canada, however, a condition of 75 F or 24 C dry bulb and relative humidity of 50 percent is typical when activity and dress of the occupants are light. ASHRAE Standards set the indoor design temperature at 78 F or 25 C with relative humidity within the comfort envelope defined in Figure 4-3. The designer should be alert for unusual circumstances that may lead to uncomfortable conditions. Certain activities may require occupants to engage in active work or require heavy protective clothing, both of which would require lower design temperatures.

8-3 THE CLTD METHOD

The methodology discussed in this section is a basic load calculation procedure that produces cooling load on an hourly basis if desired. The purpose of the cooling load is to provide a basis for system design. As discussed in Chapter 2, there are many different system types and controls. Therefore, it follows that the results of a load calculation may have to be manipulated in accordance with the building and system to be used. This will be discussed following presentation of the CLTD/CLF method to follow.

The CLTD method makes use of a temperature difference in the case of walls and roofs and *Cooling Load Factors* (*CLF*) in the case of solar gain through windows and internal heat sources. The CLTD and CLF vary with time and are a function of environmental conditions and building parameters. They have been derived from computer solutions using the transfer function procedure. A great deal of care has been taken to sample a wide variety of conditions in order to obtain reasonable accuracy. These factors have been derived for a fixed set of surface and environmental conditions; therefore, correction factors must often be applied. In general, calculations proceed as follows:

For walls and roofs

$$\dot{q}_\theta = UA(\text{CLTD})_\theta \qquad (8\text{-}10)$$

where

U = overall heat transfer coefficient, Btu/(hr-ft²-F) or W/(m²-C)
A = area, ft² or m²
CLTD = temperature difference which gives the cooling load at time θ, F or C

The CLTD accounts for the thermal response (lag) in the heat transfer through the wall or roof, as well as the response (lag) due to radiation of part of the energy from the interior surface of the wall to objects within the space.

For solar gain through glass

$$\dot{q}_\theta = A(SC)SHGF(CLF)_\theta \tag{8-11}$$

where

A = area, ft² or m²
SC = shading coefficient (internal shade)
SHGF = solar heat gain factor, Btu/(hr-ft²) or W/m²
CLF = cooling load factor for time θ

The SHGF is the maximum for a particular month, orientation, and latitude. The CLF accounts for the variation of the SHGF with time, the massiveness of the structure, and internal shade. Again the CLF accounts for the thermal response (lag) of the radiant part of the solar input.

For internal heat sources

$$\dot{q}_\theta = \dot{q}_i(CLF)_\theta \tag{8-12}$$

where

$\dot{q}_i$ = instantaneous heat gain from lights, people, and equipment, Btu/hr or W
CLF_θ = cooling load factor for time θ

The CLF accounts for the thermal response of the space to the various internal heat gains and is slightly different for each.

If the CLTD/CLF procedure has been adapted to a computer or when the transfer function method is used directly, the cooling load for each hour of the day is computed and the hours of interest may be selected. However, when the calculations are made by hand, calculations for all 24 hours in the day are too time consuming. In such a case, the time of day when the peak cooling load will occur may be estimated. In fact, two different types of peaks may need to be determined. First, the time of the peak load for each room is needed in order to compute the air quantity for that room. Second, the time of the peak load for a zone served by a central unit is required to size the unit. It is at these peak times that cooling load calculations should be made. The estimated times when the peak load will occur are determined from the tables of CLTD and CLF values together with the orientation and physical characteristics of the room or space. These tables are described in the next section. The times of the peak cooling load for walls, roofs, and windows are obvious in the tables, and the most dominant cooling load components will then determine the peak time for the entire room or zone. For example, rooms facing west with no exposed roof will experience a peak load in the late afternoon or early evening. East-facing rooms tend to peak during the morning hours. High internal loads may dominate the cooling load in some cases and cause an almost uniform load throughout the day. As mentioned earlier, further discussion will be presented later regarding the use of the cooling load results.

The details of computing the various cooling load components will be discussed in the following articles. Most of the tables to follow have been abridged and many of the footnotes have been included in the narrative to conserve space. Reference 1 should be consulted for complete data and an original source of information.

8-4 COOLING LOAD—EXTERNAL SURFACES

The calculation procedure is similar for walls, roofs, and conduction through glass. A different procedure is used for the glass solar gain.

Walls and Roofs

Tables 8-1 and 8-2 give the CLTD values in degrees F that were computed for the following conditions:

1.   Dark surface for solar radiation absorption.
2.   Inside temperature of 78 F or 26 C.
3.   Outdoor maximum temperature of 95 F or 35 C and an outdoor daily range of 21 F or 12 C. This corresponds to a mean outdoor temperature of 85 F or 29 C.
4.   Solar radiation for 40 degree north latitude on July 21.
5.   Outside convective film coefficient of 3.0 Btu/(hr-ft^2-F) or 17 W/(m^2-C).
6.   Inside convective film coefficient of 1.46 Btu/(hr-ft^2-F) or 8.3 W/(m^2-C).
7.   No forced ventilation or air ducts in the ceiling space.

To convert CLTD values from degree F to C simply multiply by 5/9. When conditions differ from the above, the CLTD (in degrees F) should be adjusted according to the following relation:

$$CLTD_{cor} = (CLTD + LM)K + (78 - t_i) + (t_{om} - 85) \qquad (8\text{-}13)$$

where

LM = is a correction for latitude and month from Table 8-3, F or C
K = color adjustment factor
t_i = room design temperature, F or C
t_{om} = outdoor mean temperature, $t_{om} = t_o - DR/2$, F or C

Table 8-1 COOLING LOAD TEMPERATURE DIFFERENCES

Roof No.	Description of Construction	Weight lb/ft²	U Btu/(hr-ft²-F)	1	2	3	4	5
						Without Suspended		
1	Steel sheet with 1-in. (or 2-in.) insulation	7 (8)	0.213 (0.124)	1	−2	−3	−3	−5
2	1-in. wood with 1-in. insulation	8	0.170	6	3	0	−1	−3
3	4-in. l.w. concrete	18	0.213	9	5	2	0	−2
4	2-in. h.w. concrete with 1-in. (or 2-in.) insulation	29	0.206 (0.122)	12	8	5	3	0
5	1-in. wood with 2-in. insulation	19	0.109	3	0	−3	−4	−5
6	6-in. l.w. concrete	24	0.158	22	17	13	9	6
7	2.5-in. wood with 1-in. insulation	13	0.130	29	24	20	16	13
8	8-in. l.w. concrete	31	0.126	35	30	26	22	18
9	4-in. h.w. concrete with 1-in. (or 2-in.) insulation	52 (52)	0.200 (0.120)	25	22	18	15	12
10	2.5-in. wood with 2-in. insulation	13	0.093	30	26	23	19	16
11	Roof terrace system	75	0.106	34	31	28	25	22
12	6-in. h.w. concrete with 1-in. (or 2-in.) insulation	75 (75)	0.192 (0.117)	31	28	25	22	20
13	4-in. wood with 1-in. (or 2-in) insulation	17 (18)	0.106 (0.078)	38	36	33	30	28
						With Suspended		
1	Steel sheet with 1-in. (or 2-in.) insulation	9 (10)	0.134 (0.092)	2	0	−2	−3	−4
2	1-in. wood with 1-in. insulation	10	0.115	20	15	11	8	5
3	4-in. l.w. concrete	20	0.134	19	14	10	7	4
4	2-in. h.w. concrete with 1-in. insulation	30	0.131	28	25	23	20	17
5	1-in. wood with 2-in. insulation	10	0.083	25	20	16	13	10
6	6-in. l.w. concrete	26	0.109	32	28	23	19	16
7	2.5 in. wood with 1-in. insulation	15	0.096	34	31	29	26	23
8	8-in. l.w. concrete	33	0.093	39	36	33	29	26
9	4-in. h.w. concrete with 1-in. (or 2-in.) insulation	53 (54)	0.128 (0.090)	30	29	27	26	24
10	2.5-in. wood with 2-in. insulation	15	0.072	35	33	30	28	26
11	Roof terrace system	77	0.082	30	29	28	27	26
12	6-in. h.w. concrete with 1-in. (or 2-in) insulation	77 (77)	0.125 (0.088)	29	28	27	26	25
13	4-in. wood with 1-in (or 2-in.) insulation	19 (20)	0.083 (0.064)	35	34	33	32	31

CLTD may be converted to degrees C multiplying by 5/9.

FOR CALCULATING COOLING LOAD FROM FLAT ROOFS[a]

					Solar Time, hr													
6	7	8	9	10	11	12	13	14	15	16	17	18	19	20	21	22	23	24
Ceiling																		
−3	6	19	34	49	61	71	78	79	77	70	59	45	30	18	12	8	5	3
−3	−2	4	14	27	39	52	62	70	74	74	70	62	51	38	28	20	14	9
−3	−3	1	9	20	32	44	55	64	70	73	71	66	57	45	34	25	18	13
−1	−1	3	11	20	30	41	51	59	65	66	66	62	54	45	36	29	22	17
−7	−6	−3	5	16	27	39	49	57	63	64	62	57	48	37	26	18	11	7
3	1	1	3	7	15	23	33	43	51	58	62	64	62	57	50	42	35	28
10	7	6	6	9	13	20	27	34	42	48	53	55	56	54	49	44	39	34
14	11	9	7	7	9	13	19	25	33	39	46	50	53	54	53	49	45	40
9	8	8	10	14	20	26	33	40	46	50	53	53	52	48	43	38	34	30
13	10	9	8	9	13	17	23	29	36	41	46	49	51	50	47	43	39	35
19	16	14	13	13	15	18	22	26	31	36	40	44	45	46	45	43	40	37
17	15	14	14	16	18	22	26	31	36	40	43	45	45	44	42	40	37	34
25	22	20	18	17	16	17	18	21	24	28	32	36	39	41	43	43	42	40
Ceiling																		
−4	−1	9	23	37	50	62	71	77	78	74	67	56	42	28	18	12	8	5
3	2	3	7	13	21	30	40	48	55	60	62	61	58	51	44	37	30	25
2	0	0	4	10	19	29	39	48	56	62	65	64	61	54	46	38	30	24
15	13	13	14	16	20	25	30	35	39	43	46	47	46	44	41	38	35	32
7	5	5	7	12	18	25	33	41	48	53	57	57	56	52	46	40	34	29
13	10	8	7	8	11	16	22	29	36	42	48	52	54	54	51	47	42	37
21	18	16	15	15	16	18	21	25	30	34	38	41	43	44	44	42	40	37
23	20	18	15	14	14	15	17	20	25	29	34	38	42	45	46	45	44	42
22	21	20	20	21	22	24	27	29	32	34	36	38	38	38	37	36	34	33
24	22	20	18	18	18	20	22	25	28	32	35	38	40	41	41	40	39	37
25	24	23	22	22	22	23	23	25	26	28	29	31	32	33	33	33	33	32
24	23	22	21	21	22	23	25	26	28	30	32	33	34	34	34	33	32	31
29	27	26	24	23	22	21	22	22	24	25	27	30	32	34	35	36	37	36

Source: Reprinted by permission from *ASHRAE Handbook, Fundamentals Volume*, 1985.

Table 8-2 COOLING LOAD TEMPERATURE DIFFERENCES

	1	2	3	4	5	6	7	8	9	10
North Latitude										
Wall Facing										
N	14	14	14	13	13	13	12	12	11	11
E	24	24	23	23	22	21	20	19	19	18
S	20	20	19	19	18	18	17	16	16	15
W	27	27	26	26	25	24	24	23	22	21
N	15	14	14	13	12	11	11	10	9	9
E	23	22	21	20	18	17	16	15	15	15
S	21	20	19	18	17	15	14	13	12	11
W	29	28	27	26	24	23	21	19	18	17
N	15	14	13	12	11	10	9	8	8	7
E	22	21	19	17	15	14	12	12	14	16
S	21	19	18	16	15	13	12	10	9	9
W	31	29	27	25	22	20	18	16	14	13
N	15	13	12	10	9	7	6	6	6	6
E	19	17	15	13	11	9	8	9	12	17
S	19	17	15	13	11	9	8	7	6	6
W	31	27	24	21	18	15	13	11	10	9
N	12	10	8	7	5	4	3	4	5	6
E	14	12	10	8	6	5	6	11	18	26
S	15	12	10	8	7	5	4	3	4	5
W	25	21	17	14	11	9	7	6	6	6
N	8	6	5	3	2	1	2	4	6	7
E	10	7	6	4	3	2	6	17	28	38
S	10	8	6	4	3	2	1	1	3	7
W	17	13	10	7	5	4	3	3	4	6
N	3	2	1	0	−1	2	7	8	9	12
E	4	2	1	0	−1	11	31	47	54	55
S	4	2	1	0	−1	0	1	5	12	22
W	6	5	3	2	1	1	2	5	8	11

[a]CLTD may be converted to degrees C by multiplying by 5/9.

Source: Reprinted by permission from *ASHRAE Handbook, Fundamentals Volume,* 1985.

Solar Time, hr													
11	12	13	14	15	16	17	18	19	20	21	22	23	24
Group A Walls													
10	10	10	10	10	10	11	11	12	12	13	13	14	14
19	19	20	21	22	23	24	24	25	25	25	25	25	25
14	14	14	14	14	15	16	17	18	19	19	20	20	20
20	19	19	18	18	18	18	19	20	22	23	25	26	26
Group B Walls													
9	8	9	9	9	10	11	12	13	14	14	15	15	15
17	19	21	22	24	25	26	26	27	27	26	26	25	24
11	11	11	12	14	15	17	19	20	21	22	22	22	21
16	15	14	14	14	15	17	19	22	25	27	29	29	30
Group C Walls													
7	8	8	9	10	12	13	14	15	16	17	17	17	16
19	22	25	27	29	29	30	30	30	29	28	27	26	24
9	10	11	14	17	20	22	24	25	26	25	25	24	22
12	12	12	13	14	16	20	24	29	32	35	35	35	33
Group D Walls													
6	7	8	10	12	13	15	17	18	19	19	19	18	16
22	27	30	32	33	33	32	32	31	30	28	26	24	22
7	9	12	16	20	24	27	29	29	29	27	26	24	22
9	9	10	11	14	18	24	30	36	40	41	40	38	34
Group E Walls													
7	9	11	13	15	17	19	20	21	23	20	18	16	14
33	36	38	37	36	34	33	32	30	28	25	22	20	17
9	13	19	24	29	32	34	33	31	29	26	23	20	17
7	9	11	14	20	27	36	43	49	49	45	40	34	29
Group F Walls													
9	11	14	17	19	21	22	23	24	23	20	16	13	11
44	45	43	39	36	34	32	30	27	24	21	17	15	12
13	20	27	34	38	39	38	35	31	26	22	18	15	12
8	11	14	20	28	39	49	57	60	54	43	34	27	21
Group G Walls													
15	18	21	23	24	24	25	26	22	15	11	9	7	5
50	40	33	31	30	29	27	24	19	15	12	10	8	6
31	39	45	46	43	37	31	25	20	15	12	10	8	5
15	19	27	41	56	67	72	67	48	29	20	15	11	8

Table 8-3   CLTD CORRECTION FOR LATITUDE AND MONTH APPLIED TO WALLS AND ROOFS, NORTH LATITUDES[a]

Lat.	Month	N	NNE NNW	NE NW	ENE WNW	E W	ESE WSW	SE SW	SSE SSW	S	HOR
24	Dec	−5	−7	−9	−10	−7	−3	3	9	13	−13
	Jan/Nov	−4	−6	−8	−9	−6	−3	3	9	13	−11
	Feb/Oct	−4	−5	−6	−6	−3	−1	3	7	10	−7
	Mar/Sept	−3	−4	−3	−3	−1	−1	1	2	4	−3
	Apr/Aug	−2	−1	0	−1	−1	−2	−1	−2	−3	0
	May/Jul	1	2	2	0	0	−3	−3	−5	−6	1
	Jun	3	3	3	1	0	−3	−4	−6	−6	1
32	Dec	−5	−7	−10	−11	−8	−5	2	9	12	−17
	Jan/Nov	−5	−7	−9	−11	−8	−4	2	9	12	−15
	Feb/Oct	−4	−6	−7	−8	−4	−2	4	8	11	−10
	Mar/Sep	−3	−4	−4	−4	−2	−1	3	5	7	−5
	Apr/Aug	−2	−2	−1	−2	0	−1	0	1	1	−1
	May/Jul	1	1	1	0	0	−1	−1	−3	−3	1
	Jun	1	2	2	1	0	−2	−2	−4	−4	2
40	Dec	−6	−8	−10	−13	−10	−7	0	7	10	−21
	Jan/Nov	−5	−7	−10	−12	−9	−6	1	8	11	−19
	Feb/Oct	−5	−7	−8	−9	−6	−3	3	8	12	−14
	Mar/Sep	−4	−5	−5	−6	−3	−1	4	7	10	−8
	Apr/Aug	−2	−3	−2	−2	0	0	2	3	4	−3
	May/Jul	0	0	0	0	0	0	0	0	1	1
	Jun	1	1	1	0	1	0	0	−1	−1	2
48	Dec	−6	−8	−11	−14	−13	−10	−3	2	6	−25
	Jan/Nov	−6	−8	−11	−13	−11	−8	−1	5	8	−24
	Feb/Oct	−5	−7	−10	−11	−8	−5	1	8	11	−18
	Mar/Sep	−4	−6	−6	−7	−4	−1	4	8	11	−11
	Apr/Aug	−3	−3	−3	−3	−1	0	4	6	7	−5
	May/Jul	0	−1	0	0	1	1	3	3	4	0
	Jun	1	1	2	1	2	1	2	2	3	2

[a]CTLD correction may be converted to degrees C by multiplying by 5/9.

Source: Reprinted by permission from the *ASHRAE Handbook, Fundamentals Volume*, 1985.

The color correction factor K should be used with caution especially for roofs. It has a value of 1.0 for dark surfaces and 0.5 for permanently light-colored surfaces. The designer must be confident that a light-colored surface will remain in that condition before using a value of K less than 1.0. Reference 1 gives a multiplying factor for Eq. 8-13 to account for attic ventilation; however, recent studies of attics and ceiling spaces have shown attic ventilation to be ineffective in reducing the cooling load. Use of this factor is not recommended for attics or ceiling air spaces.

Tables 8-4, 8-5, and 8-6 describe the roofs and walls given in Tables 8-1 and 8-2 in detail. Note, for example, in Table 8-4 the roof numbers and descriptions correspond to those in Table 8-1, and each roof is further described by use of code numbers for each layer. The details of each layer are given in Table 8-6. This same general procedure is used in Table 8-5 for walls, except a capital letter is used to denote construction groups that are thermally similar. Again the layers are described in Table 8-6. The overall heat transfer coefficients given in Table 8-1 for roofs and Table 8-5 for walls are to be used as guides. The actual value of U calculated as described in Chapter 5 should be used in calculations. When the actual roof or wall cannot be found in Tables 8-1 or 8-5, a thermally similar one should be selected. Thermally similar walls or roofs have similar mass and heat capacity.

A wall or roof may have more insulation than those given in Tables 8-1 or 8-2. For each R-7 increase in insulation, choose a roof of similar mass and heat capacity with a peak CLTD that occurs 2 hours later. For walls, simply move upward one letter in Table 8-2 to the next group for each R-7 increase in insulation. To reiterate, the actual overall heat transfer coefficient for the wall or roof should be used. It is important in selecting a wall or roof group that the mass and insulation be oriented properly. Walls with identical thermal resistance and mass will respond differently if the individual component layers are reordered. More simply stated, a wall or roof responds differently if the insulation is on the inside rather than the outside of the massive component.

Table 8-4 ROOF CONSTRUCTION CODE

Roof No.	Description	Code Number of Layers (see Table 8-6)
1	Steel sheet with 1-in. insulation	A0,E2,E3,B5,A3,E0
2	1-in. wood with 1-in. insulation	A0,E2,E3,B5,B7,E0
3	4-in. l.w. concrete	A0,E2,E3,C14,E0
4	2-in. h.w. concrete with 1-in. insulation	A0,E2,E3,B5,C12,E0
5	1-in. wood with 2-in. insulation	A0,E2,E3,B6,B7,E0
6	6-in. l.w. concrete	A0,E2,E3,C15,E0
7	2.5-in. wood with 1-in. insulation	A0,E2,E3,B5,B8,E0
8	8-in. l.w. concrete	A0,E2,E3,C16,E0
9	4-in. h.w. concrete with 1-in. insulation	A0,E2,E3,B5,C5,E0
10	2.5-in. wood with 2-in. insulation	A0,E2,E3,B6,B8,E0
11	Roof terrace system	A0,C12,B1,B6,E2,E3,C5,E0
12	6-in. h.w. concrete with 1-in. insulation	A0,E2,E3,B5,C13,E0
13	4-in. wood with 1-in. insulation	A0,E2,E3,B5,B9,E0

Source: Reprinted by permission from *ASHRAE Handbook, Fundamentals Volume,* 1985.

Table 8-5   WALL CONSTRUCTION GROUP DESCRIPTION

Group No.	Description of Construction	Weight (lb/ft²)	U-value Btu/(hr·ft²·F)	Code Numbers of Layers (see Table 8-4)
4-in. Face Brick + (Brick)				
D	4-in. Common Brick	90	0.415	A0,A2,C4,E1,E0
C	1-in. Insulation or Air space + 4-in. Common Brick	90	0.174–0.301	A0,A2,C4,B1/B2,E1,E0
B	2-in. Insulation + 4-in. Common Brick	88	0.111	A0,A2,B3,C4,E1,E0
A	Insulation or Air space + 8-in. Common Brick	130	0.154–0.243	A0,A2,C9,B1/B2,E1,E0
4-in. Face Brick + (H.W. Concrete)				
B	2-in. Insulation + 4-in. Concrete	97	0.116	A0,A2,B3,C5,E1,E0
A	Air Space or Insulation + 8-in. or more Concrete	143–190	0.110–0.112	A0,A2,B1,C10/11,E1,E0
4-in. Face Brick + (L.W. or H.W. Concrete Block)				
E	4-in. Block	62	0.319	A0,A2,C2,E1,E0
D	Air Space or Insulation + 4-in. Block	62	0.153–0.246	A0,A2,C2,B1/B2,E1,E0
C	Air Space or 1-in. Insulation + 6-in. or 8-in. Block	73–89	0.221–0.275	A0,A2,B1,C7/C8,E1,E0
B	2-in. Insulation + 8-in. Block	89	0.096–0.107	A0,A2,B3,C7/C8,E1,E0

H.W. Concrete Wall + (Finish)			
E	4-in. Concrete	63	0.585	A0,A1,C5,E1,E0
D	4-in. Concrete + 1-in. or 2-in. Insulation	63	0.119–0.200	A0,A1,C5,B2/B3,E1,E0
C	2-in. Insulation + 4-in. Concrete	63	0.119	A0,A1,B6,C5,E1,E0
C	8-in. Concrete	109	0.490	A0,A1,C10,E1,E0
B	8-in. Concrete + 1-in. or 2-in. Insulation	110	0.115–0.187	A0,A1,C10,B5/B6,E1,E0
A	2-in. Insulation + 8-in. Concrete	110	0.115	A0,A1,B3,C10,E1,E0
L.W. and H.W. Concrete Block + (Finish)			
F	4-in. Block + Air Space/Insulation	29	0.161–0.263	A0,A1,C2,B1/B2,E1,E0
E	2-in. Insulation + 4-in. Block	29–37	0.105–0.114	A0,A1,B3,C2/C3,E1,E0
E	8-in. Block	47–51	0.294–0.402	A0,A1,C7/C8,E1,E0
D	8-in. Block + Air Space/Insulation	41–57	0.149–0.173	A0,A1,C7/C8,B1/B2,E1,E0
Metal Curtain Wall			
G	With/without Air Space + 1-in./2-in./3-in. Insulation	5–6	0.091–0.230	A0,A3,B5/B6/B12,A3,E0
Frame Wall			
G	1-in. to 3-in. Insulation	16	0.081–0.178	A0,A1,B1,B2/B3/B4,E1,E0

Source: Reprinted by permission from *ASHRAE Handbook, Fundamentals Volume*, 1985.

Table 8-6 THERMAL PROPERTIES AND CODE NUMBERS OF LAYERS USED IN CALCULATIONS OF COEFFICIENTS FOR ROOF AND WALL[a]

Description	Code Number	x	Thickness and Thermal Properties				
			k	ρ	c	R	M
Outside surface resistance	A0					0.333	
1-in. stucco (asbestos cement or wood siding plaster, etc.)	A1	0.0833	0.4	116	0.20	0.208	9.7
4-in. face brick (dense concrete)	A2	0.333	0.75	130	0.22	0.444	43.3
Steel siding (aluminum or other lightweight cladding)	A3	0.005	26.0	480	0.10	0.0002	2.4
Finish	A6	0.0417	0.24	78	0.26	0.174	3.3
Air space resistance	B1					0.91	
1-in. insulation	B2	0.083	0.025	2.0	0.2	3.32	0.2
2-in. insulation	B3	0.167	0.025	2.0	0.2	6.68	0.4
3-in. insulation	B4	0.25	0.025	2.0	0.2	10.03	0.5
1-in. insulation	B5	0.0833	0.025	5.7	0.2	3.33	0.5
2-in. insulation	B6	0.167	0.025	5.7	0.2	6.68	1.0
1-in. wood	B7	0.0833	0.07	37.0	0.6	1.19	3.1
2.5-in. wood	B8	0.2083	0.07	37.0	0.6	2.98	7.7
4-in. wood	B9	0.333	0.07	37.0	0.6	4.76	12.3
2-in. wood	B10	0.167	0.07	37.0	0.6	2.39	6.2
3-in. wood	B11	0.25	0.07	37.0	0.6	3.58	9.3
3-in. insulation	B12	0.25	0.025	5.7	0.2	10.0	1.4

		x	k	ρ	c	R	M
4-in. clay tile	C1	0.333	0.33	70.0	0.2	1.01	23.3
4-in. l.w. concrete block	C2	0.333	0.22	38.0	0.2	1.51	12.7
4-in. h.w. concrete block	C3	0.333	0.47	61.0	0.2	0.71	20.3
4-in. common brick	C4	0.333	0.42	120	0.2	0.79	40.0
4-in. h.w. concrete	C5	0.333	1.0	140	0.2	0.333	46.6
8-in. clay tile	C6	0.667	0.33	70	0.2	2.02	46.7
8-in. l.w. concrete block	C7	0.667	0.33	38.0	0.2	2.02	25.4
8-in. h.w. concrete block	C8	0.667	0.6	61.0	0.2	1.11	40.7
8-in. common brick	C9	0.667	0.42	120	0.2	1.59	80.0
8-in. h.w. concrete	C10	0.667	1.0	140	0.2	0.667	93.4
12-in. h.w. concrete	C11	1.0	1.0	140	0.2	1.00	140.0
2-in. h.w. concrete	C12	0.167	1.0	140	0.2	0.167	23.4
6-in. h.w. concrete	C13	0.5	1.0	140	0.2	0.50	70.0
4-in. l.w. concrete	C14	0.333	0.1	40	0.2	3.33	13.3
6-in. l.w. concrete	C15	0.5	0.1	40	0.2	5.0	20.0
8-in. l.w. concrete	C16	0.667	0.1	40	0.2	6.67	26.7
Inside surface resistance	E0					0.685	
0.75-in. plaster; 0.75-in. gypsum or other similar finishing layer	E1	0.0625	0.42	100	0.2	0.149	6.3
0.5-in. slag or stone	E2	0.0417	0.83	55	0.40	0.050	2.3
0.375-in. felt membrane	E3	0.0313	0.11	70	0.40	0.285	2.2
Ceiling air space	E4					1.0	
Acoustic tile	E5	0.0625	0.035	30	0.20	1.786	1.9

Source: Reprinted by permission from *ASHRAE Handbook, Fundamentals Volume*, 1985

[a]Units: x = ft; c = Btu/(lbm-F); k = Btu/(hr-ft-F); R = (hr-ft²-F)/Btu; ρ = lb/ft³, M = lbm/ft².

Example 8-1. A building located in Tallahassee, Florida, has a roof described as number 1 in Table 8-4 with a suspended ceiling. There are $2\frac{1}{2}$-in. fibrous glass batts laid on top of the ceiling. Compute the cooling load per ft^2 at 10:00 A.M. and 5:00 P.M. solar time using standard design conditions for August.

Solution. The cooling load is calculated using Eq. 8-10 where U is computed for the actual roof. Data can be obtained from Chapter 5 or Table 8-6. The overall heat-transfer coefficient including the ceiling air space and acoustic tile ceiling is 0.134 Btu/(hr-ft^2-F) as given in Table 8-1. Addition of the $2\frac{1}{2}$-in. batts, which have a thermal resistance of 7 (hr-ft^2-F)/Btu, will change U to 0.07 Btu/(hr-ft^2-F). The CLTD should then be obtained from Table 8-1 for a thermally similar roof with a peak CLTD occurring 2 hours later than roof number 1. It appears that roof number 2 is the best choice. Then the uncorrected CLTD values for 10:00 A.M. and 5:00 P.M. are 13 F and 62 F, respectively. Equation 8-13 is used to correct these values as follows:

$$LM = -0.5 \text{ F; Table 8-3 at 30 deg. N. lat.}$$
$$K = 1.0; \text{ dark surface assumed}$$
$$t_i = 78 \text{ F; ASHRAE Standard}$$
$$t_{om} = 92 - (19/2) = 82.5 \text{ F; Table C-1}$$

For 10:00 A.M.

$$\text{CLTD}_{10} = (13 - 0.5)(1.0) + (78 - 78) + (82.5 - 85) = 10 \text{ F}$$

For 5:00 P.M.

$$\text{CLTD}_{17} = (62 - 0.5)(1.0) + (78 - 78) + (82.5 - 85) = 59 \text{ F}$$

Then the cooling load at 10:00 A.M. is

$$\dot{q}_c/A = (\text{CLTD}_{10}) = 0.07(10) = 0.7 \text{ Btu/(hr-ft}^2)$$

and at 5:00 P.M.

$$\dot{q}_c/A = U(\text{CLTD}_{17}) = 0.07(59) = 4.13 \text{ Btu/(hr-ft}^2)$$

Example 8-2. The building of Example 8-1 has walls consisting of 4-in. face brick plus 6-in. H.W. concrete block with 1-in. insulation between. Compute the cooling load per ft^2 at 10:00 A.M. and 5:00 P.M. solar time for the south wall using standard design conditions for August.

Solution. The wall is exactly as described in Table 8-5 as Group C. Because there is no variation from the table, the overall heat-transfer coefficient will be the same as given, 0.275 Btu/(hr-ft^2-F). Then for a south, Group C wall in Table 8-2, the uncorrected CLTD values are 9 and 22 F, respectively. Using Eq. 8-13 the corrected values become

$$\text{CLTD}_{10} = 6 \text{ F}$$
$$\text{CLTD}_{17} = 19 \text{ F}$$

The cooling load at 10:00 A.M. is then

$$\dot{q_c}/A = 0.275(6) = 1.65 \text{ Btu/(hr-ft}^2)$$

and at 5:00 P.M.

$$\dot{q_c}/A = 0.275(19) = 5.23 \text{ Btu/(hr-ft}^2)$$

Heat admission or loss through fenestration areas is affected by many factors of which the following are the most significant.

1. Solar radiation intensity and incident angle.
2. Difference between outdoor and indoor temperature.
3. Velocity and direction of air flow across the exterior and interior surfaces.
4. Low temperature radiation exchange between the surfaces of the glass and the surroundings.
5. Exterior or interior shading.

When solar radiation strikes an unshaded window (Fig. 8-4), about 8 percent of the radiant energy is typically reflected back outdoors, from 5 to 50 percent is absorbed within the glass, depending on the composition and thickness of the glass, and the remainder is transmitted directly indoors, to become part of the cooling load. The solar gain is the sum of the transmitted radiation and the portion of the absorbed radiation that flows inward. Because heat is

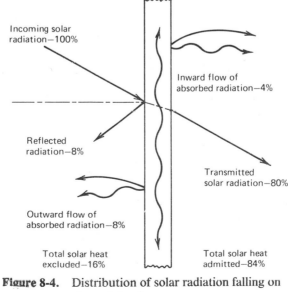

Figure 8-4.   Distribution of solar radiation falling on clear plate glass.

also conducted through the glass whenever there is an outdoor–indoor temperature difference, the total rate of heat admission is

Total heat admission through glass = Radiation transmitted through glass
  + Inward flow of absorbed solar radiation + Conduction heat gain   (8-14)

The first two quantities are related to the amount of solar radiation falling on the glass and the third quantity occurs whether or not the sun is shining. In winter the conduction heat flow may well be outward rather than inward.
   We can rewrite Eq. 8-14 to read

Total heat gain = Solar heat gain + Conduction heat gain   (8-14a)

The conduction heat gain per unit area is the product of the overall coefficient of heat transfer U for the existing fenestration and the outdoor–indoor temperature difference $(t_o - t_i)$. Values of U for a number of widely used glazing systems are given in Table 5-8. The cooling load is computed using a CLTD much the same as a wall or roof. The glass represents a much more simple situation, however, with small thermal capacity. Table 8-7 gives CLTD values for glass. The table is for 78 F or 26 C indoor temperature, 95 F or 35 C maximum outdoor temperature, and 21 degree F or 12 degree C daily range. Corrections may be made as shown for walls and roof. The cooling load due to conduction heat gain through window glass is then given by

$$\dot{q}_c = UA(\text{CLTD}) \tag{8-15}$$

The solar heat gain is much more complex because the sun's apparent motion across the sky causes the irradiation of a surface to change minute by minute. As discussed in Chapter 6, the solar heat gain may be calculated when the type of glass, latitude and location, date and time, and orientation are known. It was mentioned earlier that the cooling load due to solar heat gain is expressed in terms of a shading coefficient, a solar heat gain factor, and a cooling load factor (Eq. 8-11). We will first consider the shading coefficient.
   The ASHRAE procedure for estimating solar heat gain assumes that a constant ratio exists between the solar heat gain through any given type of fenestration system and the solar heat gain (under exactly the same solar conditions) through unshaded clear sheet glass (i.e., the *reference* glass). The reference glass is double-strength glass with a transmittance of 0.86, a reflectance of 0.08, and an 0.06 absorptance. This ratio, called the *shading coefficient,* is unique for each type of fenestration or each combination of glazing and internal shading device.

$$\text{SC} = \frac{\text{Solar heat gain of fenestration}}{\text{Solar heat gain of double-strength glass}} \tag{8-16}$$

Solar calorimeter tests are required for accurate determination of shading coefficients for combinations of glazing materials and internal shading devices. The shading coefficients for several types and combinations of glass are given in Table 8-8. The shading coefficient for any fenestration will rise above the

Table 8-7 COOLING LOAD TEMPERATURE DIFFERENCE FOR CONDUCTION THROUGH GLASS AND CONDUCTION THROUGH DOORS[a]

Solar Time, hr																							
1	2	3	4	5	6	7	8	9	10	11	12	13	14	15	16	17	18	19	20	21	22	23	24
CLTD, F																							
1	0	-1	-2	-2	-2	-2	0	2	4	7	9	12	13	14	14	13	12	10	8	6	4	3	2

[a]The CLTD may be converted to C by multiplying by 5/9.
Corrections: The values in the table were calculated for an inside temperature of 78 F and an outdoor maximum temperature of 95 F with an outdoor daily range of 21 degrees F. The table remains approximately correct for other outdoor maximums (93–102 F) and other outdoor daily ranges (16–34 F), provided the outdoor daily average temperature remains approximately 85 F. If the room air temperature is different from 78 F, and/or the outdoor daily average temperature is different from 85 F, correct as shown in Eq. 8-13.

Source: Reprinted by permission from *ASHRAE Handbook, Fundamentals Volume,* 1985.

Table 8-8   SHADING COEFFICIENTS FOR SINGLE GLASS AND INSULATING GLASS

A. Single Glass

Type of Glass	Nominal Thickness		Solar Trans.[a]	Shading Coefficient	
	in.	mm		$h_o = 4.0$ Btu $\overline{\text{hr-ft}^2\text{-F}}$	$h_o = 3.0$ Btu $\overline{\text{hr-ft}^2\text{-F}}$
Clear	$\frac{1}{8}$	3.2	0.86	1.00	1.00
	$\frac{1}{4}$	6.4	0.78	0.94	0.95
	$\frac{3}{8}$	9.5	0.72	0.90	0.92
	$\frac{1}{2}$	12.7	0.67	0.87	0.88
Heat Absorbing	$\frac{1}{8}$	3.2	0.64	0.83	0.85
	$\frac{1}{4}$	6.4	0.46	0.69	0.73
	$\frac{3}{8}$	9.5	0.33	0.60	0.64
	$\frac{1}{2}$	12.7	0.24	0.53	0.58

B. Insulating Glass[b]

Clear Out, Clear In	$\frac{1}{8}$[c]	3.2	0.71[e]	0.88	0.88
Clear Out, Clear In	$\frac{1}{4}$	6.4	0.61	0.81	0.82
Heat Absorbing[d] Out, Clear In	$\frac{1}{4}$	6.4	0.36	0.55	0.58

[a]Refer to manufacturer's literature for values.
[b]Refers to factory-fabricated units with $\frac{3}{16}$-, $\frac{1}{4}$-, or $\frac{1}{2}$-in. air space or to prime windows plus storm sash.
[c]Thickness of each pane of glass, not thickness of assembled unit.
[d]Refers to gray, bronze, and green tinted heat-absorbing float glass.
[e]Combined transmittance for assembled unit.

Source: Reprinted by permission from *ASHRAE Handbook, Fundamentals Volume,* 1985.

tabulated values when the inner surface coefficient is increased and when the outer surface coefficient is decreased. The converse is also true. Table 8-8 shows the effect for outer film coefficients of 3 and 4 Btu/(hr-ft²-F) or 17 and 23 (W/(m²-C) where the larger value is for a 7.5 mph or 3.4 m/s wind, and the smaller value is for a lower wind velocity. The effect of the film coefficient on the shading coefficient is related to energy absorbed by the glass and then transferred away by convection.

Blinds, shades, and drapes or curtains that are often installed on the inside next to windows decrease the solar heat gain. The shading coefficient is used to express this effect. Tables 8-9 and 8-10 give representative data for single sheet and insulating glass with indoor shading by blinds and shades. Note that the shading coefficient applies to the combination of glass and shading device.

Table 8-9 SHADING COEFFICIENTS FOR SINGLE GLASS WITH INDOOR SHADING BY VENETIAN BLINDS AND ROLLER SHADES

Type of Glass	Nominal Thickness in.	Nominal Thickness mm	Solar Transmittance	Type of Shading — Venetian Blinds Medium	Venetian Blinds Light	Roller Shade Opaque Dark	Roller Shade Opaque White	Roller Shade Translucent Light
Clear sheet	$\frac{3}{32}$ to $\frac{1}{4}$	2–6	0.87–0.80					
Clear	$\frac{1}{4}$ to $\frac{1}{2}$	6–13	0.80–0.71					
Clear pattern	$\frac{1}{8}$ to $\frac{3}{32}$	3–6	0.87–0.79	0.64	0.55	0.59	0.25	0.39
Heat-absorbing pattern	$\frac{1}{8}$	3	—					
Tinted	$\frac{3}{16}$, $\frac{7}{32}$	5–6	0.74, 0.71					
Heat-absorbing[a]	$\frac{3}{16}$, $\frac{1}{4}$	5–6	0.46					
Heat-absorbing pattern	$\frac{3}{16}$, $\frac{1}{4}$	5–6	—	0.57	0.53	0.45	0.30	0.36
Tinted	$\frac{1}{8}$, $\frac{7}{32}$	3–6	0.59, 0.45					
Heat-absorbing or pattern	—		0.44–0.30	0.54	0.52	0.40	0.28	0.32
Heat-absorbing[a]	$\frac{3}{8}$	10	0.34					
Heat-absorbing or pattern	—		0.29–0.15	0.42	0.40	0.36	0.28	0.31
			0.24					
Reflective coated glass (no inside shade)								
Shading coefficient = 0.30				0.25	0.23			
Shading coefficient = 0.40				0.33	0.29			
Shading coefficient = 0.50				0.42	0.38			
Shading coefficient = 0.60				0.50	0.44			

[a]Gray, bronze, and green tinted.

Source: Adapted by permission from ASHRAE Handbook, Fundamentals Volume, 1985.

Table 8-10 SHADING COEFFICIENTS FOR INSULATING GLASS WITH INDOOR SHADING BY VENETIAN BLINDS AND ROLLER SHADES

Type of Glass	Nominal Thickness, Each light in.	Nominal Thickness, Each light mm	Solar Transmittance Outer Pane	Solar Transmittance Inner Pane	Type of Shading				
					Venetian Blinds		Roller Shade		
							Opaque		Translucent
					Medium	Light	Dark	White	Light
Clear out Clear in	$\frac{3}{32}, \frac{1}{8}$	2–3	0.87	0.87	0.57	0.51	0.60	0.25	0.37
Clear out Clear in	$\frac{1}{4}$	6	0.80	0.80					
Heat-absorbing out Clear in	$\frac{1}{4}$	6	0.46	0.80	0.39	0.36	0.40	0.22	0.30
Reflective coated glass (no inside shade)									
Shading coefficient = 0.20					0.19	0.18			
Shading coefficient = 0.30					0.27	0.26			
Shading coefficient = 0.40					0.34	0.33			

Source: Adapted by permission from *ASHRAE Handbook, Fundamentals Volume*, 1985.

Shading coefficients for draperies are a complex function of color and the weave of the fabric. Although other variables also have an effect, reasonable correlation has been obtained using only color and openness of the weave. Figure 8-5 and Table 8-11 are a brief summary of information given in the *ASHRAE Handbook* (1). Many manufacturers of drapery materials furnish data on the radiation properties of their products. This is usually in the form of the reflectance and transmittance, the abscissa and ordinate for Fig. 8-5. The shading coefficient index letter is read from Fig. 8-5 (*A* to *J*) and used in Table 8-11 to determine the shading coefficient for the combination glass and drapery system. The reflectance and transmittance of the drapery are often not known. In this case the index letter may be estimated from the openness of the weave and color of the material. Openness is classified as: open, I; semiopen, II; and closed, III. Color is classified as: dark, D; medium, M; and light, L. A light-colored, closed-weave material would then be classified III$_L$. From Fig. 8-5 the index letter varies from G to J for this classification, and judgment is required in making a final selection.

There are other data available for shading coefficients of such things as shade screen, domed skylights, and hollow glass blocks. The *ASHRAE Hand-*

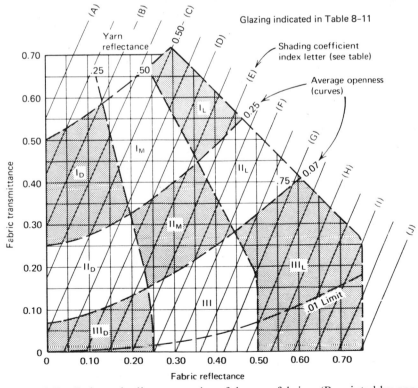

Figure 8-5. Indoor shading properties of drapery fabrics. (Reprinted by permission from *ASHRAE Handbook, Fundamentals Volume*, 1985.)

Table 8-11   SHADING COEFFICIENTS FOR SINGLE AND INSULATING GLASS WITH DRAPERIES[a]

Glazing	Glass Transmittance	Glass Shading Coefficient	A	B	C	D	E	F	G	H	I	J
Single Glass												
$\frac{1}{4}$ in. or 6 mm clear	0.80	0.95	0.80	0.75	0.70	0.65	0.60	0.55	0.50	0.45	0.40	0.35
$\frac{1}{2}$ in. or 13 mm clear	0.71	0.88	0.74	0.70	0.66	0.61	0.56	0.52	0.48	0.43	0.39	0.35
$\frac{1}{4}$ in. or 6 mm heat absorbing	0.46	0.67	0.57	0.54	0.52	0.49	0.46	0.44	0.41	0.38	0.36	0.33
$\frac{1}{2}$ in. or 13 mm heat absorbing	0.24	0.50	0.43	0.42	0.40	0.39	0.38	0.36	0.34	0.33	0.32	0.30
Reflective coated	—	0.60	0.57	0.54	0.51	0.49	0.46	0.43	0.41	0.38	0.36	0.33
(See manufacturers'	—	0.50	0.46	0.44	0.42	0.41	0.39	0.38	0.36	0.34	0.33	0.31
literature for exact values)	—	0.40	0.36	0.35	0.34	0.33	0.32	0.30	0.29	0.28	0.27	0.26
	—	0.30	0.25	0.24	0.24	0.23	0.23	0.23	0.22	0.21	0.21	0.20
Insulating glass ($\frac{1}{2}$ in. or 13 mm air space) clear out and clear in	0.64	0.83	0.66	0.62	0.58	0.56	0.52	0.48	0.45	0.42	0.37	0.35
Heat absorbing out and clear in	0.37	0.56	0.49	0.47	0.45	0.43	0.41	0.39	0.37	0.35	0.33	0.32
Reflective coated	—	0.40	0.38	0.37	0.37	0.36	0.34	0.32	0.31	0.29	0.28	0.28
(See manufacturers'	—	0.30	0.29	0.28	0.27	0.27	0.26	0.26	0.25	0.25	0.24	0.24
literature for exact values.)	—	0.20	0.19	0.19	0.18	0.18	0.17	0.17	0.16	0.16	0.15	0.15

Source: Adapted by permission from *ASHRAE Handbook, Fundamentals Volume*, 1985.

book or manufacturers' catalogs may be consulted for these cases. Heat gain through sunlit double-strength sheet glass, the reference glass, is designated as the Solar Heat Gain Factor (SHGF). Values of this quantity have been calculated for the twenty-first day of each month, for daylight hours, for various directions and various latitudes (1).

The solar heat gain factor of Eq. 8-11 is the maximum value for the month for a given orientation and latitude. The cooling load factor then represents the ratio of the actual solar heat gain, which becomes cooling load, to the maximum solar heat gain. This approach reduces the amount of tabulated data. Table 8-12 gives maximum solar heat gain factors for 32, 40, and 48 deg north latitude.

Table 8-12   MAXIMUM SOLAR HEAT GAIN FACTOR, BTU/(HR-FT^2) FOR SUNLIT GLASS, NORTH LATITUDES[a]

	32 Deg									
	N (Shade)	NNE NNW	NE NW	ENE WNW	E W	ESE WSW	SE SW	SSE SSW	S	HOR[b]
Jan.	24	24	29	105	175	229	249	250	246	176
Feb.	27	27	65	149	205	242	248	232	221	217
Mar.	32	37	107	183	227	237	227	195	176	252
Apr.	36	80	146	200	227	219	187	141	115	271
May	38	111	170	208	220	199	155	99	74	277
June	44	122	176	208	214	189	139	83	60	276
July	40	111	167	204	215	194	150	96	72	273
Aug.	37	79	141	195	219	210	181	136	111	265
Sep.	33	35	103	173	215	227	218	189	171	244
Oct.	28	28	63	143	195	234	239	225	215	213
Nov.	24	24	29	103	173	225	245	246	243	175
Dec.	22	22	22	84	162	218	246	252	252	158

	40 Deg									
	N (Shade)	NNE NNW	NE NW	ENE WNW	E W	ESE WSW	SE SW	SSE SSW	S	HOR[b]
Jan.	20	20	20	74	154	205	241	252	254	133
Feb.	24	24	50	129	186	234	246	244	241	180
Mar.	29	29	93	169	218	238	236	216	206	223
Apr.	34	71	140	190	224	223	203	170	154	252
May	37	102	165	202	220	208	175	133	113	265
June	48	113	172	205	216	199	161	116	95	267
July	38	102	163	198	216	203	170	129	109	262
Aug.	35	71	135	185	216	214	196	165	149	247
Sep.	30	30	87	160	203	227	226	209	200	215
Oct.	25	25	49	123	180	225	238	236	234	177
Nov.	20	20	20	73	151	201	237	248	250	132
Dec.	18	18	18	60	135	188	232	249	253	113

Table 8-12   MAXIMUM SOLAR HEAT GAIN FACTOR, BTU/(HR-FT²)
FOR SUNLIT GLASS, NORTH LATITUDES[a] (continued)

	N (Shade)	NNE NNW	NE NW	ENE WNW	E W	ESE WSW	SE SW	SSE SSW	S	HOR[b]
Jan.	15	15	15	53	118	175	216	239	245	85
Feb.	20	20	36	103	168	216	242	249	250	138
Mar.	26	26	80	154	204	234	239	232	228	188
Apr.	31	61	132	180	219	225	215	194	186	226
May	35	97	158	200	218	214	192	163	150	247
June	46	110	165	204	215	206	180	148	134	252
July	37	96	156	196	214	209	187	158	146	244
Aug.	33	61	128	174	211	216	208	188	180	223
Sep.	27	27	72	144	191	223	228	223	220	182
Oct.	21	21	35	96	161	207	233	241	242	136
Nov.	15	15	15	52	115	172	212	234	240	85
Dec.	13	13	13	36	91	156	195	225	233	65

48 Deg

[a]To convert to W/m² multiply by 3.155
[b]HOR = horizontal.

Source: Reprinted by permission from ASHRAE Handbook, Fundamentals Volume, 1985.

Table 8-13 gives maximum solar heat gain factors for a special case of external shade. This will be discussed later.

The cooling load factors depend on the actual solar heat gain for a particular time, the internal shade, and the building construction when there is no internal shade. At the time the cooling load factors for glass without internal shading were originally computed (1), some important variables, other than mass, were not included in the analysis. More recent research (5) has revealed the importance of additional variables such as floor covering (carpet), the ceiling, the relative amount of glass in the envelope and the size of the zone. Table 8-14, for rooms with carpeted floors, and Table 8-14a, for rooms without carpet, have been developed using transfer function coefficients from reference 5. The notes at the bottom of the tables list other variables assumed. The cooling load factors for glass with internal shade, Table 8-15, were not affected by this new work because the shades or drapes block direct solar transmission to the room. The interior shading has two effects. First, it reflects some of the solar radiation so that it never enters the space. Second, the shading device prevents the solar radiation from being absorbed by the floor, interior walls, and furnishings. Instead the shading device absorbs the radiation, which is subsequently convected to the room air.

Example 8-3.   The building of Example 8-1 has a 4 × 5 ft single glass window in the south wall. The window has light-colored venetian blinds.

Table 8-13 MAXIMUM SOLAR HEAT GAIN FACTOR FOR
EXTERNALLY SHADED GLASS, BTU/(HR-FT²)
(BASED ON GROUND REFLECTANCE OF 0.2)[a]

Use for latitudes 0–24 deg
For latitudes greater than 24, use north orientation, Table 8-12
For horizontal glass in shade, use the tabulated values for all latitudes

	N	NNE NNW	NE NW	ENE WNW	E W	ESE WSW	SE SW	SSE SSW	S	(All Latit.) HOR
Jan.	31	31	31	32	34	36	37	37	38	16
Feb.	34	34	34	35	36	37	38	38	39	16
Mar.	36	36	37	38	39	40	40	39	39	19
Apr.	40	40	41	42	42	42	41	40	40	24
May	43	44	45	46	45	43	41	40	40	28
June	45	46	47	47	46	44	41	40	40	31
July	45	45	46	47	47	45	42	41	41	31
Aug.	42	42	43	45	46	45	43	42	42	28
Sept.	37	37	38	40	41	42	42	41	41	23
Oct.	34	34	34	36	38	39	40	40	40	19
Nov.	32	32	32	32	34	36	38	38	39	17
Dec.	30	30	30	31	32	34	36	37	37	15

[a]To convert to W/m² multiply by 3.155

Source: Reprinted by permission from *ASHRAE Handbook, Fundamentals Volume,* 1985

Compute the cooling load due to the window at 10:00 A.M. and 5:00 P.M. solar time for August using standard design conditions.

Solution.   There are two components of cooling load for the window. One is due to heat conduction, the other to solar radiation. Equation 8-15 gives the cooling load due to conduction where the overall coefficient is obtained from Chapter 5 as 0.81 Btu/(hr-ft²-F). The CLTD values are given in Table 8-7 as 4 F and 13 F for 10:00 A.M. and 5:00 P.M., respectively. Because the daily average temperature is about 83 F, the table values should be reduced by 2 F. The cooling load due to conduction at 10:00 A.M. is

$$\dot{q}_c = 20(0.81)(4 - 2) = 32 \text{ Btu/hr}$$

and for 5:00 P.M.

$$\dot{q}_c = 20(0.81)(13 - 2) = 178.0 \text{ Btu/hr}$$

Considering the solar radiation and using Eq. 8-11 with a shading coefficient of 0.55 from Table 8-9; a maximum solar heat gain factor of 111 Btu/ (hr-ft²) is obtained from Table 8-12 for 32 deg north latitude. The cooling load factors from Table 8-15 are 0.58 and 0.27. The cooling load for

Table 8-14 COOLING LOAD FACTOR FOR GLASS WITHOUT LATITUDES

Direction	Room Mass	1	2	3	4	5	6	7	8	9	10
N	L	0.00	0.00	0.00	0.00	0.01	0.64	0.73	0.74	0.81	0.88
	M	0.03	0.02	0.02	0.02	0.02	0.64	0.69	0.69	0.77	0.84
	H	0.10	0.09	0.08	0.07	0.07	0.62	0.64	0.64	0.71	0.77
NE	L	0.00	0.00	0.00	0.00	0.01	0.51	0.83	0.88	0.72	0.47
	M	0.01	0.01	0.00	0.00	0.01	0.50	0.78	0.82	0.67	0.44
	H	0.03	0.03	0.03	0.02	0.03	0.47	0.71	0.72	0.59	0.40
E	L	0.00	0.00	0.00	0.00	0.00	0.42	0.76	0.91	0.90	0.75
	M	0.01	0.01	0.00	0.00	0.01	0.41	0.72	0.86	0.84	0.71
	H	0.03	0.03	0.03	0.02	0.02	0.39	0.66	0.76	0.74	0.63
SE	L	0.00	0.00	0.00	0.00	0.00	0.27	0.58	0.81	0.93	0.93
	M	0.01	0.01	0.01	0.00	0.01	0.26	0.55	0.77	0.88	0.87
	H	0.04	0.04	0.03	0.03	0.03	0.26	0.51	0.69	0.78	0.78
S	L	0.00	0.00	0.00	0.00	0.00	0.07	0.15	0.23	0.39	0.62
	M	0.01	0.01	0.01	0.01	0.01	0.07	0.14	0.22	0.38	0.59
	H	0.05	0.05	0.04	0.04	0.03	0.09	0.15	0.21	0.35	0.54
SW	L	0.00	0.00	0.00	0.00	0.00	0.04	0.09	0.13	0.16	0.19
	M	0.02	0.02	0.01	0.01	0.01	0.05	0.09	0.13	0.16	0.19
	H	0.07	0.06	0.05	0.05	0.04	0.07	0.11	0.14	0.16	0.18
W	L	0.00	0.00	0.00	0.00	0.00	0.03	0.07	0.10	0.13	0.15
	M	0.02	0.02	0.01	0.01	0.01	0.04	0.07	0.10	0.13	0.14
	H	0.06	0.06	0.05	0.04	0.04	0.06	0.09	0.11	0.13	0.15
NW	L	0.00	0.00	0.00	0.00	0.00	0.04	0.09	0.14	0.17	0.20
	M	0.02	0.02	0.01	0.01	0.01	0.05	0.10	0.13	0.17	0.19
	H	0.06	0.05	0.05	0.04	0.04	0.07	0.11	0.14	0.17	0.19
Hor.	L	0.00	0.00	0.00	0.00	0.00	0.08	0.25	0.45	0.64	0.80
	M	0.02	0.02	0.01	0.01	0.01	0.08	0.24	0.43	0.60	0.75
	H	0.07	0.06	0.05	0.05	0.04	0.11	0.25	0.41	0.56	0.68

L = Lightweight construction, 1 in. wood floor, group G wall
M = Medium-weight construction, 2 to 4 in. con. floor, group E wall
H = Heavy-weight construction, 6 to 8 in. con. floor, group C wall
Zone assumed to have 50% glass or less, a ceiling, and is small
Hor. = horizontal.

INTERIOR SHADING, WITH CARPETED FLOOR, NORTH

	Solar Time												
11	12	13	14	15	16	17	18	19	20	21	22	23	24
0.95	0.98	0.98	0.94	0.88	0.79	0.79	0.55	0.31	0.12	0.04	0.02	0.01	0.00
0.91	0.94	0.95	0.91	0.86	0.79	0.79	0.56	0.32	0.16	0.10	0.07	0.05	0.04
0.83	0.87	0.88	0.85	0.81	0.75	0.76	0.55	0.34	0.22	0.17	0.15	0.13	0.11
0.33	0.27	0.24	0.23	0.20	0.18	0.14	0.09	0.03	0.01	0.00	0.00	0.00	0.00
0.32	0.28	0.26	0.24	0.22	0.19	0.15	0.11	0.05	0.03	0.02	0.02	0.01	0.01
0.30	0.27	0.26	0.25	0.23	0.20	0.17	0.13	0.08	0.06	0.05	0.05	0.04	0.04
0.51	0.30	0.22	0.18	0.16	0.13	0.11	0.07	0.02	0.01	0.00	0.00	0.00	0.00
0.48	0.30	0.24	0.21	0.18	0.16	0.13	0.09	0.04	0.03	0.02	0.01	0.01	0.01
0.43	0.29	0.24	0.22	0.20	0.18	0.15	0.12	0.08	0.06	0.05	0.05	0.04	0.04
0.81	0.59	0.37	0.27	0.21	0.18	0.14	0.09	0.03	0.01	0.00	0.00	0.00	0.00
0.76	0.56	0.37	0.29	0.24	0.20	0.16	0.11	0.05	0.04	0.03	0.02	0.02	0.01
0.68	0.51	0.35	0.29	0.25	0.22	0.19	0.15	0.09	0.08	0.07	0.06	0.05	0.05
0.82	0.94	0.93	0.80	0.59	0.38	0.26	0.16	0.06	0.02	0.01	0.00	0.00	0.00
0.78	0.88	0.88	0.76	0.57	0.38	0.28	0.18	0.09	0.06	0.04	0.03	0.02	0.02
0.70	0.79	0.79	0.69	0.52	0.37	0.29	0.21	0.13	0.10	0.09	0.08	0.07	0.06
0.23	0.39	0.62	0.82	0.94	0.94	0.81	0.54	0.19	0.07	0.03	0.01	0.00	0.00
0.22	0.38	0.60	0.78	0.89	0.89	0.77	0.52	0.20	0.10	0.07	0.05	0.04	0.03
0.21	0.35	0.55	0.71	0.80	0.79	0.69	0.48	0.20	0.14	0.11	0.10	0.08	0.07
0.16	0.18	0.31	0.55	0.78	0.92	0.93	0.73	0.25	0.10	0.04	0.01	0.01	0.00
0.16	0.17	0.30	0.53	0.74	0.87	0.88	0.69	0.24	0.12	0.07	0.05	0.04	0.03
0.16	0.17	0.28	0.49	0.67	0.78	0.79	0.62	0.23	0.14	0.11	0.09	0.08	0.07
0.22	0.23	0.24	0.31	0.53	0.78	0.92	0.81	0.28	0.10	0.04	0.02	0.01	0.00
0.21	0.22	0.23	0.30	0.52	0.75	0.88	0.77	0.26	0.12	0.07	0.05	0.04	0.03
0.20	0.21	0.22	0.28	0.48	0.68	0.79	0.69	0.23	0.14	0.10	0.09	0.08	0.07
0.91	0.97	0.97	0.91	0.80	0.64	0.44	0.23	0.08	0.03	0.01	0.00	0.00	0.00
0.86	0.92	0.92	0.87	0.77	0.63	0.45	0.26	0.12	0.07	0.05	0.04	0.03	0.02
0.77	0.83	0.83	0.80	0.71	0.59	0.44	0.28	0.17	0.13	0.11	0.10	0.09	0.08

Table 8-14a COOLING LOAD FACTORS FOR GLASS WITHOUT LATITUDES

Direction	Room Mass	1	2	3	4	5	6	7	8	9	10
N	L	0.00	0.00	0.00	0.00	0.01	0.64	0.73	0.74	0.81	0.88
	M	0.12	0.09	0.07	0.06	0.05	0.33	0.45	0.53	0.61	0.69
	H	0.24	0.21	0.19	0.18	0.16	0.43	0.48	0.51	0.56	0.61
NE	L	0.00	0.00	0.00	0.00	0.01	0.51	0.83	0.88	0.72	0.47
	M	0.03	0.02	0.02	0.02	0.02	0.24	0.45	0.57	0.58	0.49
	H	0.08	0.07	0.07	0.06	0.06	0.27	0.43	0.49	0.45	0.37
E	L	0.00	0.00	0.00	0.00	0.00	0.42	0.76	0.91	0.90	0.75
	M	0.03	0.02	0.02	0.02	0.01	0.20	0.41	0.57	0.65	0.64
	H	0.08	0.08	0.07	0.06	0.06	0.24	0.40	0.50	0.53	0.50
SE	L	0.00	0.00	0.00	0.00	0.00	0.27	0.58	0.81	0.93	0.93
	M	0.04	0.03	0.02	0.02	0.02	0.13	0.31	0.48	0.62	0.69
	H	0.10	0.09	0.08	0.08	0.07	0.18	0.32	0.45	0.53	0.56
S	L	0.00	0.00	0.00	0.00	0.00	0.07	0.15	0.23	0.39	0.62
	M	0.05	0.04	0.04	0.03	0.02	0.05	0.09	0.14	0.24	0.38
	H	0.13	0.12	0.10	0.09	0.09	0.11	0.14	0.17	0.25	0.36
SW	L	0.00	0.00	0.00	0.00	0.00	0.04	0.09	0.13	0.16	0.19
	M	0.08	0.07	0.05	0.04	0.03	0.05	0.07	0.09	0.12	0.15
	H	0.15	0.14	0.12	0.11	0.10	0.11	0.12	0.14	0.15	0.17
W	L	0.00	0.00	0.00	0.00	0.00	0.03	0.07	0.10	0.13	0.15
	M	0.08	0.07	0.05	0.04	0.04	0.04	0.06	0.08	0.10	0.12
	H	0.14	0.13	0.12	0.11	0.10	0.10	0.11	0.12	0.13	0.14
NW	L	0.00	0.00	0.00	0.00	0.00	0.04	0.09	0.14	0.17	0.20
	M	0.08	0.06	0.05	0.04	0.03	0.05	0.07	0.10	0.13	0.15
	H	0.13	0.12	0.11	0.10	0.09	0.10	0.12	0.13	0.15	0.16
Hor.	L	0.00	0.00	0.00	0.00	0.00	0.08	0.25	0.45	0.64	0.80
	M	0.07	0.06	0.05	0.04	0.03	0.06	0.14	0.26	0.40	0.53
	H	0.16	0.15	0.13	0.12	0.11	0.13	0.20	0.29	0.39	0.48

L = Lightweight construction, 1 in. wood floor, group G wall
M = Medium-weight construction, 2 to 4 in. concrete floor, group E wall
H = Heavy-weight construction, 6 to 8 in. concrete floor, group C wall
Zone assumed to have 50% glass or less, a ceiling, and is small
Hor. = horizontal

INTERIOR SHADING, WITHOUT CARPETED FLOOR, NORTH

	Solar Time												
11	12	13	14	15	16	17	18	19	20	21	22	23	24
0.95	0.98	0.98	0.94	0.88	0.79	0.79	0.55	0.31	0.12	0.04	0.02	0.01	0.00
0.76	0.82	0.85	0.86	0.85	0.81	0.80	0.70	0.60	0.43	0.32	0.24	0.19	0.15
0.66	0.71	0.73	0.74	0.73	0.71	0.71	0.62	0.52	0.42	0.36	0.32	0.29	0.26
0.33	0.27	0.24	0.23	0.20	0.18	0.14	0.09	0.03	0.01	0.00	0.00	0.00	0.00
0.41	0.36	0.32	0.29	0.27	0.24	0.21	0.17	0.13	0.10	0.07	0.06	0.05	0.04
0.32	0.29	0.28	0.27	0.26	0.24	0.22	0.19	0.16	0.14	0.12	0.11	0.10	0.09
0.51	0.30	0.22	0.18	0.16	0.13	0.11	0.07	0.02	0.01	0.00	0.00	0.00	0.00
0.55	0.44	0.36	0.31	0.26	0.23	0.19	0.16	0.12	0.09	0.07	0.06	0.04	0.04
0.41	0.33	0.30	0.28	0.26	0.24	0.22	0.19	0.16	0.14	0.13	0.11	0.10	0.09
0.81	0.59	0.37	0.27	0.21	0.18	0.14	0.09	0.03	0.01	0.00	0.00	0.00	0.00
0.69	0.61	0.50	0.41	0.35	0.30	0.25	0.20	0.15	0.12	0.09	0.07	0.06	0.05
0.54	0.47	0.39	0.35	0.32	0.29	0.26	0.23	0.19	0.17	0.15	0.14	0.12	0.11
0.82	0.94	0.93	0.80	0.59	0.38	0.26	0.16	0.06	0.02	0.01	0.00	0.00	0.00
0.53	0.65	0.72	0.71	0.63	0.52	0.42	0.33	0.24	0.18	0.14	0.11	0.09	0.07
0.47	0.55	0.58	0.56	0.49	0.41	0.36	0.30	0.25	0.21	0.19	0.17	0.16	0.14
0.23	0.39	0.62	0.82	0.94	0.94	0.81	0.54	0.19	0.07	0.03	0.01	0.00	0.00
0.17	0.26	0.40	0.54	0.66	0.73	0.72	0.61	0.43	0.31	0.23	0.17	0.13	0.10
0.18	0.26	0.37	0.48	0.56	0.59	0.57	0.47	0.33	0.27	0.23	0.21	0.19	0.17
0.16	0.18	0.31	0.55	0.78	0.92	0.93	0.73	0.25	0.10	0.04	0.01	0.01	0.00
0.13	0.15	0.21	0.35	0.50	0.63	0.71	0.67	0.46	0.33	0.24	0.18	0.14	0.11
0.15	0.16	0.21	0.33	0.45	0.54	0.58	0.52	0.33	0.26	0.22	0.19	0.18	0.16
0.22	0.23	0.24	0.31	0.53	0.78	0.92	0.81	0.28	0.10	0.04	0.02	0.01	0.00
0.17	0.19	0.10	0.24	0.36	0.51	0.64	0.66	0.46	0.32	0.23	0.17	0.13	0.10
0.17	0.18	0.19	0.23	0.33	0.46	0.55	0.53	0.33	0.25	0.21	0.18	0.16	0.15
0.91	0.97	0.97	0.91	0.80	0.64	0.44	0.23	0.08	0.03	0.01	0.00	0.00	0.00
0.64	0.73	0.78	0.80	0.77	0.70	0.59	0.45	0.33	0.24	0.19	0.14	0.11	0.09
0.56	0.61	0.65	0.65	0.63	0.57	0.49	0.40	0.32	0.28	0.25	0.22	0.20	0.18

Table 8-15   COOLING LOAD FACTORS FOR GLASS WITH (ALL ROOM CONSTRUCTIONS)

Glass Facing	Solar											
	1	2	3	4	5	6	7	8	9	10	11	12
N	0.08	0.07	0.06	0.06	0.07	0.73	0.66	0.65	0.73	0.80	0.86	0.89
NE	0.03	0.02	0.02	0.02	0.02	0.56	0.76	0.74	0.58	0.37	0.29	0.27
E	0.03	0.02	0.02	0.02	0.02	0.47	0.72	0.80	0.76	0.62	0.41	0.27
SE	0.03	0.03	0.02	0.02	0.02	0.30	0.57	0.74	0.81	0.79	0.68	0.49
S	0.04	0.04	0.03	0.03	0.03	0.09	0.16	0.23	0.38	0.58	0.75	0.83
SW	0.05	0.05	0.04	0.04	0.03	0.07	0.11	0.14	0.16	0.19	0.22	0.38
W	0.05	0.05	0.04	0.04	0.03	0.06	0.09	0.11	0.13	0.15	0.16	0.17
NW	0.05	0.04	0.04	0.03	0.03	0.07	0.11	0.14	0.17	0.19	0.20	0.21
HOR.	0.06	0.05	0.04	0.04	0.03	0.12	0.27	0.44	0.59	0.72	0.81	0.85

Source: Reprinted by permission from *ASHRAE Handbook, Fundamentals Volume,* 1985.

10:00 A.M. is

$$\dot{q}_c = (20)(0.55)(111)0.58 = 708 \text{ Btu/hr}$$

and for 5:00 P.M. is

$$\dot{q}_c = (20)(0.55)(111)(0.27) = 330 \text{ Btu/hr}$$

The total cooling load for the window at 10:00 A.M. is then

$$\dot{q}_c = 32 + 708 = 740 \text{ Btu/hr}$$

and for 5:00 P.M. is

$$\dot{q}_c = 178 + 330 = 508 \text{ Btu/hr}$$

The shaded portion of a window is easily estimated using the methods and data presented in Chapter 6. Separate calculations are then made for the sunlit and shaded portions of the glass to obtain the cooling load; however, the cooling load due to conduction is the same for both parts. Common situations are overhangs, side projections, or setback.

At latitudes greater than 24 deg, the shaded portion of the glass is treated as a north-facing window with the SHGF and CLF read from Tables 8-12, 8-14, or 8-15 for the north orientation. The shading coefficient is the same for the sunlit and shaded parts.

For the special case of shaded horizontal glass and shaded glass for latitudes less than 24 deg, the SHGF is given in Table 8-13. CLF values are read from Tables 8-14 and 8-15 for the north orientation.

Example 8-4.   Compute the cooling load for a 4 × 6 ft clear plate glass window facing southwest at 40 deg north latitude. The solar time is 3:00

INTERIOR SHADING, NORTH LATITUDES

Time, hr

13	14	15	16	17	18	19	20	21	22	23	24
0.89	0.86	0.82	0.75	0.78	0.91	0.24	0.18	0.15	0.13	0.11	0.10
0.26	0.24	0.22	0.20	0.16	0.12	0.06	0.05	0.04	0.04	0.03	0.03
0.24	0.22	0.20	0.17	0.14	0.11	0.06	0.05	0.05	0.04	0.03	0.03
0.33	0.28	0.25	0.22	0.18	0.13	0.08	0.07	0.06	0.05	0.04	0.04
0.80	0.68	0.50	0.35	0.27	0.19	0.11	0.09	0.08	0.07	0.06	0.05
0.59	0.75	0.83	0.81	0.69	0.45	0.16	0.12	0.10	0.09	0.07	0.06
0.31	0.53	0.72	0.82	0.81	0.61	0.16	0.12	0.10	0.08	0.07	0.06
0.22	0.30	0.52	0.73	0.82	0.69	0.16	0.12	0.10	0.08	0.07	0.06
0.85	0.81	0.71	0.58	0.42	0.25	0.14	0.12	0.10	0.08	0.07	0.06

P.M. on 21 July. A drape of semiopen weave and medium color covers the window on the inside. There is a 2 ft overhang at the top of the glass, which is much wider than the window. Assume standard design conditions.

Solution.   The cooling load due to heat conduction is not affected by any external shade that may exist. Therefore, using an overall coefficient of 0.81 for single glass with internal shade and a CLTD of 14 F from Table 8-7

$$\dot{q}_c = 24(0.81)14 = 272 \text{ Btu/hr}$$

The cooling load due to solar heat gain must be done separately for the sunlit and shaded portions; however, the shading coefficient will be the same for each calculation. The drapery material has a II_M classification. Using Fig. 8-5, the index letter ranges from about D to F. To be conservative, D will be used. Then from Table 8-11 the shading coefficient is 0.65.

To find the shaded portion of the glass, the distance the shadow extends downward is computed using Eqs. 6-6 and 6-7. The solar altitude and azimuth angles are 47.2 and 76.7 deg, respectively (1), and the wall azimuth is 45 deg. Then

$$\gamma = |\phi - \psi| = |76.7 - 45| = 31.7 \text{ deg}$$

and

$$y = b \tan \beta/\cos \gamma$$
$$y = 2 \tan (47.2)/\cos (31.7) = 2.53 \text{ ft}$$

The shaded area is then

$$A_{sh} = 4(2.53) = 10.12 \text{ ft}^2$$

and the sunlit area is

$$A_{sl} = A - A_{sh} = 24 - 10.12 = 13.88 \text{ ft}^2$$

The cooling load for each part may now be computed. For the sunlit portion, the SHGF is 170 Btu/(hr-ft^2) from Table 8-12 and the CLF is 0.83 from Table 8-15. Then

$$\dot{q}_{sl} = (13.88)(0.65)170(0.83) = 1273 \text{ Btu/hr}$$

For the shaded portion, the SHGF is 38 Btu/(hr-ft^2) from Table 8-12 and the CLF is 0.82 from Table 8-15. Then

$$\dot{q}_{sh} = (10.12)(0.65)(38)(0.82) = 205 \text{ Btu/hr}$$

To summarize, the total cooling load for the window is

$$\dot{q}_t = \dot{q}_c + \dot{q}_{sl} + \dot{q}_{sh}$$
$$\dot{q}_t = 272 + 1273 + 205 = 1750 \text{ Btu/hr}$$

8-5 COOLING LOAD—INTERNAL SOURCES

Internal sources of heat energy may contribute significantly to the total cooling load of a structure, and poor judgment in the estimation of their magnitude can lead to unsatisfactory operation and/or high costs when part of the capacity is unneeded. These internal sources fall into the general categories of people, lights, and miscellaneous equipment.

People

Chapter 4 contains detailed information containing the rates at which heat and moisture are given up by occupants engaged in different levels of activity and Table 8-16 summarizes the data needed for heat gain calculations. Although the data of Table 8-16 are quite accurate, large errors are often made in the computation of heat gain from occupants because of poor estimates of the periods of occupancy or the number of occupants. Great care should be taken to be realistic about the allowance for the number of people in a structure. It should be kept in mind that rarely will a complete office staff be present or a classroom be full. On the other hand, a theater may often be completely occupied and sometimes may contain more occupants than it is designed for. Each design problem must be judged on its own merits. With the exception of theaters and other high-occupancy spaces, most spaces are designed with too large an allowance for its occupants. One should not allow for more than the equivalent full-time occupants.

The heat gain from people has two components: sensible and latent. Latent heat gain goes directly into the air in the space; therefore, this component immediately becomes cooling load with no delay. However, the sensible component from a person is delayed due to storage of a part of this energy in the

Table 8-16   RATES OF HEAT GAIN FROM OCCUPANTS OF CONDITIONED SPACES[a]

Degree of Activity	Typical Application	Total Heat Adults, Male		Total Heat Adjusted[b]		Sensible Heat		Latent Heat	
		Watts	Btu/hr	Watts	Btu/hr	Watts	Btu/hr	Watts	Btu/hr
Seated at rest	Theater, movie	115	400	100	350	60	210	40	140
Seated, very light work writing	Offices, hotels, apts	140	480	120	420	65	230	55	190
Seated, eating	Restaurant[c]	150	520	170	580[c]	75	255	95	325
Seated, light work, typing	Offices, hotels, apts	185	640	150	510	75	255	75	255
Standing, light work or walking slowly	Retail store, bank	235	800	185	640	90	315	95	325
Light bench work	Factory	255	880	230	780	100	345	130	435
Walking, 3 mph, light machine work	Factory	305	1040	305	1040	100	345	205	695
Bowling[d]	Bowling alley	350	1200	280	960	100	345	180	615
Moderate dancing	Dance hall	400	1360	375	1280	120	405	255	875
Heavy work, heavy machine work, lifting	Factory	470	1600	470	1600	165	565	300	1035
Heavy work, athletics	Gymnasium	585	2000	525	1800	185	635	340	1165

[a]Note: Tabulated values are based on 78 F room dry bulb temperature. For 80 F room dry bulb, the total heat remains the same, but the sensible heat value should be decreased by approximately 8% and the latent heat values increased accordingly.

[b]Adjusted total heat gain is based on normal percentage of men, women, and children for the application listed, with the postulate that the gain from an adult female is 85% of that for an adult male, and that the gain from a child is 75% of that for an adult male.

[c]Adjusted total heat value for eating in a restaurant, includes 60 Btu/hr for food per individual (30 Btu sensible and 30 Btu latent).

[d]For bowling figure one person per alley actually bowling, and all others as sitting (400 Btu/hr) or standing and walking slowly (790 Btu/hr).

Source: Reprinted by permission from *ASHRAE Handbook, Fundamentals Volume,* 1985.

room and furnishings. The cooling load factor is used to express this delayed effect. The CLF depends on the total hours the occupants are in the space and varies from the time of entry. Table 8-17 gives CLF values for people. The data are for continuously operating cooling equipment. When the cooling equipment is turned on and off when the occupants arrive and leave, the CLF is 1.0. This is because any energy stored in the building at closing time is still essentially there the next morning. When the cooling equipment is operated for several hours after occupancy, however, the CLF values are about the same as for continuous equipment operation. When the density of people is large such as in a theater, a CLF of 1.0 should be used.

Example 8-5. An office suite is designed with 10 private offices, a secretarial area with space for four secretaries, a reception area and waiting room, and an executive office with a connecting office for a secretary. Estimate the cooling load from the occupants at 3:00 P.M. solar time.

Solution. In the absence of additional data the following approach seems reasonable. For the private offices assume that an average of 7 out of the 10 will be occupied between 8:00 A.M. and 5:00 P.M. Assume that 3 of the 4 secretaries are always present and a receptionist is always there. The waiting room will have a transient occupancy, assume 2 people. The executive office will probably experience variable occupancy. However, an average of one occupant is about right. Assume the secretary is always present. The total number of people for which the heat gain is to be based is then 15. We will assume sedentary, very light work and use data from Tables 8-16 and 8-17. Assuming an adjusted group males and females and very light work, the sensible and latent heat gains per person are 230 and 190 Btu/hr, respectively. The latent cooling load due to people is

$$\dot{q}_l = 15(190) = 2850 \text{ Btu/hr}$$

There is a question about where the occupants go for lunch. Assume they stay in the space, then the total hours in the space are 9 and 3:00 P.M. represents the seventh hour after entry. From Table 8-17, CLF is between 0.82 and 0.83. The cooling load due to sensible heat is

$$\dot{q}_s = 15(230)0.825 = 2850 \text{ Btu/hr}$$

and the total cooling load at 3:00 P.M. is

$$\dot{q} = \dot{q}_l + \dot{q}_s = 2850 + 2850 = 5700 \text{ Btu/hr}$$

Note that the occupants could enter and leave the space on different schedules. In such a case, make a separate calculation for each group.

Lights

The cooling load due to lighting is often the major component of the space load and an accurate estimate is essential. A number of factors need to be

Table 8-17   SENSIBLE HEAT COOLING LOAD FACTORS FOR PEOPLE

Total Hours in Space	Hours after Each Entry into Space																							
	1	2	3	4	5	6	7	8	9	10	11	12	13	14	15	16	17	18	19	20	21	22	23	24
2	0.49	0.58	0.17	0.13	0.10	0.08	0.07	0.06	0.05	0.04	0.04	0.03	0.03	0.02	0.02	0.02	0.02	0.01	0.01	0.01	0.01	0.01	0.01	0.01
4	0.49	0.59	0.66	0.71	0.27	0.21	0.16	0.14	0.11	0.10	0.08	0.07	0.06	0.06	0.05	0.04	0.04	0.03	0.03	0.03	0.02	0.02	0.02	0.01
6	0.50	0.60	0.67	0.72	0.76	0.79	0.34	0.26	0.21	0.18	0.15	0.13	0.11	0.10	0.08	0.07	0.06	0.06	0.05	0.04	0.04	0.03	0.03	0.03
8	0.51	0.61	0.67	0.72	0.76	0.80	0.82	0.84	0.38	0.30	0.25	0.21	0.18	0.15	0.13	0.12	0.10	0.09	0.08	0.07	0.06	0.05	0.05	0.04
10	0.53	0.62	0.69	0.74	0.77	0.80	0.83	0.85	0.87	0.89	0.42	0.34	0.28	0.23	0.20	0.17	0.15	0.13	0.11	0.10	0.09	0.08	0.07	0.06
12	0.55	0.64	0.70	0.75	0.79	0.81	0.84	0.86	0.88	0.89	0.91	0.92	0.45	0.36	0.30	0.25	0.21	0.19	0.16	0.14	0.12	0.11	0.09	0.08
14	0.58	0.66	0.72	0.77	0.80	0.83	0.85	0.87	0.89	0.90	0.91	0.92	0.93	0.94	0.47	0.38	0.31	0.26	0.23	0.20	0.17	0.15	0.13	0.11
16	0.62	0.70	0.75	0.79	0.82	0.85	0.87	0.88	0.90	0.91	0.92	0.93	0.94	0.95	0.95	0.96	0.49	0.39	0.33	0.28	0.24	0.20	0.18	0.16
18	0.66	0.74	0.79	0.82	0.85	0.87	0.89	0.90	0.92	0.93	0.94	0.94	0.95	0.96	0.96	0.97	0.97	0.97	0.50	0.40	0.33	0.28	0.24	0.21

Source: Reprinted by permission from *ASHRAE Handbook, Fundamentals Volume*, 1985.

Table 8-18 COOLING LOAD FACTORS

"a" Classification	"b" Classification	0	1	2	3	4	5	6	7	8	Number of 9
0.45	A	0.03	0.47	0.58	0.66	0.73	0.78	0.82	0.86	0.88	0.91
	B	0.10	0.54	0.59	0.63	0.66	0.70	0.73	0.76	0.78	0.80
	C	0.15	0.59	0.61	0.64	0.66	0.68	0.70	0.72	0.73	0.75
	D	0.18	0.62	0.63	0.64	0.66	0.67	0.68	0.69	0.69	0.70
0.55	A	0.02	0.57	0.65	0.72	0.78	0.82	0.85	0.88	0.91	0.92
	B	0.08	0.62	0.66	0.69	0.73	0.75	0.78	0.80	0.82	0.84
	C	0.12	0.66	0.68	0.70	0.72	0.74	0.75	0.77	0.78	0.79
	D	0.15	0.69	0.70	0.71	0.72	0.73	0.73	0.74	0.75	0.76
0.65	A	0.02	0.66	0.73	0.78	0.83	0.86	0.89	0.91	0.93	0.94
	B	0.06	0.71	0.74	0.76	0.79	0.81	0.83	0.84	0.86	0.87
	C	0.09	0.74	0.75	0.77	0.78	0.80	0.81	0.82	0.83	0.84
	D	0.11	0.76	0.77	0.77	0.78	0.79	0.79	0.80	0.81	0.81
0.75	A	0.01	0.76	0.81	0.84	0.88	0.90	0.92	0.93	0.95	0.96
	B	0.04	0.79	0.81	0.83	0.85	0.86	0.88	0.89	0.90	0.91
	C	0.07	0.81	0.82	0.83	0.84	0.85	0.86	0.87	0.88	0.89
	D	0.08	0.83	0.83	0.84	0.84	0.85	0.85	0.86	0.86	0.87

Source: Reprinted by permission from *ASHRAE Handbook, Fundamentals Volume,* 1985.

considered because the heat gain to the air may differ significantly from the power supplied to the lights. Some of the energy emitted by the lights is in the form of radiation that is absorbed in the space. The absorbed energy is later transferred to the air by convection. The manner in which the lights are installed, the type of air distribution system, and the mass of the structure are important. Obviously a recessed light fixture will tend to transfer heat to the surrounding structure, whereas a hanging fixture will transfer more heat directly to the air. Some light fixtures are designed so that air returns through them, absorbing heat that would otherwise go into the space. Lights are often turned off and on to save energy, which makes accurate computations difficult. Lights left on 24 hours a day approach an equilibrium condition where the cooling load equals the power input.

The designer should be careful to use the correct wattage in making calculations. It is not good practice to assume nominal wattage per unit area. Additionally, all of the installed lights may not be used all the time. The instantaneous heat gain for lights may be expressed as

$$q_i = 3.412 \ W \ F_u F_s \tag{8-17}$$

where

W = summation of all installed light wattage, watts

WHEN LIGHTS ARE ON FOR 10 HOURS

Hours after Lights Are Turned On

10	11	12	13	14	15	16	17	18	19	20	21	22	23
0.93	0.49	0.39	0.32	0.26	0.21	0.17	0.13	0.11	0.09	0.07	0.06	0.05	0.04
0.82	0.39	0.35	0.32	0.28	0.26	0.23	0.21	0.19	0.17	0.15	0.14	0.12	0.11
0.76	0.33	0.31	0.29	0.27	0.26	0.24	0.23	0.21	0.20	0.19	0.18	0.17	0.16
0.71	0.27	0.26	0.26	0.25	0.24	0.23	0.23	0.22	0.21	0.21	0.20	0.19	0.19
0.94	0.40	0.32	0.26	0.21	0.17	0.14	0.11	0.09	0.07	0.06	0.05	0.04	0.03
0.85	0.32	0.29	0.26	0.23	0.21	0.19	0.17	0.15	0.14	0.12	0.11	0.10	0.09
0.81	0.27	0.25	0.24	0.22	0.21	0.20	0.19	0.17	0.16	0.15	0.14	0.14	0.13
0.76	0.22	0.22	0.21	0.20	0.20	0.19	0.18	0.18	0.17	0.17	0.16	0.16	0.15
0.95	0.31	0.25	0.20	0.16	0.13	0.11	0.08	0.07	0.05	0.04	0.04	0.03	0.02
0.89	0.25	0.22	0.20	0.18	0.16	0.15	0.13	0.12	0.11	0.10	0.09	0.08	0.07
0.85	0.21	0.20	0.18	0.17	0.16	0.15	0.14	0.14	0.13	0.12	0.11	0.11	0.10
0.82	0.17	0.17	0.16	0.16	0.15	0.15	0.14	0.14	0.14	0.13	0.13	0.12	0.12
0.97	0.22	0.18	0.14	0.12	0.09	0.08	0.06	0.05	0.04	0.03	0.03	0.02	0.02
0.92	0.18	0.16	0.14	0.13	0.12	0.10	0.09	0.08	0.08	0.07	0.06	0.06	0.05
0.89	0.15	0.14	0.13	0.12	0.12	0.11	0.10	0.10	0.09	0.09	0.08	0.08	0.07
0.87	0.12	0.12	0.12	0.11	0.11	0.11	0.10	0.10	0.10	0.09	0.09	0.09	0.09

F_u = use factor—ratio of wattage in use to that installed

F_s = special allowance factor for lights requiring more power than their rated wattage. For typical 40 W fluorescent lamps, $F_s = 1.20$

The cooling load is then given by

$$\dot{q} = \dot{q}_i(\text{CLF}) \qquad (8\text{-}18)$$

The cooling load factor is a function of the building mass, air-circulation rate, type of fixture, and time. Table 8-18 gives cooling load factors as a function of time for lights that are on for 10 hours. Data for lights on 8, 12, and 16 hours are given in the *ASHRAE Handbook*. The "a" and "b" classifications are given in Tables 8-19 and 8-20. The "a" classification depends on the nature of the light fixture, the return-air system, and the type of furnishings, whereas the "b" classification depends on the construction of the building and the type of supply and return-air system. The CLF values of Table 8-18 are for the case of continuously operating cooling equipment. If the cooling equipment is turned on and off on the same schedule as the lights, the CLF is equal to 1.0 because this condition is similar to the case when lights and cooling equipment are on continuously. When the cooling equipment is operated for a few hours after the lights are used, however, the system behaves as if the cooling equipment were operating continuously.

Table 8-19   DESIGN VALUES OF THE "a" COEFFICIENT

"a"	Furnishings	Air Supply and Return	Type of Light Fixture
0.45	Heavyweight simple furnishings, no carpet	Low rate; supply and return below ceiling $(\dot{Q} \leq 0.5)^a$	Recessed, not vented
0.55	Ordinary furniture, no carpet	Medium to high ventilation rate; supply and return below ceiling or through ceiling grill and space $(\dot{Q} \geq 0.5)^a$	Recessed, not vented
0.65	Ordinary furniture, with or without carpet	Medium to high ventilation rate or fan coil or induction type air-conditioning terminal unit; supply through ceiling or wall diffuser; return around light fixtures and through ceiling space. $(\dot{Q} \geq 0.5)^a$	Vented
0.75 or greater	Any type of furniture	Ducted returns through light fixtures	Vented or free-hanging in airstream with ducted returns

$^a\dot{Q}$ is room air supply rate in cfm/ft^2 of floor area.

Source: Reprinted by permission from *ASHRAE Handbook, Fundamentals Volume,* 1985.

In the case of light fixtures that are cooled by the return air (vented fixtures or recessed fixtures in a ceiling return plenum), the cooling load given by Eq. 8-18 using the specified "a" coefficient is not all imposed on the space. The return air carries part of the load directly back to the coil. The amount of load absorbed by the return air is a function of many variables but can be up to 50 percent of the light power.

Example 8-6.   The office suite of Example 8-5 has total installed light wattage of 8400 W. The fluorescent light fixtures are recessed with 40-W lamps. Supply air is through the ceiling with the air returning through the ceiling plenum. The lights are turned on at 8:00 A.M. and turned off at 6:00 P.M. Estimate the cooling load at 10:00 A.M. and 4:00 P.M. The floor is 6-in. concrete.

Table 8-20   THE "b" CLASSIFICATION FOR LIGHTS[a]

Room Air Circulation and Type of Supply and Return	Floor Construction and Floor Weight in Pounds per Square Foot of Floor Area				
	2-in. Wooden Floor (10) lb/ft²	3-in. Concrete Floor (40) lb/ft²	6-in. Concrete Floor (75) lb/ft²	8-in. Concrete Floor (120) lb/ft²	12-in. Concrete Floor (160) lb/ft²
Low ventilation rate— minimum required to handle cooling load. Supply through floor, wall, or ceiling diffuser. Ceiling space not vented.	B	B	C	D	D
Medium ventilation rate. Supply through floor, wall, or ceiling diffuser. Ceiling space not vented.	A	B	C	D	D
High room air circulation induced by primary air of induction unit or by fan coil unit. Return through ceiling space.	A	B	C	C	D
Very high room air circulation used to minimize room temperature gradients. Return through ceiling space.	A	A	B	C	D

[a]Based on floor covered with carpet and rubber pad. For floor covered with floor tile, use letter designation in next row down with the same floor weight.

Source: Reprinted by permission from *ASHRAE Handbook, Fundamentals Volume*, 1985.

Solution.   Assuming that about 15 percent of the lights are off all the time due to unoccupied offices, the use factor F_u is 0.85. The ballast factor F_s is 1.2. Then from Eq. 8-17

$$\dot{q}_i = 3.412(8400)0.85(1.2) = 29{,}234 \text{ Btu/hr}$$

The "a" and "b" classifications are 0.55 and C, respectively. Then from Table 8-18, CLF_{10} is 0.68 and CLF_{16} is 0.78. Then

$$\dot{q}_{10} = 29{,}234(0.68) = 19{,}880 \text{ Btu/hr}$$
$$\dot{q}_{15} = 29{,}234(0.78) = 22{,}800 \text{ Btu/hr}$$

Since the return air is flowing through the ceiling space over the light fixtures, some of the cooling load due to the lights never reaches the space.

Let us assume a 20 percent reduction in the load to the space. Then

$$\dot{q}_{10} = (1 - 0.2)\ 19{,}880 = 15{,}900\ \text{Btu/hr}$$
$$\dot{q}_{15} = (1 - 0.2)\ 22{,}800 = 18{,}240\ \text{Btu/hr}$$

One must remember, however, that the 20 percent reduction in space load must be added to the coil load.

Miscellaneous Equipment

There is an infinite variety of equipment and appliances that will provide a heat gain when installed in the conditioned space. If the equipment is completely enclosed, the heat gain equals the input to the equipment. However, this does not always happen. An electric motor may be within the space, but the device powered by the motor may be outside it. In this case the heat gain results from the inefficiency of the motor and will be only a small percentage of the rated power of the motor. Kitchen appliances usually have a hood through which air is exhausted, thus reducing the heat gain appreciably. Remember, however, that the exhausted air must be replaced by outdoor air. The *ASHRAE Handbook* (1) gives extensive data for commercial cooking equipment. Just as with people and lights, care must be taken to properly estimate the actual periods of use, so that an unreasonably large heat gain is not obtained.

After the heat gain is determined, the cooling load is computed using a cooling load factor in a fashion identical to that for people. Table 8-21 gives CLF values for the case of unhooded equipment. Typical office equipment, small computers and motors, and small electric appliances fall in this category. Similar data are given in reference 1 for hooded appliances that are mainly commercial kitchen equipment.

Table 8-21   SENSIBLE HEAT COOLING LOAD

Total Operational hours	1	2	3	4	5	6	7	8	9	10	Hours 11
2	0.56	0.64	0.15	0.11	0.08	0.07	0.06	0.05	0.04	0.04	0.03
4	0.57	0.65	0.71	0.75	0.23	0.18	0.14	0.12	0.10	0.08	0.07
6	0.57	0.65	0.71	0.76	0.79	0.82	0.29	0.22	0.18	0.15	0.13
8	0.58	0.66	0.72	0.76	0.80	0.82	0.85	0.87	0.33	0.26	0.21
10	0.60	0.68	0.73	0.77	0.81	0.83	0.85	0.87	0.89	0.90	0.36
12	0.62	0.69	0.75	0.79	0.82	0.84	0.86	0.88	0.89	0.91	0.92
14	0.64	0.71	0.76	0.80	0.83	0.85	0.87	0.89	0.90	0.92	0.93
16	0.67	0.74	0.79	0.82	0.85	0.87	0.89	0.90	0.91	0.92	0.93
18	0.71	0.78	0.82	0.85	0.87	0.89	0.90	0.92	0.93	0.94	0.94

Source: Reprinted by permission from *ASHRAE Handbook Fundamentals Volume*, 1985.

Example 8-7.   Suppose the office suite previously described has a collection of typewriters, duplicating machines, electric coffee pot, and the like, with a total nameplate rating of 3.7 kW. It is estimated that only about one-half of the equipment is in use on a continuous basis. Estimate the cooling load at 5:00 P.M.

Solution.   Assume the equipment is turned on at 8:00 A.M. and operates until 4:00 P.M. From Table 8-21 the CLF is 0.33 because 5:00 P.M. is 9 hours after startup time. The cooling load is

$$\dot{q} = 3.7(3412)(0.33)/2 = 2080 \text{ Btu/hr}$$

It should be noted that all the equipment in a space does not have to be lumped together to make a calculation. Equipment can be started and stopped at different times and separate calculations made for each.

Fans

The fans that move the air through the air distribution system are a source of cooling load. Depending on the fan location, the load will be imposed on the space or directly on the coil. In the case of a draw-through air handler, it is usually assumed that all the shaft power input is a part of the space cooling load. Remember that the fan motor may or may not be in the air stream. The load due to the return air fan and fans in blow-through air handlers is usually assumed to be imposed on the coil. Since the fan capacity and total pressure are not known at the time cooling load calculations are being made, assumed values will have to be used initially and checked later in the design process.

FACTORS FOR APPLIANCES—UNHOODED

Appliances Are On

12	13	14	15	16	17	18	19	20	21	22	23	24
0.03	0.02	0.02	0.02	0.02	0.01	0.01	0.01	0.01	0.01	0.01	0.01	0.01
0.06	0.05	0.05	0.04	0.04	0.03	0.03	0.02	0.02	0.02	0.02	0.01	0.01
0.11	0.10	0.08	0.07	0.06	0.06	0.05	0.04	0.04	0.03	0.03	0.03	0.02
0.18	0.15	0.13	0.11	0.10	0.09	0.08	0.07	0.06	0.05	0.04	0.04	0.03
0.29	0.24	0.20	0.17	0.15	0.13	0.11	0.10	0.08	0.07	0.07	0.06	0.05
0.93	0.38	0.31	0.25	0.21	0.18	0.16	0.14	0.12	0.11	0.09	0.08	0.07
0.93	0.94	0.95	0.40	0.32	0.27	0.23	0.19	0.17	0.15	0.13	0.11	0.10
0.94	0.95	0.96	0.96	0.97	0.42	0.34	0.28	0.24	0.20	0.18	0.15	0.13
0.95	0.96	0.96	0.97	0.97	0.97	0.98	0.43	0.35	0.29	0.24	0.21	0.18

Duct Heat Gain

Strictly speaking, heat gain to the air while it is moving through ducts is not an internal component; however, any heat transfer to the air in the supply system is a cooling load imposed on the space. Heat transfer to the air in the return system is additional load imposed on the central cooling unit. These load components are estimated as discussed in Chapter 7; however, extra attention must be given to thermal radiation effects when computing cooling load.

8-6 HEAT GAIN FROM INFILTRATION AND OUTDOOR VENTILATION AIR

The necessity of introducing outdoor air was discussed in Chapter 4 and current recommendations were given. Outdoor air should be held to a minimum consistent with the health and comfort of the occupants and energy conservation considerations. The outdoor air should be reduced to allow for the expected infiltration. Computation of the heat gain from outdoor make up air is properly a part of the air-quantity calculation as discussed in Chapter 3; however, it may be computed separately to obtain an estimate of the total load. In such a case the method given later for infiltration may be utilized.

The methods used to estimate the quantity of infiltration air were discussed in Chapter 7 when the heating load was considered. The same methods apply to heat-gain calculations. In this case both a sensible and latent heat gain will result and are computed as follows:

$$\dot{q}_s = \dot{m}_a c_p (t_o - t_i) = \frac{\dot{Q}_o c_p}{v_o} (t_o - t_i) \tag{8-19}$$

$$\dot{q}_l = \dot{m}_a (W_o - W_i) i_{fg} = \frac{\dot{Q}_o}{v_o} (W_o - W_i) i_{fg} \tag{8-20}$$

These heat gains may also be computed using the psychrometric chart as described in Chapter 3. Wind velocities are lower in the summer, making an appreciable difference in the computed infiltration rate.

The direction of the prevailing winds usually changes from winter to summer. This should be considered in making infiltration estimates because the load will be imposed mainly in the space where the air enters. It should also be remembered that during the summer, infiltration will enter the upper floors of high-rise buildings instead of the lower floors.

8-7 HEAT GAIN AND COOLING LOAD SUMMARY

It is important to adopt a systematic approach to the calculation and summation of the cooling load for a structure. If this is not done, some important parts of the problem may be overlooked. The computer is exceptionally good for this purpose. However, many jobs are too small to justify the use of a computer. In these situations a calculation form or worksheet is a necessity. Figure 8-6 is an example of such a worksheet designed to use the CLTD method. The worksheet has a number of advantages over a page-by-page analysis of the problem. A worksheet provides a concise record of the load estimate that can be easily checked, and a person without engineering training can be taught to make the calculation with a limited amount of supervision.

The calculations discussed so far pertain to the load estimates needed for a large commercial building that is divided into many zones with separate temperature control in each. Each zone usually has different outside exposures and internal load patterns. Therefore, as explained earlier, the peak load for each zone and the peak coincident load for all zones served by a central unit are required and hour-by-hour loads are helpful to have. A variable-volume system is an example of a system requiring such load estimates.

At the other end of the spectrum of building and system types are small detached buildings with one central, constant-flow system with one thermostat to control the temperature in all spaces. This is similar to a single-family detached house. Recent studies (4) have shown that buildings and systems of this type require design loads different from the peak values mentioned earlier. The reason for this is related to the design of the air-distribution system. For a small detached building with one central system, the calculation procedure is modified as follows:

1. Exterior walls. Use a daily average CLTD from Table 8-2.
2. Roof. Use an average CLTD for the sunlit hours of the day from Table 8-1.
3. Glass. Use an average CLF for the sunlit hours of the day from Table 8-14 or 8-15.
4. People, lights, and equipment. Compute as usual, assuming they peak in the late afternoon or when turned off.

Room-by-room calculations will then lead to a much improved proportioning of the supply air to each space, while the block load for the building will be about the same as the peak coincident load calculated on an hour-by-hour basis.

There are of course many buildings and systems that fall between the two cases discussed. The design engineer may need to be somewhat artful in using the results of his load calculations.

The next steps are to determine the air quantities and to select the equipment. These steps may be reversed depending on the type of equipment to be utilized.

8-8 AIR QUANTITIES

The preferred method of computing air quantity for cooling and dehumidification has been described in section 3-6 of Chapter 3. That method should always be used when the conditions and size of the cooling load warrant specification of special equipment. This means that the cooling and dehumidifying

				CLTD or SHGF			CLF or Δt			Area, cfm, etc.	Heat Load Btuh	Cooling Load–Btuh		
1	Identification of Room					1								
2	Length of Exposed Wall – Feet													
3	Room Dimensions – Feet													
4	Comments													
	Load Component	Direction	U or SC		Hour			Hour						Hour	
5	Gross Exposed Walls	N													
		S													
		E													
		W													
6	Windows and Doors– Convective	N													
		S													
		E													
		W													
7	Windows and Glass Doors– Solar	N													
		S													
		E													
		W													
8	Net Exposed Walls and Partitions	N													
		S													
		E													
		W													
9	Roof														
10	Floors														
11	Infiltration, Sensible		1.1												
12	Infiltration, Latent		4840												
13	People, Sensible														
14	People, Latent														
15	Internal, Sensible														
16	Internal, Latent														
17	Lights														
18	Duct Loss														
19	Total Btuh														

Figure 8-6. Cooling load worksheet.

2		Cooling Load–Btuh			3		Cooling Load–Btuh		
Area, cfm, etc.	Heat Load Btuh				Area, cfm, etc.	Heat Load Btuh	Hour		

coil is designed to match the sensible and latent heat requirements of a particular job and that the fan is sized to handle the required volume of air. The fan, cooling coil, control dampers, and the enclosure for these components are referred to as an *air handler*. These units are assembled at the factory in a wide variety of coil and fan models to suit almost any requirement. The design engineer usually specifies the entering and leaving moist-air conditions, the volume flow rate of the air, and the total pressure the fan must produce.

Specially constructed equipment cannot be justified for small commercial applications. Furthermore, these applications generally have a higher sensible heat factor, and dehumidification is not as critical as it is in large commercial buildings. Therefore, the equipment is manufactured to operate at or near one particular set of conditions. For example, typical light commercial unitary cooling equipment operates with a coil SHF of 0.75 to 0.8 with the air entering the coil at about 80 F or 27 C dry bulb and 67 F or 19 C wet bulb temperature. This equipment usually has a capacity of less than about 20 tons or 70 kW. When the peak cooling load and latent heat requirements are appropriate, this less expensive type of equipment may be used. In this case the air quantity is determined in a different way. The unit is first selected on the basis of the block sensible cooling load using the nearest available size but not less than the cooling load. Next, the latent capacity of the unit must be equal to or greater than the computed latent cooling load. This procedure assures that the unit will handle both the sensible and latent load even though an exact match does not exist. The air quantity is specified by the manufacturer for each unit and is about 400 cfm/ton or 0.0537 m³/(s-kW). The total air quantity is then divided among the various rooms according to the cooling load of each room. Table 8-22 shows performance data for some light commercial cooling equipment.

8-9 ENERGY CALCULATION BY THE BIN METHOD

The bin method of computing energy requirements was introduced in Chapter 7, when the heating load was considered. That method is equally valid for computing the energy to cool a building. However, because of solar effects a more refined load profile is desirable. A procedure will now be outlined to obtain a more realistic load profile. Reference 1 may be consulted for full details.

It is important that a building be separated into zones when part of the building requires heat while the remainder requires cooling. If the zones are combined, the heating and cooling requirements will cancel, a solution that may not be representative of the actual situation. The components of the load profile are

Solar gain through walls and roof
Solar gain through glass
Conduction gain through walls and roof
Conduction gain through glass
Lights

Table 8-22 PERFORMANCE DATA FOR SOME UNITARY AIR-CONDITIONING EQUIPMENT[a]

Outdoor Dry Bulb Temp. F	7½ Ton, 3000 cfm			10 Ton, 4000 cfm			15 Ton, 6000 cfm			20 Ton, 8000 cfm		
	Total Cap. Btu/hr × 1000	Sensible Cap. Btu/hr × 1000	EER[b] Btu/W-hr	Total Cap. Btu/hr × 1000	Sensible Cap. Btu/hr × 1000	EER[b] Btu/W-hr	Total Cap. Btu/hr × 1000	Sensible Cap. Btu/hr × 1000	EER[b] Btu/W-hr	Total Cap. Btu/hr × 1000	Sensible Cap. Btu/hr × 1000	EER[b] Btu/W-hr
75	101.2	68.4	11.0	133.4	92.4	10.6	206.8	131.5	10.4	272.5	193.0	10.4
85	96.2	66.8	9.8	126.7	89.8	9.5	195.5	127.5	9.3	257.7	187.3	9.2
95	91.0	64.8	8.7	120.0	87.4	8.5	184.0	123.4	8.3	244.0	182.0	8.2
105	85.2	62.7	7.6	112.8	84.7	7.5	171.1	118.9	7.2	229.0	176.3	7.3
115	79.5	60.7	6.7	105.5	82.1	6.6	158.2	114.5	6.2	214.1	170.7	6.4

[a]Indoor coil operating at 75 F db and 62 F wb.
[b]EER includes all fan power.

People
Equipment
Infiltration

Development of each load component as a linear function of outdoor temperature permits combination of the components into a total load profile for winter and summer conditions and determination of the balance points.

The CLTD data used for design load calculations are useful in developing the load components for walls and roofs. In order to take into account the fact that the sky is not clear all the time, it is necessary to separate the solar component from the conduction component in the CLTD. The corrected CLTD may be expressed as

$$\text{CLTD}_c = (t_o - t_i) + \text{ESTD} \tag{8-21}$$

where

ESTD = equivalent solar temperature difference, F or C
t_o = outdoor temperature, F or C
t_i = indoor temperature, F or C

now

$$\text{CLTD}_c = \text{CLTD} + (78 - t_i) + t_d - \left(\frac{DR}{2}\right) - 85 \tag{8-13}$$

where the CLTD is from Tables 8-1 or 8-2 and t_d is the outdoor design temperature. Solving for ESTD from Eq. 8-21 and (8-13)

$$\text{ESTD} = \text{CLTD} - 7 - \left(\frac{DR}{2}\right) + t_d - t_o \tag{8-22}$$

It may be shown that

$$t_d = \bar{t}_o + \frac{DR}{2} \tag{8-23}$$

where

$$t_o = \sum_{i=1}^{24} \frac{t_{oi}}{24} \tag{8-24}$$

then

$$\text{ESTD} = \text{CLTD} - 7 + \bar{t}_o - t_o \tag{8-22a}$$

and the average ESTD for a day is

$$\overline{\text{ESTD}} = \left[\sum_{i=1}^{24} \frac{\text{CLTD}_i}{24}\right] - 7 \tag{8-25}$$

The average solar transmission component for a wall or roof is then given by

$$\dot{q}_{sn} = U_n A_n \times \overline{\text{ESTD}_n} \times FPS \tag{8-26}$$

where n refers to the particular wall or roof being considered and *FPS* is the fraction of possible sunshine from Table 8-25, which accounts for cloudiness. The conduction component is given by

$$\dot{q}_c = U_n A_n (t_o - t_i) \tag{8-27}$$

In this case orientation of the surface is not a factor but the various walls and roofs may vary in composition. Equation 8-27 also gives the conduction component for glass and doors. When the $CLTD_i$ in Eq. 8-25 are averaged, the same value results for all walls and roofs. Therefore, a simplified table may be constructed, Table 8-23.

The average solar load component for glass is given by

$$\dot{q}_{sg} = \frac{\displaystyle\sum_{i=1}^{n} (\mathrm{SHGF}_i \times A_i \times \mathrm{SC}_i \times \mathrm{TCLF}_i \times FPS)}{\theta} \tag{8-28}$$

where

SHGF_i = maximum solar heat gain factor from Table 8-12, Btu/(hr-ft^2)
A_i = area of the glass, ft^2

Table 8-23   THE 24-HOUR AVERAGE SOLAR COMPONENT, ESTD

JULY

Direction				Latitude		
	16	24	32	40	48	56
North	19	6	6	5	5	5
Northeast	14	13	11	11	11	11
East	14	0	15	15	16	17
Southeast	9	11	13	14	17	19
South	3	4	7	10	14	17
Southwest	9	11	13	14	17	19
West	14	0	15	15	16	17
Northwest	14	13	12	11	11	11
Roof	22	23	23	22	22	20

JANUARY

Direction				Latitude		
North	1	1	0	0	−1	−1
Northeast	4	3	2	1	0	0
East	11	9	7	6	4	1
Southeast	18	17	16	15	14	8
South	22	23	26	21	18	12
Southwest	18	17	16	15	9	8
West	11	9	7	6	4	1
Northwest	4	3	2	1	0	0
Roof	15	11	7	3	−2	−5

Table 8-24   COOLING LOAD FACTORS, GLASS WITH INTERIOR SHADE

Direction	CLF Sum	12-Hour Average	24-Hour Average
North	11.57	0.96	0.48
Northeast	5.15	0.43	0.22
East	5.46	0.46	0.23
Southeast	6.20	0.52	0.26
South	6.43	0.54	0.27
Southwest	6.23	0.52	0.26
West	5.46	0.46	0.23
Northwest	5.15	0.43	0.22
Horizontal	8.22	0.69	0.34

$TCLF_i$ = total of the 24 cooling load factors from Table 8-14 or 8-15, Table 8-24

FPS = fraction of the possible sunshine for the location, Table 8-25

SC = classical shading coefficient, Tables 8-9, 8-10, 8-11

Θ = run time of the system, hr

In summer the actual run time for the system should be used. In winter, however, a significant load occurs at night and a run time of 24 hours is appropriate. Outside shading may have a significant effect on the energy requirements of a building. Procedures are available to estimate the average shaded area for windows with overhangs and side projections (1); however, for brevity this information is not presented here.

The internal loads may constitute a significant contribution to the total load and are important in determining the balance point. In all cases of lights, people, and equipment it is necessary to determine the average heat gain during the period under consideration. It may be necessary to plot the occupancy or lighting schedule to find a realistic average value. The internal loads are not a function of the outdoor temperature; therefore, these load components are

Table 8-25   PERCENTAGE OF POSSIBLE SUNSHINE FOR SELECTED LOCATIONS

City & State	January	July	Annual
Chicago, IL	44	73	59
Denver, CO	67	68	67
Oklahoma City, OK	57	76	68
Washington, DC	46	64	58

Source: From U.S. Weather Bureau publication. *Local Climatological Data.*

constants. Note that the occupants have a latent component that should be computed separately.

The load component due to infiltration and outdoor ventilation air is a function of outdoor dry bulb and wet bulb temperatures given by

$$\dot{q}_{vs} = \dot{Q}\rho(60)(t_o - t_i) \tag{8-29}$$

and

$$\dot{q}_{v1} = \dot{Q}\rho(60)i_{fg}(W_o - W_i) \tag{8-30}$$

where $\dot{Q}$ is in cfm, the average amount of air entering the space during the period of interest. The mean coincident wet bulb temperature should be used in evaluating W_o in Eq. 8-30, whereas W_i depends on the average indoor conditions.

The general procedure has been outlined to obtain the various components of the total load profile. Recall that most buildings operate on at least two shifts, occupied and unoccupied, and some buildings may require more than one zone. July and January usually represent the warmest and coldest months of the year in the northern hemisphere. Therefore, the usual practice is to find points on the load profile at the design conditions in January and July. An equation is then found between the two points in the form

$$\dot{q} = at_o + bt_i + c \tag{8-31}$$

Using Eq. 8-31 with $\dot{q} = 0$ and t_i known, the outdoor temperature at which the system balances may be computed. The following example will demonstrate the complete procedure.

Glass: Single pane with blinds, 40 percent of wall area, evenly distributed
Walls: $U = 0.15$ Btu/hr-ft^2-F
Roof: $U = 0.07$ Btu/hr-ft^2-F
Lights: 11.5 kW, day wd 0.12 kW, night
People: 15, day; 0, night
Equipment: 4 kW, day; 1 kW, night
Infiltration: 600 c fm, day; 300 c fm, night
Cooling equipment: Table 8-22, 15 ton,
Degradation coeff. = 0.25,
continuously operable

Figure 8-7.   Building schematic for Example 8-8.

Example 8-8. A small single-story building located in Oklahoma City, Oklahoma, is described in Fig. 8-7. The indoor operating temperatures are 75 F and 72 F for summer and winter, respectively, and there is an occupied (8:00 A.M.–7:00 P.M.) and unoccupied (7:00 P.M.–8:00 A.M. and weekends) shift. Compute the energy to operate the system, assuming the times given are solar times.

Solution. Table C-1 gives the summer and winter design temperatures as 96 F and 13 F, respectively. These temperatures will be used in determining the load profile. Let us first consider the solar component through the glass using Eq. 8-28.

Solar Through Glass, July

Surface	MSHGF	A	SC	TCLF	FPS	$\dot{q}_n$
N	36	600	0.55	11.57	0.78	107,212
E	218	200	0.55	5.46	0.78	102,126
S	131	600	0.55	6.43	0.78	216,816
W	218	200	0.55	5.46	0.78	102,126
					Total:	528,280 Btu/hr

Since the cooling equipment is operable all the time, $\theta = 24$ hr. It is also convenient to reference each component to one square foot of floor area. Then

$$(\dot{q}_{sg})_{July} = 528,280/(24)(7,500) = 2.935 \text{ Btu/(hr-ft}^2)$$

The same procedure is used for January.

Solar Through Glass, January

Surface	MSHGF	A	SC	TCLF	FPS	$\dot{q}_n$
N	22	600	0.55	11.57	0.57	47,879
E	166	200	0.55	5.46	0.57	56,829
S	252	600	0.55	6.43	0.57	304,790
W	166	200	0.55	5.46	0.57	56,829
						466,327 Btu/hr

$$(\dot{q}_{sg})_{Jan.} = 466,327/(24)(7500) = 2.591 \text{ Btu/(hr-ft}^2)$$

Now to obtain the profile for this component the design temperatures are used to derive a linear equation as follows:

$$\dot{q}_{sg} = at_o + b$$

or

$$2.935 = 96a + b$$

and

$$2.591 = 13a + b$$

from which

$$a = 0.0042 \quad \text{and} \quad b = 2.54$$

and

$$\dot{q}_{sg} = (0.0042t_o + 2.54) \text{ Btu/(hr-ft}^2) \tag{8-32}$$

The same general procedure is used for the solar component through the walls and roof using Eq. 8-26.

Solar Through Walls and Roof, July

Surface	*A*	*U*	ESTD	FPS	$\dot{q}_{sw}$
N	900	0.15	5	0.78	526.5
E	300	0.15	15	0.78	526.5
S	900	0.15	9	0.78	947.7
W	300	0.15	15	0.78	526.5
Roof	7500	0.07	22	0.78	9,009.0
					11,536.2

$$(\dot{q}_{sw})_{\text{July}} = 11{,}536/7500 = 1.538 \text{ Btu/(hr-ft}^2)$$

Solar Through Walls and Roof, January

Surface	*A*	*U*	ESTD	FPS	$\dot{q}_{sw}$
N	900	0.15	0	0.57	0
E	300	0.15	6	0.57	154
S	900	0.15	21	0.57	1,616
W	300	0.15	6	0.57	154
Roof	7500	0.07	3	0.57	898
				Total	2,822

$$(\dot{q}_{sw})_{\text{Jan.}} = 2822/7500 = 0.376 \text{ Btu/(hr-ft}^2)$$

Using the same procedure as before, the profile for this component becomes

$$\dot{q}_{sw} = 0.014t_o + 0.194 \text{ Btu/(hr-ft}^2) \tag{8-33}$$

Equation 8-27 will be used to find the load component for conduction.

Conduction for Walls, Roof, Glass

Surface	*A*	*U*	*A* × *U*
Walls	2400	0.15	360
Roof	7500	0.07	525
Glass	1600	1.00	1600
			2485 Btu/(hr-F)

$$\dot{q}_{cw} = 2,485(t_o - t_i)/7500$$

or

$$\dot{q}_{cw} = (0.331t_o - 0.331t_i) \; \text{Btu/(hr-ft}^2) \tag{8-34}$$

The load components considered so far are independent of the shift. The internal loads and infiltration to be considered next do depend on the shift.

Internal Loads, Assumed Constant the Year Round

$$\text{Lights: } \dot{q}_{1d} = 11.5(3412)/7500 = 5.23 \; \text{Btu/(hr-ft}^2) \tag{8-35}$$

$$\dot{q}_{1n} = 0.12(3412)/7500 = 0.055 \; \text{Btu/(hr-ft}^2) \tag{8-36}$$

$$\text{Equipment: } \dot{q}_{ed} = 4(3412)/7500 = 1.82 \; \text{Btu/(hr-ft}^2) \tag{8-37}$$

$$\dot{q}_{en} = 1(3412)/7500 = 0.45 \; \text{Btu/(hr-ft}^2) \tag{8-38}$$

$$\text{Occupants, sensible: } \dot{q}_{od} = 15(230)/7500 = 0.46 \; \text{Btu/(hr-ft}^2) \tag{8-39}$$
$$\dot{q}_{on} = 0.0$$

$$\text{Occupants, latent: } \dot{q}_{od} = 15(230)/7500 = 0.46 \; \text{Btu/(hr-ft}^2) \tag{8-40}$$
$$\dot{q}_{on} = 0.0$$

Infiltration (assume standard air)

$$\text{Day, sensible: } \dot{q}_{vs} = 1.1(600)(t_o - t_i)/7500 \tag{8-41}$$
$$= 0.088t_o - 0.088t_i$$

$$\text{Day, latent: } \dot{q}_{vl} = 4840(600)(0.015 - 0.009) \tag{8-42}$$
$$= 2.90 \; \text{Btu/(hr-ft}^2)$$

$$\text{Night, sensible } \dot{q}_{vs} = 1.1(300)(t_o - t_i)/7500 \tag{8-42}$$
$$= 0.044t_o - 0.044t_i \; \text{Btu/(hr-ft}^2)$$

$$\text{Night, latent } \dot{q}_{vl} = (4840)(300)(0.015 - 0.009)/7500 \tag{8-44}$$
$$= 1.45 \; \text{Btu/(hr-ft}^2)$$

Infiltration would probably be lower during the summer months than in the winter; however, it will be assumed constant in this case. The internal and infiltration loads may be summarized as follows: For the day shift sum Eqs. 8-35, 8-37, 8-39, and 8-41 to obtain the sensible component

$$\dot{q}_{is.d} = 5.23 + 1.82 + 0.46 + 0.088t_o - 0.088t_i \tag{8-45}$$
$$\dot{q}_{is.d} = (0.088t_o - 0.88t_i + 7.51) \; \text{Btu/(hr-ft}^2)$$

The latent component for the day shift is the sum of Eqs. 8-40 and 8-42:

$$\dot{q}_{il.d} = 0.46 + 2.9 = 3.36 \; \text{Btu/(hr-ft}^2) \tag{8-46}$$

The sensible component for the night shift is the sum of Eqs. 8-36, 8-38, and 8-43:

$$\dot{q}_{is,n} = 0.055 + 0.46 - 0.044t_o - 0.044t_i \qquad (8\text{-}47)$$
$$\dot{q}_{is,n} = (0.044t_o - 0.044t_i + 0.515) \text{ Btu/(hr-ft}^2) $$

The latent component for the night shift is given by Eq. 8-44:

$$\dot{q}_{il,n} = \dot{q}_{vl} = 1.45 \text{ Btu/(hr-ft}^2) \qquad (8\text{-}44)$$

The total sensible load profiles for each shift may now be determined. The latent components will be included after the balance points have been determined.

For the Day Shift

$\dot{q}_{sg} = 0.0042t_o$		$+ \ 2.54$	(8-32)
$\dot{q}_{sw} = 0.014t_o$		$+ \ 0.194$	(8-33)
$\dot{q}_{cw} = 0.331t_o$	$-0.331t_i$		(8-34)
$\dot{q}_{is} = 0.088t_o$	$-0.088t_i$	$+ \ 7.51$	(8-45)
$\dot{q}_{sd} = 0.4372t_o$	$-0.419t_i$	$+10.244$	(8-48)

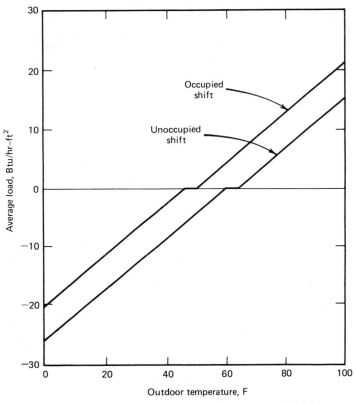

Figure 8-8.   Sensible-load profiles for Example 8-8.

For the night shift

$$\dot{q}_{sg} = 0.0042t_o \qquad\qquad\qquad\qquad + 2.54 \qquad (8\text{-}32)$$

$$\dot{q}_{sw} = 0.014t_o \qquad\qquad\qquad\qquad + 0.194 \qquad (8\text{-}33)$$

$$\dot{q}_{cw} = 0.331t_o \qquad -0.331t_i \qquad\qquad\qquad (8\text{-}34)$$

$$\dot{q}_{is} = 0.044t_o \qquad -0.044t_i \qquad + 0.515 \qquad (8\text{-}47)$$

$$\dot{q}_{sn} = 0.3932t_o \qquad -0.375t_i \qquad + 3.249 \qquad (8\text{-}49)$$

The balance point temperatures may now be computed from Eqs. 8-48 and 8-49. For the cooling season and the occupied (day) shift the indoor temperature is 75 F and for $\dot{q}_{sd} = 0$

$$t_o = [0.419(75) - 10.244]/0.4372 = 48.5 \text{ F}$$

and for the unoccupied (night) shift

$$t_o = [0.375(75) - 3.249]/0.3932 = 63.3 \text{ F}$$

These appear to be quite reasonable values. During the heating season the indoor temperature is 72 F. Using the same procedure the balance tem-

Table 8-26   BIN ENERGY CALCULATION FOR EXAMPLE 8-8

Bin Temperature	Occupied hours	Unoccupied hours	Occupied load Btu/hr	Unoccupied load Btu/hr	Cooling equipment capacity Btu/hr	Heating equipment capacity Btu/hr
1	2	3	4	5	6	6a
Table 7-6	Table 7-6	Table 7-6	Given Eq. 8-48	Given Eq. 8-49	Given Eq. 8-50	Given
102	1	1	200655	124920	174760	
97	64	40	184268	110183	180860	
92	179	117	167880	95445	186960	
87	229	178	151493	80708	193060	
82	280	338	135105	65970	199160	
77	244	532	118718	51233	205260	
72	264	545	102330	36495	211360	
67	202	545	85943	21758	217460	
62	216	426	69555	7020	223560	
57	212	389	53168	−10118	229660	200000
52	230	454	36780	−24855	235760	200000
47	166	403	20393	−39593	241860	200000
42	172	495	−11745	−54330		200000
37	156	465	−28133	−69068		200000
32	124	380	−44520	−83805		200000
27	54	175	−60908	−98543		200000
22	44	127	−77295	−113280		200000
17	21	74	−93683	−128018		200000
12	3	15	−110070	−142755		200000

peratures for the day and night shifts are 46 F and 60 F, respectively. The total sensible load profiles then appear as shown in Fig. 8-8. Note that a span of about 3 F occurs where neither heating nor cooling is required. Before proceeding with the bin calculation, the latent component should be added to Eqs. 8-48 and 8-49 for the cooling energy calculation. This latent component is usually omitted from the analysis for the heating energy. Equations 8-48 and 8-49, with $t_i = 75$ F, then become

$$\dot{q}_d = (0.437t_o - 17.82) \text{ Btu/(hr-ft}^2) \tag{8-48a}$$

and

$$\dot{q}_n = (0.393t_o - 23.43) \text{ Btu/(hr-ft}^2) \tag{8-49a}$$

These load profiles are superior to the simplified versions used in Example 7-9 of the previous chapter.

The bin analysis may now be carried out to determine the energy required. The procedure is identical to that of Example 7-9 and the occupied and unoccupied shifts are the same. Therefore, the bin hours for each shift may be taken from Table 7-6 for the bin temperatures of interest. The only

Occupied PLF	Unoccupied PLF	Occupied Run Time Hr	Unoccupied Run Time Hr	Cooling equipment Input kW	Heating Energy mcf	Cooling Energy kW-hr
7	8	9	10	11	12	13
Eq. 7-24	Eq. 7-24	$4 \times 2/(6 \times 7)$	$5 \times 3/(6 \times 8)$	Given	$0.267 \times$	$11 \times$
$Dc = 0.25$	$Dc = 0.25$			Eq. 8-51	$(9 + 10)$	$(9 + 10)$
1.00	0.93	1.1	0.8	23.38		44.8
1.00	0.90	64.9	27.0	22.68		2084.5
0.97	0.88	164.9	68.1	21.98		5121.3
0.95	0.85	189.9	87.1	21.28		5894.5
0.92	0.83	206.6	134.4	20.58		7017.5
0.89	0.81	157.8	163.4	19.88		6385.5
0.87	0.79	146.7	118.6	19.18		5090.0
0.85	0.78	94.1	70.4	18.48		3038.3
0.83	0.76	81.2	17.7	17.78		1757.3
0.81	0.76	60.8	25.8	17.08	6.9	1044.1
0.79	0.78	45.5	72.2	16.38	19.3	763.0
0.77	0.80	18.2	99.8	15.68	26.6	309.6
0.76	0.82	13.2	164.4		47.4	44.4
0.79	0.84	27.9	192.0		58.7	55.0
0.81	0.85	34.3	186.3		58.9	55.1
0.83	0.87	19.9	98.7		31.7	29.7
0.85	0.89	20.1	80.7		26.9	25.2
0.87	0.91	11.3	52.0		16.9	15.8
0.89	0.93	1.9	11.5		3.6	3.3
				Totals:	296.9	38779.0

required additional data are the air-conditioning equipment capacity and input as a function of outdoor temperature, which can be obtained from Table 8-22 for the 15-ton unit. A linear curve fit yields:

$$\text{Capacity} = (299{,}200 - 1220t_o) \text{ Btu/hr} \tag{8-50}$$

$$\text{Input} = (9.1 + 0.14t_o) \text{ kW} \tag{8-51}$$

A calculation sheet, by hand or computer, similar to Table 7-8, can now be constructed. To make a complete analysis, it will be assumed that a gas furnace with a steady-state output of 200,000 Btu/hr and efficiency of 75 percent is used to heat the building. The heating value of the natural gas is assumed to be 1000 Btu per standard cubic foot. Table 8-26 shows the results. Note that three bin temperatures occur when cooling is required during the day, whereas heating is required during the night hours. The negative loads are heating loads and the power shown in the last column includes fan energy when heating or cooling.

8-10   COOLING LOAD BY THE TRANSFER FUNCTION METHOD

The CLTD method for predicting the cooling load was discussed in the previous sections. This method is not as precise as one would like and does not supply sufficient data for detailed energy studies of large structures. Through use of the computer it is possible to solve the heat-gain–cooling-load problem, taking into account more variables to obtain solutions for every hour of the day with greater accuracy than with hand calculations. This approach also makes possible detailed studies of the energy requirements of a structure. The general procedure is referred to as the *transfer function method* and was developed by the ASHRAE Task Group on Energy Requirements (1).

A transfer function is a set of coefficients that relate an output function at some specific time to the value of one or more driving functions at that time and to previous values of both the input and output functions. For example, the heat transfer from the inside surface of a wall at a particular time is related through use of appropriate coefficients to the sol-air temperature at the same time and previous times, to the space temperature, and to the heat transfer rate at previous times. This may be written as

$$\dot{q}_{i,\theta} = A \left[\sum_{n=0} b_n(t_{e,\theta-n\Delta}) - \sum_{n=1} d_n \left(\frac{\dot{q}_{i,\theta-n\Delta}}{A} \right) - t_i \sum_{n=0} c_n \right] \tag{8-52}$$

where b_n, d_n, and c_n are the transfer function coefficients that depend on the construction of the wall or roof section. The heat gains from solar radiation on windows, heat sources within the space, ventilation, infiltration, and latent heat gains are computed as previously discussed.

The cooling load for a space at a particular time may be related through the use of appropriate coefficients to the heat gain at that and previous times and to the cooling load at previous times. This is written as

$$\dot{q}_{c,\theta} = \sum_{i=1} (v_o \dot{q}_{i,\theta} + v_1 \dot{q}_{i,\theta-\Delta} + v_2 \dot{q}_{i,\theta-2\Delta} + \cdots)$$

$$- w_1 \dot{q}_{c,\theta-\Delta} - w_2 \dot{q}_{c,\theta-2\Delta} - w_3 \dot{q}_{c,\theta-3\Delta} - \cdots \quad (8\text{-}53)$$

where the coefficients depend on the size of the time element Δ, the nature of the heat gain, and the heat storage capacity of the space.

Heat Gain Through Exterior Walls and Roofs

Transfer function coefficients have been computed for a wide range of wall and roof constructions. These calculations involve the solution of the heat conduction equation (Eq. 8-3), with the boundary conditions given by Eqs. 8-5 and 8-9 to obtain the coefficients b_n, d_n, and c_n. A sampling of the transfer function coefficients given in reference 1 are shown in Tables 8-27 and 8-28. The outside surface film coefficient used for the calculations was equal to 3.0 Btu/(hr-ft²-F) or 17 W/(m²-C), whereas the inside surface film coefficient was 1.46 Btu/(hr-ft²-F) or 8.3 W/(m²-C). The time interval Δ was one hour. Sol-air temperatures used in Eq. 8-52 should also be based on a film coefficient of 3.0 Btu/(hr-ft²-F) or 17 W/(m²-C). Table 8-29 gives some coefficients for interior partitions and Table 8-6 gives data to more adequately describe the walls, roofs, and partitions given in Tables 8-27, 8-28, and 8-29.

The sol-air temperature as a function of time is required to solve Eq. 8-52 for the heat gain through a wall or roof. Equation 8-8 expresses the sol-air temperature as a function of the dry bulb temperature, emittance and absorptance of the surface, total solar intensity, and the convective heat transfer coefficient for the surface.

$$t_e = t_o + \alpha \frac{G_t}{h_o} - \frac{\epsilon \, \Delta R}{h_o} \quad (8\text{-}8)$$

As mentioned in Sec. 8-1, the term α/h_o varies from about 0.15 (hr-ft²F)/Btu or 0.026 (m²-C)/W for a light-colored surface to a maximum of about 0.30 (hr-ft²-F)/Btu or 0.053 (m²-C)/W, whereas the term $\epsilon \, \Delta R/h_o$ varies from about zero for a vertical surface to about 7 F or 4 C for a horizontal surface. The total radiation incident on the surface is computed by the methods of Chapter 6.

The dry bulb temperature required in Eq. 8-8 is computed from the design dry bulb temperature and the daily range of temperature given in Table C-1 with the percentage of daily range given in Table 8-30 as

$$t_o = t_d - DR(X) \quad (8\text{-}54)$$

where

t_d = design dry bulb temperature, F or C
DR = daily range, F or C
X = percentage of daily range divided by 100

The last term in Eq. 8-52 is a constant because the inside design temperature is assumed to be constant. Equation 8-52 can now be solved for a particular

Table 8-27 TRANSFER FUNCTION COEFFICIENTS FOR EXTERIOR WALLS (TIME INTERVAL = 1.0 hr)

Construction Description	Code Numbers of Layers	Coefficients[a] b_n and d_n							U	$\sum_{n=0} c_n$
		$n=0$	$n=1$	$n=2$	$n=3$	$n=4$	$n=5$	$n=6$		
4-in. or 100-mm face brick, 2-in. or 50-mm insulation, and 4-in. or 100-mm lightweight concrete block	A0, A2, B3 C2, E1, E0	b 0.00000 d 1.00000	0.00046 −1.73771	0.00225 0.90936	0.00150 −0.11373	0.00016 0.00496	−0.00001		0.102	0.00437
4-in. or 100-mm face brick, air space, and 4-in. or 100-mm common brick	A0, A2, B1 C4, E1, E0	b 0.00000 d 1.00000	0.00086 −1.79201	0.00485 0.98014	0.00378 −0.16102	0.00050 0.00609	0.00001 −0.00003		0.301	0.01000
4-in. or 100-mm face brick, air space, and 4-in. or 100-mm lightweight concrete block	A0, A2, B1 C2, E1, E0	b 0.00003 d 1.00000	0.00286 −1.50943	0.01029 0.65654	0.00504 −0.07415	0.00037 0.00212			0.248	0.01859
12-in. or 300-mm heavyweight concrete	A0, A1, C11 E1, E0	b 0.00000 d 1.00000	0.00029 −1.86853	0.00303 1.09284	0.00412 −0.21487	0.00105 0.01094	0.00005 −0.00009		0.421	0.00854
Frame wall with 4-in. or 100-mm brick veneer	A0, A2, B6 A6, E0	b 0.00037 d 1.00000	0.00823 −1.03045	0.00983 0.20108	0.00125 −0.00726	0.00001			0.121	0.01969
Frame wall	A0, A6, B6 A6, E0	b 0.01977 d 1.00000	0.06317 −0.25848	0.01064 0.01072	0.00006				0.124	0.09364

Construction	Layer codes		$n=0$	$n=1$	$n=2$	$n=3$	$n=4$	$n=5$	U	$\sum c_n$
Frame wall with 3-in. or 75-mm insulation	A0, A1, B1 B4, E1, E0	b	0.00509	0.02644	0.00838	0.00010			0.081	
		d	1.00000	−0.59602	0.08757	−0.00002				0.04001
4-in. or 100-mm face brick, 2-in. or 50-mm insulation and 4-in. or 100-mm common brick	A0, A2, B3 C4, E1, E0	b	0.00000	0.00012	0.00084	0.00082	0.00014		0.111	
		d	1.00000	−1.96722	1.20279	−0.22850	0.01033	−0.00006		0.00192
8-in. or 200-mm heavyweight concrete block	A0, A1, C8 E1, E0	b	0.0004	0.0171	0.0310	0.0065	0.0001		0.402	
		d	1.0000	−1.1621	0.3132	−0.0139				0.0551
4-in. or 100-mm face brick, 8-in. or 200-mm clay tile and airspace	A0, A2, C6 B1, E1, E0	b	0.0000	0.0000	0.0007	0.0019	0.0011	0.0001	0.221	
		d	1.0000	−2.1290	1.5667	−0.4781	0.0605	−0.0029		0.0038
4-in. or 100-mm heavyweight concrete with 1-in. or 25-mm insulation	A0, A1, C5 B2, E1, E0	b	0.0005	0.0094	0.0106	0.0013			0.200	
		d	1.0000	−1.1763	0.3011	−0.0157				0.0218
4-in. or 100-mm heavyweight concrete	A0, A1, C5 E1, E0	b	0.0078	0.0705	0.0355	0.0011			0.585	
		d	1.0000	−0.8789	0.0753	−0.0001				0.1149
Sheet metal with 1-in. or 25-mm insulation	A0, A3, B2 B1, A3, E0	b	0.1424	0.0479					0.191	
		d	1.0000	−0.0013						0.1903

[a] U, b's, and c's are in Btu/(hr·ft²·°F), and d is dimensionless. To convert U, b's and c's to W/(m²·°C), multiply by 5.6783.

Source: Adapted by permission from *ASHRAE Handbook, Fundamentals Volume,* 1977.

Table 8-28 TRANSFER FUNCTION COEFFICIENTS FOR ROOFS (TIME INTERVAL = 1.0 hr)

Construction Description	Code Numbers of Layers	Coefficients[a] b_n and d_n							U	$\sum_{n=0} c_n$
		$n = 0$	$n = 1$	$n = 2$	$n = 3$	$n = 4$	$n = 5$	$n = 6$		
Roof terrace system	A0,C12,B1,B6, E2,E3,C5,E4, E5,E0	b 0.00000 d 1.0000	0.00008 −1.7304	0.00048 0.8564	0.00039 −0.1611	0.00006 0.0024			0.082	0.00101
1-in. or 25-mm wood with 1-in. or 25-mm insulation	A0,E2,E3,B5, B7,E4,E5,E0	b 0.0003 d 1.0000	0.0082 −1.0046	0.0103 0.1845	0.0014 −0.0046				0.115	0.0202
8-in. or 200-mm lightweight concrete	A0,E2,E3,C16, E4,E5,E0	b 0.00000 d 1.00000	0.00002 −1.91091	0.00046 1.22135	0.00133 −0.31019	0.00079 0.03001	0.00011 −0.00095	0.00001	0.092	0.00271
4-in. or 100-mm lightweight concrete	A0,E2,E3,C14, E4,E5,E0	b 0.0001 d 1.0000	0.0055 −1.0698	0.0141 0.2665	0.0045 −0.0143	0.0002 0.0001			0.134	0.0244
Steel sheet with 1-in. or 25-mm insulation	A0,E2,E3,B5, A3,E4,E5,E0	b 0.0085 d 1.0000	0.0505 −0.4700	0.0179 0.0476	0.0004				0.134	0.0773
2.5-in. or 64-mm wood with 1 in. or 25-mm insulation	A0,E2,E3,B5 B8,E0	b 0.0000 d 1.0000	0.0017 −1.3557	0.0068 0.5121	0.0035 −0.0634	0.0003 0.0015			0.130	0.0123
6-in. or 150-mm heavyweight concrete with 1-in. or 25-mm insulation	A0,E2,E3,B5, C13,E0	b 0.0001 d 1.0000	0.0036 −1.3001	0.0068 0.3991	0.0016 −0.0361	0.0001			0.192	0.121
4-in. or 100-mm heavyweight concrete with 1-in. or 25-mm insulation	A0,E2,E3,B5, C5,E0	b 0.0008 d 1.0000	0.0117 −1.0800	0.0100 0.2015	0.0008 −0.0051				0.200	0.0233

[a]U, b's and c's are in Btu/(hr·ft²·F), and d is dimensionless. To convert U, b's, and c's to W/(m²·C), multiply by 5.6783.

Source: Adapted by permission from ASHRAE Handbook, Fundamentals Volume, 1977.

Table 8-29   TRANSFER FUNCTION COEFFICIENTS FOR INTERIOR PARTITIONS, FLOORS, AND CEILINGS (TIME INTERVAL = 1.0 hr)

Construction Description	Code Numbers of Layers	Coefficients[a] b_n and d_n							U	$\sum\limits_{n=0} c_n$
		$n = 0$	$n = 1$	$n = 2$	$n = 3$	$n = 4$	$n = 5$	$n = 6$		
4-in. or 100-mm lightweight concrete block with ¾-in. or 19-mm plaster	E0, E1, C2, E1, E0	b 0.0048 d 1.0000	0.0514 -0.8456	0.0339 0.1397	0.0018 -0.0015				0.314	0.0919
4-in. or 100-mm heavyweight concrete block with ¾-in. or 19-mm plaster	E0, E1, C3, E1, E0	b 0.0092 d 1.0000	0.0705 -0.8203	0.0318 0.0874	0.0008 -0.0001				0.421	0.1123
8-in. or 200-mm lightweight concrete block, plastered both sides	E0, E1, C7, E1, E0	b 0.0002 d 1.0000	0.0106 -1.2098	0.0214 0.3736	0.0054 -0.0248	0.0001 0.0001			0.271	0.0377
8-in. or 200-mm heavyweight concrete block, plastered both sides	E0, E1, C8, E1, E0	b 0.0004 d 1.0000	0.0141 -1.1995	0.0231 0.3293	0.0043 -0.0132	0.0001			0.360	0.0420
12-in. or 300-mm heavyweight concrete plastered both sides	E0, E1, C11, E1, E0	b 0.0000 d 1.0000	0.0002 -1.8959	0.0023 1.1220	0.0029 -0.2203	0.0007 0.0107	-0.0001		0.372	0.0061
Frame, partition with ¾-in. or 19-mm gypsum board	E0, E1, B1, E1, E0	b 0.0729 d 1.0000	0.1526 -0.3986	0.0159 0.0208					0.388	0.2414

313

Table 8-29 TRANSFER FUNCTION COEFFICIENTS FOR INTERIOR PARTITIONS, FLOORS, AND CEILINGS (TIME INTERVAL = 1.0 hr) (continued)

		n = 0	n = 1	n = 2	n = 3		
2-in. or 50-mm heavyweight concrete floor deck	E0, A5, C12, E0	b 0.0505 d 1.0000	0.0691 −0.6662	0.0011		0.362	0.1207
4-in. or 100-mm heavyweight concrete floor deck	E0, A5, C5, E0	b 0.0111 d 1.0000	0.0405 −0.8507	0.0072 0.0229		0.341	0.0588
4-in. or 100-mm lightweight concrete floor deck	E0, A5, C2, E0	b 0.0200 d 1.0000	0.0710 −0.5878	0.0120 0.0116		0.243	0.1030
2-in. or 50-mm wood deck	E0, A5, B10, E0	b 0.0020 d 1.0000	0.0245 −0.9025	0.0174 0.1271	0.0010 −0.0008	0.201	0.0449
4-in. or 100-mm lightweight concrete deck with false ceiling	E0, A5, C2, E4, E5, E0	b 0.0020 d 1.0000	0.0195 −0.8295	0.0104 0.0534	0.0004 −0.0003	1.44	0.0323
2-in. or 50-mm wood deck with false ceiling	E0, A5, B10, E4, E5, E0	b 0.0001 d 1.0000	0.0048 −1.1372	0.0078 0.2530	0.0014 −0.0061	0.129	0.0141
Steel deck with false ceiling	E0, A5, A3, E4, E5, E0	b 0.0749 d 1.0000	0.0853 −0.1257	0.0022 0.0001		0.186	0.1624

[a] U, b's, and c's are in Btu/(hr-ft²-F), and d is dimensionless. To convert U, b's, and c's to W/(m²·C), multiply by 5.6783.

Source: Adapted by permission from ASHRAE Handbook, Fundamentals Volume, 1977.

Table 8-30 PERCENTAGE OF THE DAILY RANGE

Time, hr	Percent	Time, hr	Percent	Time, hr	Percent	Time, hr	Percent
1	87	7	93	13	11	19	34
2	92	8	84	14	3	20	47
3	96	9	71	15	0	21	58
4	99	10	56	16	3	22	68
5	100	11	39	17	10	23	76
6	98	12	23	18	21	24	82

Source: Reprinted by permission from *ASHRAE Handbook, Fundamentals Volume*, 1977. Hour 1 is 1:00 A.M. solar time.

wall or roof using Eq. 8-8 and transfer coefficient data from Tables 8-27, 8-28, or 8-29. The only practical way of doing this is by digital computer; however, some of the equations will be set up to demonstrate the procedure. With a time interval of one hour, the value for which coefficients are available, Eq. 8-52 may be expanded to give 24 separate equations for θ ranging from 1 to 24. The first equation for 1:00 A.M. is

$$\frac{\dot{q}_{i,1}}{A} = \begin{bmatrix} b_0 t_{e,1} \\ + b_1 t_{e,24} \\ + b_2 t_{e,23} \\ + b_3 t_{e,22} \\ + \cdots \end{bmatrix} - \begin{bmatrix} + d_1 \dot{q}_{i,24} \\ + d_2 \dot{q}_{i,23} \\ + d_3 \dot{q}_{i,22} \\ + d_4 \dot{q}_{i,21} \\ + \cdots \end{bmatrix} - \begin{bmatrix} t_i \sum_{n=0} c_n \end{bmatrix} \qquad (8\text{-}52a)$$

The equation for 2:00 A.M. is

$$\frac{\dot{q}_{i,2}}{A} = \begin{bmatrix} + b_0 t_{e,2} \\ + b_1 t_{e,1} \\ + b_2 t_{e,24} \\ + b_3 t_{e,23} \\ + \cdots \end{bmatrix} - \begin{bmatrix} + d_1 \dot{q}_{i,1} \\ + d_2 \dot{q}_{i,24} \\ + d_3 \dot{q}_{i,23} \\ + d_4 \dot{q}_{i,22} \\ + \cdots \end{bmatrix} - \begin{bmatrix} t_i \sum_{n=0} c_n \end{bmatrix} \qquad (8\text{-}52b)$$

and the remaining 22 equations could be written in a similar way. The complete set of equations is solved sequentially starting with Eq. 8-52a and continuing until the solution to each set does not change more than a specified amount. Three or four calculation cycles are usually required for an acceptable solution. To start the calculation the unknown heat gains $\dot{q}_{i,\theta}$ are set equal to zero. The effect of this assumption is negligible after successive calculation cycles. The final result of these calculations is the heat gain to a space for a particular wall, roof, or partition as a function of time.

When Eq. 8-52 is used to compute the heat gain for a partition, the sol-air temperation t_e must be changed to the temperature in the adjacent uncooled space t_a and its variation with time must be known. If the temperature in the uncooled space is constant, Eq. 8-52 reduces to the steady-state situation given by

$$\dot{q}_i = UA(t_a - t_i) \qquad (8\text{-}55)$$

Cooling Load by the Transfer Function Method

The cooling load depends on both the magnitude and nature of the sensible heat gain. Each component of the space heat gain gives rise to a distinct component of cooling load. For example, the cooling load resulting from heat absorbed by the floor is quite different from that absorbed by the window glass or drapes. However, the sum of all the cooling-load components for one time is equal to the total cooling load at that time.

The latent heat-gain component of the cooling load may or may not be a part of the space cooling load. The latent portion of the infiltration heat gain is a part of the room cooling load, and outdoor ventilation air may be dehumidified before it is distributed to the space. In any case the latent heat gain at a particular time results in cooling load at the same time.

Different room transfer functions are used to convert the various heat gains to cooling load. In other words, the v coefficients in Eq. 8-53 depend on the nature of the heat gain. These coefficients as well as the w coefficients also depend on the mass of the structure, the floor covering, mass of the furnishings, and size of the room. The coefficients for the room-transfer function are computed using the digital computer. The process is rather complex; however, a simple energy balance on the space forms the basis for such calculations:

$$\text{Heat gain} = \text{Heat loss} + \text{Energy storage}$$

or

$$\Sigma \dot{q}_i = \dot{q}_c + \dot{q}_o + mc \frac{dt}{d\theta} \qquad (8\text{-}56)$$

where $\dot{q}_o$ accounts for any heat transfer from the space to the environment. Rearrangement of Eq. 8-56 yields

$$\dot{q}_c = \Sigma \dot{q}_i - \dot{q}_o - mc \frac{dt}{d\theta} \qquad (8\text{-}56a)$$

The room-transfer coefficients may be derived from Eq. 8-56a and placed in four main groups according to the nature of the heat gain as shown in Table 8-31. Note that the coefficients in Table 8-31 show only a dependence on the mass of the structure. The mass is the strongest variable, but other things mentioned before also influence the value of the v and w coefficients (5). The coefficients in Table 8-31 apply when all of the heat gains eventually appear as cooling load. In most cases at least a small part of the heat gain is lost to the surroundings. This depends on the thermal conductance of the enclosure. Equation 8-57 has been prepared to estimate the fraction of the heat gain F_c that results in the cooling load:

$$F_c = 1 - 0.02K_t \qquad (8\text{-}57)$$

The v coefficients are then modified by multiplying by F_c. The unit length conductance K_t is given by

Table 8-31 COEFFICIENTS OF ROOM TRANSFER FUNCTIONS

Heat Gain Component	Room Envelope Construction	v_0	v_1 Dimensionless	v_2
Solar heat gain through glass with no interior shading and heat generated by equipment and people, which is dissipated by radiation	Light	0.224	$= 1 + w_1 - v_0$	0.0
	Medium	0.197	$= 1 + w_1 - v_0$	0.0
	Heavy	0.187	$= 1 + w_1 - v_0$	0.0
Conduction heat gain through exterior walls, roofs, partitions and doors, and windows with blinds or drapes	Light	0.703	$= 1 + w_1 - v_0$	0.0
	Medium	0.681	$= 1 + w_1 - v_0$	0.0
	Heavy	0.676	$= 1 + w_1 - v_0$	0.0
Heat generated by lights	Light	0.0	$=$ "a" in Table 8-19	$= 1 + w_1 - v_1$
	Medium	0.0	$=$ "a" in Table 8-19	$= 1 + w_1 - v_1$
	Heavy	0.0	$=$ "a" in Table 8-19	$= 1 + w_1 - v_1$
Heat generated by equipment and people, which is dissipated by convection, and energy gain due to ventilation and infiltration air	Light	1.0	0.0	0.0
	Medium	1.0	0.0	0.0
	Heavy	1.0	0.0	0.0

Source: Reprinted by permission from *ASHRAE Handbook, Fundamentals Volume,* 1985.

Table 8-31a   THE VALUE OF w_1 FOR DIFFERENT ROOM AIR CIRCULATION RATES AND ENVELOPE CONSTRUCTION

Room Envelope Construction[a]	2-in. Wood Floor	3-in. Concrete Floor	6-in. Concrete Floor	8-in. Concrete Floor	12-in. Concrete Floor	Room Air[b] Circulation & Type of Supply and Return
Specific Mass, lb/ft^2 of floor area	10	40	75	120	160	
	−0.88	−0.92	−0.95	−0.97	−0.98	Low
	−0.84	−0.90	−0.94	−0.96	−0.97	Medium
	−0.81	−0.88	−0.93	−0.95	−0.97	High
	−0.77	−0.85	−0.92	−0.95	−0.97	Very high
	−0.73	−0.83	−0.91	−0.94	−0.96	

[a]Floor covered with carpet and rubber pad; for a floor covered only with floor tile, take next w_1 value down the column.

[b]Low: Low ventilation rate—minimum required to cope with cooling load due to lights and occupants in interior zone. Supply through floor, wall, or ceiling diffuser. Ceiling space not vented.

Medium: Medium ventilation rate, supply through floor, wall, or ceiling diffuser. Ceiling space not vented.

High: Room air circulation induced by primary air of induction unit or by fan coil unit. Return through ceiling space.

Very high: High room circulation used to minimize temperature gradient in a room. Return through ceiling space.

Source: Reprinted by permission from ASHRAE Handbook, Fundamentals Volume, 1985.

$$K_t = \frac{1}{L}[(UA)_w + (UA)_{ow} + (UA)_c] \tag{8-58}$$

where

L = length of exterior wall, ft or m

U = overall heat transfer coefficient of room element, Btu/(hr-ft²-F) or W/(m²-C) (w for window, ow for outside wall, and c for corridor)

A = area of the room element, ft² or m²

The solution of Eq. 8-53 for the cooling load for a particular space is carried out in a manner analogous to that of Eq. 8-52 previously discussed. It is important to note that Eq. 8-53 must be applied to each different class of sensible heat gain shown in Table 8-31. This means that it must be solved three or four times for each space. The sum of the cooling-load components at each time then equals the total cooling load at that time.

Summary

The heat gain and cooling load calculations using the transfer function method should be carried out by a digital computer. The general procedure is outlined as follows:

1. Derive hourly values of the outdoor dry bulb temperature (Eq. 8-54). This may be part of the computer program.
2. Compute the sol-air temperatures for each surface and each hour using Eq. 8-8.
3. Compute the instantaneous sensible heat gain for each wall, partition, and roof using the procedure discussed previously and Eq. 8-52. There is a heat gain for each hour of the day for each surface.
4. Compute the instantaneous sensible heat gain for the doors. Doors are usually assumed to have negligible energy storage. Then

$$\dot{q}_{i,\theta} = (UA)_D(t_{e,\theta} - t_i) \tag{8-59}$$

5. Compute the convective heat gain for the windows.

$$\dot{q}_{i,\theta} = (UA)_w(t_{o,\theta} - t_i) \tag{8-60}$$

6. Compute the solar radiation heat gain for the windows.

$$\dot{q}_{i,\theta} = A_w(SC)(SHGF)_\theta \tag{8-61}$$

7. Compute the heat gain due to the lights, which is simply power input to the lights for the times they are on.
8. Compute the sensible heat gain due to people for the hours the space is occupied.
9. Compute the sensible heat gain due to infiltration for each hour. This may not be constant for every hour.
10. Compute the latent heat gain due to infiltration and people for each hour.

11.   Sum the instantaneous heat gains that will appear immediately as cooling load. This includes lights that are on continuously (item 7 above), sensible heat gain due to infiltration (item 9 above), and the sensible heat gain from people (item 8 above). This total corresponds to line 4 in Table 8-31.

12.   Sum the solar heat gains for glass that have no interior shading (item 6 above). This total corresponds to line 1 in Table 8-31.

13.   Sum the conduction-convection heat gains through roof, walls, windows, and solar heat gain for windows with inside blinds or drapes (3, 4, 5, 6 above). This total corresponds to line 2 of Table 8-31.

14.   Sum the heat gain due to lights that are not on all the time (item 7 above and line 3 of Table 8-31).

15.   Sum the heat gains due to equipment and people and dissipated by radiation (item 8 above). This total corresponds to line 1 of Table 8-31.

16.   Transform the separate heat gain totals of items 12, 13, 14, and 15 above to cooling load by using Eq. 8-53 and the appropriate transfer function coefficients from Table 8-31. Adjust the coefficients using Eq. 8-57 as required.

17.   Obtain the total space cooling load for each hour by summing items 10, 11, and 16 above.

8-11   HEAT EXTRACTION RATE AND ROOM TEMPERATURE

The cooling-load-calculation procedure discussed in the previous section produces data that may be used to compute the heat-extraction rate and the room temperature obtained with a particular terminal unit. This is accomplished by the use of a transfer function in a somewhat analogous manner to that used to find the cooling load. When the cooling equipment removes heat energy from the space air at a rate equal to the cooling load, the space air temperature will remain constant and the heat extraction rate $\dot{q}_x$ is equal to the cooling load $\dot{q}_c$. However, this is seldom true. A simple energy balance on the space air yields

$$\dot{q}_x - \dot{q}_c = (mc)_{\text{air}} \frac{dt_r}{d\theta} \tag{8-62}$$

Equation 8-62 shows the general behavior of the system; however, the actual process is complex because of the dependence of $\dot{q}_x$ and $\dot{q}_c$ on the air temperature and time. Therefore a transfer function has been devised, which has an obvious similarity to Eq. 8-62, to describe the process, and the digital computer has been utilized to compute the transfer function coefficients. The room air transfer function is

$$\sum_{i=0}^{1} p_i(\dot{q}_{x,\theta-i\Delta} - \dot{q}_{c,\theta-i\Delta}) = \sum_{i=0}^{2} g_i(t_i - t_{r,\theta-i\Delta}) \tag{8-63}$$

where

p_i, g_i = transfer function coefficients
$\dot{q}_c$ = cooling load at the various times
t_i = room temperature used for cooling load calculations
t_r = actual room temperature at the various times

The room air transfer function coefficients are given in Table 8-32. Note that the g coefficients have been normalized to unit floor area. The coefficients g_o and g_1 depend on the heat conductance to the surroundings UA and the infiltration and ventilation rate to the space.

$$g = g_0^* A_{fl} + [UA + \rho c_p (\dot{Q}_i + \dot{Q}_v)]p_0 \qquad (8\text{-}64)$$
$$g_1 = g_1^* A_{fl} + [UA + \rho c_p (\dot{Q}_i + \dot{Q}_v)]p_1 \qquad (8\text{-}64a)$$

where

g_i^* = normalized g coefficients, Btu/(hr-ft²-F) or W/(m²-C)
A_{fl} = floor area, ft² or m²
UA = conductance between the space and the surrounding (average), Btu/(hr-F) or W/C
$\dot{Q}$ = infiltration rate, ft³/hr or m³/s
$\dot{Q}_v$ = ventilation rate, ft³/hr or m³/s
p_i = coefficients from Table 8-32, dimensionless

The remaining g coefficient g_2 is given by

$$g_2 = g_2^* A_{fl} \qquad (8\text{-}65)$$

The next step is to describe the characteristic of the terminal unit. This can often be done by a linear expression of the form

$$\dot{q}_{x,\theta} = W + St_{r,\theta} \qquad (8\text{-}66)$$

where W and S are parameters that characterize the equipment at time θ. The equipment being modeled here is actually the cooling coil and the associated

Table 8-32   NORMALIZED COEFFICIENTS OF THE ROOM AIR TRANSFER FUNCTION[a]

Room Envelope Construction	Btu/(hr-ft²-F)			Dimensionless	
	g_0^*	g_1^*	g_2^*	p_0	p_1
Light	+1.68	−1.73	+0.05	1.0	−0.82
Medium	+1.81	−1.89	+0.08	1.0	−0.87
Heavy	+1.85	−1.95	+0.10	1.0	−0.93

[a]For all cases, the room is assumed furnished.

Source: Reprinted with permission from *ASHRAE Handbook, Fundamentals Volume*, 1985.

control system that matches the coil load to the space load. A face and bypass system would be typical.

The ability of the cooling unit to extract heat energy from the air varies from some minimum to some maximum value over the throttling range of the control system (thermostat). Equation 8-66 is valid only in the throttling range. The throttling range Δt_r may be described as the temperature at which the thermostat calls for maximum cooling minus the temperature at which the thermostat calls for minimum cooling. Figure 8-9 shows the relationship of the heat extraction rate $\dot{q}_x$ and the room air temperature according to Eq. 8-66. When the room air temperature lies outside the throttling range, the heat extraction rate is either $\dot{q}_{x,min}$ or $\dot{q}_{x,max}$ depending on whether t_r is above or below the set-point temperature t_r^*. The value of S is the slope of the characteristic in the throttling range:

$$S = \frac{[\dot{q}_{x,max} - \dot{q}_{x,min}]}{\Delta t_r} \tag{8-67}$$

The value of W is the intercept of Eq. 8-66 with the $\dot{q}_x$ axis and is defined by

$$W = \frac{[\dot{q}_{x,max} + \dot{q}_{x,min}]}{2 - St_{r,\theta}^*} \tag{8-68}$$

where $t_{r,\theta}^*$ the set-point temperature is assumed to be in the center of the throttling range. The set-point temperature may be changed during the day.

Equations 8-63 and 8-66 may be combined and solved for $\dot{q}_{x,\theta}$

$$\dot{q}_{x,\theta} = \frac{g_o W}{S + g_o} + \frac{S}{S + g_o} G_\theta \tag{8-69}$$

Figure 8-9. Cooling equipment characteristic.

where

$$G_\theta = t_i \sum_{i=0}^{2} g_i - \sum_{i=1}^{2} g_i(t_{i,\theta-i\Delta}) + \sum_{i=0}^{1} p_i(\dot{q}_{c,\theta-i\Delta}) - \sum_{i=1}^{1} p_i(\dot{q}_{x,\theta-i\Delta}) \quad (8\text{-}70)$$

When the value of $\dot{q}_{x,\theta}$ computed by Eq. 8-69 is greater than $\dot{q}_{x,\text{max}}$ it is made equal to $\dot{q}_{x,\text{max}}$, and when it is less than $\dot{q}_{x,\text{min}}$ it is made equal to $\dot{q}_{x,\text{min}}$. Finally, Eqs. 8-66 and 8-69 can be combined and solved for $t_{r,\theta}$:

$$t_{r,\theta} = \frac{(G_\theta - \dot{q}_{x,\theta})}{g_o} \quad (8\text{-}71)$$

The solution of Eq. 8-69 is carried out in the same manner as Eqs. 8-52 and 8-53 for the heat gain and cooling load discussed in Section 8-10. To summarize, Eq. 8-69 is written for each hour of the day and solved successively beginning with hour 1. Cyclic calculations are carried on until there is no significant change in $\dot{q}_{x,\theta}$ or $t_{r,\theta}$ for each hour. The transfer function coefficients are all constant for a given space unless the infiltration and ventilation rates change with time. In this case the values of g_o and g_1 must be recalculated for each set of conditions.

Figure 8-10 illustrates results obtained by using the foregoing method. Both continuous and intermittent operation are shown using somewhat undersized cooling equipment. Notice that only a small sacrifice in space temperature

Figure 8-10.   Room air temperature and heat-extraction rates for continuous and intermittent operation.

results from intermittent operation, but the total heat energy removed seems to be less, which would reduce operating costs.

The transfer function method just described is useful in finding the space-cooling load and the coil load on an hour-to-hour basis for a design day. The same method may be used to find design heating loads. In such a case the method is identical, except that the solar heat gain and the internal heat gains should be set equal to zero.

8-12   BUILDING AND EQUIPMENT THERMAL SIMULATION

Following design of the environmental control system for a building, it is often desirable to make a detailed analysis of the anticipated energy requirements of the structure for heating, cooling, lighting, and other powered equipment. This same information is often required in energy conservation studies involving existing buildings. Simulation implies that the complete system configuration is already determined; therefore, this type of analysis is distinctly different from design, previously discussed.

The transfer function method is readily adaptable to simulation analysis and works equally well for heating and cooling. The main differences as compared with the design computation are as follows:

1. Hourly historical or assumed temperature data are used instead of the design maximum temperature and the daily range.
2. Historical or assumed hourly solar radiation data are used instead of computed clear sky data so that the effect of cloudiness, haze, and so on, are taken into account.
3. More realistic occupancy and lighting schedules may be used.
4. Central cooling equipment such as chillers and boilers is modeled to compute the energy required to meet the heating or cooling demand (heat extraction rate).
5. The simulation is carried out on an hour-by-hour basis for periods of time greater than one day and generally up to a year.

Various weather data are available on magnetic tape. One type, known as the Weather Year for Energy Calculations (WYEC) was developed by ASHRAE and is in popular use. This weather tape contains data that are typical of the weather patterns for many cities in the United States for a complete year on an hourly basis.

It often happens that a building is not used or the environmental equipment operated under the same conditions for which it was designed. Therefore, these changes should be reflected in the simulation analysis.

Modeling of the central cooling and heating plant can become quite complex; however, this doesn't have to be true. The model should take into account the effect of environmental conditions and load on the operating efficiency. For example, the coefficient of performance of a water chiller depends on the chilled water temperature and the condensing water temperature. The chilled water temperature may be relatively constant, but the condensing water temperature may depend on the outdoor wet bulb temperature and the load on the chiller. Boiler performance does not depend so much on environmental conditions but the efficiency does drop rapidly with decreasing load.

A useful and simple way of modeling all types of heating and cooling equipment is to normalize the energy input and the capacity with the rated full-load input and capacity. Then the normalized input is

$$Y = E/E_{max} \qquad (8\text{-}72)$$

and the normalized capacity is

$$X = \dot{q}_x/\dot{q}_{x,max} \qquad (8\text{-}73)$$

These quantities may then be plotted and a curve fitted that comprises a model. Figure 8-11 is an example of such a model for a hot water boiler and Fig. 8-12 is for a centrifugal chiller. To construct the curves it is necessary to have performance data for partial load conditions. Most manufacturers can furnish such data. The plots of Y versus X tend to be straight lines. When this is true, it is possible to develop a model with a limited amount of performance data.

The model for the central plant must also include pumps, fans, cooling towers, and any auxiliary equipment that uses energy. The energy consumed by the lights is also often included in the overall equipment model. An estimate of the total energy consumption of the building is the overall objective. For existing buildings, the total predicted energy usage may be compared with the actual utility data.

Many computer programs are available to carry out both the design and simulation analysis previously described. Among these are TRACE, developed by the Trane Company; DOE-2, developed by the federal government; TRNSYS, developed at the University of Wisconsin; and LDCAL, developed at Oklahoma State University.

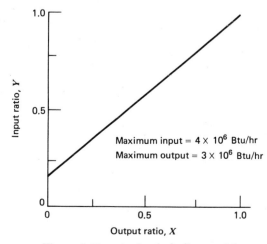

Figure 8-11.   A simple boiler model.

Figure 8-12. A simple centrifugal chiller model.

REFERENCES

1. *ASHRAE Handbook, Fundamentals Volume,* American Society of Heating, Refrigerating and Air-Conditioning Engineers, Atlanta, Ga., 1985.

2. L. W. Nelson and J. R. Tobias, "Energy Savings in Residential Buildings," *ASHRAE Journal,* Vol. 16, No. 2, February 1974.

3. *ASHRAE Cooling and Heating Load Calculation Manual,* GRP 158, American Society of Heating, Refrigerating and Air-Conditioning Engineers, New York, 1979.

4. F. C. McQuiston, "A Study and Review of Existing Data to Develop a Standard Methodology for Residential Heating and Cooling Load Calculations," *ASHRAE Transactions,* Vol. 90, Part 2, 1984.

5. Edward F. Sowell, "Updating Cooling Load Temperature Differences and Cooling Load Factors for Chapter 26 of the Handbook of Fundamentals," Final Report, RP 359, American Society of Heating, Refrigerating and Air-Conditioning, Inc., Atlanta, Ga., 1984.

PROBLEMS

8-1 Determine the ASHRAE standard design conditions for the following locations. Include the maximum outdoor temperature, the outdoor mean coincident wet bulb temperature, indoor dry bulb temperature, relative humidity, the elevation and latitude. (a) Washington, D.C., (b) San Francisco, California, (c) Denver, Colorado, (d) Dallas, Texas.

8-2 Find the adjustment to the cooling load temperature difference (CLTD) for each of the locations in Problem 8-1 assuming the color correction factor is 1.0 and the inside dry bulb temperature is 75 F.

8-3 Compute the cooling load for the south wall of a building at 40 deg north latitude in July. The solar time is 4:00 P.M. The wall is frame with an overall heat-transfer coefficient of 0.08 Btu/(hr-ft²-F). The inside design temperature is 75 F and the outside temperature is 101 F. The daily range of temperature is 22 degrees F. The wall is 8 × 16 ft with a 4 × 5 ft window. (Do not make a calculation for the window.)

8-4 The building of Problem 8-3 has a roof of light construction (No. 1, Table 8-11) with U = 0.12 Btu/(hr-ft²-F). Compute the cooling load for a 16 × 20 ft room using data from Problem 8-3 (roof only).

8-5 Compute the peak cooling load per square meter for a west wall similar to that of Table 5-7a. Assume that the wall is a dark color and located at 40 deg north latitude. The date is July 21. What time of day does the peak occur? The outdoor and indoor design temperatures are 38 C and 25 C, respectively, and the daily range is 12 deg C.

8-6 Compute the cooling load for a 8 × 40 m roof-ceiling combination similar to that of Table 5-7b at 6:00 P.M. in July. Assume 40 deg north latitude, 39 C and 24 C outdoor and indoor temperatures, and a daily range of 13 deg C.

8-7 Compute the cooling load for a frame wall like that shown in Table 8-5 with 3-in. insulation batts. The wall is on the south side of a building in Shreveport, Louisiana. Make calculations for 10:00 A.M. and 6:00 P.M. Daylight Savings Time in June, using ASHRAE standard conditions.

8-8 The building of Problem 8-7 has a typical attic-ceiling combination that can be approximated by roof No. 1 of Table 8-1 with suspended ceiling and 3½-in. insulation batts. Compute the cooling loads for the times and conditions given in Problem 8-7.

8-9 A wall for a building in Albuquerque, New Mexico, is made of 4-in. concrete with 2 in. of high-density insulation on the outside. The outside is finished with stucco and the inside is finished with gypsum board. Compute the cooling load per square foot of wall assuming standard design conditions.

8-10 A roof-ceiling combination has a metal deck with 2 in. of heavy concrete and 3 in. of rigid insulation on top. The outside layer is typical builtup membrane and stone, whereas the ceiling is acoustical tile below a ceiling air space. The building is located in Albuquerque, New Mexico. Assume standard ASHRAE design conditions and find the cooling load per square foot of roof.

8-11 Solve Problem 8-9 for a building located in (a) Tucson, Arizona; (b) Chicago, Illinois.

8-12 Solve Problem 8-10 for a building in (a) Tucson, Arizona; (b) Chicago, Illinois.

8-13 Determine the shading coefficient for clear, ¼-in. single-sheet glass under summer design conditions for (a) no internal shade, (b) shading with medium-color venetian blinds, (c) shading with white, and tightly woven drapes.

8-14 Solve Problem 8-13 using clear $\frac{1}{4}$-in. double-sheet glass.

8-15 Determine the maximum solar heat-gain factors in July for the major directions for vertical and horizontal surfaces for the following locations; (a) Las Vegas, Nevada; (b) Atlanta, Georgia; (c) Lincoln, Nebraska; (d) Spokane, Washington.

8-16 Solve Problem 8-15 for the month of December.

8-17 Consider the glass of Problem 8-13 and find the cooling-load factors (CLF) for light and heavy construction for east and west windows at 5:00 P.M. solar time.

8-18 Compute the cooling load for a window facing southeast at 40 deg north latitude at 10:00 A.M. Central Daylight Time in August. The window is regular insulating glass with a $\frac{1}{2}$-in. air space. Inside drapes have a transmittance of 0.35 and reflectance of 0.35. The indoor design temperature is 75 F and the outdoor temperature is 100 F with a 24 deg F daily range. Window dimensions are 2 ft wide by 8 ft high.

8-19 The window in Problem 8-18 has an 18-in. overhang at the top. Compute the cooling load for the window.

8-20 Compute the cooling for a window facing south at 40 deg north latitude at 1:00 P.M. sun time in October. The glass is 5 mm gray sheet with venetian blinds of medium color. The inside temperature is 25 C while the outdoor temperature is 35 C with a 13 deg daily range. The window is 1×2 m.

8-21 The window in Problem 8-20 has an overhang at the top of 0.6 m. What percentage of the window is shaded, assuming that the overhang is much wider than the window? Compute the cooling load for the window.

8-22 Compute the cooling load for the south windows of an office building that has no external shading. The windows are the insulating type with regular plate glass inside and out. Drapes with a reflectance of 0.2 and transmittance of 0.5 are fully closed. The total window area is 400 ft². Assume the indoor temperature is 78 F and the outdoor temperatures are 100 F in July and 40 F in December, with a 22 deg F daily range. Make the calculation for 12:00 noon in (a) July and (b) December at 40 deg north latitude.

8-23 A 6×6 ft regular plate glass window faces west. There is a 2-ft overhang at the top. Compute the cooling load at 3:00 P.M. solar time in August at 40 deg north latitude. Assume the outdoor-indoor temperature differential is 25 deg F and the outdoor temperature is 100 F.

8-24 A 1-m² regular plate glass window has inside venetian blinds of medium color. The window is located at 40 deg north latitude. Calculate the cooling load for (a) the window facing east at 10:00 A.M. in June with a mean outdoor temperature of 29 C and (b) the window facing west at 2:00 P.M. in June with a mean outdoor temperature of 32 C. Assume an indoor temperature of 25 C, and use solar time.

8-25 A large office space has an average occupancy of 20 people from 8:00 A.M. to 5:00 P.M. LST. Lighting is 2.5 W/ft² of recessed, unvented fluorescent fixtures. Assume a high ventilation rate and 6-in. concrete floor. Miscellaneous equipment operates intermittently and amounts to an average of 3 hp. Compute the cooling load at 4:00 P.M. for the space assuming a floor area of 4000 ft².

8-26   A space has an occupancy of 40 people engaged in sedentary activity from 8:00 A.M. to 5:00 P.M. LST. The average light level is 20 W/m² of vented fluorescent fixtures. Assume a medium ventilation rate with a ceiling plenum return and a 3-in. concrete floor. Equipment amounts to 6 kW. Estimate the cooling load for a floor area of 750 m².

8-27   A large room has 5000 watts of vented fluorescent light fixtures. The floor construction is 6-in. concrete. The air flows from the lights through a ducted return. (a) Assuming the lights have been on for 10 hr, compute the cooling load at 5:00 P.M. (b) Assuming that 20 percent of the heat from the lights is convected to the return air, what is the actual cooling load for the space?

8-28   A large office complex has a variable occupancy pattern. Twenty people arrive at 8:00 A.M. and leave at 4:00 P.M. Forty people arrive at 10:00 A.M. and leave 4:00 P.M. Ten people arrive at 1:00 P.M. and leave at 5:00 P.M. Assume seated, light activity and compute the cooling load at 4:00 P.M. and 6:00 P.M. Assume LST.

8-29   A space contains 90 people at 8:00 A.M. The people start leaving at 10:00 A.M. at the rate of 10 people per hour. What is the cooling load at 6:00 P.M. assuming seated, very light work? Assume LST.

8-30   Compute the total cooling load for the structure described by the plans and specifications furnished by the instructor using (a) the CLTD method and (b) the transfer function method using a computer program.

8-31   Estimate the energy requirements for the structure of Problem 8-30b using a computer program.

8-32   Solve Example 8-8 with the building located in (a) Denver, Colorado; (b) Chicago, Illinois; (c) Washington, D.C.

8-33   Solve Example 8-8 with two shifts where the day shift is 7:00 A.M., LST to 4:00 P.M., LST for (a) Oklahoma City, Oklahoma; (b) Denver, Colorado; (c) Chicago, Illinois; (d) Washington, D.C.

8-34   Use the bin method to compute the cooling energy for Problem 8-30a.

8-35   Solve Example 8-8 assuming that the building is rotated 90 degrees clockwise.

CHAPTER NINE

Flow, Pumps, and Piping Design

The distribution of fluids by pipes, ducts, and conduits is essential to all heating and cooling systems. The fluids encountered are gases, vapors, liquids, and mixtures of liquid and vapor (two-phase flow). From the standpoint of overall design of the building system, water and air are of greatest importance. This chapter deals with the fundamentals of incompressible flow of fluids such as air and water in conduits, considers the basics of centrifugal pumps, and develops simple design procedures for water piping systems. Basic principles of the control of fluid-circulating systems including variable flow, secondary pumping, and the relationship between thermal and hydraulic performance of the system are covered.

9-1 FLUID FLOW BASICS

The adiabatic, steady flow of a fluid in a pipe or conduit is governed by the first law of thermodynamics, which may be written

$$\frac{P_1}{\rho_1} + \frac{\overline{V}_1^2}{2g_c} + \frac{gz_1}{g_c} = \frac{P_2}{\rho_2} + \frac{\overline{V}_2^2}{2g_c} + \frac{gz_2}{g_c} + w + \frac{g}{g_c}l_f \qquad (9\text{-}1)$$

where

P = static pressure, lbf/ft^2 or N/m^2
ρ = mass density at a cross section, lbm/ft^3 or kg/m^3
$\overline{V}$ = average velocity at a cross section, ft/sec or m/s
g = local acceleration of gravity, ft/sec^2 or m/s^2
g_c = constant, 32.17 (lbm-ft)/(lbf-sec^2) or 1.0 (kg-m)/(N-s^2)
z = elevation, ft or m
w = work, (ft-lbf)/lbm or J/kg
l_f = lost head, ft or m

Each term of Eq. 9-1 has the units of energy per unit mass or specific energy. The last term on the right in Eq. 9-1 is the internal conversion of energy due to friction. The first three terms on each side of the equality are the pressure energy, kinetic energy, and potential energy, respectively. A sign convention has been selected such that work done on the fluid is negative.

Another governing relation for steady flow in a conduit is the conservation of mass. For flow along a single conduit the mass rate of flow at any two cross sections 1 and 2 is given by

$$\dot{m} = \rho_1 \overline{V}_1 A_1 = \rho_2 \overline{V}_2 A_2 \tag{9-2}$$

where

$\dot{m}$ = mass flow rate, lbm/sec or kg/s
A = cross-sectional area normal to the flow, ft² or m²

When the fluid is incompressible, Eq. 9-2 becomes

$$\dot{Q} = \overline{V}_1 A_1 = \overline{V}_2 A_2 \tag{9-3}$$

where

$$\dot{Q} = \text{volume flow rate, ft}^3/\text{sec or m}^3/\text{s}$$

Equation 9-1 has other useful forms. If it is multiplied by the mass density, an equation is obtained where each term has the units of pressure.

$$P_1 + \frac{\rho_1 \overline{V}_1^2}{2g_c} + \frac{\rho_1 g z_1}{g_c} = P_2 + \frac{\rho_2 \overline{V}_2^2}{2g_c} + \frac{\rho_2 g z_2}{g_c} + \rho w + \frac{\rho g l_f}{g_c} \tag{9-1a}$$

In this form the first three terms on each side of the equality are the static pressure, the velocity pressure, and the elevation pressure, respectively. The work term now has units of pressure and the last term on the right is the pressure lost due to friction.

Finally, if Eq. 9-1 is multiplied by g_c/g, an equation results where each term has the units of length, commonly referred to as *head*.

$$\frac{g_c}{g} \frac{P_1}{\rho_1} + \frac{\overline{V}_1^2}{2g} + z_1 = \frac{g_c}{g} \frac{P_2}{\rho_2} + \frac{\overline{V}_2^2}{2g} + z_2 + \frac{g_c w}{g} + l_f \tag{9-1b}$$

The first three terms on each side of the equality are the static head, velocity head, and elevation head, respectively. The work term is now in terms of head and the last term is the lost head due to friction.

Equations 9-1 and 9-2 are complementary because they have the common variables of velocity and density. When Eq. 9-1 is multiplied by the mass flow rate $\dot{m}$, another useful form of the energy equation results, assuming $\rho = $ constant.

$$\dot{W} = \dot{m} \left[\frac{P_1 - P_2}{\rho} + \frac{\overline{V}_1^2 - \overline{V}_2^2}{2g_c} + \frac{g(z_1 - z_2)}{g_c} - \frac{g}{g_c} l_f \right] \tag{9-4}$$

where

$$\dot{W} = \text{power, work per unit time, } \frac{\text{ft-lbf}}{\text{sec}} \text{ or W}$$

All terms on the right-hand side of the equality may be positive or negative except the lost energy, which must always be positive.

Some of the terms in Eqs. 9-1 and 9-4 may be zero or negligibly small. When the fluid flowing is a liquid, such as water, the velocity terms are usually rather small and can be neglected. In the case of flowing gases, such as air, the potential energy terms are usually very small and can be neglected; however, the kinetic energy terms may be quite important. Obviously the work term will be zero when no pump, turbine, or fan is present.

The *total pressure*, a very important concept, is the sum of the static pressure and the velocity pressure.

$$P_0 = P + \frac{\rho \overline{V}^2}{2g_c} \tag{9-5}$$

In terms of head Eq. 9-5 is written

$$\frac{g_c P_0}{g\rho} = \frac{g_c P}{g\rho} + \frac{\overline{V}^2}{2g} \tag{9-5a}$$

or

$$H_0 = H + H_v \tag{9-5b}$$

Equations 9-1 and 9-4 may be written in terms of total head and with rearrangement of terms become

$$\frac{g_c}{g} \frac{(P_{01} - P_{02})}{\rho} + (z_1 - z_2) = \frac{g_c w}{g} + l_f \tag{9-1c}$$

This form of the equation is much simpler to use with gases because the term $(z_1 - z_2)$ is negligible, and when no fan is in the system, the lost head equals the loss in total pressure head.

Lost Head

For incompressible flow in pipes and ducts the lost head is expressed as

$$l_f = f \frac{L}{D} \frac{\overline{V}^2}{2g} \tag{9-6}$$

where

f = Moody friction factor
L = length of the pipe or duct, ft or m
D = diameter of the pipe or duct, ft or m

Figure 9-1. Friction factors for pipe flow.

$\overline{V}$ = average velocity in the conduit, ft/sec or m/s
g = acceleration due to gravity, ft/sec² or m/s²

The lost head has the units of feet or meters of the fluid flowing. For conduits of noncircular cross section the diameter may be replaced by the hydraulic diameter D_h given by

$$D_h = \frac{4(\text{cross-sectional area})}{\text{wetted perimeter}} \tag{9-7}$$

Use of the hydraulic diameter is restricted to turbulent flow and cross-sectional geometries without extremely sharp corners.

Figure 9-1 shows friction data correlated by Moody (1), which is commonly referred to as the Moody diagram. Figure 9-2 is an example of relative roughness data for various types of pipe and tubing. The friction factor is a function of the Reynolds number Re, and the relative roughness e/D of the conduit in the transition zone; is a function of only the Reynolds number for laminar flow; and is a function of only relative roughness in the complete turbulence zone. Notice that the friction factor can be read directly from Fig. 9-2 if the Reynolds number and relative roughness are sufficiently large for the flow to be considered as in complete turbulence.

The Reynolds number is defined as

$$\text{Re} = \frac{\rho \overline{V} D}{\mu} = \frac{\overline{V} D}{v} \tag{9-8}$$

where

ρ = mass density of the flowing fluid, lbm/ft³ or kg/m³
μ = dynamic viscosity, lbm/(ft-sec) or (N-s)/m²
v = kinematic viscosity, ft²/sec or m²/s

The hydraulic diameter is used to calculate Re when the conduit is noncircular. Appendix B contains viscosity data for water, air, and refrigerants. The *ASHRAE Handbook* (2) has data on a wide variety of fluids.

In water systems it is often necessary to use a brine solution, usually a mixture of ethylene glycol and water. Figure 9-3 gives specific gravity and viscosity data for water and various solutions of ethylene glycol and water. Note that the viscosity is given in units of centipoise [1 lbm/(ft-sec) = 1490 centipoises and 10³ centipoise = 1 (N-s)/m²]. The following example demonstrates calculation of lost head for pipe flow.

Example 9-1.   Compare the lost head for water and a 30 percent ethylene glycol solution flowing at the rate of 110 gallons per minute (gpm) in a 3-in. standard (Schedule 40) commercial steel pipe, 200 ft in length. The temperature of the water is 50 F.

Solution.   Equation 9-6 will be used. From Table D-1 the inside diameter of 3-in. nominal-diameter Schedule 40 pipe is 3.068 in. and the inside cross-

Figure 9-2. Relative roughness of pipe materials and friction factors for complete turbulence.

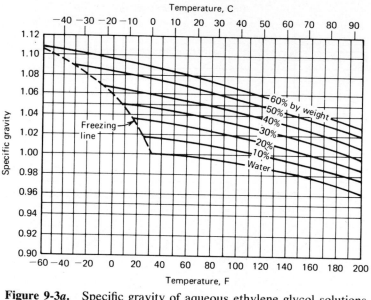

Figure 9-3a.   Specific gravity of aqueous ethylene glycol solutions. (Adapted by permission from *ASHRAE Handbook, Fundamentals Volume*, 1985.)

sectional area for flow is 0.0513 ft². The Reynolds number is given by Eq. 9-8 and the average velocity in the pipe is

$$\overline{V} = \frac{\dot{Q}}{A} = \frac{110 \text{ gal/min}}{(7.48 \text{ gal/ft}^3)(0.0513 \text{ ft}^2)} = 287 \text{ ft/min} = 4.78 \text{ ft/sec}$$

The absolute viscosity of pure water at 50 F is 1.4 centipoises or 9.4×10^{-4} lbm/(ft-sec) from Fig. 9-3b. Then

$$\text{Re} = \frac{62.4(4.78)(3.068/12)}{9.4 \times 10^{-4}} = 8.1 \times 10^4$$

From Fig. 9-2 the relative roughness e/D is 0.00058 for commercial steel pipe. Then the flow is in the transition zone and the friction factor f is 0.021 from Fig. 9-1. The lost head for pure water is then computed using Eq. 9-6.

$$l_{fw} = 0.021 \times \frac{200}{(3.068/12)} \frac{(4.78)^2}{2(32.2)} = 5.83 \text{ ft of water}$$

The absolute viscosity of the 30 percent ethylene glycol solution is 3.1 centipoises from Fig. 9-3b, and its specific gravity is 1.042 from Fig. 9-3a. The Reynolds number for this case is

$$\text{Re} = \frac{1.042(62.4)(4.78)(3.068/12)}{(3.1)/1490} = 3.8 \times 10^4$$

Figure 9-3b. Viscosity of aqueous ethylene glycol solutions. (Adapted by permission from *ASHRAE Handbook, Fundamentals Volume,* 1985.)

and the friction factor is 0.024 from Fig. 9-1. Then

$$l_{fe} = 0.024 \times \frac{200}{(3.068/12)} \frac{(4.78)^2}{2(32.2)} = 6.66 \text{ ft of E.G.S.}$$

$$= 6.94 \text{ ft of water}$$

The increase in lost head with the brine solution is

$$\text{Percent increase} = \frac{100(6.94 - 5.83)}{5.83} = 19 \text{ percent}$$

System Characteristic

The behavior of a piping system may be conveniently represented by plotting total head versus volume flow rate. Equation (9-1c) becomes

Figure 9-4.   Typical system characteristic.

$$H_p = \frac{g_c(P_{01} - P_{02})}{g\rho} + (z_1 - z_2) - l_f \qquad (9\text{-}1d)$$

where H_p represents the total head required to produce the change in static, velocity, and elevation head and to offset the lost head. If a pump is present in the system, H_p is the total head it must produce for a given volume flow rate. Since the lost head and velocity head are proportional to the square of the velocity, the plot of total head versus flow rate is approximately parabolic, as shown in Fig. 9-4. Note that the elevation head is the same regardless of the flow rate. System characteristics are useful in analyzing complex circuits such as the parallel arrangement of Fig. 9-5. Circuit 1a2 and 1b2 each have a characteristic as shown in Fig. 9-6. The total flow rate is equal to the sum of $\dot{Q}_a$ and $\dot{Q}_b$ and the total head is the same for both circuits; therefore the characteristics are summed for various values of H_p to obtain the curve for the complete system shown as $(a + b)$. Series circuits have a common flow rate and the total heads are additive (Fig. 9-7). More discussion of system characteristics will follow introduction of pumps in Sec. 9-2.

Flow Measurement

It is important to make provisions for the measurement of flow rate in piping and duct systems. The most common devices for making these measurements are the pitot tube, orifice meter, or other restrictive device, and anemometers, which are generally direct velocity-measuring instruments. The pitot tube and orifice meter will be discussed here because they have greatest application in duct and piping systems. Figure 9-8 shows a pitot tube installed

Figure 9-5.   Arbitrary parallel flow circuit.

Figure 9-6. System characteristic for parallel circuits.

in a duct. The pitot tube senses both total and static head. The difference, the velocity head, is measured with a manometer. If the pitot tube is very small relative to the duct size, then local measurements and traverses may be made. When Eq. 9-1 is applied to a streamline between the tip of the pitot tube and a point a short distance upstream, the following equation results (the lost energy is assumed to be negligibly small and the mass density constant):

$$\frac{P_1}{\rho} + \frac{V_1^2}{2g_c} = \frac{P_2}{\rho} = \frac{P_{02}}{\rho} \tag{9-9}$$

or

$$\frac{P_{02} - P_1}{\rho} = \frac{V_1^2}{2g_c} \tag{9-9a}$$

Solving for V_1, we get

Figure 9-7. System characteristic for series circuits.

Figure 9-8. Pitot tube in a duct.

$$V_1 = \left[2g_c \frac{(P_{02} - P_1)}{\rho} \right]^{1/2}$$ (9-10)

Equation 9-10 yields the velocity upstream of the pitot tube. The velocity may be different at other transverse locations in the pipe. For laminar flow the velocity distribution is parabolic while turbulent flow exhibits a more nearly uniform velocity distribution except near the wall.

Figure 9-9 shows a profile that might be typical of turbulent flow. It is generally necessary to traverse the pipe or duct and to integrate either graphically or numerically to find the average velocity in the duct. Equations 9-2 and 9-3 are then used to find the mass or volume flow rate.

The average velocity for incompressible steady flow is expressed by

$$\bar{V}A = \int_A V(x) \, dA$$ (9-11)

which is simply an adaptation of Eq. 9-3. For a circular duct the cross-sectional area A is given by

$$A = \pi x^2 \quad \text{and} \quad dA = 2\pi x \, dx$$ (9-12)

Figure 9-9. Velocity profile in a pipe or duct.

Then, letting x equal r at the wall, Eq. 9-11 becomes

$$\overline{V} = \frac{2}{r^2} \int_0^r V(x)x \, dx \tag{9-13}$$

In fully developed turbulent flow, the most common situation, the velocity profile can be well represented by

$$\frac{V}{V_{max}} = \left(1 - \frac{x}{r}\right)^{1/n} \tag{9-14}$$

where n varies from 6 to 8 over the Reynolds number range of 4×10^3 to 5×10^5 (6). For practical cases of air flow in ducts, $\overline{V}/V_{max}$ is about 0.82 for fully developed flow. In practice it is usually necessary to determine the velocity profile using the pitot tube and to then integrate graphically or numerically. In Fig. 9-10 and Eq. 9-13, the average velocity in the duct is given by

$$\overline{V} = \frac{2}{r^2} \sum_{i=1}^{m} x_i V_i(x) \, \Delta x_i \tag{9-13a}$$

The number of elements m required in Eq. 9-13a depends on how $V(x)$ varies with x. There is usually a central core over which the velocity is relatively constant. There are cases when the velocity profile is not symmetrical, that is, where the velocity is not constant throughout the circular element for a particular radius. In this case velocity profiles should be measured along at least two diameters, which are oriented at 90 degrees. Then for each circular element in the summation of Eq. 9-13a the velocity $V_i(x)$ is the average of the four measured values at the radius x_i.

Pitot tubes are more frequently used to measure the velocity of gases than liquids. With gases such as air the available velocity head is often small, making precise measurements quite important. Micromanometers and inclined or Hook gages are used to measure $(P_{02} - P_1)$ in these cases. A form of the pitot tube is often used in liquid piping circuits to sense the flow rate for control purposes.

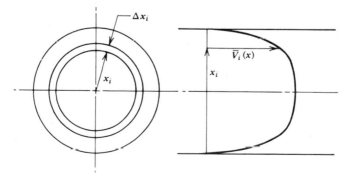

Figure 9-10. Integrating the velocity profile over the pipe cross-sectional area.

Example 9-2.   A pitot tube is installed in an air duct on the center line. The velocity head as indicated by an inclined gage is 0.32 in. of water, the static head is 0.25 in. of water, the air temperature is 60 F, and barometric pressure is 29.92 in. of mercury. Assuming fully developed turbulent flow exists where the average velocity is approximately 82 percent of the center-line value, compute the volume and mass flow rates for a 10-in. diameter duct.

Solution.   The mass and volume flow rates are obtained from the average velocity, using Eqs. 9-2 and 9-3. The average velocity is fixed by the center-line velocity in this case, which is computed by using Eq. 9-10. Since the fluid flowing is air, the density term in Eq. 9-10 is that for air, ρ_a. The pressure difference $P_{02} - P_1$ in Eq. 9-10 is the measured pressure indicated by the inclined gage as 0.32 in. of water. The pressure equivalent of this column of water is given by

$$P_{02} - P_1 = \text{(ft. of water)}\ \frac{g}{g_c}\ \rho_w$$

$$P_{02} - P_1 = \left(\frac{0.32}{12}\right) \text{ft} \left(\frac{0.32}{32.2}\right) \frac{\text{lb}_f}{\text{lbmw}} (62.4) \frac{\text{lbmw}}{\text{ft}^3} = 1.664 \frac{\text{lb}_f}{\text{ft}^2}$$

To get the density of the air we assume an ideal gas

$$\rho_a = \frac{P_a}{R_a T_a} = \frac{(29.92)(0.491)(144)}{(53.35)(60 + 460)} = 0.076 \frac{\text{lbma}}{\text{ft}^3}$$

which neglects the slight pressurization of the air in the duct. The center-line velocity is given by Eq. 9-10

$$V_{ce} = \left[\frac{(2)(32.2)(1.664)}{(0.076)}\right]^{1/2} = 37.6 \text{ ft/sec}$$

and the average velocity is

$$\overline{V} = 0.82\ V_{ce} = (0.82)(37.6) = 30.8 \text{ ft/sec}$$

The mass flow rate is given by Eq. 9-2 with the area given by

$$A = \frac{\pi}{4}\left(\frac{10}{12}\right)^2 = 0.545 \text{ ft}^2$$

$$\dot{m} = \rho_a \overline{V} A = 0.076(30.8)0.545 = 1.28 \text{ lbm/sec}$$

The volume flow rate is

$$\dot{Q} = \overline{V} A = 30.8(0.545)60 = 1007 \text{ ft}^3/\text{min}$$

using Eq. 9-3.

Example 9-3.   Air flows in a 0.3-m diameter circular duct. At the end of a long straight run the center-line velocity is measured to be 12 m/s. Estimate the average velocity and mass flow rate of the air assuming standard conditions for the air.

Solution.   Because the measurement follows a long straight run of pipe, we shall assume that the flow is fully developed and that the velocity profile is symmetrical. It will be necessary to integrate Eq. 9-13 using Eq. 9-14 to describe the velocity profile. The exponent n should be about 7 for this case and can be checked later. Substitution of Eq. 9-14 in Eq. 9-13 and integration with $n = 7$ yields

$$\overline{V} = \frac{2V_{max}}{r^{15/7}} \int_0^r x^{8/7}\, dx = \frac{14}{15} V_{max}$$

and

$$\overline{V} = \frac{14}{15}(12) = 11.2 \text{ m/s}$$

The mass flow rate is given by Eq. 9-2

$$\dot{m} = \rho \overline{V} A = 1.2(11.2)\frac{\pi}{4}(0.3)^2 = 0.95 \text{ kg/s}$$

where the mass density of standard air, 1.2 kg/m³, has been used. To check the value of the exponent n, it is necessary to compute the Reynolds number.

$$\text{Re} = \frac{\rho \overline{V} D}{\mu} = \frac{1.2(11.2)0.3}{20.4 \times 10^{-6}} = 197{,}650$$

where the dynamic viscosity is from Table B-4a. Reference 6 indicates that n equal to 7 is valid for the Reynolds number just computed.

Flow measuring devices of the restrictive type use the pressure drop across an orifice, nozzle, or venturi to predict flow rate. The square-edged orifice is widely used because of its simplicity. Figure 9-11 shows such a meter with the location of the pressure taps (3). The flange-type pressure taps are widely used in HVAC piping systems and are standard fittings available commercially. The orifice plate may be fabricated locally or may be purchased. Reference 3 outlines the manufacturing procedure in detail.

The orifice meter is far from being an ideal flow device and introduces an appreciable loss in total pressure. An empirical *discharge coefficient* is

$$C = \frac{\dot{Q}_{actual}}{\dot{Q}_{ideal}} \tag{9-15}$$

The ideal flow rate may be derived from Eq. 9-1 with the lost energy equal to zero. Applying Eq. 9-1 between the cross sections defined by the pressure taps gives

Figure 9-11.   Recommended location of pressure taps for use with concentric thin-plate and square-edged orifices according to Reference 3.

$$\frac{P_1}{\rho} + \frac{\overline{V_1^2}}{2g_c} = \frac{P_2}{\rho} + \frac{\overline{V_2^2}}{2g_c} \tag{9-16}$$

To eliminate the velocity $\overline{V}_1$ from Eq. 9-16, Eq. 9-3 is recalled and

$$\overline{V}_1 = \overline{V}_2 \frac{A_2}{A_1} \tag{9-3a}$$

Substitution of Eq. 9-3a into Eq. 9-16 and rearrangement yields

$$\overline{V}_2 = \frac{1}{[1 - (A_2/A_1)^2]^{1/2}} \left[2g_c \frac{(P_1 - P_2)}{\rho} \right]^{1/2} \tag{9-17}$$

Then by using Eqs. 9-15 and 9-17 we get

$$\dot{Q}_{\text{actual}} = \frac{CA_2}{[1 - (A_2/A_1)^2]^{1/2}} \left[2g_c \frac{(P_1 - P_2)}{\rho} \right]^{1/2} \tag{9-18}$$

The term $[1 - (A_2/A_1)^2]^{1/2}$ is referred to as the *velocity of approach factor*. In practice the discharge coefficient and velocity of approach factor are often combined and called the *flow coefficient C_d*:

$$C_d = \frac{C}{[1 - (A_2/A_1)^2]^{1/2}} \tag{9-19}$$

This is merely a convenience. For precise measurements other corrections and factors may be applied, especially for compressible fluids (3). Figure 9-12 shows representative values of the flow coefficient C_d. The data apply to pipe diameters over a wide range (1 to 8 in.) and to flange or radius taps within about 5 percent. When precise flow measurement is required, reference 3 should be consulted for more accurate flow coefficients.

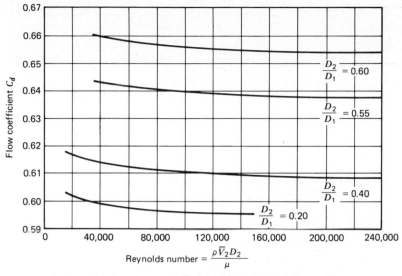

Figure 9-12.   Flow coefficients for square-edge orifices.

In complex piping systems it is necessary to install meters at strategic locations so that the system can be balanced when it is put into operation. The designer must specify an orifice that, when the desired flow rate is attained, will have a reasonable pressure drop but one not large enough to be wasteful of pumping power. The following example illustrates an orifice design procedure.

Example 9-4.   Design an orifice to be used in a 4-in. standard (schedule 40) commercial steel pipe that has a design flow rate of 200 gal per minute (gpm). The pipe will supply chilled water in summer at 45 F and in winter hot water at 150 F. A change in head, H_{12} of 5 to 10 in. of mercury is acceptable.

Solution.   Referring to Fig. 9-12, we note that the flow coefficient curves become independent of Reynolds number at high Reynolds numbers. It would therefore be desirable to operate in this region. It is also desirable to operate with a diameter ratio of 0.5 to 0.6 because the flow coefficient data seem to be most accurate in this range. Therefore, it is assumed that $D_2/D_1 = 0.55$ and $C_d = 0.637$. The change in head $\Delta P/\rho$ is calculated from Eq. 9-18.

$$\frac{P_1 - P_2}{\rho} = \left(\frac{\dot{Q}}{C_d A_2}\right)^2 \frac{1}{2g_c}$$

$$D_2 = 0.55 D_1 = 0.55(4.026) = 2.21 \text{ in.}$$

where the inside diameter of 4-in. pipe is 4.026 in.

$$A_2 = \frac{\pi}{4} \left(\frac{2.21}{12} \right)^2 = 0.027 \text{ ft}^2$$

$$\dot{Q} = \frac{200}{60(7.48)} = 0.446 \text{ ft}^3/\text{sec}$$

where 7.48 gal equals 1 ft³

$$\frac{P_1 - P_2}{\rho} = \left[\frac{0.446}{(0.637)0.027} \right]^2 \frac{1}{2(32.2)} = 10.38 \frac{\text{ft-lbf}}{\text{lbmw}}$$

or

$$H_{12} = \frac{10.38}{12.55} \times 12 = 9.92 \text{ in. of mercury}$$

where the effective specific gravity of the manometer fluid is 12.55. This pressure drop is satisfactory. The Reynolds number will now be computed to check the assumed flow coefficient. The largest expected value of the viscosity is used to compute the Reynolds number because this yields the lowest Re number. If we use Eq. 9-8 and data from Fig. 9-3.

$$\text{Re} = \frac{\rho \bar{V}_2 D_2}{\mu} = \frac{62.4(0.446/0.027)(2.21/12)}{(1.6/1490)} = 1.8 \times 10^5$$

Figure 9-13a. A single-inlet, flexible-coupled centrifugal pump. (Courtesy of ITT Bell and Gossett, Skokie, Illinois.)

From Fig. 9-12 note that the flow coefficient C_d is nearly constant at Reynolds numbers greater than about 2×10^5. For this particular case the Reynolds number is sufficiently large to be acceptable.

9-2 CENTRIFUGAL PUMPS

The centrifugal pump is by far the most frequently used type of pump in HVAC systems. The essential parts of a centrifugal pump are the rotating member or impeller and the surrounding case. The impeller is usually driven by an electric motor that may be close-coupled (on the same shaft as the impeller) or flexible coupled (Fig. 9-13a). The fluid enters the center of the rotating impeller, is thrown into the volute, and flows outward through the diffuser (Fig. 9-13b). The fluid leaving the impeller has high kinetic energy that is converted to static pressure in the volute and diffuser as efficiently as possible. Although there are various types of impellers and casings (4), the principle

Figure 9-13b. Cutaway of single-inlet, flexible-coupled centrifugal pump. (Courtesy of ITT Bell and Gossett, Skokie, Illinois.)

Figure 9-14a. Centrifugal pump performance data.

Figure 9-14b. Centrifugal pump performance data.

349

of operation is the same. The pump shown in Fig. 9-13 is a single suction pump because the fluid enters the impeller from only one side. The double-suction type has fluid entering from both sides. A pump may be staged with more than one impeller on the same shaft with one casing. The fluid leaves the first stage and enters the impeller of the second stage before leaving the casing.

Pump performance is most commonly given in the form of curves. Figure 9-14 is an example of such data for a pump that may be operated at two different speeds with several different impellers. For each speed a different curve is given for each impeller. These curves give the total dynamic head, efficiency, shaft power, and the net positive suction head as a function of capacity.

The *total dynamic head* furnished by a pump can be understood by applying Eq. 9-1b to the fluid entering and leaving the pump.

$$H_p = \frac{wg_c}{g} = \frac{g_c(P_1 - P_2)}{g\rho} + \frac{\overline{V}_1^2 - \overline{V}_2^2}{2g} + (z_1 - z_2) \tag{9-20}$$

The lost head is unavailable as useful energy and is omitted from the equation. Losses are typically accounted for by the *efficiency*, defined as the ratio of the useful power actually imparted to the fluid to the shaft power input:

$$\eta_p = \frac{\dot{W}}{\dot{W}_s} = \frac{\dot{m}w}{\dot{W}_s} = \frac{\rho\dot{Q}w}{\dot{W}_s} \tag{9-21}$$

The shaft power may be obtained from Eq. 9-21

$$\dot{W}_s = \frac{\dot{m}w}{\eta_p} = \frac{\rho\dot{Q}w}{\eta_p} = \frac{\rho\dot{Q}H_pg}{\eta_pg_c} \tag{9-22}$$

Therefore a definite relationship exists between the curves for total head, efficiency, and shaft power in Fig. 9-14.

If the static pressure of the fluid entering a pump approaches the vapor pressure of the liquid too closely, vapor bubbles will form in the impeller passages. This condition is detrimental to pump performance and the collapse of the bubbles is noisy and may damage the pump. This phenomenon is known as cavitation. The amount of pressure in excess of the vapor pressure required to prevent cavitation, known as the *required net positive suction head (NPSHR)*, is a characteristic of a given pump and varies considerably with speed and capacity. NPSHR is determined by the actual testing of each model.

Whereas each pump has its own NPSHR, each system has its own *available net positive suction head (NPSHA)*.

$$\text{NPSHA} = \frac{P_sg_c}{\rho g} + \frac{\overline{V}_s^2}{2g} - \frac{P_vg_c}{\rho g} \tag{9-23}$$

where

$g_cP_s/\rho g$ = static head at the pump inlet, ft or m, absolute
$\overline{V}_s^2/2g$ = velocity head at the pump inlet, ft or m
$g_cP_v/\rho g$ = static vapor pressure head of the liquid at the pumping temperature, ft or m, absolute

The net positive suction head available must always be greater than the NPSHR or noise and cavitation will result.

Example 9-5.   Suppose the pump of Fig. 9-14 is installed in a system as shown in Fig. 9-15. The pump is operating at 3500 rpm with the 6-in. impeller and delivering 200 gpm. The suction line is standard 4-in. pipe that has an inside diameter of 4.026 in. Compute the NPSHA and compare it with the NPSHR. The water temperature is 60 F.

Solution.   From Fig. 9-14 the NPSHR is 10 ft of head. The available net positive suction head is computed from Eq. 9-23; however, the form will be changed slightly through the application of Eq. 9-1b between the water surface and the pump inlet.

$$\frac{P_B g_c}{\rho g} = \frac{P_s g_c}{\rho g} + \frac{\overline{V}_s^2}{2g} + z_s + l_f$$

or

$$\frac{P_s g_c}{\rho g} + \frac{\overline{V}_s^2}{2g} = \frac{P_B g_c}{\rho g} - z_s - l_f$$

Then Eq. 9-23 becomes

$$\text{NPSHA} = \frac{P_B g_c}{\rho g} - z_s - l_f - \frac{P_v g_c}{\rho g} \tag{9-23a}$$

Assuming standard barometric pressure

$$\frac{P_B g_c}{\rho g} = \frac{29.92}{12} \times 13.55 = 33.78 \text{ ft of water}$$

$$\frac{P_v g_c}{\rho g} = \frac{0.2562 \times 144}{62.4} = 0.59 \text{ ft of water}$$

where P_v is read from Table A-1. Then from Eq. 9-23a

$$\text{NPSHA} = 33.78 - 10 - 5 - 0.59 = 18.19 \text{ ft of water}$$

which is almost twice as large as the NPSHR. If the water temperature is increased to 160 F and other factors remain constant, the NPSHA becomes

$$\text{NPSHA} = 33.78 - 10 - 5 - \left(\frac{4.74 \times 144}{61}\right) = 7.6 \text{ ft}$$

which is less than the NSPHR of 10 ft. Cavitation will undoubtedly result.

In an open system such as a cooling tower the pump suction (inlet) should be flooded if possible; that is, the inlet should be lower than the free water surface. This is particularly important if the fluid temperature is high such as in the case of condensate from a steam condenser. Long runs of suction piping should be eliminated whenever possible, and care should be taken to eliminate

Figure 9-15.   Open system with suction lift.

air traps on the suction side of the pump. In an open reservoir or sump the end of the suction pipe should be adequately covered with fluid to prevent entrainment of air from the vortex formed at the pipe entrance. An inlet velocity of less than 3 ft/sec or 1 m/s will minimize vortex formation. Care must be taken to locate the pump in a space where freezing will not occur and where maintenance may be easily performed.

The pump foundation, usually concrete, should be sufficiently rigid to support the pump base plate. This is particularly important for flexible-coupled pumps to maintain alignment between the pump and motor. The pump foundation should weigh from $1\frac{1}{2}$ to 3 times the total pump and motor weight for vibration and sound control.

It is extremely important to install the suction and discharge piping such that forces are not transmitted to the pump housing. Expansion joints are required on both the suction and discharge sides of the pump to isolate expansion and contraction forces, and the piping must be supported independent of the pump housing. In-line pumps, as in Fig. 2-12b, are an exception to this rule since they are designed to be supported by the piping. However, the piping must still be supported on both sides of the pump.

9-3   COMBINED SYSTEM AND PUMP CHARACTERISTICS

The combination of the system and pump characteristics (head versus capacity) is very useful in the analysis and design of piping systems. Figure 9-16 is an example of how a system with parallel circuits behaves with a pump installed. Recall that the total head H_p furnished by the pump is given by Eq. 9-20. Note that the combination operates at point t, where the characteristics cross. The pump and system must both operate on their characteristics; therefore, the point where they cross is the only possible operating condition. This concept is very important in understanding more complex systems. The flow rate for each of the parallel circuits in Fig. 9-16 is quite obvious because the required change in total head from 1 to 2 is equal for both circuits.

Figure 9-17 illustrates a series-type circuit. When the valve is open, the operating point is at point a with flow rate $\dot{Q}_a$ and total head H_a. Partial closing

Figure 9-16. Combination of system and pump characteristics for parallel circuits.

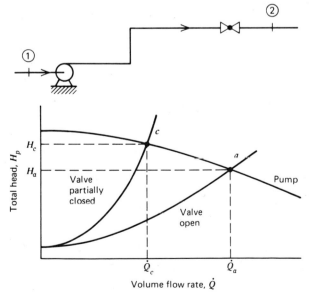

Figure 9-17. Combination of system and pump characteristics for series circuits.

353

of the valve introduces additional flow resistance (head loss) and is similar to adding series resistance in an electrical circuit. The new system characteristic crosses the pump curve at point c and the flow rate is $\dot{Q}_c$ with total head H_c. All piping systems should contain valves for control and adjustment purposes.

A typical design problem is one of pump selection. The following example illustrates the procedure.

Example 9-6.   A water piping system has been designed to distribute 150 gpm and the total head requirement is 36 ft. Select a pump using the data of Fig. 9-14 and specify the power rating for the electric motor.

Solution.   Figure 9-18 shows the characteristic for the piping system as it was designed. Point 0 denotes the operating capacity desired. Examination of Fig. 9-14 indicates that the low-speed version of the given pump covers the desired range. The desired operating point lies between the curves for the $6\frac{1}{2}$ and 7 in. impellers. The curves are sketched on Fig. 9-18. Obviously, the pump with the 7-in. impeller must be selected, but the flow rate will be about 160 gpm as indicated by point 1. Therefore a valve must be adjusted (closed slightly) to modify the system characteristic as shown, to obtain 150 gpm at point 2. Again referring to Fig. 9-14 the shaft power requirement is read as about 2.3 hp. Electric motors usually have an efficiency of 85 to 90 percent and a 3-hp motor should be specified.

Variable Pump Speed

It is common to vary the impellor rotational speed in controlling a water-distribution system. The flow rate, head, and shaft power are related to the

Figure 9-18.   Example of a pump selection for a given system.

new and old speeds. The laws governing this relationship are known as the *Affinity Laws for Pumps*. They may be stated as

$$\dot{Q}_n = \dot{Q}_o \frac{\text{rpm}_n}{\text{rpm}_o} \tag{9-24}$$

$$H_{pn} = H_{po} \left[\frac{\text{rpm}_n}{\text{rpm}_o} \right]^2 \tag{9-25}$$

$$\dot{W}_{sn} = \dot{W}_{so} \left[\frac{\text{rpm}_n}{\text{rpm}_o} \right]^3 \tag{9-26}$$

The affinity laws may be used in conjunction with the system characteristic to generate a new pump head characteristic. The total system will operate where the new pump characteristic and old system characteristic cross.

Example 9-7. The 1750 rpm pump with 7-in. impeller of Fig. 9-14 is operating in a system as shown as point 1 of Fig. 9-19. It is desired to reduce the pump speed until the flow rate is 100 gpm. Find the new pump head, shaft power, and efficiency.

Solution: From the system characteristic it may be observed that the pump must produce 25 ft of heat at a flow rate of 100 gpm. This is one

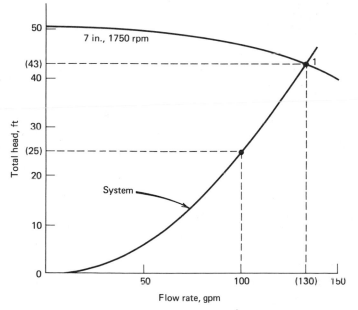

Figure 9-19. Pump and system characteristics for Example 9-7.

point on the new pump characteristic. The new pump speed can be found from either Eq. 9-24 or 9-25. Using Eq. 9-24

$$\text{rpm}_n = \text{rpm}_o \, (\dot{Q}_n/\dot{Q}_o)$$
$$= 1750 \, (100/130) = 1346$$

The new shaft power is given by Eq. 9-26 with $W_{so} = 2.1$ hp from Fig. 9-14.

$$\dot{W}_{sn} = 2.1 \, (1346/1750)^3 = 0.96 \text{ hp}$$

The pump efficiency could be recalculated using Eq. 9-21. However, it may be deduced from the affinity laws that the efficiency will remain constant at about 69.4 percent.

$$\frac{\eta_{pn}}{\eta_{po}} = \frac{\dot{Q}_n H_{pn}/\dot{W}_{sn}}{\dot{Q}_o H_{po}/\dot{W}_{so}} = 1$$

9-4 PIPING SYSTEM DESIGN

There are many different types of piping systems used with HVAC components and there are many specialty items and refinements that make up these systems. Chapters 13, 14, and 15 of reference 5 give a detailed description of various arrangements of the components making up the complete system. The main thrust of the discussion to follow is to develop methods for the design of basic piping systems used to distribute hot and chilled water. Chapter 32 of reference 2 pertains to the sizing of pipe. The basic concepts will first be covered and applied to simple open- and closed-loop systems. The principles involved in designing larger variable-flow systems using secondary pumping will then be discussed in Sec. 9-5.

Open-loop System

A typical open-loop piping system is shown in Fig. 9-20. Characteristically an open-loop system will have some part of the circuit open to the atmosphere. The cooling tower of Fig. 9-20 shows the usual valves, filters, and fittings installed in this type of circuit. It is important to protect the pump with a filter. The isolation valves provide for maintenance without complete drainage of the system, whereas a ball or plug valve should be provided at the pump outlet for adjustment of the flow rate. Expansion joints and a rigid base support are shown to isolate the pump as previously discussed.

Closed-loop System

Two closed-loop systems are shown in Fig. 9-21 and are known as two-pipe and four-pipe systems. Although it is not necessary to install a filter in a

Figure 9-20. Typical open-loop water system.

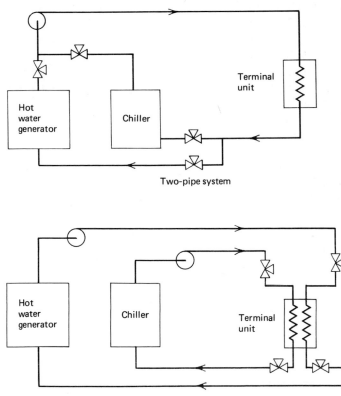

Two-pipe system

Four-pipe system

Figure 9-21. Schematic diagrams of two-pipe and four-pipe flow systems.

closed-loop system, an expansion tank and air separator are required. The expansion tank protects the system from being damaged by a change in volume due to temperature variations; provides a collecting space for air removed by the separator; and creates a point of constant pressure in the system. The expansion tank is covered fully later. Closed-loop systems are normally pressurized so that the complete loop will be above atmospheric pressure. A pressure regulator is installed in the makeup water line, which is connected to the system at the same place as the expansion tank. Isolation and control values are required, and instrumentation to measure flow rates and temperatures at strategic locations is essential. This is especially important when parallel circuits are employed.

Pipe Sizing Criteria

Piping systems often pass through or near occupied spaces where noise generated by the flowing fluid may be objectionable. Therefore, a velocity limit of 4 ft/sec or 1.2 m/s for 2-in. pipe and smaller is generally imposed. For larger sizes a limit on the head loss of 4 ft per 100 ft of pipe is imposed. This corresponds to about 0.4 kPa/m in SI units. These criteria should not be treated as hard rules but rather as guides. Noise is caused by entrained air, abrupt pressure drops, and turbulence in general. If these factors can be minimized, the given criteria can be relaxed. Open systems such as cooling tower circuits are usually remote from occupied spaces. Therefore, somewhat higher velocities may be used in such a case. A reasonable effort to design a balanced system will prevent drastic valve adjustments and will contribute to a quieter system.

The piping layout for a heating and air-conditioning system depends on the location of the central and terminal equipment and the type of system to be used. All of the piping may be located in the central equipment room or may run throughout the building to terminal units in every room. In this case the available space may be a controlling factor. It must be kept in mind that piping for domestic hot and cold water, sewage, and other services must be provided in addition to the heating and air-conditioning requirements. The designer must constantly check to make sure the piping will fit into the allowed space.

Pipe Sizing

After the piping layout has been completed, the problem of sizing the pipe consists mostly of applying the design criteria discussed before. Where possible the pipes should be sized so that drastic adjustments are not required. Often an ingenious layout helps in this respect. The system and pump characteristics are also useful in the design process.

To facilitate the actual pipe sizing and computation of head loss, charts such as those shown in Figs. 9-22 and 9-23 for pipe and copper tubing have been developed. Figures 9-22 and 9-23 are based on 60 F or 16 C water and give head losses that are about 10 percent high for hot water. Examination of Figs. 9-22, and 9-23 shows that head loss may be obtained directly from the flow rate and nominal pipe size or from flow rate and water velocity. When the head loss and flow rate are known, a pipe size and velocity may be obtained.

Pipe fittings and valves also introduce losses in head. These losses are usually accounted for by use of a *resistance coefficient K*, which is the number of velocity heads lost because of the valve or fitting. Thus,

$$l_f = K \frac{\overline{V}^2}{2g} \tag{9-27}$$

Comparing this definition with Eq. 9-6 it can be seen that

$$K = f \frac{L}{D} \tag{9-27a}$$

The ratio L/D is the equivalent length in pipe diameters of straight pipe that will cause the same pressure drop as the valve or fitting under the same flow conditions. This is a convenient concept to use when one is computing head loss in a piping system. Representative values of resistance coefficients for some common valves and fittings are given in Fig. 9-24a. Conversions between K, L/D, and L can be obtained for various pipe sizes by the use of Fig. 9-24b.

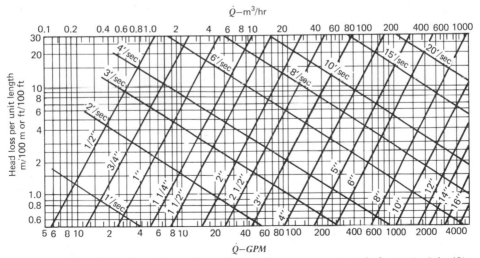

Figure 9-22. Friction loss due to flow of water in commercial steel pipe (schedule 40). (Reprinted by permission from *ASHRAE Handbook, Fundamentals Volume*, 1977.)

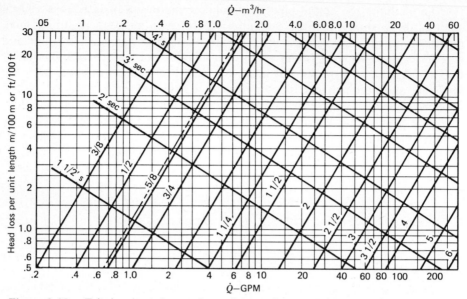

Figure 9-23. Friction loss due to flow of water in type L copper tube. (Reprinted by permission from *ASHRAE Handbook, Fundamentals Volume*, 1977.)

When using SI units it is suggested that the *L/D* ratio be determined from Fig. 9-24*b*, using the nominal pipe size. The equivalent length in meters may then be determined using the inside diameter *D* in meters. The lost head for a given length of pipe of constant diameter and containing fittings is computed as the product of the lost head per foot from Figs. 9-22 and 9-23 and the total equivalent length of the pipe and fittings.

> **Example 9-8.**   Compute the lost head for a 150 ft run of standard pipe, having a diameter of 3 in. The pipe run has 3 standard 90 degree elbows, a globe valve, and a gate valve. One hundred gpm of water flows in the pipe.
>
> *Solution.*   The equivalent length of the various fittings will first be determined by using Figs. 9-24*a* and 9-24*b*.
>
> $$\text{Globe Valve: } K_1 = 340\ f_t, \ f_t = 0.018;\ \text{Fig. 9-24}a\ \text{and Table 9-1}$$
> $$K_1 = 340(0.018) = 6.1$$
> $$L = 86\ \text{ft; Fig. 9-24}b$$
> $$\text{Elbow: } K = 30\ f_t, \ f_t = 0.018$$
> $$K = 30(0.018) = 0.54$$
> $$L = 8\ \text{ft}$$
> $$\text{Gate Valve: } K_1 = 8\ f_t, \ f_t = 0.018$$
> $$K_1 = 8(0.018) = 0.14$$
> $$L = 2\ \text{ft}$$

Figure 9-24a. Resistance coefficients K for various valves and fittings. (Courtesy of the Crane Company, Technical Paper No. 410.)

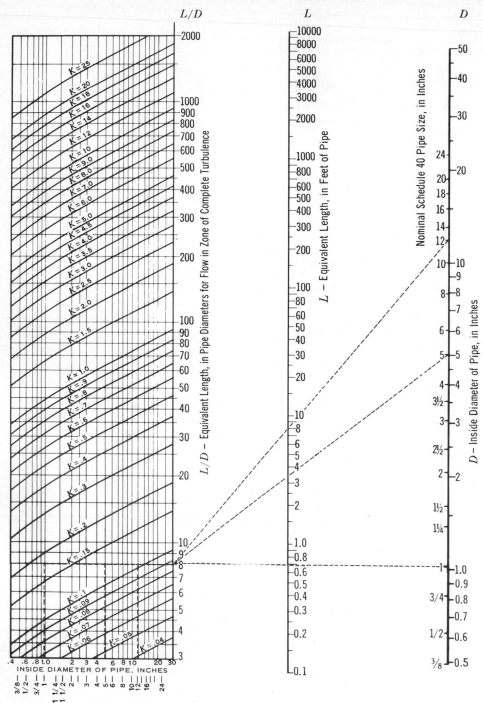

Figure 9-24b. Equivalent lengths L and L/D and resistance coefficient K. (Courtesy of the Crane Company, Technical Paper No. 410.)

Table 9-1 FORMULAS, DEFINITION OF TERMS, AND VALUES OF f_t FOR FIGURE 9-24

Formula 1 $K_2 = \dfrac{K_1 + \left(\sin\dfrac{\theta}{2}\right)0.8(1 - \beta^2) + 2.6(1 - \beta^2)^2}{\beta^4}$

Formula 2 $K_2 = \dfrac{K_1 + 0.5\left(\sin\dfrac{\theta}{2}\right)(1 - \beta^2) + (1 - \beta^2)^2}{\beta^4}$

$\beta = \dfrac{D_1}{D_2}; \quad \beta^2 = \left(\dfrac{D_1}{D_2}\right)^2 = \dfrac{A_1}{A_2}; \quad D_1 = \text{smaller diameter} \\ A_1 = \text{smaller area}$

Nominal Size	$\frac{1}{2}$ in.	$\frac{3}{4}$ in.	1 in.	$1\frac{1}{4}$ in.	$1\frac{1}{2}$ in.	2 in.
Friction Factor (f_t)	0.027	0.025	0.023	0.022	0.021	0.019
$2\frac{1}{2}$, 3 in.	4 in.	5 in.	6 in.	8–10 in.	12–16 in.	18–24 in.
0.018	0.017	0.016	0.015	0.014	0.013	0.012

The total equivalent length is then

Actual length of pipe	150 ft
One globe valve	86 ft
Three elbows	24 ft
One gate valve	2 ft
Total	262 ft

From Fig. 9-22 the lost head l'_f is 2.3 ft per 100 ft of length or

$$l'_f = 2.3 \times 10^{-2} \text{ ft/ft of length}$$

The lost head for the complete pipe run is then given by

$$l_f = L_e l'_f = (262)2.3 \times 10^{-2} = 6.0 \text{ ft}$$

The lost head for control valves, check valves, strainers, and other such devices is often given in terms of a coefficient C_v. The coefficient is numerically equal to the flow rate of water at 60 F in gpm, which will give a pressure drop of one lbf/in.^{2} (2.31 ft of water). Because the head loss is proportional to the square of the velocity, the pressure drop or lost head may be computed at other flow rates:

$$\frac{l_{f1}}{l_{f2}} = \left(\frac{\dot{Q}_1}{\dot{Q}_2}\right)^2 \tag{9-28}$$

In terms of the coefficient C_v:

$$l_f = 2.31 \left(\frac{\dot{Q}}{C_v}\right)^2 \tag{9-29}$$

where $\dot{Q}$ and C_v are both in gpm and l_f is in ft of water.

It may be shown that the flow rate of any fluid is given by

$$\dot{Q} = C_v \left[\frac{\Delta P(62.4)}{\rho}\right]^{1/2} \tag{9-30}$$

where ΔP is in lbf/in.^{2} and ρ is in lbm/ft^3.

There is a relationship between C_v and the resistance coefficient K. By using Eqs. 9-3 and 9-6, we can show that

$$C_v = \frac{0.208\, D^2}{\sqrt{K}}$$

where D is in feet. In SI units a flow coefficient C_{vs} is defined as the flow rate of water at 15 C in m^3/s with a pressure loss of 1 kPa given by

$$C_{vs} = 1.11\, \frac{D^2}{\sqrt{K}} \tag{9-31}$$

where D is in meters.

Example 9-9. A strainer has a C_v rating of 60. It is to be used in a system to filter 50 gpm of water. What pressure drop and head loss can be expected?

Solution. Equation (9-29) will yield the desired result:

$$l_f = 2.31 \left(\frac{50}{60}\right)^2 = 1.6 \text{ ft of water}$$

The pressure drop ΔP is given by

$$\Delta P = \frac{1.6(62.4)}{144}\, \frac{g}{g_c} = 0.7 \text{ lbf/in.}^2$$

Heating and cooling units and terminal devices usually have head-loss information furnished by the manufacturer. The head loss is often used to indicate the flow rate for the adjustment of the system. Equation 9-28 may be used to estimate head loss at other than specified conditions.

There is no one set procedure for pipe sizing. The following example will demonstrate some approaches to the problem.

Figure 9-25. Two-pipe system-design example.

Example 9-10. Figure 9-25 shows a two-pipe water system such as might be found in a central equipment room. The terminal units *a*, *b*, and *c* are air-handling units that contain air-to-water finned tube heat exchangers. An actual system would probably contain both a heater and a chiller; only one or the other is to be considered here. Size the piping and specify the pumping requirements.

Solution. The first step is to select criteria for sizing of the pipe. Because the complete system is confined to a central equipment room, the velocity and head loss criteria may be relaxed somewhat. Let the maximum velocity be 5 ft/sec and the maximum head loss be about 7 ft per 100 foot in the main run. Somewhat higher values may be used in the parallel circuits. By using Fig. 9-22 we select pipe sizes as follows:

$$D_{23} = D_{81} = 2\tfrac{1}{2} \text{ in.}; \quad \overline{V} = 4 \text{ ft/sec}; \quad l'_f = 2.6 \text{ ft/100 ft}$$
$$D_{34} = D_{78} = 1\tfrac{1}{2} \text{ in.}; \quad \overline{V} = 4\tfrac{3}{4} \text{ ft/sec}; \quad l'_f = 6.5 \text{ ft/100 ft}$$
$$D_{45} = D_{67} = 1 \text{ in.}; \quad \overline{V} = 3\tfrac{3}{4} \text{ ft/sec}; \quad l'_f = 7.4 \text{ ft/100 ft}$$

The lost head for the various sections is now computed. The equivalent lengths used for the various fittings are approximate in the following analysis and are summed with the actual pipe section lengths. In actual practice the fittings should be identified and the equivalent lengths found as shown in Examples 9-8 and 9-9. A three-dimensional piping diagram would also be used to describe the piping layout in detail.

Section 8 to 3 (exclusive of the heater or chiller)

$$L_{83} = 140 \text{ ft}$$
$$l_{f83} = 2.6(140)/100 = 3.6 \text{ ft}$$

Sections 3 to 4 and 7 to 8

$$L_e = L_{43} + L_{78} = 2(15) = 30 \text{ ft}$$
$$l_{f34} + l_{f78} = 30(6.5)/100 = 2.0 \text{ ft}$$

Sections 4 to 5 and 6 to 7

$$L_e = L_{45} + L_{67} = 60 \text{ ft}$$
$$l_{f45} + l_{f67} = 60(7.4)/100 = 4.4 \text{ ft}$$

Chiller, Unit c and Orifice

$$l_f = 14 + 10 + 5 = 29 \text{ ft}$$

The total lost energy for this pipe run including fittings and equipment is

$$\frac{(P_{02} - P_{01})}{\rho} = (l_f)_t \frac{g}{g_c} = 3.6 + 2.0 + 4.4 + 29 = 39.0 \text{ (ft-lbf)/lbm}$$

Applying Eq. 9-1c to just the pump, we get

$$w = \frac{(P_{01} - P_{02})}{\rho}$$

Therefore

$$w = -39.0 \text{ (ft-lbf)/lbm}$$

which is equivalent to stating that the pump must develop 39 ft of head.
The two remaining parallel circuits will now be sized to balance the system within reason.

Section 4 to 7

$$(l_{f47})_t = l_{f4567} = 4.4 + 10 + 5 = 19.4 \text{ ft}$$
$$L_{47} = 45 \text{ ft}$$
$$l_{f47} = 19.4 - 15 = 4.4 \text{ ft}$$

$$(l'_f)_{47} = \frac{4.4(100)}{45} = 9.8 \text{ ft/100 ft}$$

From Fig. 9-22

$$D_{47} = 1\tfrac{1}{4} \text{ in.}; \qquad \overline{V} = 4\tfrac{1}{3} \text{ ft/sec}$$

Section 3 to 8

$$(l_{f38})_t = l_{f3478} = (l_{f47})_t + l_{f34} + l_{f78} = 19.4 + 2.0 = 21.4 \text{ ft}$$
$$L_{38} = 40 \text{ ft}$$
$$l_{f38} = 21.4 - 20 = 1.4 \text{ foot}$$

$$(l'_f)_{38} = \frac{1.4(100)}{40} = 3.5 \text{ ft/100 ft}$$

From Fig. 9.22

$$D_{38} = 2 \text{ in.;} \qquad \overline{V} = 3 \text{ ft/sec}$$

Only minor adjustments should be required when the system is put into operation.

The sizing of pipe and calculation of lost head follows the same procedure for larger and more complex systems. In the sections to follow this will become evident.

The Expansion Tank

The expansion tank is a much more important element of a piping system than generally assumed. It was mentioned earlier that the expansion tank provides for changes in volume and may be part of the air elimination system, and it establishes a point of constant pressure in the system. This latter purpose is most important. A point of constant pressure is necessary to establish the pressure at other points of the closed-loop system; otherwise the system would be like an electrical circuit without a *ground*. The location of the expansion tank then becomes an important design consideration. One rule can be stated that has no exceptions. *A system, no matter how large or complex, must have only one expansion tank.* Consider the piping system shown in Fig. 9-25. The pressure regulator in the makeup water line establishes the pressure in the expansion tank and the pipe at point 1, except for a small amount of elevation head. The pressure at any other point in the system may then be computed relative to point 1 using Eq. 9-1. Note that the arrangement shown in Fig. 9-25 will produce positive pressures throughout the system, assuming that the pressure at point 1 is positive, the tank pressure is usually between about 10 and 50 psig. If the expansion tank were located at point 2 in Fig. 9-25, it would be possible to have negative pressures in the system depending on the lost head for the system. The pressure would be lowest at point 1. It is not possible to state one fixed rule for location of the expansion tank; however, it is usually best to locate the tank and pump as shown in Fig. 9-25 in a chilled water loop so that the pump is discharging into the system. A hot water boiler requires a different approach because it must be equipped with a safety relief valve and improper location of the expansion tank and pump may cause unnecessary opening of the relief valve. Therefore, the expansion tank should be located at the boiler outlet with the pump located either just upstream or just downstream of the boiler. Again the pressures in the system should be analyzed to ensure that positive pressures occur throughout.

Sizing of the expansion tank is important and depends on the total volume of the system, the maximum and minimum system pressures and temperatures, the piping material, the type of tank, and how it is installed. Expansion tanks are of two types. The first type is simply a tank where air is compressed above the free liquid–air interface by system pressure. The second type has a balloonlike bladder within the tank that contains the air. The bladder does not fill the complete tank and is inflated, prior to filling the system, to the pressure setting of the makeup water pressure regulator. Either tank can be used in hot or chilled water systems; however, the first type is usually used in hot water systems because it provides a convenient place for air to collect when released from the heated water in the boiler. The second, bladder type, is usually applied with chilled water systems because cold water tends to absorb the air in the free surface type of tank and release it elsewhere in the system where it is removed. This process may eventually lead to a *water-logged* system where no compressible volume exists. Drastic structural damage can occur with a water-logged system.

Relations may be derived for sizing of the expansion tanks by assuming the air behaves as an ideal gas. The type of tank and the way it is employed in the system then influence the results. Consider the free liquid–air interface type where the water in the tank always remains at its initial temperature (uninsulated and connected by a small pipe); the expansion and compression of the air in the tank is isothermal; and the air in the tank is initially at atmospheric pressure. The resulting relation for the tank volume is

$$V_T = \frac{V_w \left[\left(\dfrac{v_2}{v_1} - 1 \right) - 3\alpha\, \Delta t \right]}{\dfrac{P_a}{P_1} - \dfrac{P_a}{P_2}} \tag{9-32}$$

where

V_T = expansion tank volume, ft³ or m³
V_w = volume of water in the system, ft³ or m³
P_a = local barometric pressure, psia or kPa
P_1 = pressure at lower temperature, t_1 (regulated system pressure), psia or kPa
P_2 = pressure at higher temperature, t_2 (some maximum acceptable pressure), psia or kPa
t_1 = lower temperature (initial fill temperature for hot water system or operating temperature for chilled water system), F or C

t_2 = higher temperature (some maximum temperature for both hot and chilled water systems), F or C

v_1 = specific volume of water at t_1, ft³/lbm or m³/kgm

v_2 = specific volume of water at t_2, ft³/lbm or m³/kgm

α = linear coefficient of thermal expansion for the piping, F^{-1} or C^{-1}, $6.5 \times 10^{-6} F^{-1}$ or $11.7 \times 10^{-6} C^{-1}$ for steel pipe, and 9.3×10^{-5} F^{-1} or $16.74 \times 10^{-6} C^{-1}$ for copper pipe

Δt = higher temperature minus the lower temperature, F or C

If the initial air charge in the tank is not compressed from atmospheric pressure but rather is forced into the tank at the design operating pressure, as with a bladder-type tank, and then expands or compresses isothermally, the following relation results:

$$V_T = \frac{V_w \left[\left(\dfrac{v_2}{v_1} - 1 \right) - 3\alpha \, \Delta t \right]}{1 - \dfrac{P_1}{P_2}} \tag{9-33}$$

where the variables are defined as for Eq. 9-32.

The expansion tank must be installed so that the assumptions outlined before for the size calculations, Eqs. 9-32 and 9-33, are valid. This generally means that the expansion tank is not insulated and is connected to the main system by a relative long, small-diameter pipe so that water from the system does not circulate into the expansion tank. The following example demonstrates the expansion tank problem.

Example 9-11. Compute the expansion tank volume for a chilled water system that contains 2000 gal of water. The system is regulated to 10 psig at the tank with an operating temperature of 45 F. It is estimated that the maximum water temperature during extended shutdown would be 100 F and a safety relief valve in the system is set for 35 psia. Assume standard barometric pressure and steel pipe.

Solution: A bladder-type would be the best choice; however, calculations will be made for both types. Equation 9-32 will give the volume of the free liquid–air interface type tank where $v_2 = 0.01613$ ft³/lbm and $v_1 = 0.01602$ ft³/lbm from Table A-1.

$$V_{TF} = \frac{2000 \left(\dfrac{0.01613}{0.01602} - 1 \right) - 3(6.5 \times 10^{-6})(55)}{\left(\dfrac{14.696}{24.696} - \dfrac{14.696}{49.696} \right)}$$

$$V_T = 38.7 \text{ gal or } 5.2 \text{ ft}^3$$

Equation 9-33 will give the volume of the bladder-type tank

$$V_{TB} = \frac{2000\left[\left(\dfrac{0.01613}{0.01602} - 1\right) - 3(6.5 \times 10^{-6})(55)\right]}{1 - \dfrac{24.696}{49.696}}$$

$V_{TB} = 23.0$ gal or 3.1 ft^3

Note that the volume of the bladder-type tank is less than the free-surface type. This is an advantage in large systems.

9-5 CONTROL OF HYDRONIC SYSTEMS

The small system shown in Fig. 9-25 has no provisions for control other than the hand-operated balancing valves and the expansion tank. As shown, the flow rate of water is constant everywhere in the system, the chiller cycles off at some selected minimum water temperature, and the air side of the system is controlled by a space thermostat. The cooling coils in such a system are said to *run wild*. The need to control the flow of water in the coils in response to the load so that the partial load characteristics of the space can be met was discussed in Chapter 3. The most feasible way of matching the water-side to the air-side load is to regulate the amount of water flowing through the coil. Two ways to do this are shown in Fig. 9-26. (a) A two-way valve may be used to throttle the flow to maintain a relatively fixed water temperature leaving the coil or (b) a three-way mixing valve may be used to bypass some of the flow with remixing downstream of the coil. In both cases (a) and (b) the coil receives the same flow of water and the temperature leaving the coil at T is the same; however, the overall effect of the two different control methods on the system is different. The two-way valve produces a variable flow rate with a fixed water temperature differential, whereas the three-way valve produces a fixed flow rate (downstream of the valve) and a variable water temperature differential. We will see later that the two-way valve control method is preferred although the three-way valve method has been the most popular. High energy costs have caused designers to reevaluate their design methods and as a consequence many advances in hydronic system design have recently evolved. One reason for the use of three-way valves is the requirement to maintain a constant flow rate of water through the source unit, the chiller or boiler. Note in Fig. 9-25 that if two-way valve control is used on each coil, the flow rate through the chiller will decrease as the load becomes lower and lower. This is intolerable and may cause severe damage to the chiller. Therefore, the three-way valve control method, which maintains a constant flow rate, is a better choice for a small system with a single chiller or boiler. As systems become large in capacity and have extensive piping systems that may serve several buildings from one

(a)

Two–way valve

(b)

Three–way valve

Figure 9-26.   Basic control valve types.

central plant, it is not feasible to use one pump to serve the complete system. In such a case it is desirable to interconnect many subsystems into one integrated system containing several pumps. This type of system can use the two-way valve control, which results in water flow rates proportional to the load and more economical operation.

Example 9-12.   The coils in Fig. 9-26 are identical and require 20 gpm of water at full load. The water enters at 45 F and the flow controllers are set for 55 F. A partial-load condition exists where the flow rate through the coil is reduced to 12 gpm. Find the temperature of the water being returned to the chiller for each type of control valve.

Solution: The two-way valve system will return the water at 55 F assuming the controller can maintain the water temperature leaving the coil at exactly 55 F. In the case of the three-way valve system, water at 45 F is mixed with 55 F water leaving the coil. An energy balance on the valve yields

$$\dot{Q}_1 T_1 + \dot{Q}_2 T_2 = \dot{Q}_3 T_3$$

$$T_3 = \frac{\dot{Q}_1 T_1 + \dot{Q}_2 T_2}{\dot{Q}_3} = \frac{8(45) + 12(55)}{20} = 51 \text{ F}$$

It is clear that the two-way valve leads to variable flow and a fixed temperature differential, whereas the three-way valve gives a constant flow rate with a variable temperature differential.

9-6 LARGE SYSTEM DESIGN

A piping system operating with an incompressible fluid like water is sensitive to disturbances such as valves opening or closing or pumps cycling on and off. Therefore, with several pumps all in the same system, ranging from large to small, care must be taken that valve action or pump cycling in one part of the system does not cause a problem in another section, such as overloading a pump or an idle pump being driven like a turbine. This leads to the conclusion that interconnected circuits each containing a pump must be dynamically and hydraulically isolated. This is known as *the principle of the common pipe*.

Consider Fig. 9-27a, which shows a secondary load circuit connected to a primary circuit. Under full design load the three-way valve is in its normal position and there is no flow in Section C. This clearly places the secondary pump in series with the primary pump and sets up a potential for trouble. To overcome this situation a pipe common to both circuits is installed as shown in Fig. 9-27b. The common pipe is always designed to have lost head approaching zero by making it very short in length and of ample size. There is now no possibility of the primary circuit affecting the secondary pump because the secondary circuit is dynamically isolated from the primary circuit. To understand this concept, remember that the secondary pump must operate at the

Figure 9-27a. Potentially troublesome secondary piping circuit.

Figure 9-27b.   Common pipe used to isolate secondary from primary
circuit.

point where its characteristic crosses the characteristic of the secondary circuit
and continuity must exist at every tee. Consider junction A with the three-way
valve in its normal position. The secondary circuit is designed for a given flow
rate, say 100 gpm, using three-way valve control of the coils. The system is
balanced so that the flow rate from the primary circuit is 100 gpm. Therefore,
a mass balance at junction A shows that the common pipe has no flow and the
100 gpm leaves the secondary circuit at point B through the normally open
(NO) port of the three-way valve. At a partial load, some of the 100 gpm flow
in the secondary circuit, say 40 gpm, will recirculate in the common pipe and
mix with 60 gpm of water from the primary supply circuit while the remaining
40 gpm of the 100 gpm from the supply circuit will flow through section C and
the three-way valve, mixing with 60 gpm from the secondary circuit. Remember
the three-way valve will only let water out of the secondary circuit when it
reaches a set temperature, say 55 F.

   If the coils in the secondary circuit of Fig. 9-27b were controlled using
two-way valves, it would be a variable-flow secondary circuit. In this case there
would be no flow in the common pipe and the excess flow from the primary
circuit would flow through section C. The primary supply circuit is still a
constant flow system. The arrangement of Fig. 9-27b is not intended to be a
model system but only a means of presenting the common pipe principle.

   The systems of Fig. 9-27 using coils with either three-way or two-way
valve control returns water to the source at a variable temperature depending
on the total system load. This is generally not desirable in large chilled water
systems because the chillers cannot operate at their full capacity unless the
water returns at some reasonably high temperature compatible with their de-

sign. Because two-way valve control produces a fixed temperature rise in the water and a variable flow rate, the primary circuit should be of this type.

Variable Volume Flow

Let us consider a secondary load circuit using two-way valves as shown in Fig. 9-28a with the primary circuit three-way valve replaced with a two-way valve. Suppose the secondary circuit is designed for 100 gpm and for the present assume $t_s = t_p$. Then when balanced at full load, the design flow will enter the tee at A, the secondary pump will take it all and there will be no flow in the common pipe. Likewise at B, the design flow will exit the secondary circuit and flow through the two-way valve into the primary return. Now consider a partial load condition with a 60 gpm flow to the coils, caused by the response of the two-way coil valves to decreased space load. The secondary pump will respond as shown in Fig. 9-28b, backing up on its characteristic. Then, 60 gpm will enter the tee at A and flow to the pump. The same amount will exit at B with no flow in the common pipe. Note that this arrangement will produce a fixed temperature differential across the primary circuit and a variable flow rate to the primary supply circuit.

It is often the case that the primary supply water temperature t_p is lower than the secondary circuit requires, t_s. For this case the temperature sensor is moved from point C to point D in Fig. 9-28a. This arrangement will cause some secondary water to flow in the common pipe and mix at point A to obtain water at point D at a temperature t_s. We must now provide a suitable primary circuit to accommodate the chillers or boilers.

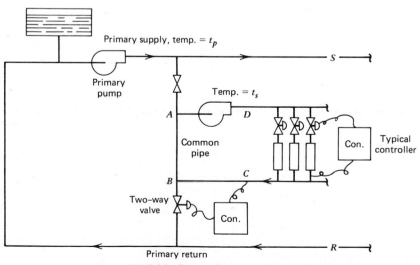

Figure 9-28a.   Variable-flow primary and secondary circuits.

Figure 9-28b.   Secondary system and pump characteristics.

The primary distribution system shown in Fig. 9-28 is simply another piping and pumping system. As the coil control valves throttle back the flow in the secondary circuits, the primary pump will back up on its characteristic curve in the same way as the secondary pump, as shown in Fig. 9-28b. Methods of controlling the pumps to achieve additional energy savings will be covered later. It is interesting to note the sequence of control events. The load on a coil dictates its flow requirement and its two-way valve is the primary control. The secondary pump is the next response followed by the response of the primary pump. Note that only one simple control system is required, the two-way valves.

The next problem is to add the source system apparatus to the primary supply circuit. One method of configuration is shown in Fig. 9-29, where two chillers are shown, each with its own pump. There is no limit on the number of source elements (chillers or boilers), but there should be at least two and they do not have to be of the same capacity. Each source pump should be designed to match the circuit from B to C and A to D, whereas the primary pump is sized to match the primary distribution circuit including the two-way valves but excluding the secondary circuits and the source (chiller or boiler) circuits. Note that the common pipe shown in the primary distribution circuit, Fig. 9-29, decouples the source pumps from the primary pump.

To understand the flow dynamics of the system shown in Fig. 9-29, assume a total design flow of 500 gpm, equal-sized chillers, and supply and return water temperatures of 42 F and 55 F, respectively. Fig. 9-28a shows the type of secondary circuits. Assume that the secondary circuits use 42 F water and the two-way valves control the flow so that 55 F water leaves the coils. Then at

Figure 9-29.   Variable-flow system with constant flow chillers.

full design load the total flow rate of 500 gpm is divided evenly between the two chillers; the primary pump is operating at 500 gpm and the total flow is divided among the various secondary circuits, which are operating at their full design load. The primary circuit common pipe has no flow in either direction. Now suppose the various loads have decreased and the coil two-way valves have reduced the flow in the secondary circuits so that the total required flow is 400 gpm. The primary pump will back up on its characteristic to 400 gpm; however, the chiller pumps are unaffected by the actions of the secondary and primary pumps because of the common pipes and continue to move 250 gpm each. Consider the tee at A. Continuity requires that 250 gpm go to chiller 1 and 150 gpm flow to the tee at B. Chiller 2 requires 250 gpm; therefore, continuity requires that 100 gpm flow from the common pipe to the tee at B. In other words, 100 gpm of the 250 gpm entering the tee at C must be returned through the common pipe to the tee at B. The chillers are controlled by thermostats at A and B. At this partial-load condition note that chiller 1 receives water at 55 F and remains fully loaded. Chiller 2, however, now receives water at a lower temperature (about 50 F) and is not fully loaded. Chiller 2 is said to *unload,* which means it will operate at less than its full capacity using less power input. As the coil loads continue to decrease, the primary circuit flow will continue to decrease. When the total flow reaches 250 gpm, chiller 2 and its pump will cycle off because all of the flow through it will be diverted to the common pipe and reenter it at B. Further reduction in flow below 250 gpm will cause chiller 1 to unload and eventually cycle off when the total coil load reaches zero. As the coil loads increase from zero, the primary flow will increase, causing warm water to flow toward points A and B. Thermostat A will activate chiller 1 and its pump, which operates until the total flow exceeds 250 gpm, when thermostat B starts chiller 2 and its pump. Note that no matter how many chillers are used, they will unload from left to right and load from right to left. Also note that all the chillers that are operating are fully loaded except one that may be partially loaded. This permits maximum operating efficiency. Further, this type system provides the minimum flow of water to meet the space load, which leads to low pumping costs. Finally notice that only a bare minimum of controls are needed for a rather extensive distribution system. The common pipes dramatically reduce the complexity because any one circuit can be designed and analyzed independent of the others.

Figure 9-30 shows a variation on the location of the common pipe in the primary distribution circuit. Analysis of this arrangement shows that the chillers will load and unload equally, which means that most of the time none are fully loaded unless some extra controls are used to cycle one or more chillers off and on. This type set-up is sometimes used where the load is relatively constant and operating personnel are present to monitor the system and start and stop one or more units if necessary.

Figure 9-31 shows how the system of Fig. 9-29 can utilize thermal storage. Note that under partial load, the extra chiller capacity cools the water in the storage tank and chiller 2 will not unload until water leaving the tank is at a temperature less than the system return water temperature. At some other time when the total system load exceeds the total chiller capacity, water flows through both chillers and through the storage tank out into the distribution system. Note that the primary pump has a capacity greater than the total capacity of the chiller pumps.

The design and sizing of the piping and pumps for the large variable-flow system follow the same general procedures given for constant-flow systems in

Figure 9-30.   Chillers arranged to share the load equally.

Figure 9-31.   Variable-flow system adapted to a storage system.

Sec. 9-4. Each part of the variable-flow system is designed for full load. Partial-load operation is then controlled as described previously.

Pump Control

The primary and secondary pumps in a variable-flow system may be controlled in some way rather than allowing them to back up on their characteristics. This is desirable for operating economy and also prevents overpressuring of the two-way valves, which may vibrate or leak if too large a pressure differential exists. The control method most frequently used is to reduce pump speed in response to a critical pressure differential some place in the circuit. For example, the path to and from one particular coil in a secondary circuit will require the greatest pressure differential of all the coils in that circuit. Therefore, the differential pressure sensor for the pump motor speed control should be located across that coil and control valve and set so that the pump will always produce enough head for that coil. Frequently the critical coil is the one located farthest from the pump. The primary pump will be controlled in the same general way. In this case, the critical secondary circuit must be identified and the pressure sensor located accordingly.

The foregoing discussions of water system control and design have generally referred to chilled water systems. However, the concepts apply to all kinds of source elements for both heating and cooling.

Control Valve Characteristics and Selection

The selection of control valves is an important step in hydronic system design. There are two main considerations. These are the size or head loss at design flow and the relation of flow to valve plug lift, the *valve characteristic*.

In sizing control valves care must be taken that the valve is not so large that its control range is very small. That is, it is undesirable for a large change in flow to result from a small lift of the valve plug. To prevent this the valve should be selected to have about the same head loss when fully open as the element being controlled. For example, a two-way valve for a coil with a head loss of 10 ft of water should have a head loss of about 10 ft of water. The valve head loss is determined from its rated C_v coefficient, as discussed earlier in the chapter.

The design of the valve plug depends on the liquid medium for the application such as hot water, chilled water, or steam. A valve plug designed as shown in Fig. 9-32b is said to be linear, as shown by curve A in Fig. 9-32a, whereas a plug shaped as shown in Fig. 9-32c is for an equal percentage valve as shown by curve B in Fig. 9-32a. The requirement for different characteristics relates to the temperature changes for the fluids at decreased loads.

For steam the heat exchanger load is directly proportional to the flow rate because the condensing vapor is at about the same temperature for all flow rates and a linear valve is quite satisfactory (curve A Fig. 9-32a).

Hot water presents a different problem because a decrease in flow rate is accompanied by an increase in the temperature change of the water. The net result may be only a small reduction in heat exchange for a large reduction in flow. To obtain a better relation between lift and output for this case, an equal percentage valve should be used (curve B of Fig. 9-32a).

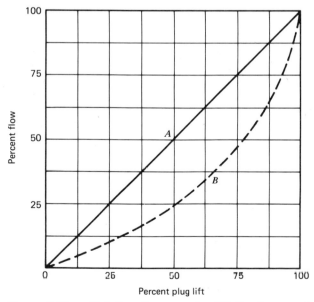

Figure 9-32a. Relative flow vs. plug lift for linear and equal percentage valves.

Figure 9-32b.   Linear or V-port valve.

Figure 9-32c.   Equal percentage valve.

Chilled water coils have a limited water temperature range of 10 to 15 F and can be adequately controlled by a properly sized linear valve.

REFERENCES

1. L. F. Moody, "Friction Factors for Pipe Flow," *Transactions of ASME*, Vol. 66, 1944.

2. *ASHRAE Handbook, Fundamentals Volume*, American Society of Heating, Refrigerating and Air-Conditioning Engineers, Atlanta, Ga., 1985.

3. *Fluid Meters, Their Theory and Application*, American Society of Mechanical Engineers, New York, 1959.

4. *ASHRAE Handbook, Equipment Volume*, American Society of Heating, Refrigerating and Air-Conditioning Engineers, Atlanta, Ga., 1983.

5. *ASHRAE Handbook, HVAC Systems and Applications*, American Society of Heating, Refrigerating and Air-Conditioning Engineers, Atlanta, Ga., 1987.

6. H. Schlichting, *Boundary Layer Theory*, 4th ed., McGraw-Hill, New York, 1960.

7. "Flow of Fluids Through Valves, Fittings, and Pipes," Technical Paper No. 410, The Crane Co., Chicago, Ill., 1976.

PROBLEMS

9-1   Consider the system shown in Fig. 9-33, which has a nozzle at point 2. Compute (a) the work done on the water in (ft-lbf)/lbm, ft of head and lbf/in.^{2}, (b) the power delivered to the water, and (c) sketch the system characteristic.

Figure 9-33.   Sketch for Problem 9-1.

Figure 9-34. Sketch for Problem 9-2.

9-2 The system shown in Fig. 9-34 transfers water to the tank at a rate of 0.015 m³/s through standard commercial steel 4-in. pipe (ID = 103 mm). The total equivalent length of the pipe is 100 m. The increase in elevation is 40 m. Compute (a) the work done on the water in kJ/kg, (b) the power delivered to the water in kW, and (c) sketch the system characteristic.

9-3 The piping of Fig. 9-35 is part of a larger water distribution system. The pressure at point 1 is 10 psig (70 kPa), and the pipe is all the same size. (a) Compute the pressure at points 2, 3 and 4. (b) Sketch the system characteristic for the complete run of pipe.

9-4 A pump located at ground level circulates water in the chilled water system for a 20-story building. Consider a vertical riser from the pump to an equipment room on the twentieth floor. The lost head in the riser is 25 ft (8 m) of water and the pump produces 120 ft (37 m) of head. What must the pressure be on the suction side of the pump for a pressure of 5 psig (35 kPa) to exist in the riser on the twentieth floor? Assume 12 ft (3.7 m) of elevation per floor.

9-5 For the building of Problem 9-4 it is required that the domestic service water pressure be the same on the twentieth floor as supplied by the city water main. Assuming a lost head of 20 ft (6 m) in the distribution riser to the twentieth floor, how much head must a booster pump produce?

$(l_f)_{23} = 20$ ft (6 m)
$(l_f)_{34} = 15$ ft (4.6 m)
$H_p = 80$ ft (24 m)

Figure 9-35. Sketch for Problem 9-3.

Figure 9-36. Schematic for Problem 9-6.

9-6 Consider the piping system shown in Fig. 9-36. Sketch the characteristics for each separate part of the system and combine them to obtain the characteristic for the complete system.

9-7 Compute the lost heat for 100 gpm of 20 percent ethylene glycol solution flowing through 300 ft of $2\frac{1}{2}$-in. commercial steel pipe. The temperature of the solution is 60 F.

9-8 Compute the lost head for 0.006 m³/s of 10 C water flowing in $2\frac{1}{2}$-in. (62 mm) standard commercial steel pipe. The total equivalent length is 100 m.

9-9 A piping system has three parallel circuits. Circuit A requires 20 ft (6 m) of head with a flow rate of 50 gpm (3.2 L/s); circuit B requires 25 ft (7.5 m) of head with a flow rate of 30 gpm (1.9 L/s); and circuit C requires 30 ft (9 m) of head with a flow rate of 45 gpm (2.8 L/s). (a) Construct the characteristic for each circuit and find the characteristic for the combination of A, B, and C. (b) What is the flow rate in each circuit when the total flow rate is 100 gpm (6.3 L/s)? (c) How much head is required to produce a total flow rate of 125 gpm (7.9 L/s)? (d) What is the flow rate in each circuit of part (c)?

9-10 A pitot tube is being used to measure the flow rate of air in a 6-in. diameter duct. The pressure indicated by the inclined gage (velocity head) is 0.25 in. of water. The static pressure is essentially standard atmospheric and the air temperature is 120 F. Assume the average velocity is 80 percent of the center-line velocity. Compute the volume flow rate in cubic feet per minute.

9-11 Assume the measurement made in Problem 9-10 is on the center line and the velocity profile can be described by Eq. 9-14 with $n = 7.5$. Find (a) the average velocity in the duct, (b) the volume flow rate, (c) the mass flow rate, and (d) the ratio of the average to the maximum velocity.

9-12 Saturated water vapor at 101.35 kPa flows in a standard 8-in. pipe (203 mm). A pitot tube located at the center of the pipe shows a velocity head of 10 mm of mercury. Find (a) the velocity of the water vapor at this location and (b) the mass flow rate assuming the average velocity is 82 percent of the maximum velocity.

9-13 Integrate the velocity profile of Problem 9-12, using Eq. 9-14 and assuming $n = 8$. Find the average velocity of the steam. Compute the Reynolds number. Does this agree with the assumed value of n?

9-14 Design an orifice to be used in 2-in. standard commercial steel pipe to measure 60 gpm of 30 percent ethylene glycol solution at 50 F. A pressure drop of 5 to 10 in. of mercury is required.

9-15 A square-edged orifice is installed in standard 4-in. water pipe (103 mm ID). The orifice diameter is 50 mm and a head differential across the orifice of 98 mm of mercury is observed. Compute the volume flow rate of the water assuming a temperature of 10 C. What is the Reynolds number based on the orifice diameter? Does the Reynolds number agree with the flow coefficient?

9-16 The piping system of Problem 9-7 has an increase in elevation of 18 ft from inlet to outlet. (a) Select a pump using Fig. 9-14 and sketch both the system and pump characteristics. (b) How much power is delivered to the fluid in horsepower? (c) How much shaft power is required? (d) If the electric motor has an efficiency of 90 percent, what is the size of the motor required?

9-17 Two $6\frac{1}{2}$-in. 1750 rpm pumps as shown in Fig. 9-14 are used in parallel to deliver 240 gpm at 35 ft of head. (a) Sketch the system and pump characteristics. (b) What is the shaft power requirement of each pump? (c) If one pump fails, what is the flow rate and shaft power requirement of the pump still in operation? (d) Could this type of failure cause a problem in general?

9-18 A 7-in. 3500-rpm pump, shown in Fig. 9-14, is to be used to transfer lake water to a water-treatment plant. The flow rate is to be 300 gpm. What is the maximum height that the pump can be located above the lake surface to prevent cavitation? Assume the water has a maximum temperature of 80 F, the lost head in the suction line is 2 ft of water, and barometric pressure is 28 in. of mercury.

9-19 Select a pump using Fig. 9-14 for a system that requires a flow rate of 250 gpm (15.8 L/s) at a head of 150 ft (45 m) of water. (a) Sketch the pump and system characteristics and show the operating flow rate, efficiency, and power assuming no adjustments. (b) Assume the system has been adjusted to 250 gpm (15.8 L/s) and find the efficiency and power.

9-20 A system requires a flow rate of 210 gpm (13.25 L/s) and a head of 120 ft (37 m). (a) Select a pump from Fig. 9-14 that most closely matches the required flow rate and head and list its shaft power and efficiency. (b) Suppose a 7-in., 3500-rpm pump was selected for the system. At what flow rate and head would the system operate assuming no adjustments? Find the efficiency and power. (c) The system of part (b) is adjusted to a flow rate of 210 gpm (13.25 L/s). What is the efficiency and power? (d) Show the pump and system characteristics of (a), (b), and (c) above on the same graph.

9-21 Refer to Problem 9-20b. Suppose that the pump speed were reduced to obtain 210 gpm (13.25 L/s) and find the rpm, head, efficiency, and shaft power.

9-22 Use the data of Fig. 9-14 and the pump affinity laws and plot a characteristic for a pump with a 7-in. impeller rotating at 2500 rpm.

9-23 Size schedule 40 commercial steel pipe for the given flow rates. Comment on your selections. (a) 40 gpm (2.5 L/s), (b) 30 gpm (1.9 L/s), (c) 15 gpm (0.95 L/s), (d) 50 gpm (3.2 L/s), (e) 100 gpm (6.3 L/s), (f) 1500 gpm (95 L/s).

9-24 Find the lost head for each of the following fittings. (a) 2-in. standard elbow with flow rate of 50 gpm (3.2 L/s), (b) 4-in. globe valve with flow rate of 300 gpm (19 L/s), (c) branch of 3-in. standard tee with 150 gpm (9.5 L/s).

9-25 A control valve has a C_v of 80. It has been selected to control the flow in a coil that requires 120 gpm. What head loss can be expected for the valve?

9-26 Size the piping for the cooling tower circuit shown in Fig. 9-37. The water flow rate is 0.03 m³/s and the total equivalent length of the pipe and fittings is 200 m. Pressure loss for the condenser coil is 35 kPa and the strainer has a C_{vs} of 1.14×10^{-2} m³/s per kPa pressure loss. What is the head requirement for the pump?

9-27 Size the piping for the layout shown in Fig. 9-38 and specify the pump requirements. Assume all the turns and fittings are as shown on the diagram. The pipe is commercial steel. Table 9-2 gives the required data.

9-28 Size the piping and specify pump requirements for a cooling tower installation similar to that shown in Fig. 9-20. The volume flow rate of the water is 300 gpm. The piping is commercial steel. Assume fittings are as shown. The head loss in the condenser is 20 ft of water. C_v for the strainer is 200. The horizontal distance from the condenser to the cooling tower is 80 ft. The vertical distance from the pump to the top of the tower is 30 ft. The tower sump is 12 ft above the pump.

9-29 Size the piping for the layout shown in Fig. 9-38. Assume that fittings are as shown and the pipe is standard commercial steel. Table 9-3 gives pertinent data. Specify pump requirements.

9-30 Determine the volume of a free surface expansion tank for a system similar to that shown in Fig. 9-38. The system volume is 500 gal (1.9 m³). Assume a system gage pressure of 12 psi (83 kPa) and an operating temperature of 42 F (6 C). A maximum temperature and pressure of 110 F (43 C) and 50 psig (345 kPa) are specified. Assume steel pipe.

9-31 Rework Problem 9-30 for a bladder-type expansion tank.

9-32 Find the volume of a free surface expansion tank for a hot water system with a volume of 1200 gal (4.6 m³). The system gage pressure is regulated to 20 psi (140 kPa) at the tank and is initially filled with water at 60 F (16 C). The pressure

Figure 9-37. Sketch for Problem 9-26.

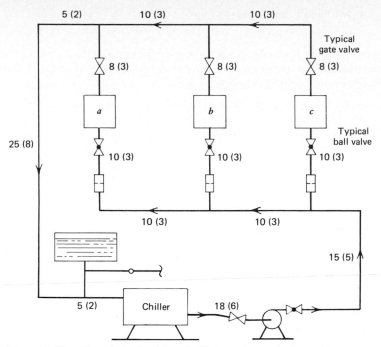

Figure 9-38. Sketch for Problem 9-27. Lengths are in feet with meters in parentheses.

Table 9-2 DATA FOR PROBLEM 9-27

| Unit | gpm | Head Loss in Feet | |
		Coil	Orifice
a	30	15	6
b	40	12	6
c	50	10	6
Chiller	120	20	—

Table 9-3 DATA FOR PROBLEM 9-29

| Unit | m³/s | Head Loss in Meters | |
		Coil	Orifice
a	0.002	5	2
b	0.003	4	2
c	0.0033	4	2
Chiller	0.0083	10	—

385

relief valve on the boiler is set for a gage pressure of 50 psi (345 kPa) and the maximum water temperature is expected to be 220 F (104 C). The system is predominantly copper tubing.

9-33   Consider the system shown in Fig. 9-27b operating at one-third load. Water is supplied at 42 F (6 C) and the control valve is set for 57 F (14 C). The secondary circuit uses three-way valves on its coils and the design full-load water flow rate is 150 gpm (9.5 L/s). (a) How much water is flowing into the tee at A and how much water is flowing in section C? (b) What is the temperature of the water entering the secondary pump? (c) What is the temperature of the water leaving the three-way valve at point D? (d) What size pipe should be used for the main sections of the secondary circuit and the connections to the primary circuit?

9-34   Consider the system shown in Fig. 9-28a. The primary supply water temperature is 40 F (4.5 C) and the controller, with sensor located at D, is set for 47 F (8.3

Figure 9-39.   Preliminary sketch of a central chilled water system.

C). The controllers on the coil valves are set for 57 F (14 C). (a) If the full-load secondary circuit flow rate is 100 gpm (6.3 L/s), how much water must recirculate in the common pipe? (b) How much water is supplied and returned to the primary circuit? (c) Size the main sections of the secondary circuit (*C* and *D*), the common pipe, and the connections to the primary circuit.

9-35   Consider the system shown in Fig. 9-29, where the chillers are of equal size. Assume the system is designed to circulate 1000 gpm (63 L/s) under full-load

Note: Piping is type L copper
        All lengths are total equivalent lengths

Coil	Flow rate gpm (L/s)	Lost head ft (m) Coil	Con. valve
A	40 (2.5)	12 (3.7)	10 (3)
B	40 (2.5)	15 (4.6)	12 (3.7)
C	50 (3.2)	18 (5.5)	15 (4.6)

Figure 9-40.   Schematic of secondary circuit for a variable flow system for Problem 9-38.

Notes: Pipe is schedule 40, commercial steel
All lengths are total equivalent lengths excluding control valves

Circuit	Flow rate gpm (L/s)	Control valve head loss ft (m)
A	60 (3.8)	40 (12)
B	70 (4.4)	50 (15)
C	70 (4.4)	50 (15)

Figure 9-41. Schematic of a primary water circuit for a variable flow system for Problem 9-39.

design conditions at 45 F (7 C) and the secondary circuits utilize water at the same temperature. Water is returned from the secondary circuits at 60 F (15.6 C). At a part-load condition, 650 gpm (41 L/s) of water flows to the secondary circuits. (a) What is the flow rate of the water in the common pipe? (b) What is the temperature of the water at point B? (c) What is the load ratio (load/capacity) for chiller 2? (d) Size the pipe, based on full-load design conditions (except the secondary circuits). (e) The primary pump operates at 3500 rpm when fully loaded. Approximately what speed is required at the part-load condition? (f) What is the power reduction at part load?

9-36   Consider the system in Fig. 9-30, where the chillers are of equal size. Design and part-load operating conditions are the same as Problem 9-35. (a) What is the flow rate of the water in the common pipe? (b) What is the temperature of the water entering both chillers? (c) What is the load ratio for each chiller?

9-37   Consider the preliminary system layout in Fig. 9-39. The final design is to be a variable flow system similar to Fig. 9-29 with two-way valve control throughout. The primary system has three chillers of equal size and each secondary system has two or three air handlers. (a) Sketch the system in more detail showing pumps, common pipes, control valves, and so on. (b) The primary supply water is at 42 F (5.6 C); however, half the buildings require water at about 45 F (7 C). Show how the primary controller sensors are located in all cases. (c) The coil valve controllers are set for 58 F (14 C). Show the flow rates and water temperatures at strategic points in the system at full load. (d) Variable-speed pumps are to be used in the primary and secondary circuits. Show where you would install the differential pressure sensors for the pump speed controllers.

9-38 Size the pipe for the secondary water circuit shown in Fig. 9-40. The pipe is type L copper. Notice that the lengths given are the *total* equivalent lengths excluding the coil and control valves. Select a pump from Fig. 9-14 and sketch the system and pump characteristics.

9-39 Size the pipe for the primary circuit shown in Fig. 9-41. The lengths shown are the *total equivalent lengths* for the section exclusive of the control valve. Specify the primary pump performance requirement.

CHAPTER TEN

Room Air Distribution

The object of air distribution in heating, ventilating, and air-conditioning systems is to create the proper temperature, humidity, air motion, and air quality in the occupied portion of the conditioned room. To obtain comfort conditions within this space, standard limits for an acceptable effective draft temperature have been established. This term comprises air temperature, air motion, relative humidity, and their physiological effect on the human body. Any variation from accepted standards of one of these elements may result in discomfort to the occupants. Discomfort also may be caused by lack of uniform conditions within the space or by excessive fluctuation of conditions in the same part of the space. Such discomfort may arise due to excessive room air temperature variations (horizontally, vertically, or both), excessive air motion (draft), failure to deliver or distribute the air according to the load requirements at the different locations, or rapid fluctuation of room temperature or air motion (gusts).

10-1 AIR MOTION AND COMFORT

A measure of the effective temperature difference between any point in the occupied space and the control conditions is called the *effective draft temperature*. It is defined by the equation proposed by Rydberg and Norback (1).

$$EDT = (t_x - t_r) - M(\overline{V}_x - \overline{V}_r) \tag{10-1}$$

where

t_r = 76 F or 24 C
$\overline{V}_r$ = 30 ft/min or 0.15 m/s
t_x = local air stream dry bulb temperature, F or C
$\overline{V}_x$ = local air stream velocity, ft/min or m/s
M = 0.077 (F-min)/ft or 7.66 (C-s)/m

Equation 10-1 takes into account the feeling of coolness produced by air motion. It also shows that the effect of a one degree F temperature change is equivalent to a 15 ft/min velocity change. In summer the local air stream temperature t_x is usually below the control temperature. Hence both temperature and velocity terms are negative when the velocity $\overline{V}_x$ is greater than $\overline{V}_r$ and both of them add to the feeling of coolness. If in winter $\overline{V}_x$ is above $\overline{V}_r$, it will reduce the feeling of warmth produced by t_x. Therefore, it is usually possible to have zero difference in effective temperature between location x and the control point in winter but not in summer. Research indicates that a high percentage of people in sedentary occupations are comfortable where the effective draft temperature is between -3 F $(-1.7$ C$)$ and $+2$ F $(1.1$ C$)$ and the air velocity is less than 70 ft per minute (0.36 m/s). These conditions are used as criteria for developing the Air Distribution Performance Index (ADPI) to be discussed later.

Conditioned air is normally supplied to air outlets at velocities much higher than would be acceptable in the occupied space. The conditioned air temperature may be above, below, or equal to the temperature of the air in the occupied space. Proper air distribution causes entrainment of room air by the primary air stream and the resultant mixing reduces the temperature differences to acceptable limits before the air enters the occupied space. It also counteracts the natural convection and radiation effects within the room.

10-2 BEHAVIOR OF JETS FROM OUTLETS

The air projection from free round openings, grilles, perforated panels, ceiling diffusers, and other outlets is related to the average velocity at the face of the air supply opening. A free jet has four zones of expansion, and the center-line velocity in any of the zones is related to the initial velocity as shown in Fig. 10-1. Regardless of the type of opening the jet will tend to assume a circular shape. The effect of nearby surfaces will be considered later. In zone III, the most important zone from the point of view of room air distribution, the relation between the jet center-line velocity and the initial velocity is given by

$$\frac{\overline{V}_x}{\overline{V}_0} = K \frac{\sqrt{A_0}}{x} \tag{10-2}$$

or

$$\overline{V}_x = \frac{K\dot{Q}_0}{(\sqrt{A_0}x)} \tag{10-2a}$$

where

  $\overline{V}_x$ = center-line velocity at any x, ft/min or m/s
  $\overline{V}_0$ = initial velocity, ft/min or m/s
  A_0 = area corresponding to initial velocity, ft^2 or m^2
   x = distance from outlet to point of measurement of $\overline{V}_x$, ft or m

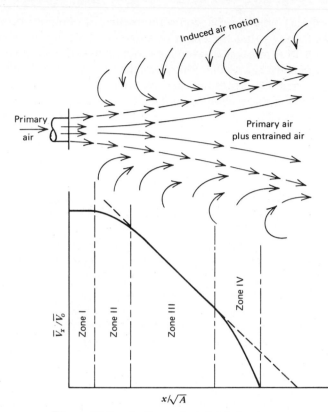

Figure 10-1. Isothermal air-jet behavior.

$\dot{Q}_0$ = air flow rate at outlet, cfm or m³/s
K = constant of proportionality, dimensionless

Equations 10-2 and 10-2a strictly pertain to isothermal free jets, but with the proper A and K the equations define the throw for any type of outlet. The throw is the distance from the outlet to where the maximum velocity in the jet has decreased to some specified value such as 50, 100, or 150 ft/min. The constant K varies from about 6 for free jets to about 1 for ceiling diffusers.

The jet expands because of entrainment of room air; the air beyond zone II is a mixture of primary and induced air. The ratio of the total volume of the jet to the initial volume of the jet at a given distance from the origin depends mainly on the ratio of initial velocity $\bar{V}_0$ to the terminal velocity $\bar{V}_x$. The *induction ratio* is

$$\frac{\dot{Q}_x}{\dot{Q}_0} = C\frac{\bar{V}_0}{\bar{V}_x} \tag{10-3}$$

where

$\dot{Q}_x$ = total air mixture at distance x from the outlet, cfm or m³/s
C = entrainment coefficient, 2 for round, free jet, dimensionless

In zone IV where the terminal velocity is low, Eq. 10-3 will give values about 20 percent high.

When a jet is projected parallel to and within a few inches of a surface, the induction or entrainment is limited on the surface side of the jet. A low-pressure region is created between the surface and the jet, and the jet attaches itself to the surface. This phenomenon results if the angle of discharge between the jet and the surface is less than about 40 degrees and if the jet is within about one foot of the surface. The jet from a floor outlet is drawn to the wall and the jet from a ceiling outlet is drawn to the ceiling. This surface effect increases the throw for all types of outlets and decreases the drop for horizontal jets. The drop is illustrated in Fig. 10-2.

In nonisothermal jets, buoyant forces cause the jet to rise when the air is warm and drop when cool. These conditions result in shorter throws for jet velocities less than 150 ft/min or 0.76 m/s. The lower sketch of Fig. 10-2 shows the drop for a cool jet of air.

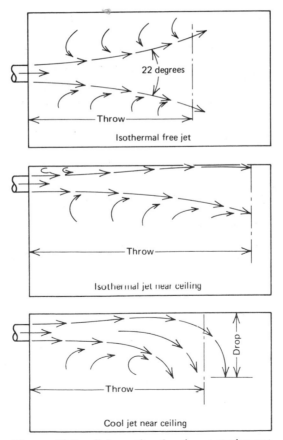

Figure 10-2. Schematic showing entrainment, surface effect, and drop.

The following general statements may be made concerning the characteristics of air jets.

1. Surface effect increases the throw and decreases the drop compared to free space conditions.
2. Increased surface effect may be obtained by moving the outlet away from the surface somewhat so that the jet spreads over the surface after impact.
3. Increased surface effect may be obtained by spreading the jet when it is discharged.
4. Spreading the air stream reduces the throw and drop.
5. Drop primarily depends on the quantity of air and only partially on the outlet size or velocity. Thus the use of more outlets with less air per outlet reduces drop.

Room Air Motion

Room air near the jet is entrained and must then be replaced by other room air. The room air always moves toward the supply and thus sets all the room air into motion. Whenever the average room air velocity is less than about 50 ft/min or 0.25 m/s, buoyancy effects may be significant. In general, about 8 to 10 air changes per hour are required to prevent stagnant regions (velocity less than 15 ft/min or 0.08 m/s). However, stagnant regions are not necessarily a serious condition. The general approach is to supply air in such a way that the high-velocity air from the outlet does not enter the occupied space. The region within 1 ft of the wall and above about 6 ft from the floor is out of the occupied space for practical purposes.

Figure 10-3 shows velocity envelopes for a high sidewall outlet. Equation 10-2 has been used to estimate the throw for the terminal velocities shown. In order to interpret the air motion shown in terms of comfort, it is necessary to

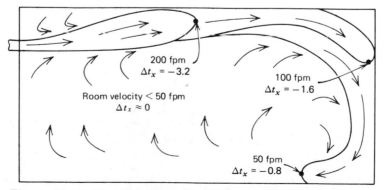

Figure 10-3. Jet and room air velocities and temperatures for $\bar{V}_o = 1000$ ft/min and $\Delta t_o = -20$ F.

estimate the local air temperatures corresponding to the terminal velocities. The relationship between the center-line velocities and the temperature differences is given approximately (3) by

$$\Delta t_x = 0.8 \, \Delta t_0 \, \frac{\overline{V}_x}{\overline{V}_0} \tag{10-4}$$

where Δt_x and Δt_0 are the differences in temperature between the local stream temperature and the room $(t_x - t_r)$ and the outlet air and the room $(t_0 - t_r)$. Equation 10-4 is similar to Eq. 10-1 but slightly more general. Temperatures calculated using Eq. 10-4 are shown in Fig. 10-3. On the opposite wall where the terminal velocity is 100 ft/min the air temperature is 1.6 F below the room temperature. The temperature difference for the 50 ft/min envelope shows that within nearly the entire occupied space the temperature is less than about 0.8 F below the room temperature and the room air motion is under 50 ft/min.

Basic Flow Patterns

The basic flow patterns for the most often used types of outlets are shown in Figs. 10-4 to 10-7 (4). The high-velocity primary air is shown by the dark shading, and the total air is represented by the light shading. These areas represent the high-momentum regions of the room air motion. Natural convection (buoyancy) effects are evident in all cases. Note that stagnant zones always have a large temperature gradient. When this occurs in the occupied space, air needs to be projected into the stagnant region to enhance mixing. An ideal condition would be uniform room temperature from the floor to about

Figure 10-4. Room air distribution patterns for perimeter floor diffuser—spreading jet. (Courtesy of National Environmental Systems Contractors Association, 1501 Wilson Blvd., Arlington, Virginia 22209.)

Figure 10-5. Room air distribution patterns for floor register—nonspreading jet. (Courtesy of National Environmental Systems Contractors Association, 1501 Wilson Blvd., Arlington, Va. 22209.)

6 ft above the floor. However, a gradient of about 4 F or 2 C should be acceptable to about 85 percent of the occupants.

The perimeter-type outlets shown in Fig. 10-4 are generally regarded as superior for heating applications. This is particularly true when the floor is over an unheated space or a slab and where considerable glass area exists in the wall. Diffusers with a wide spread are usually best for heating because buoyancy

Figure 10-6. Room air distribution patterns for high sidewall register. (Courtesy of National Environmental Systems Contractors Association, 1501 Wilson Blvd., Arlington, Va. 22209.)

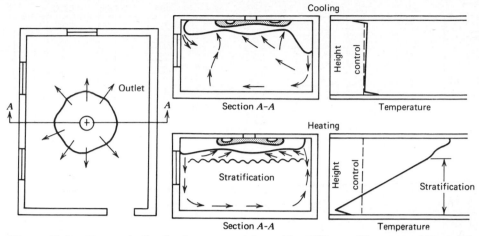

Figure 10-7.   Room air distribution patterns for ceiling diffuser. (Courtesy of National Environmental Systems Contractors Association, 1501 Wilson Blvd., Arlington, Va. 22209.)

tends to increase the throw. For the same reason the spreading jet is not as good for cooling applications because the throw may not be adequate to mix the room air thoroughly. However, the perimeter outlet with a nonspreading jet is quite satisfactory for cooling. Figure 10-5 shows a typical cooling application of the nonspreading perimeter diffuser. It can be seen that the nonspreading jet is less desirable for heating because a larger stratified zone will usually result. Diffusers are available that may be changed from the spreading to nonspreading type according to the season.

The high sidewall type of register shown in Fig. 10-6 is often used in mild climates and on the second and succeeding floors of multistory buildings. This type of outlet is not recommended for cold climates or with unheated floors. Figure 10-6 shows that a considerable temperature gradient may exist between floor and ceiling when heating; however, this type outlet gives good air motion and uniform temperatures in the occupied zone for cooling application.

The ceiling diffuser shown in Fig. 10-7 is very popular in commercial applications and many variations of it are available. The air patterns shown in Fig. 10-7 are typical. Because the primary air is projected radially in all directions, the rate of entrainment is large, causing the high-momentum jet to diffuse quickly. This feature enables the ceiling diffuser to handle larger quantities of air at higher velocities than most other types. Figure 10-7 shows that the ceiling diffuser is quite effective for cooling applications but generally poor for heating. However, satisfactory results may be obtained in commercial structures when the floor is above a heated space.

The return air intake generally has very little effect on the room air motion, but the location may have a significant effect on the performance of the heating and cooling equipment. Because it is desirable to return the coolest air to the

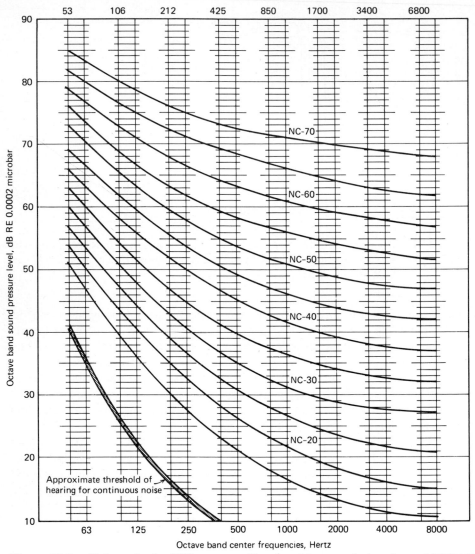

Figure 10-8.   Noise criteria (NC) curves. (Reprinted by permission from *ASHRAE Handbook, Fundamentals Volume*, 1977.)

furnace and the warmest air to the cooling coil, the return air intake should be located in a stagnant region.

Noise

Noise produced by the air diffuser and air can be annoying to the occupants of the conditioned space. Noise associated with air motion usually does not have distinguishable frequency characteristics and the level or loudness is basically a statistically representative sample of human reactions. Loudness contours or curves of equal loudness versus frequency can be established from such reactions.

A method of providing information on the spectrum content of noise is the use of the *noise criteria* (NC) curves and numbers. The NC curves are shown in Fig. 10-8. These are a series of curves constructed using loudness contours, and the speech-interfering properties of noise and are used as a simple means of specifying sound-level limits for an environment by a simple, single-number rating. They have been found to be quite generally applicable for conditions of comfort. In general, levels below a NC of 30 are considered to be quiet, whereas levels above a NC of 50 or 55 are considered noisy. The activity within the space is a major consideration in determining an acceptable level. Table 10-1 gives recommended noise criteria numbers for various applications (5). To determine the acceptability of a given space for a given specification, sound pressure level must be measured at several octave band center frequencies and compared with the specified NC curve of Fig. 10-8. To meet the particular NC rating, the actual octave band reading should lie on or below the NC curve.

Table 10-1   RECOMMENDED NOISE
CRITERIA FOR ROOMS

Application	
Broadcasting studios	[a]
Concert halls	[a]
Legitimate theaters	[a]
Schoolrooms	NC 25–35
Apartments and hotels	NC 30–35
Assembly halls (amplification)	NC 30–35
Homes (sleeping area)	NC 25–30
Conference rooms	NC 25–30
Motion picture theaters	NC 30–35
Hospitals	NC 25–30
Churches	NC 20–30[a]
Courtrooms	NC 30–35
Libraries	NC 30–35
Small private offices	NC 30–35
Restaurants	NC 35–40
Coliseums for sports only	NC 30–40
Stenographic offices	NC 35–45

[a]An acoustics expert should be consulted.

Source: Reprinted by permission from *ASHRAE Handbook, Systems Volume,* 1980.

Some manufacturers of air-diffusing equipment rate their products using the noise criteria concept. An example of these data is presented in the next section.

10-3   AIR-DISTRIBUTION SYSTEM DESIGN

This section discusses the selection and placement of the air outlets. If this is done purely on the basis of comfort, the preceding discussions on room air motion dictate the type of system and the location of the air inlets. However, the architectural design and the functional requirements of the building often override comfort.

When the designer is free to select the type of air-distribution system based on comfort, the perimeter type of system with vertical discharge of the supply air is to be preferred for exterior spaces when the heating requirements exceed 2000 degrees (F) days. This type of system is excellent for heating and satisfactory for cooling when adequate throw is provided. When the floors are warmed and the degree (F) day requirement is between about 3500 and 2000, the high sidewall outlet with horizontal discharge toward the exterior wall is acceptable for heating and quite effective for cooling. When the heating requirement falls below about 2000 degree (F) days, the overhead ceiling outlet or high sidewall diffuser are recommended because cooling is the predominant mode. Interior spaces in commercial structures are usually provided with overhead systems because cooling is required most of the time.

Commercial structures often are constructed in such a way that ducts cannot be installed to serve the desired air-distribution system. Floor space is very valuable and the floor area required for outlets may be covered by shelving or other fixtures, making a perimeter system impractical. In this case an overhead system must be used. In some cases the system may be a mixture of the perimeter and overhead type.

Renovation of commercial structures may represent a large portion of a design engineer's work. Compromises are almost always required in this case, and the air distribution system is often dictated by the nature of the existing structure.

In all cases where an ideal system cannot be used it is particularly important that the air-diffusing equipment be carefully selected and located. Although most manufacturers of air diffusers and grilles furnish extensive data on the performance of their products, there is no substitute for experience and good judgment in designing the air-distribution system.

Table 10-2 gives performance data for a type of diffuser that may be used for perimeter systems having a vertical discharge from floor outlets or as a linear diffuser in the ceiling or sidewall. Note that the data pertain to the capacity, throw, total pressure loss, noise criteria, and free area as a function of the size. It is important to read the notes given with these catalog data. Notice that throw values for three different terminal velocities are given. The diffuser may be almost any length, but its capacity is based on a length of 1 ft, whereas the throw is based on a 4-ft active length, and the NC is based on 10 ft of length. Some corrections are also required when the diffuser is used as a return intake.

Performance data for one type of round ceiling diffuser are shown in Table 10-3, and Table 10-4 shows data for an adjustable diffuser that would generally be used for high sidewall applications. The same general data are given. Note that the diffuser of Table 10-4 has adjustable vanes and throw data are given for three different settings, 0, $22\frac{1}{2}$, and 45 degrees. Figure 10-9 shows a T-bar type diffuser, which is used extensively with modular ceilings. These diffusers are often associated with variable air volume systems and sometimes have automatic flow control built into the diffuser itself. The diffuser shown produces horizontal throw parallel to the ceiling in opposing directions. Table 10-5 gives performance data for the T-bar diffuser.

Return grilles are quite varied in design. The construction of the grille has very little to do with the overall performance of the system except to introduce some loss in pressure and noise if not properly sized. The appearance of a return grille is important, and the louver design is usually selected on this basis. Table 10-6 gives data for one style of grille. Note that the capacity, pressure loss, and noise criteria are the main performance data given.

The *Air Distribution Performance Index* (ADPI) was mentioned in connection with Eq. 10-1. The ADPI is defined as the percentage of measurements taken at many locations in the occupied zone of a space that meet the -3 F to 2 F effective draft temperature criteria. The objective is to select and place the air diffusers so that an ADPI approaching 100 percent is achieved. Note that ADPI is based only on air velocity and effective draft temperature, a local temperature difference from the room average, and is not directly related to the level of dry bulb temperature or relative humidity. These effects and other factors such as mean radiant temperature must be accounted for as discussed in Chapter 3. The ADPI provides a means of selecting air diffusers in a rational way. There are no specific criteria for selection of a particular type of diffuser except as discussed before, but within a given type the ADPI is the basis for selecting the throw. The space cooling load per unit area is an important consideration. Heavy loading tends to lower the ADPI. However, loading does not influence design of the diffuser system significantly. Each type of diffuser has a characteristic room length, as shown in Table 10-7. Table 10-8 is the ADPI selection guide. Table 10-8 gives the recommended ratio of throw to characteristic length that should maximize the ADPI. A range of throw-to-length ratios are also shown that should give a minimum ADPI. Note that the throw is based on a terminal velocity of 50 ft/min for all diffusers except the ceiling slot type. The general procedure for use of Table 10-8 is as follows:

1. Determine the air flow requirements and the room size.
2. Select the type of diffuser to be used.
3. Determine the room characteristic length.
4. Select the recommended throw-to-length ratio from Table 10-8.

Table 10-2 PERFORMANCE DATA FOR A TYPICAL LINEAR DIFFUSER

Size/Area	Total Pressure	0.009	0.020	0.036	0.057	0.080	0.109	0.143	0.182	0.225
2 in.	Flow, cfm/ft	22	33	44	55	66	77	88	99	110
	NC	—	—	12	18	23	27	31	34	37
0.055	Throw, ft—sill or floor	1-1-1	4-4-4	7-7-7	9-9-10	11-11-12	13-14-16	14-16-18	15-17-20	17-19-21
3 in.	Flow, cfm/ft	38	58	77	96	115	134	154	173	192
	NC	—	—	11	17	22	26	30	33	36
0.096	Throw, ft—sill or floor	2-2-2	7-7-7	10-10-11	12-13-14	15-16-17	18-19-20	20-21-23	23-24-25	25-25-26
4 in.	Flow, cfm/ft	56	83	111	139	167	195	222	250	278
	NC	—	—	12	18	23	27	31	34	37
0.139	Throw, ft—sill or floor	3-3-3	9-9-9	13-13-13	16-16-17	20-20-21	22-23-24	24-25-26	27-27-27	30-30-30
5 in.	Flow, cfm/ft	72	107	143	179	215	250	286	322	358
	NC	—	—	12	18	23	27	31	34	37
0.179	Throw, ft—sill or floor	4-4-4	10-10-10	14-14-14	18-18-18	22-22-23	24-24-24	27-27-28	30-30-31	32-32-32

	6 in.								
Flow, cfm/ft	88	133	177	221	265	310	354	398	442
NC	—	—	13	19	24	28	32	35	38
0.221 Throw, ft—sill or floor	5-5-5	10-10-10	15-15-15	18-18-18	23-23-23	25-25-25	28-28-28	31-31-31	32-32-32

Area is given in square feet per foot of length.

1. All pressures are in in. H_2O.

2. Minimum throw values refer to a terminal velocity of 150 ft/min, middle to 100 ft/min and maximum to 50 ft/min, for a 4-ft active section with a cooling temperature differential of 20 F. The multiplier factors listed in the table are applicable for other lengths.

Terminal Velocity

Active Length	150 ft/min	100 ft/min	50 ft/min
1 ft	0.5	0.6	0.7
10 ft or continuous	1.6	1.4	1.2

3. The NC values are based on a room absorption of 80 Db, re 10^{-12} watts and a 10-ft active section.

NC Correction for Length

Active length, ft	1	2	4	6	8	10	15	20	25	30
Correction	-10	-7	-4	-2	-1	0	+2	+3	+4	+5

Source: Reprinted by permission of Environmental Elements Corporation, Dallas, Texas.

Table 10-3 PERFORMANCE DATA FOR A TYPICAL ROUND CEILING DIFFUSER

Size	Neck Velocity, ft/min Velocity Pressure	400 0.010	500 0.016	600 0.023	700 0.031	800 0.040	900 0.051	1000 0.063	1200 0.090
6 in.	Total pressure	0.026	0.041	0.059	0.079	0.102	0.130	0.161	0.230
	Flow rate, cfm	80	100	120	140	160	180	200	235
	Radius of diffusion, ft	2-2-4	2-3-5	2-4-6	3-4-7	3-5-8	4-5-9	4-6-10	5-7-11
	NC	—	—	14	19	23	26	30	35
8 in.	Total pressure	0.033	0.052	0.075	0.101	0.130	0.166	0.205	0.292
	Flow rate, cfm	140	175	210	245	280	315	350	420
	Radius of diffusion, ft	2-4-6	3-4-7	4-5-9	4-6-10	5-7-11	5-8-13	6-9-14	7-11-17
	NC	—	15	21	26	31	34	37	44
10 in.	Total pressure	0.027	0.043	0.062	0.084	0.108	0.138	0.170	0.243
	Flow rate, cfm	220	270	330	380	435	490	545	655
	Radius of diffusion, ft	3-4-7	3-5-8	4-6-10	5-7-11	5-8-13	6-9-15	7-10-16	8-12-20
	NC	—	11	17	21	26	30	33	39
12 in.	Total pressure	0.026	0.042	0.060	0.081	0.105	0.134	0.166	0.236
	Flow rate, cfm	315	390	470	550	630	705	785	940
	Radius of diffusion, ft	3-5-8	4-6-10	5-7-12	6-8-13	6-10-15	7-11-17	8-12-19	10-14-23
	NC	—	11	17	22	26	30	33	39

24 in.							
Total pressure	0.024	0.038	0.054	0.073	0.094	0.120	0.148	0.211
Flow rate, cfm	1260	1570	1880	2200	2510	2820	3140	3770
Radius of diffusion, ft	6-9-15	8-12-19	9-14-22	11-16-26	12-19-30	14-21-34	16-23-37	19-28-45
NC	—	13	19	24	28	32	35	41
36 in.							
Total pressure	0.024	0.038	0.055	0.074	0.096	0.122	0.151	0.216
Flow rate, cfm	2820	3520	4230	4930	5630	6340	7040	8450
Radius of diffusion, ft	9-14-22	12-18-28	14-21-34	16-25-39	19-28-45	21-32-50	23-35-56	28-42-67
NC	—	15	21	26	30	34	38	43

1. All pressures are in in. H_2O.
2. Minimum radii of diffusion are to a terminal velocity of 150 ft/min, middle to 100 ft/min, and maximum to 50 ft/min.
3. The NC values are based on a room absorption of 18 Db, re 10^{-13} watts or 8 Db, re 10^{-12} watts.

Dimensions

A	Size			
	B	C	D	E
6	6½	11⅛	1¾	1⅛
8	8½	14¾	2⅛	1½
10	10½	18¼	2⅞	2⅛
12	12½	22	3⅜	2⅜
24	24½	43¼	7¾	6⅝
36	36½	64½	10⅛	8⅜

Source: Reprinted by permission of Environmental Elements Corporation, Dallas, Texas.

Table 10-4 PERFORMANCE DATA FOR AN ADJUSTABLE TYPE, HIGH SIDEWALL DIFFUSER

Size, inches		Velocity, ft/min	300	400	500	600	700	800	1000	1200
		Velocity Pressure	0.006	0.010	0.016	0.022	0.030	0.040	0.062	0.090
	Total Pressure	0	0.010	0.017	0.028	0.038	0.052	0.069	0.107	0.156
		$22\frac{1}{2}$	0.011	0.019	0.031	0.043	0.058	0.078	0.120	0.175
		45	0.016	0.029	0.047	0.064	0.088	0.117	0.181	0.263
8 × 4	cfm		55	70	90	110	125	145	180	215
7 × 5	NC					10	15	19	25	31
6 × 6	Throw, ft	0	4-7-13	6-8-15	7-11-17	9-13-19	10-15-20	11-16-22	14-17-24	15-19-26
$A_c = 0.18\ \text{ft}^2$		$22\frac{1}{2}$	3-6-10	5-6-12	6-9-14	7-10-15	8-12-16	9-13-18	11-14-19	12-15-21
		45	2-3-7	3-4-8	4-5-9	4-7-10	5-7-10	6-8-11	7-9-12	8-10-13
10 × 4	cfm		65	90	110	130	155	175	220	265
8 × 5	NC					10	15	19	25	31
7 × 6	Throw, ft	0	4-7-14	7-10-17	8-12-19	9-15-21	11-16-23	13-17-24	16-19-27	17-21-29
$A_c = 0.22\ \text{ft}^2$		$22\frac{1}{2}$	3-6-11	6-8-14	6-10-15	7-12-17	9-13-18	10-14-19	13-15-22	14-17-23
		45	2-4-7	3-5-9	4-6-10	5-7-10	6-8-11	6-9-12	8-10-13	9-11-15
12 × 4	cfm		80	105	130	155	180	210	260	310
10 × 5	NC					11	16	20	26	32

		0	100	135	170	205	240	270	340	410
8 × 6	Throw,	0	5-8-16	7-11-19	9-13-21	10-16-23	12-17-24	14-19-26	17-21-29	19-23-32
$A_c = 0.26$ ft²	ft	22½	4-6-13	6-9-15	7-10-17	8-13-18	10-14-19	11-15-21	14-17-23	15-18-26
		45	3-4-8	4-5-9	4-7-10	5-8-11	6-9-12	7-9-13	8-11-15	9-12-16
16 × 4	cfm		100	135	170	205	240	270	340	410
12 × 5	NC					12	17	21	27	33
10 × 6	Throw,	0	5-9-18	8-12-21	10-15-24	12-19-26	14-20-28	16-22-30	20-24-33	22-26-37
$A_c = 0.34$ ft²	ft	22½	4-7-14	6-10-17	8-12-19	10-15-21	11-16-22	13-18-24	16-19-26	18-21-30
		45	3-4-9	4-6-11	5-8-11	6-9-13	7-10-14	8-11-15	10-12-17	11-13-18
18 × 4	cfm		115	155	195	235	275	310	390	470
14 × 5	NC					13	18	22	28	34
12 × 6	Throw,	0	6-9-19	9-13-23	11-16-25	13-19-28	15-22-30	17-23-32	21-26-36	23-27-40
8 × 8	ft	22½	5-7-15	7-10-18	9-13-20	10-15-22	12-18-24	14-18-26	17-21-29	18-22-32
$A_c = 0.39$ ft²		45	3-5-10	4-6-11	5-18-13	7-10-14	8-11-15	9-12-16	11-13-18	12-14-20

Source: Reprinted by permission of Environmental Elements Corporation, Dallas, Texas.

0 deg.
0 deg.
Deflection

22½ deg.
0 deg.
22½ deg.
22½ deg.
Deflection

45 deg.
0 deg.
45 deg.
45 deg.
Deflection

Model	A	H	B	C	D	E
27	½	24	4	12	5	5⅞
		48			7	3⅞
28	¾	24	4½	12	6	4⅞
		48			8	2⅞

Figure 10-9. A typical T-bar type diffuser assembly. (Courtesy of Environmental Corporation, Dallas, Tex.)

408

Table 10-5 PERFORMANCE DATA FOR THE T-BAR DIFFUSERS OF FIGURE 10-9

Model 27

H-24	Cfm	55	62	68	80	95	110	120	135	150
	Horiz. proj.	2-3-4	2-3-4	2-3-5	2-4-6	3-5-7	3-6-8	3-6-9	4-7-10	4-8-11
	Total press.	.04	.06	.07	.10	.14	.18	.22	.28	.34
	NC	—	11	14	19	24	28	32	35	38

H-48	Cfm	104	120	135	160	185	215	240	270	295
	Horiz. proj.	2-4-5	2-4-6	3-5-7	3-6-8	4-6-9	4-7-10	5-8-12	5-9-13	6-10-14
	Total press.	.04	.05	.07	.10	.13	.18	.22	.28	.34
	NC		14	17	22	27	31	35	38	41

Model 28

H-24	Cfm	80	90	100	120	140	160	180	200	215
	Horiz. proj.	2-3-5	2-3-5	2-4-6	3-5-7	3-6-8	4-7-9	4-6-10	5-8-12	5-8-12
	Total press.	.05	.06	.08	.11	.15	.20	.25	.31	.36
	NC	17	21	24	29	34	38	42	45	48

H-48	Cfm	140	155	175	210	245	280	315	350	385
	Horiz. proj.	2-4-6	3-4-6	3-5-7	4-6-8	5-7-10	5-8-11	5-9-13	6-10-14	7-12-16
	Total press.	.04	.05	.06	.08	.11	.15	.19	.23	.28
	NC	15	19	22	27	32	36	40	43	46

1. All pressures are in in. H_2O.
2. Minimum projection is to a terminal velocity of 150 fpm, middle to 100 fpm, and maximum to 50 fpm.
3. NC values are based on room absorption of 10 dB, re 10^{-12} watts.

Source: Reprinted by permission of Environmental Elements Corporation, Dallas, Texas.

5. Calculate the throw.
6. Select the appropriate diffuser from catalog data such as that in Tables 10-2, 10-3, 10-4, or 10-5.
7. Make sure any other specifications are met (noise, total pressure, etc.).

To illustrate the use of diffuser and grille performance data to design air distribution systems, let us consider a few examples.

Example 10-1. The room shown in Fig. 10-10 is part of a single-story office building located in the central United States. A perimeter type of air distribution system is used since heating will be important. The air quantity required for the room is 250 cfm. Select diffusers for the room.

Solution. Diffusers of the type shown in Table 10-2 should be used for this application. Because the room has two exposed walls, an air outlet should be placed under each window in the floor near the wall (Fig. 10-

Table 10-6 PERFORMANCE DATA FOR ONE TYPE OF RETURN GRILLE

		Core Velocity fpm	200	300	400	500	600	700	800
		Velocity Pressure	0.002	0.006	0.010	0.016	0.023	0.031	0.040
A_c	Size, inches	Negative Static Pressure	0.011	0.033	0.055	0.088	0.126	0.170	0.220
0.34 ft²	16 × 4 10 × 6 / 12 × 5	cfm	70	100	135	170	205	240	270
		NC			13	20	25	30	33
0.39 ft²	18 × 4 12 × 6 / 14 × 5 8 × 8	cfm	80	115	155	195	235	275	310
		NC			14	21	26	31	34
0.46 ft²	20 × 4 14 × 6 / 16 × 5 10 × 8	cfm	90	140	185	230	275	320	370
		NC			15	22	27	32	35
0.52 ft²	24 × 4 16 × 6 / 18 × 5	cfm	105	155	210	260	310	365	415
		NC			16	23	28	33	36
0.60 ft²	28 × 4 18 × 6 10 × 10 / 20 × 5 12 × 8	cfm	120	180	240	300	360	420	480
		NC			17	24	29	34	37
0.69 ft²	30 × 4 20 × 6 12 × 10 / 24 × 5 14 × 8	cfm	140	205	275	345	415	485	550
		NC			17	24	29	34	37
0.81 ft²	36 × 4 22 × 6 14 × 10 / 28 × 5 16 × 8	cfm	160	245	325	405	485	565	650
		NC		10	18	25	30	35	38
0.90 ft²	40 × 4 26 × 6 16 × 10 / 30 × 5 18 × 8 12 × 12	cfm	180	270	360	450	540	630	720
		NC		11	19	26	31	36	39
1.07 ft²	48 × 4 30 × 6 14 × 12 / 36 × 5 18 × 10	cfm	215	320	430	535	640	750	855
		NC		12	20	27	32	37	40

Area	Sizes		cfm / NC						
1.18 ft²	34 × 6, 24 × 8	20 × 10, 16 × 12	14 × 14						
			cfm	235	355	470	590	710	825	945
			NC		13	21	28	33	38	41
1.34 ft²	60 × 4, 48 × 5	36 × 6, 18 × 12	16 × 14						
			cfm	270	400	535	670	805	940	1070
			NC		13	21	28	33	38	41
1.60 ft²	72 × 2, 30 × 8	24 × 10, 22 × 12	18 × 14, 16 × 16						
			cfm	320	480	640	800	960	1120	1280
			NC		14	22	29	34	39	42
1.80 ft²	60 × 5, 48 × 6	36 × 12, 30 × 10	24 × 12, 20 × 14	18 × 16					
			cfm	360	540	720	900	1080	1260	1440
			NC		15	23	30	35	40	43
2.08 ft²	72 × 5, 60 × 6	40 × 8, 36 × 10	30 × 12, 24 × 14	20 × 16, 18 × 18					
			cfm	415	625	830	1040	1250	1460	1660
			NC		16	24	31	36	41	44
2.45 ft²	72 × 6, 48 × 8	32 × 12, 26 × 14	24 × 16, 20 × 18						
			cfm	490	735	980	1220	1470	1720	1960
			NC		17	25	32	37	42	45
2.78 ft²	36 × 12, 30 × 14	26 × 16, 24 × 18	22 × 20						
			cfm	555	835	1110	1390	1670	1950	2220
			NC		18	26	33	38	43	46
3.11 ft²	60 × 8, 40 × 10	40 × 12, 36 × 14	30 × 16, 26 × 18	24 × 20					
			cfm	620	935	1240	1560	1870	2180	2490
			NC		19	27	34	39	44	47
3.61 ft²	72 × 8, 60 × 10	48 × 12, 36 × 16	30 × 18, 24 × 24						
			cfm	720	1080	1440	1800	2170	2530	2890
			NC		20	28	35	40	45	48

30 deg. $1\frac{1}{16}$ $\frac{1}{2}$

1. All pressures are in in. H₂O.
2. The NC values are based on a room absorption of 8 dB, Re 10^{-12} watts, and one return.

Source: Reprinted by permission of Environmental Elements Corporation, Dallas, Texas.

Table 10-7 CHARACTERISTIC ROOM LENGTH FOR SEVERAL DIFFUSER TYPES

Diffuser Type	Characteristic Length, L
High sidewall grille	Distance to wall perpendicular to jet
Circular ceiling diffuser	Distance to closest wall or intersecting jet
Sill grille	Length of room in the direction of the jet
Ceiling slot diffuser	Distance to wall or midplane between outlets

Source: Reprinted by permission from *ASHRAE Handbook, Fundamentals Volume,* 1977.

Table 10-8 ADPI SELECTION GUIDE

Terminal Device	Room Load Btu/(hr-ft²)	x_{50}/L for Max. ADPI	Maximum ADPI	For ADPI Greater Than	Range of x_{50}/L
High	80	1.8	68	—	—
sidewall	60	1.8	72	70	1.5–2.2
grilles	40	1.6	78	70	1.2–2.3
	20	1.5	85	80	1.0–1.9
Circular	80	0.8	76	70	0.7–1.3
ceiling	60	0.8	83	80	0.7–1.2
diffusers	40	0.8	88	80	0.5–1.5
	20	0.8	93	90	0.7–1.3
Sill grille	80	1.7	61	60	1.5–1.7
straight	60	1.7	72	70	1.4–1.7
vanes	40	1.3	86	80	1.2–1.8
	20	0.9	95	90	0.8–1.3
Sill grille	80	0.7	94	90	0.8–1.5
spread	60	0.7	94	80	0.6–1.7
vanes	40	0.7	94	—	—
	20	0.7	94	—	—
Ceiling	80	0.3[a]	85	80	0.3–0.7
slot	60	0.3[a]	88	80	0.3–0.8
diffusers	40	0.3[a]	91	80	0.3–1.1
	20	0.3[a]	92	80	0.3–1.5

[a]x_{100}/L.

Source: Reprinted by permission from *ASHRAE Handbook, Fundamentals Volume,* 1977.

Figure 10-10.   Plan view of a room showing location of different types of outlets.

10c). This will help to counteract the cold air moving downward from the window as a result of natural convection. The total air quantity is divided equally between the two diffusers. According to Table 10-1 the NC should be about 30 to 40. If we assume that the room has an 8-ft ceiling, the room characteristic length is 8 ft. Table 10-8 gives a throw-to-length ratio ranging from 1.7 to 0.9 for a straight vane diffuser. We will assume a room load of 40 Btu/(hr-ft^2), then

$$x_{50}/L = 1.3$$

and

$$x_{50} = 1.3(8) = 10.4 \text{ ft}$$

From Table 10-2 a 4 × 12 in. diffuser with 125 cfm has a throw between

$$x_{50} = 13(0.7) = 9.1 \text{ ft}$$

and

$$x_{50} = 17(0.7) = 11.9 \text{ ft}$$

because 125 cfm lies between 111 cfm and 139 cfm. The throw given in the table has been corrected for diffuser length. The NC is quite acceptable and is between 12 and 18 uncorrected for length. The total pressure required by the diffuser is between 0.036 and 0.057 in. H$_2$O and is about

$$\Delta P = (125/111)^2(0.036) = 0.046 \text{ in. H}_2\text{O}$$

An acceptable solution is listed as follows:

Size, in.	Capacity, cfm	Throw, ft	NC	ΔP_0, in. H$_2$O
4 × 12	125	10.5	5	0.046

The loss in total pressure for the diffuser is an important consideration. The value shown above would be acceptable for a light commercial system.

Example 10-2.   Suppose the room of Fig. 10-10 is located in the southern latitudes where overhead systems are recommended. Select a round ceiling diffuser system and a high sidewall system. Also select a return grille.

Solution.   The data of Table 10-3 with information from Tables 10-7 and 10-8 will be used to select a ceiling diffuser. The characteristic length is 7 or 8 ft and the throw-to-length ratio is 0.8; then

$$x_{50} = 0.8(7.0) = 5.6 \text{ ft}$$

The best choice would be

Size, in.	Throw, ft	NC	ΔP_0, in. of H_2O
10	$7\frac{1}{2}$	10	0.035

The throw is larger than desired but the throw-to-length ratio is within the range to give a minimum ADPI of 80 percent. Figure 10-10 shows this application. A high sidewall diffuser may be selected from Table 10-4. In this case the throw-to-length ratio should be about 1.6 and the characteristic length is 14 ft; then

$$x_{50} = 1.6(14) = 22.4 \text{ ft}$$

The following units using the $22\frac{1}{2}$ degree spread would be acceptable:

Size, in.	Throw, ft	NC	ΔP_0, in. H_2O
16 × 4 12 × 5 10 × 6	$22\frac{1}{2}$	18	0.063

Figure 10-10b shows the diffuser location. It would be desirable to locate the return air intake near the floor for heating purposes and in the ceiling for cooling. However, two different intakes are not generally used except in extreme cases, and the return will be located to favor the cooling case or to accommodate the building structure. For the room shown in Figure 10-10 it will be assumed that the building design prevents practical location of the return near the floor and the return is located in the ceiling as shown. We may select the following grilles from Table 10-6:

Size, in.	NC	ΔP_0, in. H_2O
24 × 4		
16 × 6		
18 × 5	23	0.072
12 × 6		
8 × 8	27	0.11

Example 10-3. Figure 10-11 shows a sketch of a recreational facility with pertinent data on ceiling height, air quantity, and building dimensions. The elevated seating rises 6 ft from the floor. The floor area and walls are not available for air outlets in the locker rooms. The structure is located in Topeka, Kansas. Select an air diffuser system for the complete structure.

Solution.   It would be desirable to use a perimeter type of system throughout the structure; floor area is not available in all of the spaces, however, and air motion will be enhanced in the central part of the gymnasium by an overhead system.

   The entry area is subject to large infiltration loads and has a great deal of glass area. Therefore, outlets should be located in the floor around the perimeter. There is 50 ft of perimeter wall with 12 ft taken up by doors. Then about 38 ft of linear diffuser could be used if required. Noise is not a limiting factor and the throw should be about 12 ft based on the ADPI (Table 10-8). If we refer to Table 10-2, the 2-in. size has a throw of 12 ft,

Figure 10-11.   A single-story recreational facility.

total pressure loss of 0.08 in. H_2O, a NC of 23, and a capacity of 66 cfm/ft. The total length of the required diffusers would then be

$$L_d = \frac{1000}{66} = 15 \text{ or } 16 \text{ ft}$$

This total length should be divided into four equal sections and located as shown in Figure 10-11.

The office and classrooms should also be equipped with perimeter air inlets. The throw should be about 12 to 15 ft and a NC of about 30 would be acceptable. Referring to Table 10-2, a 3-in. size may be used with a capacity of 115 cfm/ft. The NC is 22 and the throw is 15 ft with a loss in total pressure of 0.08 in. H_2O. The total length of diffuser is then computed as

$$L_d = \frac{600}{115} = 5.2 \text{ or } 5 \text{ ft}$$

The total length may be divided into two sections, or a single 5-ft length will function adequately as shown in Fig. 10-11. The corner classroom should have two outlets.

The elevated seating on each side of the gym should also be equipped with perimeter upflow air outlets because of the exposed walls and glass. A throw of 10 ft would be acceptable because the seating is elevated about 6 ft. Noise is not a major factor. There is about 145 ft of exposed wall on each side and 7500 cfm is required. Therefore, a capacity of at least 52 cfm/ft is required. From Table 10-2, a 2-in. size with a capacity of 55 cfm/ft will give a throw of 10 ft with a loss in total pressure of 0.057 in. H_2O. The total length of diffuser is computed as

$$L_d = \frac{7500}{55} = 136.4 \text{ or } 136 \text{ ft}$$

The total length should be divided into at least five sections and located beneath each window as shown in Fig. 10-11.

The central portion of the gymnasium should be equipped with round ceiling diffusers. Table 10-3 has data for this type of outlet. The total floor area is divided into imaginary squares and a diffuser selected with a capacity to serve that area with a throw just sufficient to reach the boundary of the area. It is better for the throw to be a little short than too long. If the total area is divided into 12 equal squares of about 25 × 25 ft, a 12-in. diffuser in each area with a capacity of about 630 cfm would be in the acceptable range, although the throw of 15 ft is slightly large. Since this is a gym area, some sacrifice of comfort is acceptable and a more economical system will result if large diffusers are used. Imagine that the area is divided into three equal squares of about 50 × 50 ft. Then each diffuser should provide about 2500 cfm and have a throw of about 25 ft. A 24-in.

size, which has a capacity of 2510 cfm and a throw of 30 ft, would be acceptable. The loss in total pressure is about 0.094 in. H$_2$O. The throw is slightly high, but is within the range given in Table 10-8. Three diffusers should be located as shown in Fig. 10-11.

The locker room areas may be equipped either with ceiling-type diffusers or with high sidewall outlets because the floor area is all covered near the walls. Since moisture will tend to collect near the ceiling and condense on ceiling diffusers, a high sidewall system will be used. If four 16 × 4 in. diffusers with capacity of 300 cfm are selected from Table 10-4, a throw of 31 ft (zero deflection) will result in a loss in total pressure of 0.085 in. of water with a NC of 24. This throw is less than that required by the ADPI of Table 10-8. A different diffuser type should be used in a critical situation. The diffusers should be equally spaced about 6 to 12 in. below the ceiling as shown in Fig. 10-11.

The air return grilles should all be placed in the ceiling unless the structure has a basement, which would make placement of grilles near the floor feasible if desired. Because cooling and ventilation will be important factors in the gym and locker room area, a ceiling type of return air system will be utilized. Return grilles may be selected from Table 10-6 as follows:

Figure 10-12.   Schematic of VAV air-distribution system for a room.

No.	Size, in.	Capacity, cfm	ΔP_0, in. H_2O	NC	Location
1	24 × 12	900	0.07	30	Entry
1	24 × 8	590	0.07	28	Office
1	24 × 8	590	0.07	28	Classroom 1
1	24 × 8	590	0.07	28	Classroom 2
10	30 × 24	2320	0.07	36	Gym
1	24 × 16	1220	0.07	32	Men's L.R.
1	24 × 16	1220	0.07	32	Women's L.R.

Figure 10-13. A single-duct VAV terminal box.

It has been assumed that all of the air will flow back through the air return before any of it is exhausted.

Variable air volume air-distribution systems usually involve the use of linear or T-Bar diffusers and a thermostat-controlled metering device, referred to as a VAV terminal box. Figure 10-12 shows how such a device is used in relation to the main air supply and the diffusers. Figure 10-13 is a photo and schematic of a typical terminal box. There are almost infinite variations in these devices depending on the manufacturer. Some are self-powered, using energy from the flowing air, whereas others use power from an external source. Many of the self-powered boxes require a relatively high static pressure and therefore are adaptable only to high-velocity systems. However, there are models available that operate with pressures compatible with low-velocity systems. The layout and selection of the diffusers follow the principles and methods previously discussed.

REFERENCES

1. J. Rydberg and P. Norback, "ASHVE Research Report No. 1362—Air Distribution and Draft," *ASHVE Transactions*, Vol. 65, 1949.

2. F. C. Houghten, Carl Gutberlet, and Edward Witkowski, "Draft Temperatures and Velocities in Relation to Skin Temperatures and Feeling of Warmth," *ASHVE Transactions*, Vol. 44, 1938.

3. Alfred Koestel, "Computing Temperature and Velocities in Vertical Jets of Hot or Cold Air," *ASHVE Transactions*, Vol. 60, 1954.

4. H. E. Straub, "Principles of Room Air Distribution," *Heating, Piping, and Air Conditioning*, 1969.

5. *ASHRAE Handbook, Systems Volume*, American Society of Heating, Refrigerating and Air-Conditioning Engineers, Atlanta, Ga., 1987.

PROBLEMS

10-1   A free isothermal jet is discharged horizontally from a circular opening. There is no nearby surface. The initial velocity and volume flow rate are 700 ft/min and 200 cfm, respectively. Compute (a) the throw for terminal velocities of 50, 100, and 150 ft/min, (b) the total volume flow rate of the jet for each terminal velocity in (a) above.

10-2   Air is discharged from a circular pipe 100 mm in diameter to an open space. The discharge velocity is 5 m/s. Assuming standard air, compute (a) the throw for terminal velocities of 0.25 and 0.5 m/s, and (b) the total volume flow rate of the air at each terminal velocity in (a) above.

10-3   A free jet is discharged horizontally below a ceiling. The initial velocity and volume flow rate are 1000 ft/min and 300 cfm. The initial jet temperature is 100

F, whereas the room is to be maintained at 75 F. If the coefficient K in Eq. 10-2 has a value of 9, compute the throw and the difference in temperature between the center line of the jet and the room at terminal velocities of 50, 100, and 150 ft/min.

10-4 Air at 49 C is discharged into a room at 20 C with a velocity of 4 m/s. Estimate the stream temperature where the jet velocities are 0.25 and 0.5 m/s.

10-5 Suppose a given space requires a very large quantity of circulated air for cooling purposes. What type of diffuser system would be best? Why?

10-6 A space has a low but essentially constant occupancy with a moderate cooling load. What type of air diffuser system would be best for heating and cooling? Explain. Assume that the space is on the ground floor.

10-7 Consider a single-story structure with many windows. What would be the best all-around air-distribution system for (a) the northern part of the United States, and (b) the southern states? Explain.

10-8 Consider a relatively large open space with a small cooling load and low occupancy located in the southern part of the United States. What type of air-distribution system would be best? Explain.

10-9 Select a perimeter-type diffuser system for the building shown in Fig. 10-14. It is general office space. Dimensions given are in feet.

10-10 Select a round ceiling diffuser system for the building in Problem 10-9.

10-11 Select a high sidewall diffuser system for the buulding in Problem 10-9.

10-12 Select return air grilles for the building in Problem 10-9. Assume the return system must be placed in the attic and each room must have a return.

10-13 Select an air-distribution system and diffusers and grilles for the structure described by plans and specifications furnished by the instructor.

Figure 10-14.   Floor plan for Problem 10-9.

10-14 A linear floor diffuser is required for a space with an air supply rate of 480 cfm. The room has a 12-ft ceiling and cooling load of 40 Btu/(hr-ft²). (a) Select a diffuser from Table 10-2 for this application. (b) Determine the total pressure and NC for your selection.

10-15 Suppose a round ceiling diffuser is to be used in the situation described in Problem 10-14. The room has plan dimensions of 16 × 18 ft. (a) Select a diffuser from Table 10-3 for this application. (b) Determine the total pressure and NC for the diffuser.

10-16 Assume that two high side wall diffusers are to be used for the room described in Problems 10-14 and 10-15, and they are to be installed in the wall with the longest dimension. (a) Select suitable diffusers from Table 10-4. (b) Determine the total pressure and NC for the diffusers.

10-17 Select a suitable return grille from Table 10-5 for the room described in Problem 10-14. Total pressure for the grille should be less than 0.10 in. H₂O and one dimension should be 12. in.

10-18 Consider a room with a 20-ft exposed wall that has two windows. The other dimension is 30 ft. The room is part of a variable air volume system. (a) Lay out and select T-Bar diffusers from Table 10-5 if the room requires a total air quantity of 800 cfm and the maximum total pressure available is 0.10 in. H₂O. (b) Note the total pressure, the throw to where the maximum velocity has decreased to 100 ft/min, and the NC for each diffuser.

10-19 Consider a 18 × 30 ft room in the southwest corner of a zone. There are windows on both exterior walls and the peak air quantity for the room is 1000 cfm. (a)

Figure 10-15.   Floor plan for a large office space.

Lay out and select T-Bar diffusers from Table 10-6 using a maximum total pressure of 0.15 in. H_2O. (b) Note the total pressure, the throw to where the maximum velocity has decreased to 100 ft/min, and the NC for each diffuser.

10-20 Select perimeter-type diffusers for the room shown in Fig. 10-15. The perimeter distribution is for the heating system that is secondary to the VAV cooling system. The perimeter system requires 1800 cfm (850 L/s) evenly distributed along the exterior walls. Locate the diffusers on the floor plan. Limit the total pressure to 0.10 in. H_2O.

10-21 Select round ceiling diffusers for the room shown in Fig. 10-15. The room has a cooling load of 80,000 Btu/hr (23.5 kW) and a design air supply rate of 3000 cfm (1416 L/s). Locate the diffusers on the floor plan. A maximum total pressure of 0.12 in. H_2O is allowed.

10-22 Select T-Bar (slot) type diffusers for the room of Fig. 10-15 and locate them on the floor plan. The cooling load is 90,000 Btu/hr (26.4 kW) and the design air supply rate is 3200 cfm (1500 L/s). The maximum allowable total pressure is 0.10 in. H_2O.

10-23 Select and locate a return grille(s) for the room of Problem 10-21. A quiet system is desirable.

10-24 Select and locate return grilles for the room of Problem 10-22. Limit the NC to less than 30.

CHAPTER ELEVEN

Fans and Building Air Distribution

The previous chapter considered the distribution and movement of the air within the conditioned space. It was assumed that the proper amount of air at the required total pressure was delivered to each diffuser. This chapter discusses the details of distributing the air to the various spaces in the structure. Proper design of the duct system and the selection of appropriate fans and accessories are essential. A poorly designed system may be noisy and inefficient and lead to discomfort of occupants. Correction of faulty design is expensive and sometimes practically impossible.

11-1 FANS

The fan is an essential component of almost all heating and air-conditioning systems. Except in those cases where free convection creates air motion, a fan is used to move air through ducts and to induce air motion in the space. An understanding of the fan and its performance is necessary if one is to design a satisfactory duct system.

The *centrifugal fan* is the most widely used because it can efficiently move large or small quantities of air over a wide range of pressures. The principle of operation is similar to the centrifugal pump in that a rotating impeller mounted inside a scroll type of housing imparts energy to the air or gas being moved. Figure 11-1 shows the various components of a centrifugal fan. The impeller blades may be forward-curved, backward-curved, or radial. The blade design influences the fan characteristics and will be considered later.

The *vaneaxial fan* is mounted on the center line of the duct and produces an axial flow of the air. Guide vanes are provided before and after the wheel to reduce rotation of the air stream.

The *tubeaxial fan* is quite similar to the vaneaxial fan but does not have the guide vanes. Figure 11-2 illustrates both types.

Figure 11-1.   Exploded view of a centrifugal fan. (Reprinted
by permission from *AHSRAE Handbook,
Equipment Volume,* 1983.)

Axial flow fans are not capable of producing pressures as high as those of
the centrifugal fan but can move large quantities of air at low pressure. Axial
flow fans generally produce higher noise levels than centrifugal fans.

11-2   FAN PERFORMANCE

The performance of fans is generally given in the form of a graph showing
pressure, efficiency, and power as a function of capacity. The energy transferred to the air by the impeller results in an increase in static and velocity
pressure; the sum of the two pressures gives the total pressure. Although these
quantities are commonly referred to as pressure, they are usually expressed in
terms of head such as inches or millimeters of water. When Eq. 9-1b is applied
to a fan and constant density is assumed, the following result is obtained:

$$\frac{g_c w}{g} = \frac{g_c}{g}\left(\frac{P_1}{\rho} - \frac{P_2}{\rho}\right) + \frac{1}{2g}(\bar{V}_1^2 - \bar{V}_2^2) = \frac{g_c}{g}\frac{(P_{01} - P_{02})}{\rho} \qquad (11\text{-}1)$$

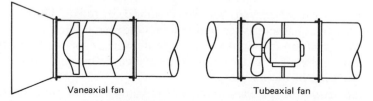

Figure 11-2.   Axial flow fans.

In this form the equation expresses the increase in total pressure of the air. Multiplying Eq. 11-1 by g/g_c gives

$$w = \frac{P_{01} - P_{02}}{\rho} \qquad (11\text{-}1a)$$

which is an expression for the energy imparted to the air. Multiplication of Eq. 11-1a by the mass flow rate of the air produces an expression for the *total power* imparted to the air:

$$\dot{W}_t = \frac{\dot{m}(P_{01} - P_{02})}{\rho} \qquad (11\text{-}2)$$

The *static power* is the part of the total power that is used to produce the change in static head:

$$\dot{W}_s = \frac{\dot{m}(P_1 - P_2)}{\rho} \qquad (11\text{-}3)$$

Fan efficiency may be expressed in two ways. The *total fan efficiency* is the ratio of total air power to the shaft power input:

$$\eta_t = \frac{\dot{W}_t}{\dot{W}_{sh}} = \frac{\dot{m}(P_{01} - P_{02})}{\rho \dot{W}_{sh}} \qquad (11\text{-}4)$$

In terms of volume flow rate Eq. 11-4 becomes

$$\eta_t = \frac{\dot{Q}(P_{01} - P_{02})}{\dot{W}_{sh}} \qquad (11\text{-}4a)$$

where

$$\dot{Q} = \text{volume flow rate, ft}^3/\text{min or m}^3/\text{s}$$
$$P_{01} - P_{02} = \text{change in total pressure, lbf/ft}^2 \text{ or Pa}$$
$$\dot{W}_{sh} = \text{shaft power, (ft-lbf)/min or W}$$

It has been common practice in the United States for $\dot{Q}$ to be in ft^3/min., $(P_{01} - P_{02})$ to be in in. H$_2$O, and $\dot{W}_{sh}$ to be in horsepower. In this special case

$$\eta_t = \frac{\dot{Q}(P_{01} - P_{02})}{6350 \, \dot{W}_{sh}} \qquad (11\text{-}4b)$$

The *static fan efficiency* is the ratio of the static air power to the shaft power input:

$$\eta_s = \frac{\dot{W}_s}{\dot{W}_{sh}} = \frac{\dot{m}(P_1 - P_2)}{\rho \dot{W}_{sh}} = \frac{\dot{Q}(P_1 - P_2)}{\dot{W}_{sh}} \qquad (11\text{-}5)$$

where the same consistent units are used as with Eq. 11-4a. Using the units of Eq. 11-4b, we get

$$\eta_s = \frac{\dot{Q}(P_1 - P_2)}{6350 \dot{W}_{sh}} \qquad (11\text{-}5a)$$

Figures 11-3, 11-4, 11-5, and 11-6 illustrate typical performance curves for centrifugal and vaneaxial fans. Note the difference in the pressure characteristics for the different types of blade. Also note the point of maximum efficiency with respect to the point of maximum pressure. The power characteristics of vaneaxial fans are distinctly different from centrifugal fans. Note that the power increases as the flow rate approaches zero for a vaneaxial fan, which is opposite

Figure 11-3.   Forward-tip fan characteristics.

Figure 11-4.   Backward-tip fan characteristics.

Figure 11-5.   Radial-tip fan characteristics.

Figure 11-6. Vaneaxial fan characteristics.

to that of a centrifugal fan. Otherwise the general behavior is similar. Emphasis will be given to the centrifugal fan in discussion to follow with comments related to vaneaxial fans when appropriate. Fan characteristics are discussed in greater detail later.

A conventional representation of fan performance is shown in Fig. 11-7 for a specific backward-curved blade fan. In this case total pressure and total efficiency are also given. Note that the area for desired application is marked.

Figure 11-7. Conventional curves for backward-curved blade fan. (Reprinted by permission from *ASHRAE Journal*, Vol. 14, Part I, No. 1, 1972.)

Figure 11-8. Performance data for a typical backward-curved blade fan. (Reprinted by permission from ASHRAE Journal, 14, Part I, No. 1, 1972.)

When data from this zone are plotted on a logarithmic scale, the curves appear as shown in Fig. 11-8. This plot has some advantages over the conventional representation (6). Many different fan speeds can be conveniently shown, and the system characteristic is a straight line parallel to the efficiency lines.

Table 11-1 compares some of the more important characteristics of centrifugal fans.

Table 11-1 COMPARISON OF CENTRIFUGAL FAN TYPES

Item	Forward-curved Blades	Radial Blades	Backward-curved Blades
Efficiency	Medium	Medium	High
Space required	Small	Medium	Medium
Speed for given pressure rise	Low	Medium	High
Noise	Poor	Fair	Good

The noise emitted by a fan is of great importance in many applications. For a given pressure the noise level is proportional to the tip speed of the impeller and to the air velocity leaving the wheel. Furthermore, fan noise is roughly proportional to the pressure developed regardless of the blade type. However, backward-curved fan blades are generally considered to have the better (lower) noise characteristics.

The pressure developed by a fan is limited by the maximum allowable speed. If noise is not a factor, the straight radial blade is superior. Fans may be operated in series to develop higher pressures, and multistage fans are also constructed. However, difficulties may arise when fans are used in parallel. Surging back and forth between fans may develop, particularly if the system demand is changing. Forward-curved blades are particularly unstable when operated at the point of maximum efficiency.

Combining both the system and fan characteristics on one plot is very useful in matching a fan to a system and to ensure fan operation at the desired conditions. Figure 11-9 illustrates the desired operating range for a forward-curved blade fan. The range is to the right of the point of maximum efficiency. The backward-curved blade fan has a selection range that brackets the range of maximum efficiency and is not so critical to the point of operation; however this type should always be operated to the right of the point of maximum pressure. Figure 11-10 shows the combined system and fan characteristics using the logarithmic plot. The system characteristic is line $S - S$. For a given system the efficiency does not change with speed; however, capacity, total pressure, and power all depend on the speed. Changing the fan speed will not change the relative point of intersection between the system and fan characteristics. This can be done only by changing fans.

Figure 11-11 shows fan characteristics for a forward-curved blade fan using SI units, except that capacity is in m^3/min instead of m^3/s.

There are several simple relationships between fan capacity, pressure, speed, and power, which are referred to as the *fan laws*. The first three fan laws are the most useful and are stated as follows:

Figure 11-9. Optimum match between system and forward-curved blade fan.

Figure 11-10.   Performance chart showing combination of fan and system. (Reprinted by permission from *ASHRAE Journal*, Vol. 14, Part I, No. 1, 1972.)

1. The capacity is directly proportional to the fan speed.
2. The pressure (static, total, or velocity) is proportional to the square of the fan speed.
3. The power required is proportional to the cube of the fan speed.

The last three fan laws are of less utility but are nevertheless useful.

4. The pressure and power are proportional to the density of the air at constant speed and capacity.
5. Speed, capacity, and power are inversely proportional to the square root of the density at constant pressure.
6. The capacity, speed, and pressure are inversely proportional to the density, and the power is inversely proportional to the square of the density at a constant mass flow rate.

Example 11-1.   A centrifugal fan is operating as shown in Fig. 11-12 at point 1. Estimate the capacity, total pressure, and power requirement when the speed is increased to 1050 rpm. The initial power requirement is 2 hp.

Figure 11-11. Performance data for a forward-curved blade fan.

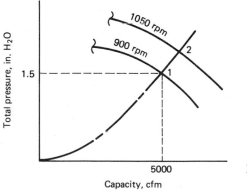

Figure 11-12. Fan and system characteristics for Example 11-1.

Solution. The first three fan laws may be used to estimate the new capacity, total pressure, and power.

Capacity:

$$\frac{\dot{Q}_1}{\dot{Q}_2} = \frac{\text{rpm}_1}{\text{rpm}_2}$$

or

$$\dot{Q}_2 = \dot{Q}_1 \frac{\text{rpm}_2}{\text{rpm}_1} = 5000 \left(\frac{1050}{900} \right) = 5833 \text{ ft}^3/\text{min}.$$

Total pressure:

$$\frac{P_{01}}{P_{02}} = \left(\frac{\text{rpm}_1}{\text{rpm}_2} \right)^2$$

$$P_{02} = P_{01} \left(\frac{\text{rpm}_2}{\text{rpm}_1} \right)^2 = 1.5 \left(\frac{1050}{900} \right)^2 = 2.04 \text{ in. H}_2\text{O}$$

Power:

$$\frac{\dot{W}_1}{\dot{W}_2} = \left(\frac{\text{rpm}_1}{\text{rpm}_2} \right)^3$$

$$\dot{W}_2 = \dot{W}_1 \left(\frac{\text{rpm}_2}{\text{rpm}_1} \right)^3 = 2 \left(\frac{1050}{900} \right)^3 = 3.2 \text{ hp}$$

11-3   FAN SELECTION

It is not practical to satisfy all of the many different fan applications with one fan design. Therefore, the engineer is faced with selecting the right fan for an application. The following discussion explains the specific characteristics of different types of fans as they relate to their selection. Radial-bladed fans are not usually used in heating, ventilating, and air-conditioning systems and are not discussed.

Backward-curved Bladed Fans

This type is used for general heating, ventilating, and air-conditioning systems, especially where system size permits significant horsepower savings. Such fans can be used in low-, medium-, and high-pressure HVAC systems. These are the highest efficiency designs of all centrifugal fan types. For a given duty, these fans will operate at the highest speed of the different centrifugal fans.

The performance curve is stable, and this type of fan has a load-limiting horsepower characteristic (Fig. 11-7). The horsepower curve reaches a maximum near the peak efficiency area and becomes lower for free delivery. Since there is a maximum horsepower requirement, if the fan is equipped with a motor of such size that this maximum requirement is met, there will be no danger of overloading the motor.

These fans are also used in industrial applications where power savings will be significant. The airfoil type blade must be used only in those applications where the air is clean and the airfoil blade is not subject to erosion or corrosion.

Forward-curved Blade Fan

This type of fan is usually used in low-pressure heating, ventilation, and air-conditioning applications, such as domestic furnaces, central station units, and packaged air-conditioning equipment. This design tends to have the lowest efficiency of the centrifugal fans and will operate at the lowest speed of the various centrifugal fans.

The pressure curve is less steep than that of the other designs. There is a dip in the pressure curve to the left of peak pressure and the highest efficiency occurs just to the right of peak pressure. The fan should be rated well to the right of the peak pressure point. The horsepower curve rises continually toward free delivery, and this must be taken into account when the fan is applied and the motor is selected.

Vaneaxial Fan

This type of fan is becoming more commonly used in heating, ventilating, and air-conditioning systems in low-, medium, and high-pressure applications and is particularly advantageous where straight-through flow is required.

Vaneaxial fans usually have blades of airfoil design, which permits medium- to high-pressure capability at relatively high efficiency.

The performance curve (Fig. 11-6) shows the highest pressure character- istics of the axial designs at medium volume flow rate. The performance curve includes a break to the left of peak pressure, which is caused by dynamic stall. Application on this part of the curve should be avoided.

Some fans of this design have the capability of changing the pitch of the blade to meet different application requirements. In some cases this is accom- plished by shutting the fan down and changing the blade angle to a new position and restarting the fan. In other cases, the pitch of the fan blade can be changed with the fan in operation. This latter method provides good control character- istics for the fan in VAV systems.

To select a fan for a given system it is necessary to know the capacity and total pressure requirement of the system. To assist in the actual fan selection, manufacturers furnish graphs such as those shown in Figs. 11-7 and 11-8 with the areas of preferred operation shown. The static pressure is often given but not the total pressure. The total pressure may be computed from the capacity and the fan outlet dimensions. Data pertaining to noise are also available from most manufacturers and are generally similar to those discussed in Chapter 10.

In many cases manufacturers present their fan performance data in the form of tables. Tables 11-2 and 11-2a are examples of such data for two forward-curved blade fans. Note that the static pressure is given instead of the total pressure; however, the outlet velocity is given, which makes it convenient to calculate the velocity pressure to find the total pressure.

It is important that the fan be quiet. Generally a fan will generate the least noise when operated near the peak efficiency. Operation considerably beyond the point of maximum efficiency will be noisy. Forward-curved blades operated at high speeds will be noisy and straight blades are generally noisy, especially at high speed. Backward-curved blades may be operated on both sides of the peak efficiency at relatively high speeds with less noise than the other types of fans.

Example 11-2.   Comment on the suitability of using the fan described by Fig. 11-8 to move 15,000 cfm at 3. 5 in. H_2O total pressure. Estimate the speed and power requirement.

Solution.   Examination of Fig. 11-8 shows that the fan would be quite suitable. The operating point would be just to the right of the point of maximum efficiency and the fan speed is between 800 and 900 rpm. Therefore, the fan will operate in a relatively quiet manner. The speed and power required may be estimated directly from the graph as 830 rpm and 9.5 hp, respectively.

Example 11-3.   Determine whether the fan given in Table 11-2 is suitable for use with a system requiring 1250 cfm at 1.8 in. H_2O total pressure.

Solution.   There is a possibility that the fan could be used. At 1250 cfm the outlet velocity is about 1750 ft/min. To utilize velocity in ft/min and to give the velocity pressure in in. H_2O we modify Eq. 9-5:

$$H_v = \frac{\overline{V}^2}{2g} = \frac{\overline{V}}{2g}\frac{\rho_a}{\rho_w}\frac{12}{3600} = \left(\frac{\overline{V}}{4005}\right)^2$$

where

$\overline{V}$ = average velocity, ft/min
H_v = velocity pressure, in. H_2O

Then

$$H_v = \left(\frac{1750}{4005}\right)^2 = 0.19 \text{ in. } H_2O$$

The static pressure at the fan outlet is then computed as

$$H_s = H_t - H_v = 1.8 - 0.19 = 1.61 \text{ in. } H_2O$$

Table 11-2 PRESSURE-CAPACITY TABLE FOR A FORWARD-CURVED BLADE FAN

Volume Flow Rate cfm	Outlet Velocity ft/min	$\frac{1}{2}$ in. H$_2$O[a]		$\frac{5}{8}$ in. H$_2$O		$\frac{3}{4}$ in. H$_2$O		1 in. H$_2$O		$1\frac{1}{4}$ in. H$_2$O		$1\frac{1}{2}$ in. H$_2$O	
		rpm	bhp[b]	rpm	bhp	rpm	bhp	rpm	bhp	rpm	bhp	rpm	bhp
851	1200	848	0.13	933	0.16	1018	0.19	—	—	—	—	—	—
922	1300	866	0.15	945	0.18	1019	0.21	—	—	—	—	—	—
993	1400	884	0.17	957	0.20	1030	0.23	1175	0.30	—	—	—	—
1064	1500	901	0.19	973	0.22	1039	0.26	1182	0.32	—	—	—	—
1134	1600	926	0.22	997	0.24	1057	0.29	1190	0.35	1320	0.43	—	—
1205	1700	954	0.25	1020	0.27	1078	0.31	1200	0.38	1325	0.46	1436	0.55
1276	1800	983	0.28	1044	0.31	1100	0.34	1210	0.42	1330	0.50	1440	0.59
1347	1900	1011	0.31	1068	0.35	1126	0.38	1230	0.46	1341	0.54	1447	0.63
1418	2000	1039	0.35	1092	0.39	1152	0.42	1250	0.50	1352	0.59	1458	0.66
1489	2100	1068	0.39	1115	0.43	1178	0.47	1275	0.54	1370	0.62	1470	0.72
1560	2200	1096	0.44	1147	0.47	1204	0.51	1300	0.59	1390	0.67	1482	0.77
1631	2300	1124	0.48	1179	0.52	1230	0.56	1325	0.64	1420	0.73	1500	0.83
1702	2400	1152	0.53	1210	0.58	1256	0.62	1350	0.70	1448	0.78	1525	0.88

[a]Static pressure.
[b]bhp = shaft power in horsepower.
Note. Data are for a 9 in. wheel diameter and an outlet of 0.71 ft.[2]

Table 11-2a   PRESSURE-CAPACITY TABLE FOR A FORWARD-CURVED BLADE FAN

Volume Flow Rate m³/s	Outlet Velocity m/s	0.7 kPa[a]		0.8 kPa		0.9 kPa		1.0 kPa		1.1 kPa		1.2 kPa	
		rpm	kW	rpm	kW	rpm	kW	rpm	kW	rpm	kW	rpm	kW
3.35	7	692	4.33										
3.83	8	688	4.79	737	5.44								
4.32	9	679	5.20	732	6.06	778	6.90	825	7.68				
4.78	10	664	5.48	721	6.48	770	7.46	819	8.43	864	9.47		
5.27	11	654	5.82	704	6.82	755	7.98	808	9.02	855	10.1	900	11.2
5.75	12	656	6.38	699	7.31	743	8.43	790	9.47	840	10.5	887	11.7
6.23	13	663	7.12	702	7.98	741	8.87	781	9.84	825	11.0	871	12.3
6.72	14	674	7.90	710	8.72	747	9.62	781	10.6	817	11.6	855	12.7
7.18	15	686	8.95	720	9.77	755	10.7	787	11.6	820	12.5	853	13.5
7.67	16	702	10.1	733	10.8	765	11.6	797	12.6	828	13.6	860	14.5
8.13	17			748	12.0	778	12.9	808	13.9	839	14.8	869	15.8
8.62	18					793	14.3	822	15.4	851	16.3	880	17.3
9.10	19											891	18.9

[a]Static pressure.
[b]Outlet area = 0.479 m². Wheel diameter = 660 mm. Tip speed = rpm × 2.07 m/s.

From Table 11-2 the maximum static pressure recommended for the fan is 1.5 in. H_2O and the speed is relatively high even at that condition. If the fan speed were to be further increased, fan noise would probably be unacceptable. A larger fan should be selected. The fan shown in Table 11-2 would be suitable for use with 1250 cfm and total pressures of 0.75 to 1 in. H_2O. The speed would range from about 975 to 1100 rpm.

Example 11-4.   A duct system requires a fan that will deliver 6 m³/s of air at 1.2 kPa total pressure. Is the fan of Table 11-2a suitable? If so, determine the speed, shaft power, and total efficiency.

Solution.   The required volume flow rate falls between 5.75 and 6.23 m³/s in the left-hand column of Table 11-2a. The corresponding outlet velocities are 12 and 13 m/s and the velocity pressure for each case is

$$(P_v)_{5.75} = \rho_a \frac{\overline{V}^2}{2} = 1.2 \frac{(12)^2}{2} = 86.4 \text{ Pa}$$

$$(P_v)_{6.23} = \frac{1.2(13)^2}{2} = 101.4 \text{ Pa}$$

Moving across the table to the 1.1 kPa static pressure column, the total pressure at 5.75 m³/s is

$$(P_0)_{5.75} = 1100 + 86.4 = 1186.4 \text{ Pa}$$

and at 6.23 m³/s

$$(P_0)_{6.23} = 1100 + 101.4 = 1201.4 \text{ Pa}$$

By interpolation the total pressure at 6 m³/s is

$$(P_0)_{6.0} = 1186.4 + \frac{6 - 5.75}{6.23 - 5.75}(1201.4 - 1186.4)$$

$$= 1190 \text{ Pa} \quad \text{or} \quad 1.19 \text{ kPa}$$

Although the total pressure at 6 m³/s is barely adequate, the fan speed can be increased to obtain total pressures up to almost 1.3 kPa at a capacity of 5.75 to 6.23 m³/s.

The fan speed may be determined by interpolation to be

$$\text{rpm} = 840 - \frac{6 - 5.75}{6.23 - 5.75}(840 - 825) = 832$$

and the shaft power is likewise found to be

$$\dot{W}_{sh} = 10.5 + \frac{6 - 5.75}{6.23 - 5.75}(0.5) = 10.76 \text{ kW}$$

The total power imparted to the air is given by Eq. 11-2:

$$\dot{W}_t = \frac{\dot{m}}{\rho}(P_{01} - P_{02}) = \dot{Q}(P_{01} - P_{02}) \tag{11-2}$$

A
Poor
Restrictive duct and
transformation

B
Good
Open inlet or full size
straight approach duct

C
Poor
Spin with wheel—
reduced volume and pressure
spin against wheel—
power increase required

D
Good
Turning vanes for uniform
distribution, no swirl at inlet

E
Poor
High duct entry loss

F
Good
Low duct entry loss

A
Poor

B
Better

C
Better

D
Poor

E
Better

F
Poor

G
Poor

H
Better

15 deg.
maximum

Figure 11-13. Comparison of good and poor fan connections. (Reprinted by permission from *ASHRAE Handbook, Equipment Volume*, 1983.)

where $\dot{Q}$ is in m³/s, $(P_{01} - P_{02})$ is in N/m² or Pa, and W_t is in watts. Then

$$\dot{W}_t = (6)(1.2)(1000)/(1000) = 7.2 \text{ kW}$$

The total efficiency is then given by

$$\eta_t = \frac{\dot{W}_t}{\dot{W}_{sh}} = 7.2/10.76 = 0.67$$

11-4   FAN INSTALLATION

The performance of a fan can be reduced drastically by improper connection to the duct system. In general, the duct connections should be such that the air may enter and leave the fan as uniformly as possible with no abrupt changes in direction or velocity (Fig. 11-13). The designer must rely on good judgment and ingenuity in laying out the system. Space is often limited for the fan installation, and a less than optimum connection may have to be used. In this case the designer must be aware of the penalties (loss in total pressure and efficiency).

If a fan and system combination do not seem to be operating at the volume flow rate and pressure specified, the difficulty may be that the system is not the same as specified in the design. The point of operation will not be at the design point of rating on the fan curve. In Fig. 11-14, point B is the specified

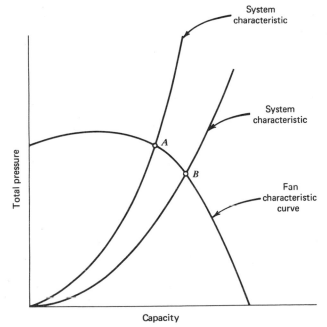

Figure 11-14.   Fan and system characteristics showing deficient operation.

point of operation but tests may show that the actual point of operation is point *A*. The important thing to notice in this case is that the difference between the specified point of rating is due to a change in the system characteristic curve and not the fan. The fan curve is in its original position and the problem is simply to get the system characteristic curve to cross the fan curve at the appropriate point. This problem is discussed in Sec. 11-5 to follow.

An entirely different change in the point of operation between the fan and the system can come about by an actual change in the fan performance curve. For example, fans are designed for uniform and straight flow into the inlet of the fan. If this flow is upset in any way, the fan will not perform on the original performance curve but will produce a new performance curve. Typical inlet and outlet connections that are detrimental to performance are shown in Fig. 11-13. This change in fan performance is discussed next.

The Air Movement and Control Association, Inc. (AMCA) has published, *system effect factors* in their *Fan Application Manual* (8), which express the effect of various fan connections on system performance. These factors are in the form of flow resistances that are added to the computed system resistance.

System Effect Factors

The total pressure requirements of a fan are calculated by methods discussed in Sec. 11-7 and are the result of pressure losses in ductwork, fittings, heating and cooling coils, dampers, filters, process equipment, and similar sources. All of these sources of pressure loss are based on uniform velocity profiles. The velocity profile at the fan outlet is not uniform, and fittings at or near the fan outlet will develop pressure losses greater than the rated value.

Fans are normally tested with open inlets and the flow to the wheel is uniform. In actual installations many other inlet configurations are encountered, and these will adversely affect the catalogued performance. This effect on fan performance is not the usual pressure loss of ductwork, fittings, and equipment as normally computed but is an additional loss.

In order to properly rate the fan, the inlet and outlet effects must be taken into account and the pressure requirements of the fan as normally calculated must be increased. These effects, identified as *system effect factors,* may be estimated by using the procedure outlined next.

Fan Outlet Condition

As shown in Fig. 11-15, the outlet velocity profiles of both centrifugal and axial flow fans are not uniformly distributed across the outlet duct until the air has traveled through a certain length of the duct. This length is identified as *one effective duct length.* To make best use of energy developed by the fan, this length of duct should be provided at the fan outlet. Preferably, the outlet duct should be the same size as the fan outlet, but good flow can be obtained if the duct is not greater in area by about 110 percent nor less in area than

Figure 11-15. Fan-outlet velocity profiles.

about 85 percent of the fan outlet. The slope of transition elements should not be greater than 15 degrees for the converging elements or greater than 7 degrees for the diverging elements.

One effective duct length is a function of fan outlet velocity as shown in Table 11-3.

Table 11-3   EFFECTIVE DUCT LENGTH

Duct Velocity		One Effective Duct Length
fpm	**m/s**	**Duct Diameters**
2500	12.5	2.5
3000	15.0	3.0
4000	20.0	4.0
5000	25.0	5.0
6000	30.0	6.0
7000	35.0	7.0
8000	40.0	8.0

If the duct is rectangular, the equivalent duct diameter is given by

$$D = (4 \times H \times W/\pi)^{1/2} \tag{11-6}$$

where

D = equivalent duct diameter, ft or m
H = rectangular duct height, ft or m
W = rectangular duct width, ft or m

In those cases where a shorter discharge duct length is used, an additional pressure loss will result, and this additional pressure must be added to the fan total pressure requirements. The additional pressure loss may be calculated by

$$\Delta P_o = K \times \left(\frac{\overline{V}}{1000}\right)^2 \tag{11-7}$$

where

ΔP_o = pressure loss, in. H_2O
$\overline{V}$ = velocity at inlet plane, ft/min
K = factor from Table 11-4

The value of blast area ratio is defined as

blast area ratio = blast area/outlet area

The blast area should be obtained from the fan manufacturer for the particular fan being considered. For estimating purposes values of blast area ratio are given in Table 11-5.

To obtain the rated performance from the fan, the first elbow fitting should be at least one effective duct length from the fan outlet (Fig. 11-16). If this length cannot be provided, an additional pressure loss will result and must be added to the fan total pressure requirements. The additional pressure loss may be determined from Eq. 11-7 with a loss factor from Table 11-6.
The factors in Table 11-6 are for single-wheel-single-inlet (SWSI) fans. For double-wheel-double-inlet (DWDI) fans, apply multipliers of 1.25 for position B, 0.85 for position D, and 1.0 for positions A and C.

Table 11-4   DISCHARGE DUCT LOSS FACTORS

Blast Area Ratio	Effective Duct Length—Percent			
	0	12	25	50
0.5	0.13	0.08	0.025	0.01
0.6	0.08	0.03	0.015	0.01
0.7	0.05	0.025	0.01	0.01
0.8	0.03	0.015	0.01	
0.9	0.015	0.01		

Table 11-5   BLAST AREA RATIOS

Fan Type		Blast Area Ratio
Centrifugal:		
Backward-curved		0.70
Radial		0.80
Forward-curved		0.90
Axial:		
Hub Ratio	0.3	0.90
	0.4	0.85
	0.5	0.75
	0.6	0.65
	0.7	0.50

There are other types of fittings that have an effect similar to the outlet elbow that are not covered here. Reference 8 should be consulted for full details.

Inlet Conditions.   If it is necessary to install an elbow on the fan inlet, a straight run of duct is recommended between the elbow and the fan, and a long–radius elbow should be used. Inlet elbows create an additional loss, which must be

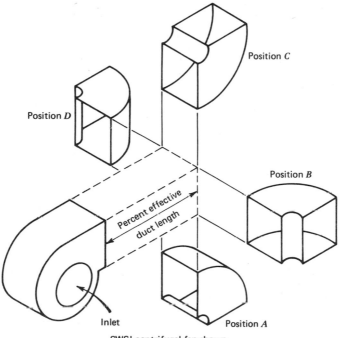

Position C

Position D

Position B

Percent effective duct length

Inlet

Position A

SWSI centrifugal fan shown

Figure 11-16.   Outlet-duct elbow positions.

Table 11-6   OUTLET DUCT ELBOW LOSS FACTORS

Blast Area Ratio	Outlet Elbow Position	Effective Duct Length—Percent			
		0	**12**	**25**	**50**
	A	0.20	0.16	0.13	0.05
0.4	B	0.29	0.29	0.16	0.08
	C & D	0.29	0.29	0.20	0.10
	A	0.13	0.10	0.08	0.13
0.5	B	0.20	0.16	0.13	0.05
	C & D	0.20	0.20	0.13	0.08
	A	0.10	0.08	0.05	0.025
0.6	B	0.13	0.10	0.08	0.03
	C & D	0.16	0.13	0.10	0.05
0.7	A	0.05	0.03	0.025	0.01
	B, C & D	0.08	0.05	0.03	0.015
	A	0.05	0.04	0.03	0.01
0.8	B	0.08	0.06	0.04	0.02
	C	0.10	0.09	0.06	0.03
	D	0.09	0.08	0.05	0.02
	A	0.04	0.03	0.02	0.01
0.9	B & D	0.06	0.05	0.03	0.02
	C	0.08	0.06	0.04	0.02

added to the fan total pressure requirements. Table 11-7 shows loss factors for both vaned and unvaned elbows. The additional loss may be calculated from Eq. 11-7.

Enclosure Restrictions.   In those cases where a fan (or several fans) is built into a fan cabinet construction or is installed in a plenum, it is recommended that the walls be at least one inlet diameter from the fan housing and that a space of at least two inlet diameters be provided between fan inlets. If these recommendations cannot be met, additional pressure losses will result and must be added to the fan total pressure requirements. Every effort must be made to keep the inlet of the fan free of obstructions. The fan inlet should be located so that the inlet is not obstructed by other equipment, walls, pipes, beams, columns, and so on, since such obstructions will degrade the performance of the fan (8).

Example 11-5.   A double-wheel-double-inlet (DWDI) backward-curved blade fan is operating with both inlet and outlet duct elbows. The outlet duct

Table 11-7 INLET DUCT ELBOW LOSS FACTORS

Figure No.	Duct Radius Ratio R/D	No Duct	Duct Length	
			2D	5D
11-17a	0.75	0.10	0.05	0.025
	1.0	0.08	0.05	0.015
	2.0	0.05	0.03	0.015
	3.0	0.03	0.125	0.015
11-17b		0.20	0.13	0.08
11-17c	0.5	0.16	0.10	0.05
	0.75	0.10	0.08	0.025
	1.0	0.08	0.05	0.025
	2.0	0.05	0.03	0.015
	3.0	0.05	0.025	0.015
	R/H			
11-17d	0.5	0.16	0.10	0.05
	0.75	0.13	0.08	0.03
	1.0	0.08	0.05	0.025
	2.0	0.05	0.03	0.015
11-17e	0.5	0.05	0.03	0.015
	1.0	0.03	0.025	0.01
	2.0	0.015	0.015	0.01

elbow is in position C, Fig. 11-16, and is located one duct diameter from the fan outlet. The average velocity in the duct is 4000 ft/min. The fan inlet is configured as shown in Fig. 11-17d, with a duct length of two diameters and R/H of 0.75.

Solution.   The first consideration is the effective duct length for the outlet. From Table 11-3, one effective duct length is four duct diameters for a duct velocity of 4000 ft/min. However, the elbow is located at one duct diameter; therefore, an additional pressure loss will result for both the outlet duct and the elbow. The percent effective duct length is $\frac{1}{4}$, or 25 percent. The blast area ratio is 0.7 from Table 11-5. The discharge duct loss factor K is then 0.01 from Table 11-4, and the additional lost pressure for the duct is

$$\Delta P_{od} = 0.01 \ (4000/1000)^2 = 0.16 \ \text{in. H}_2\text{O from Eq. 11-7}$$

Table 11-6 gives the outlet duct elbow loss factor as 0.03 for 25 percent effective duct length, position C, and blast area ratio of 0.7. The multiplier for the DWDI fan is 1.0 in this case. Then from Eq. 11-7

Figure 11-17.   Inlet-duct elbow configurations.

$$\Delta P_{oe} = 0.03 \ (4000/1000)^2 = 0.48 \text{ in. } H_2O$$

The inlet duct elbow loss factor is given as 0.08 in Table 11-7 for the Fig. 11-17*d* configuration with a duct length of 2*D* and *R/H* of 0.75. Then using Eq. 11-7 and assuming the fan inlet velocity is one half the outlet velocity

$$\Delta P_{oi} = 0.08 \ (2000/1000)^2 = 0.32 \text{ in. } H_2O$$

Finally the total lost pressure for all the inlet and outlet system effects is

$$\Delta P_o = \Delta P_{od} + \Delta P_{oe} + \Delta P_{oi}$$
$$= 0.16 + 0.48 + 0.32 = 0.96 \text{ in. } H_2O$$

This must be added to the computed system total pressure to obtain the actual total pressure that the fan must produce. This is illustrated in Fig.

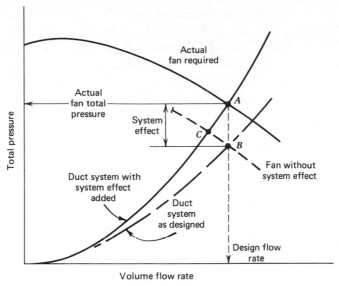

Figure 11-18.   Illustration of system effect for Example 11-5.

11-18. Notice that a fan selected on the basis of zero system effect would operate at point *C* instead of point *B*. The fan selected, taking into account the system effect, operates at point *A*, producing the desired flow rate.

11-5   FIELD PERFORMANCE TESTING

The design engineer is often responsible for checking the fan installation when it is put into operation. In cases of malfunction or a drastic change in performance, the engineer must find and recommend corrective action. The logarithmic plot of fan performance is again quite convenient. From the original system design the specified capacity and total pressure are known and the fan model number and description establishes the fan characteristics as shown in Fig. 11-19. The system characteristic is shown as line *S-D-A-S*. The system shown was designed to operate at about 13,000 cfm and 3 in. H_2O total pressure. To check the system, measurements of capacity and total pressure are made in the field using a pitot tube and an inclined manometer. The use of these instruments was discussed in Chapter 9.

Several different conditions may be indicated by capacity and pressure measurements. First, if the measurements indicate operation at point *A* in Fig. 11-19, the system and fan are performing as designed. Operation at points *B* or *C* indicates that the fan is performing satisfactorily, but that the system is not operating as designed. At point *B* the system has more pressure loss than anticipated, and at point *C* the system has less pressure loss than desired. To obtain the desired capacity the fan speed must be increased to about 900 rpm for point *B* and reduced to about 650 rpm for point *C*. Operation at point *D*

Figure 11-19.   Performance curves showing field test combinations. (Reprinted by permission from *ASHRAE Journal*, Vol. 14, Part I, No. 1, 1972.)

indicates that the fan is not performing as it should. This may be because an incorrect belt drive or belt slippage has caused the fan to operate at the incorrect speed. Poorly designed inlet and outlet connections to the fan may have altered fan performance as previously discussed. The correct capacity may be obtained by correcting the fan speed or by eliminating the undesirable inlet and outlet connections. Operation at point *E* indicates that neither the system nor the fan is operating as designed, which is the usual case found in the field. Although in this situation corrective action could be made by decreasing fan speed to about 700 rpm, any undesirable features of the fan inlet or outlet should first be eliminated to maintain a high fan efficiency. After any increase in fan speed the change in power requirements should be carefully ascertained, because fan power is proportional to the cube of the fan speed.

Example 11-5.   A duct system was designed to handle 2.5 m³/s of air with a total pressure requirement of 465 Pa. The fan of Fig. 11-11 was selected for the system. Field measurements indicate that the system is operating at 2.4 m³/s at 490 Pa total pressure. Recommend corrective action to bring the system up to the design capacity.

Solution. The system characteristics for the design condition and the actual condition may be sketched on Fig. 11-11 and are parallel to the efficiency lines. The fan is performing as specified; however, the system has more flow resistance that it was designed for. To obtain the desired volume flow rate, we must reduce the system flow resistance or increase the fan speed. A check should first be made for unnecessary flow restrictions or closed dampers and, where practical, adjustments should be made to lower the flow resistance. As a last resort the fan speed must be increased, keeping in mind that the power requirements and noise level will increase.

Assuming that the duct system cannot be altered, the fan speed for this example must be increased from 900 rpm to 975 rpm to obtain 2.5 m²/s of flow. The total pressure produced by the fan will increase from 490 to 564 Pa. The shaft power requirement at the design condition was 2100 W and will be increased to 2400 W at the higher speed. In this case the required increase in fan speed is moderate and the increase in noise level should be minimal. The motor must be checked, however, to be sure that an overload will not occur at the higher speed.

11-6   FANS AND VARIABLE AIR VOLUME SYSTEMS

The variable volume air distribution system is usually designed to supply air to a large number of spaces with the total amount of circulated air varying between some minimum and the full-load design quantity. Normally the minimum is about 20 to 25 percent of the maximum. The volume flow rate of the air is controlled independently of the fan by the terminal boxes, and the fan must respond to the system. Because the fan capacity is directly proportional to fan speed, and power is proportional to the cube of the speed, the fan speed should be decreased as volume flow rate decreases. There are practical and economic considerations involved, however. A variable-speed electric motor is ideal. Fan drives that make use of magnetic couplings have been developed and are referred to as *eddy current drives*. These devices have almost infinite adjustment of fan speed. Their disadvantage is high cost. Another approach is to change the fan speed by changing the diameter of the V-belt drive pulley by adjusting the pulley shives. This requires a mechanism that will operate while the drive is turning. The disadvantage of this approach is maintenance. The motor speed system that has emerged as the best from the standpoint of cost, reliability, and efficiency is the *adjustable frequency control system*. This type of controller will operate with most alternating current motors, although motors of high quality are desired.

Another approach to control of the fan is to throttle and introduce a swirling component to the air entering the fan, which alters the fan characteristic in such a way that less power is required at the lower flow rates. This is done with variable inlet vanes, which are a radial damper system located at the inlet

Figure 11-20. Centrifugal fan inlet vanes. (Reproduced by permission of the Trane Company, LaCrosse, Wisc.)

to the fan. Figure 11-20 illustrates the use of fan inlet vanes. Gradual closing
of the vanes reduces the volume flow rate of air and changes the fan characteristic as shown in Fig. 11-21. This approach is not as effective in reducing
fan power as fan speed reduction, but the cost and maintenance are low.

The fan speed or inlet vane position is normally controlled to maintain a
constant static pressure at some location in the duct system. Static pressure
could be sensed at the fan outlet; however, this will lead to abnormally low
pressure at some terminal boxes located a large distance from the fan. Therefore, the static pressure sensor should be located downstream at a location
such that the most distant terminal box will have an acceptable level of static
pressure. This will result in a minimum pressure level throughout the system
and the best operating economy.

Consider the response of a fan in a VAV system with static pressure control
as discussed earlier. The minimum and maximum flow rates are shown in Figs.
11-21 and 11-22 for variable-speed and inlet vanes, respectively. Without any
fan control, the operating point must move along the constant-speed and the
full-open vane characteristics. This results in high static pressure, low efficiency, and wasted fan power. Further, it may not be possible to have stable
operation of the fan at the minimum flow rate. There will also be a very high
static pressure in the system, which is undesirable. When the fan speed is
reduced or inlet vanes are closed to maintain a fixed static pressure at some
downstream location in the duct system, the fan static pressure actually decreases to that shown at point 2. This occurs because the lost pressure between
the fan and the sensing point decreases as the flow rate decreases. This is

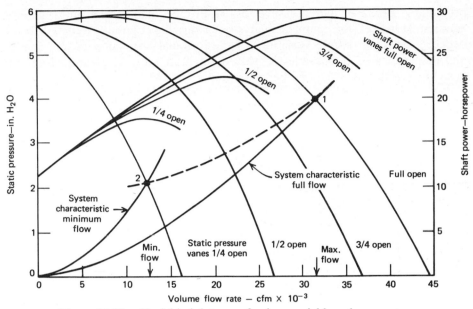

Figure 11-21. Variable inlet vane fan in a variable-volume system.

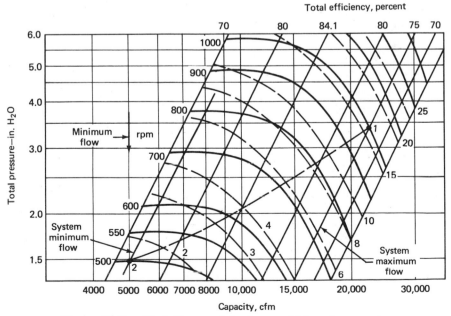

Figure 11-22. Variable-speed fan in a variable-volume system.

predictable because duct pressure loss is proportional to the duct velocity squared. However, the complete analysis of a variable volume system is difficult because there are infinite variations of the terminal box dampers, fan speed, and inlet vanes. It is possible to locate point 2 in Figs. 11-21 and 11-22 and system operation will then be between points 1 and 2.

11-7 AIR FLOW IN DUCTS

The general subject of fluid flow in ducts and pipes was discussed in Chapter 9. The special topic of air flow is treated in this section. Although the basic theory is the same, certain simplifications and computational procedures will be adopted to aid in the design of air ducts.

Equation 9-1b applies to the flow of air in a duct. Neglecting the elevation head terms, assuming that the flow is adiabatic, and no fan is present, Eq. 9-1b becomes

$$\frac{g_c}{g} \frac{P_1}{\rho_1} + \frac{\overline{V_1^2}}{2g} = \frac{g_c}{g} \frac{P_2}{\rho_2} + \frac{\overline{V_2^2}}{2g} + l_f \tag{11-8}$$

and in terms of the total head with ρ constant

$$\frac{g_c}{g} \frac{P_{01}}{\rho} = \frac{g_c}{g} \frac{P_{02}}{\rho} + l_f \tag{11-8a}$$

Equations 11-8 provide a great deal of insight into the duct flow problem. The only important terms remaining in the energy equation are the static head, the velocity head, and the lost head. The static and velocity heads are interchangeable and may increase or decrease in the direction of flow depending on the

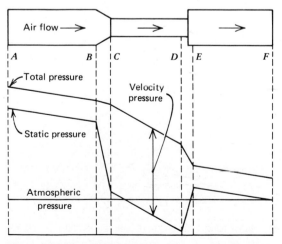

Figure 11-23.   Pressure changes during flow in ducts.

duct cross-sectional area. Because the lost head l_f must be positive, the total pressure always decreases in the direction of flow. Figure 11-23 illustrates these principles.

For duct flow the units of each term in Eq. 11-8 are usually in in. H_2O because of their small size. When SI units are used, each term has units of N/m^2 or Pa. Equation 11-8 then takes the following form for constant density:

$$P_1 + \frac{\rho \overline{V}_1^2}{2g_c} = P_2 + \frac{\rho \overline{V}_2^2}{2g_c} + \frac{\rho g l_f}{2g_c} \tag{11-8b}$$

where l_f has the units of meters as defined in Eq. 9-6. To simplify the notation, the equations may be written

$$H_{s1} + H_{v1} = H_{s2} + H_{v2} + l_f \tag{11-8c}$$

and

$$H_{01} = H_{02} + l_f \tag{11-8d}$$

In this form each term has the units of length in any system of units. For air at standard conditions

$$H_v = \left(\frac{\overline{V}}{4005}\right)^2 \text{ in. } H_2O \tag{11-9}$$

where $\overline{V}$ is in ft/min, and

$$P_v = \left(\frac{\overline{V}}{1.29}\right)^2 \text{ Pa} \tag{11-10}$$

where $\overline{V}$ is in m/s.

The lost head due to friction in a straight, constant area duct is given by Eq. 9-6 and the computational procedure is the same as discussed in Sec. 9-2. Because this approach becomes tedious when designing ducts, special charts have been prepared. Figures 11-24, 11-24a, 11-24b, and 11-24c are examples of such charts for air flowing in galvanized steel ducts with approximately 40 joints per 100 ft or 30 m. The charts are based on standard air and fully developed flow. For the temperature range of 50 F or 10 C to about 100 F or 38 C there is no need to correct for viscosity and density changes. Above 100 F or 38 C, however, a correction should be made. The density correction is also small for moderate pressure changes. For elevations below about 2000 ft or 610 m the correction is small. The correction for density and viscosity will normally be less than one. For example, dry air at 100 F at an elevation of 2000 ft would exhibit a pressure loss about 10 percent less than given in Fig. 11-24. For average-to-rough ducts, a correction factor for density and viscosity may be expressed as

$$C = \left(\frac{\rho_a}{\rho_s}\right)^{0.9} \left(\frac{\mu_a}{\mu_s}\right)^{0.10} \tag{11-11}$$

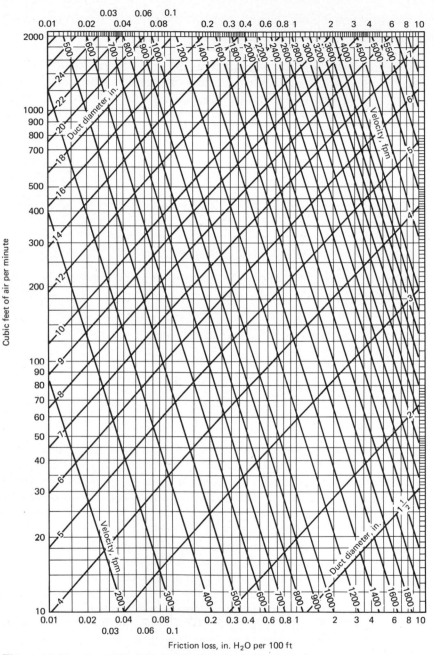

Figure 11-24. Lost head due to friction for galvanized steel ducts. (Reprinted by permission from *ASHRAE Handbook, Fundamentals Volume*, 1977.)

Figure 11-24a. Lost head due to friction for galvanized steel ducts. (Reprinted by permission from *ASHRAE Handbook, Fundamentals Volume,* 1977.)

455

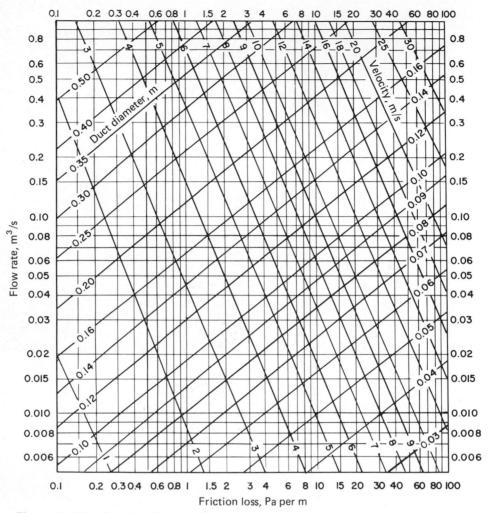

Figure 11-24b.   Lost head due to friction for galvanized steel ducts. (*Using SI units in Heating, Air Conditioning and Refrigeration* by W.F. Stoecker, 1975, Business News Publishing Company.)

where

ρ = air density
μ = air viscosity

and subscripts a and s refer to actual and standard conditions, respectively. The actual lost pressure is then given by

$$\Delta P_{oa} = C\Delta P_{os} \qquad (11\text{-}12)$$

where ΔP_{os} is from Figures 11-24.

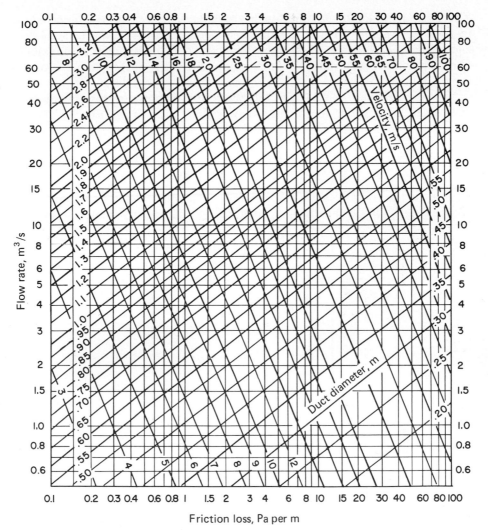

Figure 11-24c. Lost head due to friction for galvanized steel ducts. (*Using SI units in Heating, Air Conditioning and Refrigeration* by W.F. Stoecker, 1975, Business News Publishing Company.)

The effect of roughness is a more important consideration. A common problem to designers is determination of the roughness effect of fibrous glass duct liners and fibrous ducts. This material is manufactured in several grades with various degrees of absolute roughness. Further, the joints and fasteners necessary to install the material affect the overall pressure loss. Smooth galvanized ducts typically have a friction factor of about 0.02, whereas fibrous liners and duct materials will have friction factors varying from about 0.03 to 0.06 depending on the quality of the material and joints, and the duct diameter.

The usual approach to account for this roughness effect is to use a correction factor that is applied to the pressure loss obtained for galvanized metal duct such as Fig. 11-24. Figure 11-25 shows a range of data for commercially available fibrous duct liner materials. These correction factors probably do not allow for typical joints and fasteners.

A more refined approach to the prediction of pressure loss in rough or lined ducts is to generate a chart, such as Fig. 11-24, using Eq. 9-6 and the Colebrook function (2)

$$\frac{1}{\sqrt{f}} = -2 \log_{10} \left[\frac{e}{3.7D} + \frac{2.51}{Re_D \sqrt{f}} \right] \tag{11-13}$$

to express the friction factor. Equation 11-13 is valid in the transition region where f depends on the absolute roughness e, the duct diameter D, and the

Figure 11-25.   Range of roughness correction factors for commercially available duct liners.

Reynolds number Re_D. Equations 9-6 and 11-13 and the ideal gas property relation may be easily programmed for a small computer to calculate the lost pressure for a wide range of temperatures, pressures, and roughness. This general approach eliminates the need for corrections of any kind.

The head loss due to friction is greater for a rectangular duct than for a circular duct of the same cross-sectional area and capacity. For most practical purposes ducts of aspect ratio not exceeding 8:1 will have the same lost head for equal length and mean velocity of flow as a circular duct of the same hydraulic diameter. When the duct sizes are expressed in terms of hydraulic diameter D_h and when the equations for friction loss in round and rectangular ducts are equated for equal length and capacity, Eq. 11-14 for the circular equivalent of a rectangular duct is obtained.

$$D_e = 1.3 \frac{(ab)^{5/8}}{(a + b)^{1/4}} \tag{11-14}$$

where a and b are the rectangular duct dimensions in any consistent units. Table 11-8 has been compiled using Eq. 11-14. A more complete table is given in the *ASHRAE Handbook* (2).

Oval ducts are sometimes used in commercial duct systems. The frictional pressure loss may be treated in the same manner as rectangular ducts using the circular equivalent of the oval duct as defined by

$$D_e = \frac{1.55A^{0.625}}{P^{0.25}} \tag{11-15}$$

with

$$A = \frac{b^2}{4} + b(a - b) \tag{11-15a}$$

and

$$P = b + 2(a - b) \tag{11-15b}$$

where

a = major diameter of oval duct, in. or m
b = minor diameter of oval duct, in. or m

Equations 11-15 are valid for aspect ratios ranging from 2 to 4 (2).

11-8 AIR FLOW IN FITTINGS

Whenever a change in area or direction occurs in a duct or when the flow is divided and diverted into a branch, substantial losses in total pressure may occur. These losses are usually of greater magnitude than the losses in the straight pipe and are referred to as *dynamic losses*.

Table 11-8 CIRCULAR EQUIVALENTS OF RECTANGULAR DUCTS FOR EQUAL FRICTION AND CAPACITY—DIMENSIONS IN INCHES, FEET, OR METERS

Side Rectangular Duct	6	7	8	9	10	11	12	13	14	15	16	17	18	19	20	22	24
6	6.6																
7	7.1	7.7															
8	7.5	8.2	8.8														
9	8.0	8.6	9.3	9.9													
10	8.4	9.1	9.8	10.4	10.9												
11	8.8	9.5	10.2	10.8	11.4	12.0											
12	9.1	9.9	10.7	11.3	11.9	12.5	13.1										
13	9.5	10.3	11.1	11.8	12.4	13.0	13.6	14.2									
14	9.8	10.7	11.5	12.2	12.9	13.5	14.2	14.7	15.3								
15	10.1	11.0	11.8	12.6	13.3	14.0	14.6	15.3	15.8	16.4							
16	10.4	11.4	12.2	13.0	13.7	14.4	15.1	15.7	16.3	16.9	17.5						
17	10.7	11.7	12.5	13.4	14.1	14.9	15.5	16.1	16.8	17.4	18.0	18.6					
18	11.0	11.9	12.9	13.7	14.5	15.3	16.0	16.6	17.3	17.9	18.5	19.1	19.7				
19	11.2	12.2	13.2	14.1	14.9	15.6	16.4	17.1	17.8	18.4	19.0	19.6	20.2	20.8			
20	11.5	12.5	13.5	14.4	15.2	15.9	16.8	17.5	18.2	18.8	19.5	20.1	20.7	21.3	21.9		
22	12.0	13.1	14.1	15.0	15.9	16.7	17.6	18.3	19.1	19.7	20.4	21.0	21.7	22.3	22.9	24.1	
24	12.4	13.6	14.6	15.6	16.6	17.5	18.3	19.1	19.8	20.6	21.3	21.9	22.6	23.2	23.9	25.1	26.2
26	12.8	14.1	15.2	16.2	17.2	18.1	19.0	19.8	20.6	21.4	22.1	22.8	23.5	24.1	24.8	26.1	27.2
28	13.2	14.5	15.6	16.7	17.7	18.7	19.6	20.5	21.3	22.1	22.9	23.6	24.4	25.0	25.7	27.1	28.2
30	13.6	14.9	16.1	17.2	18.3	19.3	20.2	21.1	22.0	22.9	23.7	24.4	25.2	25.9	26.7	28.0	29.3
32	14.0	15.3	16.5	17.7	18.8	19.8	20.8	21.8	22.7	23.6	24.4	25.2	26.0	26.7	27.5	28.9	30.1
34	14.4	15.7	17.0	18.2	19.3	20.4	21.4	22.4	23.3	24.2	25.1	25.9	26.7	27.5	28.3	29.7	31.0
36	14.7	16.1	17.4	18.6	19.8	20.9	21.9	23.0	23.9	24.8	25.8	26.6	27.4	28.3	29.0	30.5	32.0
38	15.0	16.4	17.8	19.0	20.3	21.4	22.5	23.5	24.5	25.4	26.4	27.3	28.1	29.0	29.8	31.4	32.8
40	15.3	16.8	18.2	19.4	20.7	21.9	23.0	24.0	25.1	26.0	27.0	27.9	28.8	29.7	30.5	32.1	33.6

Source: Reprinted by permission from ASHRAE Handbook, Fundamentals Volume, 1977.

Dynamic losses vary as the square of the velocity and are conveniently represented by

$$H_0 = C(H_v) \tag{11-16}$$

where the loss coefficient C is a constant and Eqs. 11-9 or 11-10 express H_v. When different upstream and downstream areas are involved as in an expansion or contraction, either the upstream or downstream value of H_v may be used in Eq. 11-16 and C will be different in each case.

Fittings are classified as either *constant flow*, such as an elbow or transition, or as *divided flow*, such as a wye or tee. Tables 11-9 through 11-13, at the end of this section, give loss coefficients for many different types of constant flow fittings. It should be kept in mind that the quality and type of construction may vary considerably for a particular type of fitting. Some manufacturers provide data for their own products.

Example 11-6. Compute the lost pressure in a 6-in., 90 degree 4-piece elbow that has 150 cfm of air flowing through it. The ratio of turning radius to diameter is 1.5.

Solution. The lost pressure will be computed from Eq. 11-16. From Table 11-9 the loss coefficient is read as 0.27 and the average velocity in the elbow is computed as

$$\bar{V} = \frac{\dot{Q}}{A} = \frac{\dot{Q}}{(\pi/4)D^2} = \frac{(150)4(144)}{\pi(36)} = 764 \text{ ft/min}$$

then

$$H_0 = C\left(\frac{\bar{V}}{4005}\right)^2 = 0.27\left(\frac{764}{4005}\right)^2 = 0.010 \text{ in. } H_2O$$

In SI units the elbow diameter is 15.24 cm and the flow rate is 4.25 m³/min. The average velocity is then

$$\bar{V} = \frac{\dot{Q}}{A} = \frac{4.25}{(\pi/4)(0.1524)^2(60)} = 3.88 \text{ m/s}$$

The loss coefficient C is dimensionless and is therefore unchanged. Using Eq. 11-10, we get

$$H_0 = C\left(\frac{\bar{V}}{1.29}\right)^2 = 0.27\left(\frac{3.88}{1.29}\right)^2 = 2.44 \text{ Pa}$$

This result compares favorably with the result obtained in English units since 0.010 in. H_2O may be converted directly to 2.44 Pa.

Elbows are generally efficient fittings in that their losses are small when the turn is gradual. When an abrupt turn is used without turning vanes, the lost pressure will be four or five times larger.

When considering the lost pressure in divided flow fittings, the loss in the straight-through section as well as the loss through the branch outlet must be considered. Tables 11-12 and 11-13 give data for branch-type fittings. The angle of the branch takeoff has a great influence on the loss coefficient. Converging flow (Table 11-13) differs from diverging flow (Table 11-12). Finally, the loss coefficient is usually based on the velocity in the upstream section of the fitting.

Tables 11-12 and 11-13 also give representative data for the straight-through flow for the common types of divided flow fittings. The velocity may increase, decrease, or remain constant through the fitting. In every case there will be some loss in total pressure. At velocities below about 1000 ft/min or 5 m/s, the loss can be neglected for practical purposes; however, it is good practice to allow for a small loss.

Example 11-7.   Compute the loss in total pressure for a 90 degree branch takeoff as well as for the straight-through section. The main section is 10 × 12 in. with a flow rate of 1100 cfm. The branch flow rate is 250 cfm through a 7-in. duct. The main duct does not reduce following the branch.

Solution.   It is first necessary to compute the average velocity in each section of the fitting.

$$\overline{V}_u = \frac{\dot{Q}_u}{A_u} = \frac{1100}{(10 \times 12/144)} = 1320 \text{ ft/min}$$

$$\overline{V}_b = \frac{\dot{Q}_b}{A_b} = \frac{250}{(\pi/4)(7/12)^2} = 935 \text{ ft/min}$$

$$V_d = \frac{\dot{Q}_d}{A_d} = \frac{850}{(10 \times 12/144)} = 1020 \text{ ft/min}$$

The ratio of the branch to upstream velocity is

$$\frac{\overline{V}_b}{\overline{V}_u} = \frac{935}{1320} = 0.71$$

and the ratio of the downstream to upstream velocity is

$$\frac{\overline{V}_d}{\overline{V}_u} = \frac{1020}{1320} = 0.77$$

The ratio of the branch to upstream flow rate is

$$\frac{\dot{Q}_b}{\dot{Q}_u} = \frac{250}{1100} = 0.23$$

The loss coefficient for the branch is then read from Table 11-12 as 1.28 and

$$H_{ob} = C_b \left[\frac{\overline{V}_u}{4005} \right]^2 = 1.28 \left(\frac{1320}{4005} \right)^2 = 0.139 \text{ in. } H_2O$$

The lost pressure for the straight-through section is also determined from Table 11-12. The loss coefficient is about 0.03 and

$$H_{od} = 0.03 \left(\frac{1320}{4005} \right)^2 = 0.003 \text{ in. } H_2O$$

Some of the data just discussed for branch-type fittings are for pipes of circular cross section. When rectangular duct is used, the pressure losses will vary depending on the design of the fitting. Extensive data for fittings of all kinds are given in reference 2. These data should be used in real design situations.

It is often convenient to express the effect of fittings in terms of *equivalent length* as was done in Chapter 9 for pipes. The equivalent length is a function of the air velocity and the size (diameter) of the fitting. In the case of pipes with water flowing, the Reynolds number is sufficiently large to assume the fully turbulent friction factor is always valid. This is not generally true for air flowing in ducts unless the duct is large. However, for low-velocity systems a friction factor can be assumed, based on expected flow conditions, to derive equivalent lengths from the loss coefficient data of Tables 11-8 through 11-12. Recall that

$$\Delta P_f = f \frac{L}{D} \frac{\overline{V}^2}{2g} = C \frac{\overline{V}^2}{2g} \tag{11-17}$$

and

$$\frac{L}{D} = \frac{C}{f} \tag{11-18}$$

Table 11-14 gives friction factors that will generally be valid for various duct diameters.

Figures 11-26 and 11-27 give some equivalent lengths commonly used for residential system design. The lengths shown have an unknown amount of conservatism built in. This approach simplifies calculations and is helpful in the design of simple low-velocity systems.

Example 11-8.   Compute the equivalent lengths for the fittings in the duct system of Fig. 11-28. The fittings are an entrance, a 45 degree wye, the straight-through section of the wye fitting, a 45 degree elbow, and a 90 degree elbow.

Figure 11-26. Equivalent lengths of some plenum and branch fittings in feet, with meters in parentheses. (Courtesy of National Environmental Systems Contractors Association, 1501 Wilson Blvd., Arlington, Va. 22209.)

Figure 11-27. Equivalent lengths of common duct fittings in feet, with meters in parentheses. (Courtesy of National Environmental Systems Contractors Association, 1501 Wilson Blvd., Arlington, Va. 22209.)

Figure 11-28.   A simple duct system.

Solution.   Table 11-11A gives the loss coefficient for an entrance. In this case, θ is either 0 or 180 degrees and C_o is 0.5. Then using Eq. 11-18 and Table 11-14 for f, and a 10-in. diameter

$$\frac{L_i}{D} = \frac{0.5}{0.022} = 22.7$$

and

$$L_i = 22.7 \left(\frac{10}{12}\right) = 19 \text{ ft}$$

Table 11-12A gives the loss coefficients for the branch of a wye. For this case

$$\frac{\dot{Q}_b}{\dot{Q}_u} = \frac{120}{400} = 0.3$$

and

$$\frac{A_b}{A_u} = \left(\frac{6}{10}\right)^2 = 0.36$$

Then for θ of 45 degrees, $C_b = 0.45$. There is a question about what diameter should be used for the equivalent length calculation. Since C_b is based on the upstream condition, use the diameter of the upstream section of 10 in. Then

$$\frac{L_b}{D} = \frac{0.45}{0.022} = 20.5$$

and

$$L_b = 20.5 \left(\frac{10}{12}\right) = 17 \text{ ft}$$

Data for the straight-through section of the branch fitting is given in Table 11-12B.

$$\frac{\overline{V}_d}{\overline{V}_u} = \frac{\dot{Q}_d}{\dot{Q}_u} \left(\frac{D_u}{D_d}\right)^2 = \frac{280}{400} \left(\frac{10}{9}\right)^2 = 0.864$$

then $C_d = 0.014$ and

$$\frac{L_d}{D} = \frac{0.014}{0.022} = 0.64$$

and

$$L_d = 0.64 \left(\frac{10}{12}\right) = 0.53 \text{ ft}$$

which is negligible for all practical purposes.

Assume that the 90 degree elbow is the 4-piece type with a r/D ratio of 1.0. Then the loss coefficient is 0.37 from Table 11-9A and

$$\frac{L_e}{D} = \frac{0.37}{0.028} = 13.2$$

and

$$L_e = 13.2 \left(\frac{6}{12}\right) = 7 \text{ ft}$$

A 45 degree elbow will have about one-half the equivalent length of a 90 degree elbow.

To compute the lost pressure, we add the equivalent length of the fitting to the adjacent pipe length. The following example shows the procedure.

Example 11-9.   Compute the lost pressure for each branch of the simple duct system shown in Fig. 11-28 using the equivalent length approach with data from Figs. 11-26 and 11-27 and SI units.

Solution.   The lost pressure will first be computed from 1 to a; then sections a to 2 and a to 3 will be handled separately.

The equivalent length of section 1 to a consists of the actual length of the 25-cm duct plus the equivalent length for the entrance from the plenum of 11 m given in Fig. 11-26. Then

$$L_{1a} = 15 + 11 = 26 \text{ m}$$

From Fig. 11-24b at 0.19 m^3/s and for a pipe diameter of 25 cm, the lost pressure is 0.85 Pa/m of pipe. Then

$$\Delta P_{01a} = (0.85)(26) = 22.1 \text{ Pa}$$

Section a to 3 has an equivalent length equal to the sum of the actual length, and the equivalent length for the 45 degree branch takeoff, one 45 degree elbow, and one 90 degree elbow.

$$L_{a3} = 12 + 11 + 1.5 + 3 = 27.5 \text{ m}$$

From Fig. 11-24b, at 0.057 m³/s and for a 15-cm diameter pipe, the lost pressure is 1.0 Pa/m of pipe. Then

$$\Delta P_{0a3} = (1.0)(27.5) = 27.5 \text{ Pa}$$

Section a to 2 has an equivalent length equal to the sum of the actual length and the equivalent length for the straight-through section of the branch fitting. A length of 2 m is assumed for this case.

$$L_{a2} = 15 + 2 = 17 \text{ m}$$

From Fig. 11-24b, the loss per meter of length is 0.6 Pa and

$$\Delta P_{0a2} = (0.6)(17) = 10.2 \text{ Pa}$$

Then the lost pressure for section 1 to 2 is

$$\Delta P_{012} = \Delta P_{01a} + \Delta P_{0a2} = 22.1 + 10.2 = 32.3 \text{ Pa}$$

and for section 1 to 3

$$\Delta P_{013} = \Delta P_{01a} + \Delta P_{0a3} = 22.1 + 27.5 = 49.6 \text{ Pa}$$

Table 11-9 TOTAL PRESSURE LOSS COEFFICIENTS FOR ELBOWS

A. Elbow, 90 Degrees; 3, 4, & 5-Pieces, Round

5-piece   4-piece   3-piece

Coefficients for 90 Degrees Elbows:[a]

C_o

No. of Pieces	r/D				
	0.5	0.75	1.0	1.5	2.0
5	—	0.46	0.33	0.24	0.19
4	—	0.50	0.37	0.27	0.24
3	0.90	0.54	0.42	0.34	0.33

Table 11-9   TOTAL PRESSURE LOSS COEFFICIENTS FOR ELBOWS
(Continued)

B. Elbow, Mitered, with Single-Thickness Vanes, Rectangular

C_o

| Design | Dimensions, in. | | | |
No.	r	s	L	C_o
1	2.0	1.5	0.75	0.12
2	4.5	2.25	0	0.15
3	4.5	3.25	1.60	0.18

C. Elbow, Mitered, Rectangular

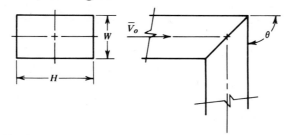

C_o

| θ, deg. | H/W | | | | | | | | | | |
	0.25	0.5	0.75	1.0	1.5	2.0	3.0	4.0	5.0	6.0	8.0
20	0.08	0.08	0.08	0.07	0.07	0.07	0.06	0.06	0.05	0.05	0.05
30	0.18	0.17	0.17	0.16	0.15	0.15	0.13	0.13	0.12	0.12	0.11
45	0.38	0.37	0.36	0.34	0.33	0.31	0.28	0.27	0.26	0.25	0.24
60	0.60	0.59	0.57	0.55	0.52	0.49	0.46	0.43	0.41	0.39	0.38
75	0.89	0.87	0.84	0.81	0.77	0.73	0.67	0.63	0.61	0.58	0.57
90	1.3	1.3	1.2	1.2	1.1	1.1	0.98	0.92	0.89	0.85	0.83

[a]For 45° elbow, multiply by 0.6

Source: Reprinted by permission from *ASHRAE Handbook, Fundamentals Volume,* 1985.

Table 11-10 TOTAL PRESSURE LOSS COEFFICIENTS FOR TRANSITIONS

A. Contraction, Round, & Rectangular

$\theta = 180°$

C_o

A_1/A_o	θ, degrees						
	10	**15–40**	**50–60**	**90**	**120**	**150**	**180**
2	0.05	0.05	0.06	0.12	0.18	0.24	0.26
4	0.05	0.04	0.07	0.17	0.27	0.35	0.41
6	0.05	0.04	0.07	0.18	0.28	0.36	0.42
10	0.05	0.05	0.08	0.19	0.29	0.37	0.43

B. Conical Diffuser, Round

$\theta = 180°$

C_o

A_1/A_o	θ, degrees									
	8	**12**	**16**	**20**	**30**	**45**	**60**	**90**	**120**	**180**
2	0.11	0.11	0.14	0.19	0.32	0.33	0.33	0.32	0.31	0.30
4	0.15	0.17	0.23	0.30	0.46	0.61	0.68	0.64	0.63	0.62
6	0.17	0.20	0.27	0.33	0.48	0.66	0.77	0.74	0.73	0.72
10	0.19	0.23	0.29	0.38	0.59	0.76	0.80	0.83	0.84	0.83
16	0.19	0.22	0.31	0.38	0.60	0.84	0.88	0.88	0.88	0.88

Source: Reprinted by permission from *ASHRAE Handbook, Fundamentals Volume,* 1985.

Table 11-11 TOTAL PRESSURE LOSS COEFFICIENTS FOR DUCT ENTRANCES

A. Conical Converging Bellmouth with End Wall, Round and Rectangular

Rectangular: $D = 2 HW/(H + W)$

C_o

$\dfrac{L}{D}$	θ, degrees								
	0	10	20	30	40	60	100	140	180
0.025	0.50	0.47	0.45	0.43	0.41	0.40	0.42	0.45	0.50
0.05	0.50	0.45	0.41	0.36	0.33	0.30	0.35	0.42	0.50
0.075	0.50	0.42	0.35	0.30	0.26	0.23	0.30	0.40	0.50
0.10	0.50	0.39	0.32	0.25	0.22	0.18	0.27	0.38	0.50
0.15	0.50	0.37	0.27	0.20	0.16	0.15	0.25	0.37	0.50
0.60	0.50	0.27	0.18	0.13	0.11	0.12	0.23	0.36	0.50

B. Smooth Converging Bellmouth with End Wall

r/D	0	0.01	0.02	0.03	0.04	0.05
C_o	0.50	0.43	0.36	0.31	0.26	0.22

r/D	0.06	0.08	0.10	0.12	0.16	≥ 0.20
C_o	0.20	0.15	0.12	0.09	0.06	0.03

Source: Reprinted by permission from *ASHRAE Handbook, Fundamentals Volume,* 1985.

Table 11-12   TOTAL PRESSURE LOSS COEFFICIENTS FOR DIVERGING FLOW FITTINGS

A. Diverging Wye, Rectangular and Round

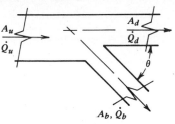

$A_u = A_d$; $H_b = H_u$, where H is height of rectangular duct

a. $\theta = 30$ degrees

Branch, $C_{u,b}$

A_b/A_u	$\dot{Q}_b/\dot{Q}_u$								
	0.1	0.2	0.3	0.4	0.5	0.6	0.7	0.8	0.9
0.8	0.75	0.55	0.40	0.28	0.21	0.16	0.15	0.16	0.19
0.7	0.72	0.51	0.36	0.25	0.18	0.15	0.16	0.20	0.26
0.6	0.69	0.46	0.31	0.21	0.17	0.16	0.20	0.28	0.39
0.5	0.65	0.41	0.26	0.19	0.18	0.22	0.32	0.47	0.67
0.4	0.59	0.33	0.21	0.20	0.27	0.40	0.62	0.92	1.3
0.3	0.55	0.28	0.24	0.38	0.76	1.3	2.0	—	—
0.2	0.40	0.26	0.58	1.3	2.5	—	—	—	—
0.1	0.28	1.5	—	—	—	—	—	—	—

b. $\theta = 45$ degrees

Branch, $C_{u,b}$

A_b/A_u	$\dot{Q}_b/\dot{Q}_u$								
	0.1	0.2	0.3	0.4	0.5	0.6	0.7	0.8	0.9
0.8	0.78	0.62	0.49	0.40	0.34	0.31	0.32	0.35	0.40
0.7	0.77	0.59	0.47	0.38	0.34	0.32	0.35	0.41	0.50
0.6	0.74	0.56	0.44	0.37	0.35	0.36	0.43	0.54	0.68
0.5	0.71	0.52	0.41	0.38	0.40	0.45	0.59	0.78	1.0
0.4	0.66	0.47	0.40	0.43	0.54	0.69	0.95	1.3	1.7
0.3	0.66	0.48	0.52	0.73	1.2	1.8	2.7	—	—
0.2	0.56	0.56	1.0	1.8	—	—	—	—	—
0.1	0.60	2.1	—	—	—	—	—	—	—

Table 11-12   TOTAL PRESSURE LOSS COEFFICIENTS FOR DIVERGING FLOW FITTINGS (Continued)

B. Diverging Tee, Rectangular Main to Round Tap

$$A_u = A_d$$

Branch, $C_{u,b}$

$\bar{V}_b/\bar{V}_u$	$\dot{Q}_b/\dot{Q}_u$								
	0.1	**0.2**	**0.3**	**0.4**	**0.5**	**0.6**	**0.7**	**0.8**	**0.9**
0.2	1.00								
0.4	1.01	1.07							
0.6	1.14	1.10	1.08						
0.8	1.18	1.31	1.12	1.13					
1.0	1.30	1.38	1.20	1.23	1.26				
1.2	1.46	1.58	1.45	1.31	1.39	1.48			
1.4	1.70	1.82	1.65	1.51	1.56	1.64	1.71		
1.6	1.93	2.06	2.00	1.85	1.70	1.76	1.80	1.88	
1.8	2.06	2.17	2.20	2.13	2.06	1.98	1.99	2.00	2.07

Main

$\bar{V}_d/\bar{V}_u$	0	0.1	0.2	0.3	0.4	0.5	0.6	0.8	1.0
$C_{u,d}$	0.35	0.28	0.22	0.17	0.13	0.09	0.06	0.02	0

Source: Reprinted by permission from *ASHRAE Handbook, Fundamentals Volume,* 1985.

Table 11-13 TOTAL PRESSURE LOSS COEFFICIENTS FOR CONVERGING FLOW FITTINGS

A. Converging Wye (45°), Round

Branch, $C_{d,b}$

$\dfrac{\bar{V_b}}{\bar{V_d}}$	A_b/A_d						
	0.1	0.2	0.3	0.4	0.6	0.8	1.0
0.4	− .56	− .44	− .35	− .28	− .15	− .04	0.05
0.5	− .48	− .37	− .28	− .21	− .09	0.02	0.11
0.6	− .38	− .27	− .19	− .12	0	0.10	0.18
0.7	− .26	− .16	− .08	− .01	0.10	0.20	0.28
0.8	− .21	− .02	0.05	0.12	0.23	0.32	0.40
0.9	0.04	0.13	0.21	0.27	0.37	0.46	0.53
1.0	0.22	0.31	0.38	0.44	0.53	0.62	0.69
1.5	1.4	1.5	1.5	1.6	1.7	1.7	1.8
2.0	3.1	3.2	3.2	3.2	3.3	3.3	3.3
2.5	5.3	5.3	5.3	5.4	5.4	5.4	5.4
3.0	8.0	8.0	8.0	8.0	8.0	8.0	8.0

Main, $C_{d,u}$

$\dfrac{\bar{V_u}}{\bar{V_d}}$	A_b/A_d						
	0.1	0.2	0.3	0.4	0.6	0.8	1.0
	−8.6	−4.1	−2.5	−1.7	−.97	−.58	−.34
0.2	−6.7	−3.1	−1.9	−1.3	−.67	−.36	−.18
0.3	−5.0	−2.2	−1.3	−.88	−.42	−.19	−.05
0.4	−3.5	−1.5	−.88	−.55	−.21	−.05	0.05
0.5	2.3	−.95	−.51	−.28	−.06	0.06	0.13
0.6	−1.3	−.50	−.22	−.09	0.05	0.12	0.17
0.7	−.63	−.18	−.03	0.04	0.12	0.16	0.18
0.8	−.18	0.01	0.07	0.10	0.13	0.15	0.17
0.9	0.03	0.07	0.08	0.09	0.10	0.11	0.13
1.0	0.01	0	0	0.10	0.02	0.04	0.05

Table 11-13   TOTAL PRESSURE LOSS COEFFICIENTS FOR
CONVERGING FLOW FITTINGS (Continued)

B. Converging Tee, Round

Branch, $C_{d,b}$

$\dot{Q}_b/\dot{Q}_d$	A_b/A_d						
	0.1	0.2	0.3	0.4	0.6	0.8	1.0
0	-1.0	-1.0	-1.0	$-.90$	$-.90$	$-.90$	$-.90$
0.1	0.40	$-.37$	$-.51$	$-.46$	$-.50$	$-.51$	$-.52$
0.2	3.8	0.72	0.17	$-.02$	$-.14$	$-.18$	$-.24$
0.3	9.2	2.3	1.0	0.44	0.21	0.11	$-.08$
0.4	16	4.3	2.1	0.94	0.54	0.40	0.32
0.5	26	6.8	3.2	1.1	0.66	0.49	0.42
0.6	37	9.7	4.7	1.6	0.92	0.69	0.57
0.7	43	13	6.3	2.1	1.2	0.88	0.72
0.8	65	17	7.9	2.7	1.5	1.1	0.86
0.9	82	21	9.7	3.4	1.8	1.2	0.99
1.0	101	26	12	4.0	2.1	1.4	1.1

Main

$\dot{Q}_b/\dot{Q}_d$	0	0.1	0.2	0.3	0.4	0.5	0.6	0.7	0.8	0.9	1.0
$C_{d,u}$	0	0.16	0.27	0.38	0.46	0.53	0.57	0.59	0.60	0.59	0.55

Source: Reprinted by permission from *ASHRAE Handbook, Fundamentals Volume*, 1985.

Table 11-14   FRICTION
FACTORS FOR
VARIOUS
GALVANIZED
STEEL DUCTS

Diameter		Friction
in.	mm	Factor
4	10	0.035
6	15	0.028
8	20	0.023
10	25	0.022
12	30	0.019
14	36	0.016
16	40	0.016
20	50	0.015
24	60	0.014

11-9   TURNING VANES AND DAMPERS

The two main accessory devices used in duct systems are vanes and dampers. It is the responsibility of the design engineer to specify the location and use of the devices.

Turning vanes have the purpose of preventing turbulence and consequent high loss in total pressure where turns are necessary in rectangular ducts.

Vane rail

Vane

Figure 11-29.   Typical use of vanes in a rectangular duct.

Although large-radius turns may be used for the same purpose, this requires more space. When turning vanes are used, an abrupt 90 degree turn is made by the duct, but the air is turned smoothly by the vanes, as shown in Fig. 11-29. Turning vanes are of two basic designs (Fig. 11-30). The airfoil type is the more efficient of the two, but they are more expensive to fabricate than the single-piece flat vane. The vanes shown in Figs. 11-30a and 11-30b are used in ducts having a maximum width of 36 in.; the vanes of Fig. 11-30c are used in larger ducts.

Dampers are necessary to balance a system and to control makeup and exhaust air. The dampers may be hand operated and locked in position after

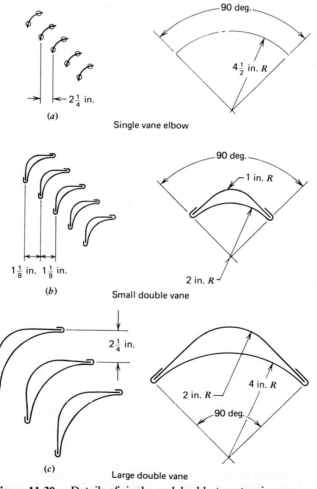

Figure 11-30. Details of single- and double-type turning vanes.

Figure 11-31. Typical opposed blade damper assembly.

adjustment or may be motor operated and controlled by temperature sensors or by other remote signals. The damper may be a single blade on a shaft or a multiblade arrangement as shown in Fig. 11-31. The blades may also be connected to operate in parallel.

A combination damper and turning assembly, called an extractor, is shown in Fig. 11-32. The extractor is adjustable from outside the duct and may be extended into the air stream to regulate flow to a branch. Most designs completely close off flow to the branch when retracted. A four-bar linkage keeps the vanes aligned with the flow.

Pressure losses for turns equipped with turning vanes were discussed in Sec. 11-8. The pressure loss for an extractor will be about the same as an elbow with single plate vanes.

Figure 11-32. Combination turning vane and damper assembly.

11-10   DUCT DESIGN—GENERAL CONSIDERATIONS

The purpose of the duct system is to deliver a specified amount of air to each diffuser in the conditioned space at a specified total pressure. This is to ensure that the space load will be absorbed and the proper air motion within the space will be realized. The method used to lay out and size the duct system must result in a reasonably quiet system and must not require unusual adjustments to achieve the proper distribution of air to each space. A low noise level is achieved by limiting the air velocity, by using sound-absorbing duct materials or liners, and avoiding drastic restrictions in the duct such as nearly closed dampers. Figure 11-33 gives recommended velocities and pressure losses for duct systems. A low-velocity duct system will generally have a pressure loss of less than 0.15 in. H_2O per 100 ft (1.23 Pa/m), whereas high-velocity systems may have pressure losses up to about 0.7 in. H_2O per 100 ft (5.7 Pa/m).

The use of fibrous glass duct materials has gained wide acceptance in recent times because they are very effective for noise control. These ducts are also attractive from the fabrication point of view because the duct, insulation, and reflective vapor barrier are all the same piece of material. Metal ducts are usually lined with fibrous glass material in the vicinity of the air distribution equipment and for some distance away from the equipment. The remainder of the metal duct is then wrapped or covered with insulation and a vapor barrier. Insulation on the outside of the duct also reduces noise.

The duct system should be relatively free of leaks, especially when the ducts are outside the conditioned space. Air leaks from the system to the outdoors result in a direct loss that is proportional to the amount of leakage and the difference in enthalpy between the outdoor air and the air leaving the conditioner. Presently 1 percent of the total volume flow of the duct system is an accepted maximum leakage rate in high-velocity systems. No generally accepted level has been set up for low-velocity systems. However, care should be taken to tape or otherwise seal all joints to minimize leakage in all duct systems and the sealing material should have a projected life of 20 to 30 years.

The layout of the duct system is very important to the final design of the system. Generally the location of the air diffusers and air-moving equipment is first selected with some attention given to how a duct system may be installed. The ducts are then laid out with attention given to space and ease of construction. It is very important to design a duct system that can be constructed and installed in the allocated space. If this is not done, the installer may make changes in the field that lead to unsatisfactory operation.

The total pressure requirements of a duct system are an important consideration. From the standpoint of first cost, the ducts should be small; however, small ducts tend to give high air velocities, high noise levels, and large losses in total pressure. Therefore, a reasonable compromise between first cost, operating cost, and practice must be reached. The cost of owning and operating an air-distribution system can be expressed in terms of system parameters, energy cost, life of the system, and interest rates such that an optimum velocity

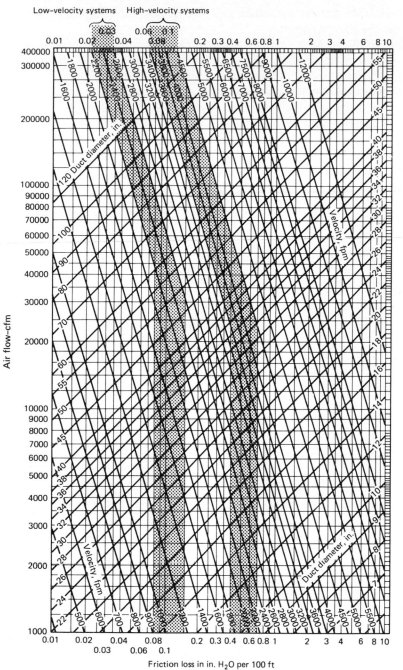

Figure 11-33. Recommended velocity and friction rate design limits (Reprinted by permission from *ASHRAE Handbook, Fundamentals Volume*, 1985.)

or friction rate can be established (2). A number of computer programs are available for this purpose.

The total pressure requirements of a duct system are determined in two main ways. For residential and light commercial applications, all of the heating, cooling, and air-moving equipment is determined by the heating and/or cooling load. Therefore, the fan characteristics are known before the duct design is begun. Furthermore, the pressure losses in all other elements of the system except the supply and return ducts are known. The total pressure available for the ducts is then the difference between the total pressure characteristic of the fan and the sum of the total pressure losses of all of the other elements in the system excluding the ducts. Figure 11-34 shows a typical total pressure profile for a unitary type system. In this case the fan is capable of developing 0.6 in.

Element	Total pressure loss, in. H_2O
Return grille	0.04
Return duct	0.08
Filter	0.08
Heat and cool coils	0.23
Supply ducts	0.14
Diffusers	0.03
Fan total pressure-in. H_2O	0.60

Figure 11-34.   Total pressure profile for a typical unitary system.

H_2O at the rated capacity. The return grille, filter, coils, and diffusers have a combined loss in total pressure of 0.38 in. H_2O. Therefore, the available total pressure for which the ducts must be designed is 0.22 in. H_2O. This is usually divided for low-velocity systems so that the supply-duct system has about twice the total pressure loss of the return ducts.

Large commercial and industrial duct systems are usually designed using velocity as a limiting criterion and the fan requirements are determined after the design is complete. For these larger systems the fan characteristics are specified and the correct fan is installed in the air handler at the factory or on the job.

It has been the practice of some designers to neglect velocity pressure. This assumption does not simplify the duct design procedure and is unnecessary. When the air velocities are high, the velocity pressure must be considered to achieve reasonable accuracy. If static and velocity pressure are computed separately as required by some design methods, the problem becomes very complex. The trend is to use total pressure exclusively in duct design procedures because it is simpler and accounts for all of the flow energy.

It is possible to lay out and size a duct system that cannot possibly function properly because the designer did not understand or failed to recognize the system pressure dynamics. In this connection, the concept of pressure and energy grade lines is useful in understanding a complex air distribution system.

Pressure Gradient Diagrams

As stated earlier, the total pressure in a duct system at any location is the sum of the static pressure and the velocity pressure. As frictional and dynamic effects occur in the airstream in the course of flow, an energy loss occurs that appears as a reduction in total pressure. Therefore, in any real duct system, except where energy is added with a fan, the total pressure will decrease in the direction of flow. This, however, is not necessarily true of the static and velocity pressures since the configuration of the air passages may cause trade-offs between the two.

Figure 11-35 is a pressure gradient diagram for a simple fan system, where

P_o = total pressure (in. or mm H_2O)
P_s = static pressure (in. or mm H_2O)
P_v = velocity pressure (in. or mm H_2O)
P_b = barometric pressure

The line connecting all of the respective total pressure points along the duct system is called the *energy grade line* (EGL). The line connecting all of the static pressure points along the duct is called the *hydraulic grade line* (HGL). The vertical distance between these lines at any section is the velocity head. The reference pressure may be any value, although for simplicity local atmospheric pressure is normally used. For fans, fittings, entrances, exits, and the like, the loss is generally considered to be concentrated at the location of these

Figure 11-35.   A simple fan system.

items, although the effects of these elements extend to some distance down- and upstream. Several points should be noted regarding the relationships de- picted on the diagram. First, the only location at which there is a total pressure increase is at the point of external energy input, namely the fan. At all other locations, total pressure will decrease, in straight equal-area ducts at a rate equal to the pressure drop per unit length, and in fittings at a rate determined by the applicable dynamic losses. Second, the total pressure always equals or exceeds the static pressure by a quantity equal to the velocity pressure. Third, though the total pressure may not increase except at the fan, the static pressure may increase or decrease in relation to the arrangement of the system, at times in unusual manners.

Though this application may seem simple, similar applications occur fre- quently and are misunderstood because the pressure gradient diagram of the proposed system was not considered. Consider the simple industrial exhaust system shown in Figure 11-36. In this instance, the total pressure loss in section AB is 0.12 in. H_2O. However, the conversion of static pressure to velocity pressure at section B has caused another apparent loss in static pressure of one velocity head. A manometer placed at point B would read a negative static pressure of -0.37 in. H_2O relative to the room pressure. This effect, in addition to the free discharge loss of an additional velocity head, has contributed sig- nificantly to the losses in this system in a manner difficult to visualize without the use of the pressure gradient diagram.

The pressure in any building served by an air system is dictated by the location of the fan and the duct system arrangement. The problem of deter- mining, understanding, or controlling the relationship of the pressure in a build- ing to the ambient or surrounding pressure is best understood through the use of the pressure gradient diagram. Figure 11-37 will begin to develop this concept

Figure 11-36.   A simple exhaust system.

more clearly. In this system, one additional element has been added: an out-door-air intake. This system would be operable in this manner only if an exhaust system were operating to exhaust the outdoor air. Otherwise, the space pressure would have to be positive to permit the outdoor air to be exfiltrated from the building. There is no guarantee that this system will operate satisfactorily.

In many cases the return-air duct system is used as the route for the excess exhaust or relief air. Figure 11-38 displays such a pressure gradient diagram. In this case the pressure in the duct system must be above barometric pressure

Figure 11-37.   A simple duct system with outdoor air intake.

Figure 11-38.   A simple duct system with outdoor
air intake and relief.

upstream of the return damper and below barometric pressure downstream of
the return damper. In addition, and more importantly, the room pressure in
this condition must exceed the barometric pressure by the static pressure losses
in the return duct system plus the relief duct system. This condition will cause
the familiar scene of exterior doors standing open with no possible means of
correcting the problem short of modifying the return/relief duct system. In most
cases the problem cannot be corrected without the use of an additional fan.

The use of a second fan can hold the building pressure closer to the bar-

Figure 11-39.   A duct system with outdoor air intake
and relief with two fans.

ometric pressure. This system is shown in Figure 11-39. In this case, the total energy required is still the same as in the case of Figure 11-38 except that it is divided between the two fans. The return fan acts to negate the losses in the return/relief air system, furnishing the positive pressure necessary at the entrance to the relief point such that the outdoor air that has been introduced may be effectively exhausted. The pressure gradient diagram is invaluable as a tool for studying such systems to evaluate the methods and components for achieving proper system operation.

11-11   DESIGN OF LOW-VELOCITY DUCT SYSTEMS

The methods described in this section pertain to low-velocity systems as defined in Fig. 11-33. These methods can be used for high-velocity system design, but the results will not be satisfactory in some cases.

Equal-friction Method

The principle of this method is to make the pressure loss per foot of duct length the same for the entire system. If the layout is symmetrical with all runs from fan to diffuser about the same length, this method will produce a good balanced design. However, most duct systems have a variety of duct runs ranging from long to short. The short runs will have to be dampered, which can cause considerable noise.

The usual procedure is to select the velocity in the main duct adjacent to the fan in accordance with Fig. 11-33. The known flow rate then establishes the duct size and the lost pressure per unit of length using Fig. 11-24. This same pressure loss per unit length is then used throughout the system. A desirable feature of this method is the gradual reduction of air velocity from fan to outlet, thereby reducing noise problems. After sizing the system, the designer must compute the total pressure loss of the longest run (largest flow resistance), taking care to include all fittings and transitions. When the total pressure available for the system is known in advance, the design loss value may be established by estimating the equivalent length of the longest run and computing the lost pressure per unit length.

> **Example 11-10.**   Select duct sizes for the simple duct system of Fig. 11-40 using the equal friction method and SI units. The total pressure available for the duct system is 0.12 in. H_2O or 30 Pa and the loss in total pressure for each diffuser at the specified flow rate is 0.02 in. H_2O or 5 Pa.
>
> *Solution.*   Because the system is small, the velocity in the main supply duct should not exceed about 1000 ft/min or 5 m/s and the branch duct velocities should not exceed about 600 ft/min or 3 m/s. The total pressure available for the ducts, excluding the diffusers, is 25 Pa and the longest run is 1-2-3. The equivalent-length method will be used to account for

Figure 11-40. A simple duct layout.

losses in the fittings. Then for simplicity if we use Figs. 11-26 and 11-27

$$L_{123} = (L_1 + L_{ent}) + (L_2 + L_{st}) + (L_3 + L_{wye} + L_{el} + L_{boot})$$
$$L_{123} = (11 + 6) + (1.5 + 4.6) + (3 + 11 + 3 + 9) = 49.1 \text{ m}$$

and

$$\Delta P_0' = \frac{\Delta P_0}{L_{123}} = \frac{25}{49.1} = 0.509 \text{ Pa/m}$$

This value will be used to size the complete system using Fig. 11-24. Table 11-15 summarizes the results showing the duct sizes, velocity in each section, and the loss in total pressure in each section. The rectangular sizes selected are rather arbitrary in this case.

It is of interest to check the actual loss in total pressure from the plenum to each outlet.

$$(\Delta P_0)_{123} = 8.67 + 3.11 + 13.26 = 25.04 \text{ Pa}$$
$$(\Delta P_0)_{124} = 8.67 + 3.11 + 14.54 = 26.32 \text{ Pa}$$
$$(\Delta P_0)_{15} = 8.67 + 18.87 = 27.54 \text{ Pa}$$

Table 11-15   SOLUTION TO EXAMPLE 11-10

Section Number	$\dot{Q}$ m³/s	D cm	w × h cm × cm	Velocity m/s	$\Delta P_0'$ Pa/m	L_e m	ΔP_0 Pa
1	0.237	30	40 × 18	3.5	0.51	17	8.67
2	0.166	26.5	38 × 16	3.2	0.51	6.1	3.11
3	0.071	19	28 × 11	2.6	0.51	26	13.26
4	0.095	21.5	32 × 13	2.8	0.51	28.5	14.54
5	0.071	19	28 × 11	2.6	0.51	37	18.87

The loss in total pressure for the three different runs are unequal when it is assumed that the proper amount of air is flowing in each. However, the actual physical situation is such that the loss in total pressure from the plenum to the conditioned space is equal for all runs of duct. That is, if the actual energy grade lines for each duct run were constructed, they would all begin at the plenum pressure and end at the room pressure. Therefore, the total flow rate from the plenum will divide itself among the three branches in order to satisfy the lost pressure requirement. If no adjustments are made to increase the lost pressure in sections 3 and 4, the flow rates in these sections will increase relative to section 5 and the total flow rate from the plenum will increase slightly because of the decreased system resistance. However, dampers in sections 3 and 4 could be adjusted to balance the system. Nevertheless this duct sizing method is used extensively. It is not always necessary that the system be designed to balance without adjustments.

Balanced-capacity Method

This method of duct design has been referred to as the "balanced pressure loss method" (3). However, it is the flow rate or capacity of each outlet that is balanced and not the pressure. As discussed earlier, the loss in total pressure automatically balances regardless of the duct sizes. The basic principle of this method of design is to make the loss in total pressure equal for all duct runs from fan to outlet when the required amount of air is flowing in each. In general all runs will have a different equivalent length and the pressure loss per unit length for each run will be different. It is theoretically possible to design every duct system to be balanced. This may be shown by combining Eqs. 9-6 and 11-8d to obtain

$$H_{01} - H_{02} = l_f = f \frac{L_e}{D} H_v \qquad (11\text{-}19)$$

Then for a given duct and fluid flowing

$$H_{01} - H_{02} = l_f(L_e, D, \overline{V}) \qquad (11\text{-}20)$$

Because the volume flow rate $\dot{Q}$ is a function of the velocity $\overline{V}$ and the diameter D, Eq. 11-20 may be written as

$$H_{01} - H_{02} = l_f(\dot{Q}, L_e) \qquad (11\text{-}20a)$$

For a given equivalent length the diameter can always be adjusted to obtain the necessary velocity that will produce the required loss in total pressure. There may be cases, however, when the required velocity may be too high to satisfy noise limitations and a damper or other means of increasing the equivalent length will be required.

The design procedure for the balanced capacity method is the same as the equal friction method in that the design pressure loss per unit length for the run of longest equivalent length is determined in the same way depending on whether the fan characteristics are known in advance. The procedure then changes to one of determining the required total pressure loss per unit length in the remaining sections to balance the flow as required. The method shows where dampers may be needed and provides a record of the total pressure requirements of each part of the duct system. Example 11-11 demonstrates the main features of the procedure.

Example 11-11.   Design the duct system in Fig. 11-40 and Example 11-10 by using the balanced capacity method. Use SI units.

Solution.   The total pressure available and the equivalent lengths will be the same as those in Example 11-10. In addition, the procedure for the design of the longest run L_{123} is exactly the same. Sections 4 and 5 must then be sized to balance the system. It is obvious from Fig. 11-40 that the lost pressure in section 4 must equal that of section 3. Then from Table 11-15

$$\Delta P_{04} = \Delta P_{03} = 13.26 \text{ Pa}$$

The equivalent length of section 4 is 28.5 m; therefore,

$$\Delta P'_{04} = \frac{\Delta P_{04}}{L_4} = \frac{13.26}{28.5} = 0.465 \text{ Pa/m}$$

From Fig. 11-24*b*, at $\Delta P'_{04} = 0.465$ and $\dot{Q} = 0.095 \text{ m}^3/\text{s}$

$$D_4 = 22 \text{ cm} \quad (26 \times 15 \text{ cm rectangle})$$
$$\overline{V}_4 = 2.7 \text{ m/s}$$

The loss in total pressure for section 5 is

$$\Delta P_{05} = \Delta P_{02} + \Delta P_{03} = 3.11 + 13.26 = 16.37 \text{ Pa}$$

The equivalent length of section 5 is 37 m; therefore

$$\Delta P'_{05} = \frac{\Delta P_{05}}{L_5} = \frac{16.37}{37} = 0.442 \text{ Pa/m}$$

From Fig. 11-24*b*

$$D_5 = 19.5 \text{ cm} \quad (22 \times 15 \text{ cm rectangle})$$
$$\overline{V}_5 = 2.4 \text{ m/s}$$

Comparison of the diameters or rectangular sizes obtained for sections 4 and 5 using the two different methods discussed thus far shows the sizes to be different, as suggested in the discussion following Example 11-10. For these simple examples the differences were not dramatic. In real systems the equivalent lengths of the various runs vary considerably and the balanced-capacity method is superior to the equal-friction method when system balance is critical.

The only limitation of the balanced capacity method is the use of equivalent lengths for the fittings. Experience has shown this to be a minor error for low-velocity systems. Further, when the equivalent lengths are derived as illustrated earlier, the method is quite accurate within one's ability to describe the system.

Example 11-12.   Design the duct system of Fig. 11-41 by using the balanced-capacity method. The velocity in the duct attached to the plenum must not exceed 900 ft/min and the overall loss in total pressure should not exceed about 0.32 in. H_2O. Total pressure losses for the diffusers are all equal to 0.04 in. H_2O. Rectangular ducts are required. The lengths shown are the *total* equivalent lengths of each section. Use English units.

Solution.   In order to hold the duct sizes to a minimum, the maximum velocity criteria will be used to establish the design pressure loss for the longest run of duct. Using Fig. 11-24 the pressure per 100 ft and equivalent diameter of section 1 is

$$H'_{01} = 0.096 \text{ in. } H_2O/100 \text{ ft}$$
$$D_1 = 12.9 \text{ in.}$$

If this pressure loss per 100 ft is used to design sections 1-2-3-4-5, the longest run, the loss in total pressure will be

$$H_{015} = 0.096(55 + 25 + 40 + 45 + 95)/100 = 0.250 \text{ in. } H_2O$$

When the diffuser loss is added, the maximum lost pressure criterion is still satisfied ($H_0 = 0.29$ in. H_2O); therefore, the longest run will be sized using $H'_{01} = 0.096$ in. $H_2O/100$ ft. Table 11-16 summarizes this procedure.

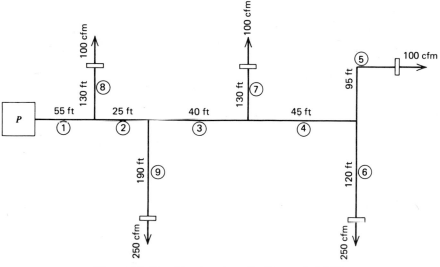

Figure 11-41.   Duct system for Example 11-12.

Table 11-16   DATA FOR EXAMPLE 11-12

Section Number	Q cfm	D_e in.	$w \times h$ in. $\times$ in.	Velocity fpm	L_e ft	H_0 in. H_2O
1	800	12.9	18 $\times$ 8	900	55	0.053
2	700	12.1	16 $\times$ 8	880	25	0.024
3	450	10.2	15 $\times$ 6	780	40	0.038
4	350	9.4	13 $\times$ 6	740	45	0.044
5	100	5.9	6 $\times$ 5	540	95	0.091
Diffuser 5	100	—	—	—	—	0.040
				Total		0.290

Branches 6, 7, 8, and 9 must now be sized to balance the system. Because all of the diffusers have the same lost pressure, they need not be considered in the remainder of the analysis.

$$H_{06} = H_{05} = 0.091 \text{ in. } H_2O$$
$$L_6 = 120 \text{ ft}$$
$$H'_{06} = 0.091(100)/120 = 0.076 \text{ in. } H_2O/100 \text{ ft}$$
$$D_6 = 8.7 \text{ in.}   (11 \times 6 \text{ rectangle})$$
$$\overline{V}_6 = 610 \text{ ft/min}$$
$$H_{07} = H_{04} + H_{05} = 0.044 + 0.091 = 0.135 \text{ in. } H_2O$$
$$L_7 = 130 \text{ ft}$$

$$H'_{07} = \frac{0.135(100)}{130} = 0.100 \text{ in. } H_2O/100 \text{ ft}$$
$$D_7 = 5.8 \text{ in.}   (6 \times 5 \text{ rectangle})$$
$$\overline{V}_7 = 540 \text{ ft/min}$$
$$H_{09} = H_{03} + H_{07} = 0.038 + 0.135 = 0.173 \text{ in. } H_2O$$
$$L_9 = 190 \text{ ft}$$
$$H'_{09} = \frac{0.173(100)}{190} = 0.091 \text{ in. } H_2O/100 \text{ ft}$$
$$D_9 = 8.4 \text{ in.}   (10 \times 6 \text{ rectangle})$$

$$\overline{V}_9 = 650 \text{ ft/min}$$
$$H_{08} = H_{09} + H_{02} = 0.173 + 0.024 = 0.197 \text{ in. } H_2O$$
$$L_8 = 130 \text{ ft}$$
$$H'_{08} = \frac{0.197(100)}{130} = 0.152 \text{ in. } H_2O/100 \text{ ft}$$
$$D_8 = 5.4 \text{ in.}   (6 \times 4 \text{ rectangle})$$
$$\overline{V}_8 = 630 \text{ ft/min}$$

The resulting velocities for sections 8 and 9 are slightly high but are probably acceptable.

The previous examples have shown the balanced-capacity method to be straight-forward and to produce much detailed information about the duct system, particularly with respect to the duct velocities and the placement of dampers. However, the calculation procedure is cumbersome and time-consuming when carried out for a complex system as shown. To alleviate these undesirable features, a worksheet has been devised. An example has been solved using the worksheet (Fig. 11-42). A full-sized worksheet is enclosed in the packet at the end of the text, Chart 6. Note that the worksheet has provisions for recording information concerning the various elements of the air-distribution system including the fan, filters, cooling and heating coils, and the diffusers. One added feature of the worksheet is the provision for subbranches, which sometimes occur in complex systems. The following procedure is used with the duct design worksheet.

1. Fill in the heading on the worksheet based on the equipment selected for the job.
2. Select diffusers for each outlet and record the total pressure requirements as shown.

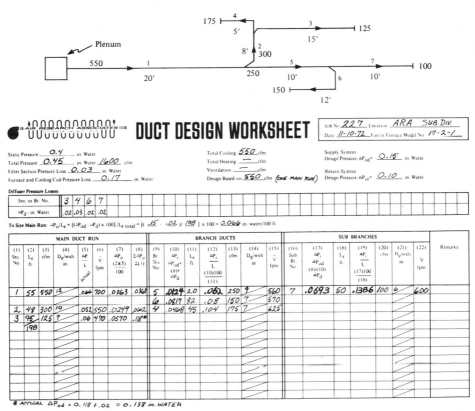

Figure 11-42. Duct design example using worksheet and balanced capacity method. (Reprinted with permission of Air Research Associates, Inc.)

3. Determine the main duct run (longest equivalent length). This is run 1 to 2 to 3 in the example; then section 4 is a branch, section 5 to 6 is a branch, and section 7 is a subbranch. Section 5 to 6 is selected as the branch because it has a larger equivalent length than section 5 to 7.

4. Enter the section numbers, equivalent lengths, and volume flow rate for each section as shown. Notice the arrangement of the section numbers with respect to the duct system. Reading across the first line, section 5 to 6 branches off section 1 and section 7 branches off section 5. Similarly, section 4 branches off section 2.

5. Calculate the loss in total pressure per 100 ft for the main duct run in the space provided and use the value and Fig. 11-24 to size the various sections. If rectangular ducts are to be used, record the exact duct diameter from the chart. If circular ducts are to be used, round off to the nearest standard diameter as shown for the example.

6. Record the actual loss per 100 ft in column 5, the velocity in column 6, and compute the actual loss in total pressure for each section: record in column 7. The accumulated loss in total pressure is recorded in column 8. The last number in column 8 is the actual loss in total pressure for the main duct run exclusive of the diffuser and should be less than or equal to the design value.

7. Size the branch ducts by comparing the loss in total pressure for each branch as shown in column 10. Find $\Delta P/L$ in column 12 and select the duct diameter using Fig. 11-24. When a branch has two or more sections as shown, the branch is sized for the loss in total pressure and equivalent length for the complete branch. For this example, sections 5 and 6 must have a loss of 0.0817 in. H_2O and the total equivalent length is 102 ft. Then $\Delta P/L = 0.08$. From the friction chart $D_6 = 7$ in. and $D_5 = 8.5$ in. D_5 must be rounded up to 9 in. and then $\Delta P/L = 0.062$. The loss in total pressure for section 5 is finally calculated as 0.0124 in. H_2O.

8. The subbranches are sized in the same way as the branches. Only a few systems have more than one or two subbranches. Most systems have none.

9. If the ducts are to be rectangular, Table 11-8 may be used to convert the diameters to rectangular dimensions.

The completed worksheet makes a good record and provides a check on the velocities in each section. A check for computational errors may be made by summing the the pressure losses horizontally in columns 8, 10, and 17 plus the lost pressure for the diffuser.

Return Air Systems

The design of the return system may be carried out using the methods described earlier. In this case the air flows through the branches into the main duct and back to the plenum. Although the losses in constant-flow fittings are the same regardless of the flow direction, divided-flow fittings behave differently and different equivalent lengths or loss coefficients must be used. Table 11-13 gives loss coefficients for typical divided-flow fittings used in return systems. Reference 2 gives considerable data for converging-type fittings of both circular and rectangular cross section. It should be noted that for low velocity ratios the loss coefficient can become negative with converging flow streams. This seems contrary to natural laws; however, this behavior is a result of a high-velocity stream mixing with a low-velocity stream. Kinetic energy is transferred from the higher to the lower velocity air, which results in an increase in energy or total pressure of the slower stream. Low-velocity return systems are usually designed using the equal-friction method. The total pressure loss for the system is then estimated as discussed for supply duct systems. Dampers may be required just as with supply systems. In large commercial systems a separate fan for the return air may be required.

11-12   HIGH-VELOCITY DUCT DESIGN

Because space allocated for ducts in large commercial structures is very limited due to the high cost of building construction, alternatives to the low-velocity central system are often sought. One approach is to use hot and cold water, which is piped to the various spaces where small central units or fan coils may be used. However, it is sometimes desirable to use rather extensive duct systems without taking up too much space. The only way this can be done is to move the air at much higher velocities. Figure 11-33 indicates that high-velocity systems may use velocities as high as 4500 ft/min (23 m/s) and lost pressures as high as 0.7 in. H_2O per 100 ft (5.7 Pa/m). The use of high velocities reduces the duct sizes dramatically, but introduces some new problems.

Noise is probably the most serious consequence of high-velocity air movement. Special attention must be given to the design and installation of sound-attenuating equipment in the system. Generally, a sound absorber is installed just downstream of the fan and is similar to an automobile's muffler. Because the air cannot be introduced to the conditioned space at a high velocity, a device called a *terminal box* is used to throttle the air to a low velocity, control the flow rate, and attenuate the noise. The terminal box is located in the general vicinity of the space it serves and may distribute air to several outlets. These boxes take many forms.

The energy required to move the air through the duct system at high velocity is also an important consideration. The total pressure requirement is typically on the order of several inches of water and is due to the large losses associated with the high velocities. To partially offset the high fan power requirements, variable-speed fans are often used to take advantage of low volume flow requirements of variable-volume systems at off-design conditions. Careful selection of duct fittings and the duct layout can also reduce the power requirement.

The higher static and total pressures required by high-velocity systems aggravate the duct leakage problem. Ordinary duct materials such as light gauge steel and ductboard do not have suitable joints to prevent leakage and are also too weak to withstand the forces arising from the high pressure differentials. Therefore, an improved duct fabrication system has been developed for high-velocity systems. The duct is generally referred to as *spiral duct* and has either a round or oval cross section. The fittings are machine formed and are especially designed to have low pressure losses and close fitting joints to prevent leakage. The pressure losses in spiral pipe and fittings are generally different from those in ordinary galvanized duct and fittings. There is also a much wider variety of fittings available for high-velocity systems. For brevity, the data on pressure losses presented in Secs. 11-7 and 11-8 will be used in this section. In actual practice, however, engineering data should be obtained from the manufacturer.

The criterion for designing high-velocity duct systems is somewhat different from that used for low-velocity systems. Emphasis is shifted from a self-balancing system to one that has minimum losses in total pressure. The high-velocity system achieves the proper flow rate at each outlet through the use of the terminal box. The terminal box is designed to operate with a minimum pressure loss of about 0.25 to 1.0 in. H_2O or 60 to 245 Pa. Some of these devices maintain a constant volume flow rate, whereas others vary the flow rate in proportion to the space heating or cooling requirement. A double-duct type may also be used where warm and cool air are blended according to the space requirement. The run of duct that has the largest potential flow resistance should be designed to have a total pressure loss as low as possible. This run will then determine the system total pressure. The remaining runs will presumably result in a lower lost pressure but will be balanced by the terminal box.

Total pressure losses must be estimated carefully using the loss coefficient method (Eq. 11-16), because the equivalent length method is not reliable at the velocities used with high-velocity systems. The most efficient fittings from the standpoint of lost pressure should be selected. For example, an ordinary square entrance to a duct from a plenum may be used in a low-velocity system but, when used with high-velocity air, the lost pressure becomes prohibitive. Table 11-11 indicates that the loss coefficient for a formed entrance is about one-tenth that of a square entrance. Similar savings in total pressure may be achieved in the selection of branch fittings. Manufacturer's data should be consulted and the best fitting selected. The designer should become acquainted with the standard practices for duct construction (7) before attempting an actual design problem.

There are a number of duct design procedures that may be used for high-velocity systems. These will be described and demonstrated separately. The design procedures are relatively independent of the type of air-distribution system involved. For example, a variable air volume duct system should be designed for the full-load condition when air flow is at a maximum. At full load the VAV system operates the same as a constant-volume system. However, a VAV system may be very extensive, serving zones that peak at much different times. In such a case, each section of the duct system must be sized to handle the maximum amount of air required.

Static Regain Method

This method systematically reduces the air velocity in the direction of flow in such a way that the increase (regain) in static pressure in each transition just balances the pressure losses in the following section. This method is suitable for high-velocity, constant volume systems having long runs of duct with many takeoffs. With this procedure approximately the same static pressure exists at the entrance to each branch, which simplifies terminal unit selection and system balancing. The main disadvantages of the method are (1) the very low velocities and large duct sizes that may result at the end of long runs; (2) the tedious bookkeeping and trial-and-error aspects of the method; and (3) the total pressure requirements of each part of the duct system that are not readily apparent.

The general procedure for use of the static regain method is to first select a velocity for the duct attached to the fan or supply plenum. With the capacity this establishes the size of this main duct. The run of duct that appears to have the largest flow resistance is then designed first, using the most efficient fittings and layout possible. A velocity is assumed for the next section in the run and the static pressure regain is compared to the lost pressure for that section. Usually about two velocities must be checked to find a reasonable balance between the static pressure regain and the losses of a section. It must also be kept in mind that standard duct sizes must be used, which usually prevents an exact balance between regain and losses. The spiral duct is available in diameters of 3 to 24 in. using increments of 1 in. and from 24 to 50 in. using increments of 2 in. Standard metric pipe ranges from about 8 to 60 cm in 1-cm increments and from 60 to 120 cm in 2-cm increments. The following example demonstrates the procedure for a simple duct system.

Example 11-13.   Design the duct system shown in Fig. 11-43 using the static regain method. Each outlet has a terminal box that requires a minimum of 0.5 in. H_2O total pressure. Other pertinent data are shown on the sketch.

Solution.   The first step in the solution is to select a velocity for the section of duct connected to the plenum. Figure 11-33 indicates that a maximum velocity of 2500 fpm should be used when the duct carries 2000 cfm. The section may then be sized using Fig. 11-24. A 12-in. duct will result in a velocity of about 2550 fpm, which is acceptable. Then

$$D_1 = 12 \text{ in.}; \qquad \overline{V}_1 = 2550 \text{ fpm}; \qquad H_{v1} = 0.405 \text{ in. } H_2O$$

From Table 11-11B assume a loss coefficient for the formed entrance of 0.03 and from Fig. 11-24 a pressure loss per 100 ft of duct of $H'_{01} = 0.75$ in. $H_2O/100$ ft. Then

$$(H_0)_p = 0.03 H_{v1} = 0.03(0.405) = 0.012 \text{ in. } H_2O$$

Figure 11-43.   Duct system for Examples 11-13 and 11-14.

$$(H_0)_{pa} = \frac{H'_{01}L}{100} = \frac{0.75(50)}{100} = 0.375 \text{ in. } H_2O$$

$$H_{01} = 0.387 \text{ in. } H_2O$$

The total pressure required at the plenum to meet the needs of section 1 is then given by

$$H_{0p} = H_{01} + H_{v1} = 0.387 + 0.405 = 0.792 \text{ in. } H_2O$$

If the static region method could be followed exactly, this would be the system total pressure, exclusive of the terminal box requirement, because velocity pressure is to be converted to static pressure to offset the lost pressure.

The run of duct with the largest flow resistance appears to be sections 1-2-3-4, but this must be checked at the conclusion of the design. To size section 2 it is required that the increase in static pressure (decrease in velocity pressure) from point a to point b must equal the lost pressure from point b to point c. This is equivalent to

$$H_a - H_c = H_{vc} - H_{va} + (H_0)_{ab} + (H_0)_{bc} = 0 \qquad (11\text{-}21)$$

or

$$H_a - H_c = H_{v2} - H_{v1} + \Sigma(H_0)_{ac} = 0 \qquad (11\text{-}21a)$$

and this may be generalized as

$$H_u - H_d = H_{vd} - H_{vu} + \Sigma(H_0) = 0 \qquad (11\text{-}21b)$$

where the subscripts u and d refer to upstream and downstream, respectively. To size section 2 a diameter of 11 in. will be assumed and calculations made to check against Eq. 11-21b.

$$D_2 = 11 \text{ in.;} \qquad \bar{V}_2 = 2000 \text{ fpm;} \qquad H'_{02} = 0.53 \text{ in. } H_2O/100 \text{ ft}$$
$$H_{v2} = 0.25 \text{ in. of water;} \qquad \dot{Q}_2 = 1500 \text{ cfm}$$
$$(H_0)_{ab} = 0.03 \text{ in. of water} \quad \text{(assumed, see Table 11-12B)}$$
$$(H_0)_{bc} = 0.53(20)/100 = 0.106 \text{ in. } H_2O$$
$$\Sigma(H_0)_{ac} = 0.136 \text{ in. } H_2O$$

Then using Eq. 11-21b

$$H_a - H_c = 0.25 - 0.405 + 0.136 = -0.019 \text{ in. } H_2O$$

This indicates that a larger increase in static pressure than required was achieved in the transition. Although the next smaller pipe size could be tried, the 11-in. duct would prove to be the best solution.

The same general procedure is used to size section 3 assuming a 10-in. diameter duct.

$$D_3 = 10 \text{ in.;} \qquad \bar{V}_3 = 1850 \text{ fpm;} \qquad H'_{03} = 0.5 \text{ in. } H_2O/100 \text{ ft}$$
$$H_{v3} = 0.213 \text{ in. } H_2O; \qquad \dot{Q}_3 = 1000 \text{ cfm}$$
$$(H_0)_{ca} = 0.015 \text{ in. } H_2O \quad \text{(assumed, see Table 11-12B)}$$
$$(H_0)_{de} = \frac{0.5(30)}{100} = 0.15 \text{ in. } H_2O$$
$$\Sigma(H_0)_{ce} = 0.165 \text{ in. } H_2O$$
$$H_c - H_e = 0.2313 - 0.25 + 0.165 = 0.128 \text{ in. } H_2O$$

Assume another diameter for section 3.

$$D_3 = 11 \text{ in.;} \qquad \bar{V}_3 = 1520 \text{ cfm;} \qquad H'_{03} = 0.31 \text{ in. } H_2O/100 \text{ ft}$$
$$H_{v3} = 0.144 \text{ in. } H_2O; \qquad \dot{Q}_3 = 1000 \text{ cfm}$$
$$(H_0)_{cd} = 0.017 \text{ in. } H_2O \quad \text{(assumed, see Table 11-12B)}$$
$$(H_0)_{de} = 0.31(30)/100 = 0.093 \text{ in. } H_2O$$
$$\Sigma(H_0)_{ce} = 0.11 \text{ in. } H_2O$$
$$H_c - H_e = 0.004 \text{ in. } H_2O$$

This is a very good solution for section 3.

It is obvious that the design of a system using the foregoing procedure can become quite tedious. A tabular form reduces the work considerably and will be used to conclude this design. Table 11-17 contains all of the design calculations for this example.

The total pressure requirement for the system is determined by the run with the maximum cumulative static pressure loss plus the velocity pressure in section 1.

$$H_0 = (H_0)_{1237} + H_{v1} = 0.392 + 0.405 = 0.797 \text{ in. } H_2O$$

Table 11-17 STATIC REGAIN DESIGN METHOD

Section Number	Item	D in.	$\dot{Q}$ cfm	$\bar{V}$ fpm	H_v in. H_2O	L ft	H'_0 or C in. H_2O per 100 ft	H_0 in. H_2O	ΣH_0 in. H_2O	$(H_u - H_d)^a$ in. H_2O	$\Sigma(H_u - H_d)$ in. H_2O
1	Duct	12	2000	2550	0.405	50	0.75	0.375			
	Entrance						0.03	0.012	0.387	0.387	0.387
2	Duct	11	1500	2000	0.250	20	0.53	0.106			
	Fitting							0.030	0.136	−0.019	0.368
3	Duct	10	1000	1850	0.213	30	0.50	0.150			
	Fitting							0.015	0.165	0.128	
	Duct	11	1000	1520	0.144	30	0.31	0.093			
	Fitting							0.017	0.110	0.004	0.372
4	Duct	9	600	1350	0.114	30	0.32	0.096			
	Fitting						0.45	0.051	0.147	0.117	
	Duct	10	600	1100	0.075	30	0.19	0.057			
	Fitting						0.40	0.030	0.087	0.018	0.390
	Branch 7 off Section 3										
7	Duct	9	400	910	0.052	60	0.15	0.090			
	Fitting							0.012			
	Fitting						0.200	0.010	0.112	0.020	0.392
	Branch 6 off Section 2										
6	Duct	8	500	1450	0.131	40	0.41	0.164			
	Fitting						0.40	0.052	0.216	0.097	
	Duct	9	500	1140	0.081	40	0.23	0.092			
	Fitting						0.50	0.041	0.133	−0.036	0.332
	Branch 5 off Section 1										
5	Duct	8	500	1450	0.131	30	0.41	0.123			
	Fitting						0.50	0.066	0.189	−0.085	0.302

[a]Eq. 11-21b.

where $H_{1237} = \Sigma(H_u - H_d)$ from Table 11-17. However, this does not include the pressure loss for the terminal box. In allowing for this pressure, remember that the total pressure just upstream of the box is equal to the velocity pressure. Section 7 has a velocity pressure of 0.052 in. H_2O. Because the terminal box requires at least 0.5 in. H_2O, the system total pressure requirement should be increased to

$$H_0 = 0.797 + (0.5 - 0.052) = 1.245 \text{ in. } H_2O$$

The required total pressure for this example is relatively low because of its small size. As high-velocity systems become larger, the maximum velocity may be increased according to Fig. 11-33 with a corresponding increase in required total pressure.

Example 11-14.   Suppose the duct system of Example 11-13 handles a total volume flow of 23 m³/s or 49,000 cfm. Design section 1 for the maximum velocity allowable from Fig. 11-33 and estimate the total pressure required for the system assuming static regain design. Assume a minimum total pressure of 245 Pa for the terminal boxes.

Solution.   Figure 11-33 shows a maximum velocity of 4500 ft/min or 23 m/s for this case. Then using Fig. 11-24 the duct (section 1) is sized as

$$D_1 = 1.2 \text{ m}; \qquad \overline{V}_1 = 22.5 \text{ m/s}; \qquad \Delta P'_{01} = 3.2 \text{ Pa/m}$$

Assume the loss coefficient for the formed entrance is 0.03 and the velocity pressure in section 1 is

$$P_{v1} = \left(\frac{22.5}{1.29}\right)^2 = 304.2 \text{ Pa}$$

Then

$$(\Delta P_o)_p = 0.03(304.2) = 9.1 \text{ Pa}$$
$$(\Delta P_o)_{pa} = 3.2(15.2) = 48.6 \text{ Pa}$$

The total pressure loss for section 1 is then 58 Pa. A conservative estimate of the system total pressure is given by

$$P_o = P_{o1} + P_{v1} + (P_o)_{box}$$
$$P_o = 58 + 304 + 245 = 607 \text{ Pa}$$

The total pressure method is an adaptation of the static regain method. The method provides insight into the actual energy requirements of each part of the duct system. The objective is to size each section so that the total pressure available at the beginning of a section is just enough to balance the loss in total pressure in the section.

Velocity-reduction Method

This method evolved following general acceptance of the total pressure concept in duct design. Appreciation of the total energy requirements of the duct system has placed emphasis on control of the local velocities and total pressure losses during the design of the system. At the same time almost no effort is made to design a self-balancing system. The method is based on the selection of acceptable air velocities at the inlet and in the branches of the system and then on a gradual reduction of the velocity in between. The total pressure losses are then calculated for each section proceeding from either end. The various air velocities are selected such that standard duct sizes will result in order to avoid double effort on the part of the designer.

This design method is recommended for the VAV duct system with zones that peak at widely varying times. Because the air flow rates in the various parts of the system change continually, there is no advantage to using a sophisticated design method. Careful attention to the maximum velocities attained in each section is of greatest importance.

With the ducts sized and the fittings known, the total pressure losses can be calculated, the pressure or energy grade lines plotted, and the maximum pressure loss or critical path of the system determined. The use of the energy grade line analysis, previously discussed, is particularly useful to make sure the system will function in an efficient manner.

A solution to the duct system of Fig. 11-43 using the assumed velocity method is given in Table 11-18 in SI units. The known characteristics, flow rate and duct length, were first entered in the table. The maximum velocities for sections 1, 4, 5, 6, and 7 were then selected along with the duct diameters for the same sections and entered in the table. The velocities and duct diameters were then selected for the intermediate sections. The reductions in velocity were divided as evenly as possible. The various losses were then computed and summed as shown for each run of duct. After completion of the table the results can be checked for any abnormally high losses; in such a case the duct diameter or a fitting may be changed to lower the overall total pressure requirement of the system. The system as designed in Table 11-18 requires about 503 Pa total pressure.

Accessories

The main accessories used with high-velocity duct systems are sound absorbers and dampers. Sound absorbers were mentioned earlier and may take many forms. Although the design of these elements is beyond the scope of this book, manufacturer's data or other references on acoustics and noise should be consulted.

Balancing dampers may be used in high-velocity systems; however, this function is usually handled by the terminal box. Fire dampers are of great importance in this system. The function of the fire damper is to divert the flow

Table 11-18   VELOCITY REDUCTION DESIGN METHOD

Section Number	Item	D cm	$\dot{Q}$ m³/s	$\bar{V}$ m/s	L m	$\Delta P_0'$ Pa/m	C	H_0 Pa	ΣH_0 Pa
1	Duct	30	0.945	14	15.2	7.4		112.5	
	Entrance						0.03	3.5	
2	Duct	28	0.70	11.2	6.1	5.4		32.9	
	Fitting							8.7	
3	Duct	23	0.47	11.0	9.1	6.2		56.4	
	Fitting							3.5	
4	Duct	19	0.283	9.4	9.1	6.0		54.6	
	Fitting						0.45	32.7	
	Box							124.5	429.3
	Branch 7 off Section 3								
7	Duct	16	0.188	9.8	18.3	8.0		146.4	
	Fitting							2.5	
	Fitting						0.20	14.5	
	Box							124.5	505.4
	Branch 6 off Section 2								
6	Duct	18	0.237	9.0	12.2	6.0		73.2	
	Fitting						0.40	30.2	
	Box							124.5	385.5
	Branch 5 off Section 1								
5	Duct	18	0.237	9.0	9.1	6.0		54.6	
	Fitting						0.45	53.0	
	Box							124.5	348.1

of air to or from a particular area if a fire should occur. Large commercial structures have partitions, floors, and ceilings that confine a fire to a given area for some specified time. When an air duct passes through one of these fire barriers, a fire damper is generally required. Some of these dampers are held open by fusible links, whereas others are controlled by smoke detectors or other such devices.

The location and control of the fire dampers are the responsibility of the design engineer, who must become acquainted with the governing codes for the particular application and city and state in which he or she is practicing.

REFERENCES

1. *ASHRAE Handbook, Equipment Volume,* American Society of Heating, Refrigerating and Air-Conditioning Engineers, Atlanta, Ga., 1983.

2. *ASHRAE Handbook, Fundamentals Volume,* American Society of Heating, Refrigerating and Air-Conditioning Engineers, Atlanta, Ga., 1985.

3. Burgess H. Jennings, *Environmental Engineering Analysis and Practice,* International Textbook Company, Scranton, Pa., 1970.

4. F. C. McQuiston, "Duct Design for Balanced Air Distribution in Low Velocity Systems," *Proceedings,* Conference on Improved Efficiency in HVAC
Components, Purdue University, Lafayette, Ind., 1974.

5. "Low Velocity Duct Construction Standards," 4th ed., Sheet Metal and
Air-Conditioning Contractors National Association, Washington, D.C., 1969.

6. J. B. Graham, "Methods of Selecting and Rating Fans," *ASHRAE Journal,*
1972.

7. "High Pressure Duct Construction Standards," 3rd ed., Sheet Metal and
Air-Conditioning Contractors National Association, Washington D.C., 1975.

8. "AMCA Fan Application Manual," Parts 1, 2, and 3, Air Movement and
Control Association, Inc., 30 West University Drive, Arlington Heights, Ill.

PROBLEMS

11-1 A centrifugal fan is delivering 1700 cfm of air at a static pressure differential
(across the fan) of 1.0 in. H_2O. The fan has an outlet area of 0.71 ft^2 and requires
0.7 horsepower shaft input. Compute (a) the static power, (b) the static efficiency,
(c) the total efficiency, and (d) the fan total pressure.

11-2 A fan is delivering 2.1 m^3/s of air at a total pressure of 280 Pa. The fan has an
outlet area of 0.19 m^2 and requires 1100 W of shaft power. Compute (a) the total
efficiency, (b) the static efficiency, and (c) the static pressure differential.

11-3 The fan of Problem 11-1 is operating at 1350 rpm. The fan speed is reduced to
1200 rpm. Compute the capacity, static and total pressure, and the shaft power
at the lower speed.

11-4 The fan of Problem 11-2 is operating at 700 rpm. The speed is increased to 800
rpm. Compute the capacity, static and total pressure, and the shaft power at
the higher speed.

11-5 Consider the fan curves shown in Fig. 11-7 and assume the data given are for
a speed of 900 rpm. (a) Construct a new total pressure characteristic for a speed
of 750 rpm for the good selection zone. (b) Construct a new shaft power curve
for 750 rpm. (c) Sketch the total pressure and power characteristics for both
speeds on the same graph with a system characteristic for 10,000 cfm (900 rpm).
(d) What is the total pressure, volume flow rate, and shaft power at 750 rpm?

11-6 The fan of Fig. 11-7 is rated at standard sea level density. Suppose it is to be
used in Denver, Colorado, where the elevation is 5280 ft. (a) Construct total
pressure and shaft power characteristics for the selection zone for the new
condition. (b) Compute the percent change in power between sea level and 5280
ft elevation for a volume flow rate of 10,000 cfm.

11-7 The fan of Fig. 11-11 is rated at sea level. Suppose it is to be used in Albuquerque,
New Mexico, where the elevation is 1618 m. (a) Construct the total pressure
characteristic for 800 rpm. (b) Compute the percent change in power from sea
level to 1618 m for 800 rpm and a volume flow rate of 155 m^3/s.

11-8 Comment on the desirability of using the fan described in Fig. 11-7 to circulate (a) 10,000 cfm at about 1.8 in. H_2O total pressure, (b) 15,000 cfm at about 0.9 in. H_2O total pressure, (c) 2500 cfm at 2.0 in. H_2O total pressure.

11-9 Would the fan shown in Fig. 11-8 be suitable for use in a system requiring (a) 5000 cfm at 1.5 in. H_2O total pressure? (b) 30,000 cfm at 5.0 in. H_2O total pressure? (c) 15,000 cfm at 4.0 in. H_2O total pressure? Explain.

11-10 A duct system has been designed for 150 m^3/min at 500 Pa total pressure. Would the fan shown in Fig. 11-11 be suitable for this application? Explain and estimate the total efficiency, fan speed, and shaft power.

11-11 A small system requires 0.88 in. H_2O total pressure at a flow rate of 1420 cfm. Select a suitable fan using the data of Table 11-2. (a) Sketch the system and fan characteristics showing the operating point. (b) What is the fan speed and power?

11-12 A system requires 1.0 kPa total pressure at a flow rate of 6.25 m^3/s. (a) Select a fan using Table 11-2a. (b) Sketch the system and fan characteristics showing the operating point. (c) What is the fan speed and power?

11-13 A duct system has been designed to have 2.5 in. H_2O total pressure loss. It is necessary to use an elbow at the fan outlet that turns the air 90 degrees to the side. Reference 8 gives a system effect factor of 0.2 in. H_2O for this situation. The fan inlet also has an elbow with a system effect factor of 0.15 in. H_2O. (a) What total pressure should the fan for this system produce to circulate the desired amount of air? (b) Sketch the system and fan characteristics with and without the system effect factors. (c) The design volume flow rate is 15,000 cfm. Using Fig. 11-8 estimate the resulting flow rate when the system effect factors are not used.

11-14 The fan shown in Fig. 11-11 is operating in a system at 900 rpm with a flow rate of 150 m^3/s. The system was designed to circulate 170 m^3/min. Assuming that the duct pressure losses were accurately calculated, estimate the system effect factor for the fan.

11-15 A SWSI, backward-curved blade fan discharges air into a 12 × 16 in. rectangular duct at the rate of 6700 cfm. An elbow located 20 in. from the fan outlet turns to the right. (a) Estimate the system effect for the outlet duct. (b) Estimate the system effect for the elbow.

11-16 The fan of Problem 11-15 also has an inlet duct and elbow as shown in Fig. 11-17b. The diameter is 16 in. and the duct length is 32 in. Estimate the system effect.

11-17 A forward-curved blade fan with a 12 × 12 in. outlet is to be installed in a system to deliver 2500 cfm. What length of outlet duct is required to prevent any system effect?

11-18 The fan of Problem 11-17 has an inlet duct configured as shown in Fig. 11-17e with an R/H ratio of 1.0 and H of 12 in. About how long must the duct be so that the system effect will not exceed 0.16 in. H_2O?

11-19 The fan shown in Fig. 11-8 was selected for a constant-volume system requiring 15,000 cfm and a total pressure of 4.5 in. H_2O. The operating speed was expected to be about 900 rpm. (a) When the system was started, measurements were made and a flow rate of 9000 cfm and total pressure of 2.0 in. H_2O were observed. Explain the probable cause of the discrepancy. (b) At another time, after the problem of part (a) was corrected, a flow rate of 10,000 cfm and a total pressure of 4.8 in. H_2O was observed. Explain the possible problem in this case. (c) Several years later, assuming the problem of part (b) was solved, measurements show a flow rate of 13,000 cfm and a total pressure of 4.3 in. H_2O. Explain the probable cause of the decreased flow rate.

11-20 Refer to Fig. 11-22 and compute the percent difference in shaft power between flow conditions 1 and 2.

11-21 A variable-volume system using a fan as shown in Fig. 11-22 will operate at about 15,000 cfm a majority of the time. (a) Estimate the power saved, in kW-hr for one day, as compared with a constant-volume system operating at point 1. (b) Estimate the power saving as compared with a VAV system with no fan speed control.

11-22 Consider the fan of Fig. 11-21, operating as shown in a VAV system. Compute the percent decrease in shaft power between flow conditions 1 and 2.

11-23 Assume that the fan and VAV system shown in Fig. 11-21 operate at an average capacity of 25,000 cfm over a given 24-hr period. (a) Estimate the power savings as compared with a constant-volume system operating at point 1. (b) Estimate the power saving as compared with a VAV system with no fan control, that is, where point 1 will move along the full open characteristic.

11-24 Estimate the lost pressure in 50 ft of 22 × 20 in. metal duct with an air flow rate of 3000 cfm. The duct is lined with 1 in. of the type B liner shown in Fig. 11-25.

11-25 Assuming the duct of Problem 11-24 is operating with standard sea level air, estimate the lost pressure for air at the same temperature but at an altitude of 5000 ft.

11-26 Compute the lost pressure for a 16 m length of 0.5 × 0.6 m duct lined with 25 mm of the type C liner shown in Fig. 11-25. The volume flow rate is 3 m^3/s.

11-27 Compute the lost pressure for the duct of Problem 11-26 for an altitude of 1500 m.

11-28 A circular metal duct 20 ft in length has an abrupt contraction at the inlet and an abrupt expansion at the exit. Both have an area ratio of 0.6. The duct has a diameter of 10 in. with a flow rate of 600 cfm. Estimate the loss in total pressure for the duct including the contraction and expansion.

11-29 Consider a 6-in. circular branch duct that is tapped into the side of a large rectangular duct. The branch is a flexible fiberglass material with roughness similar to the type C liner of Fig. 11-25. The branch is 20 ft in length with a flow rate of 150 cfm. Estimate the lost pressure for the takeoff and duct assuming $\overline{V}_b/\overline{V}_u = 0.75$.

11-30 Compare the lost pressure for a bellmouth entrance, $r/D = 0.06$, Fig. 11-11b with an abrupt entrance, $\theta = 180$ degrees, Fig. 11-11a for (a) duct velocity of 1000 ft/min (5 m/s) and (b) 4000 ft/min (20 m/s).

Figure 11-44. Schematic for Problem 11-35.

11-31 Compare the lost pressure for a 90 degree mitered rectangular elbow, 16×12 in., Table 11-9C and the same elbow with turning vanes No. 3, Fig. 11-9b. The volume flow rate of air is (a) 1600 cfm (0.76 m³/m) and (b) 3500 cfm (1.65 m³/s).

11-32 Compute the lost pressure for a diverging wye fitting such as those in Fig. 11-12. The flow rate in the 12-in. (30 cm) upstream section is 800 cfm (0.38 m³/s) and the flow rate in the 6-in. (15 cm) branch is 250 cfm (0.12 m³/s). Assume (a) a 45 degree wye and (b) a 90 degree wye.

11-33 Compute the lost pressure for a converging wye type fitting such as those of Fig. 11-13. The flow rate in the 12-in. (30 cm) upstream section is 550 cfm (0.26 m³/s) and the flow rate in the 6-in. (15 cm) branch is 250 cfm (0.12 m³/s). Assume (a) a 45 degree wye and (b) a 90 degree wye.

11-34 The computed lost pressure for a fitting is 0.2 in. H_2O (50 Pa) based on sea level conditions. What is the lost pressure based on conditions in Denver, Colorado?

11-35 Compute the loss in total pressure for each run of the duct system shown in Fig. 11-44. The ducts are of round cross section. Turns and fittings are as shown. Use the loss coefficient and the equivalent length approaches (Fig. 11-26 and 11-27) and compare the answers. Use English units.

11-36 Refer to Problem 11-30 and compute the equivalent lengths for the two different entrances assuming a duct diameter of 12 in. (30 cm). Compute the lost pressure using the equivalent lengths.

Figure 11-45. Schematic for Problem 11-38.

11-37 Refer to Problem 11-31 and compute the equivalent lengths for the elbows. Compute the lost pressure using the equivalent lengths.

11-38 The duct system shown in Fig. 11-45 is one branch of a complete air-distribution system. The system is a perimeter type located below the floor. The diffuser boots are shown in Fig. 11-27. Size the various sections of the system using the equal friction method and round pipe. A total pressure of 0.13 in. H_2O is available at the plenum. Compute the actual loss in total pressure for each run assuming the proper amount of air is flowing.

11-39 Consider the duct layout shown in Fig. 11-46. The system is supplied air by a rooftop unit that develops 0.25 in. H_2O total pressure external to the unit. The return air system requires 0.10 in. H_2O. The ducts are to be of round cross section and the maximum velocity in the main run is 850 ft/min, whereas the branch velocities must not exceed 650 ft/min. (a) Size the ducts using the equal-friction method. Show the location of any required dampers. Compute the total pressure loss for the system. (b) Size the ducts using the balanced-capacity method and the duct design worksheet.

11-40 Design the duct system shown in Fig. 11-47 for circular ducts. The fan produces a total pressure of 0.70 in. H_2O at 1000 cfm. The lost pressure in the filter, furnace, and evaporator is 0.35 in. H_2O. The remaining total pressure should be divided between the supply and return with 65 percent used for the supply system. Diffuser and grille losses are shown on the diagram. Use equivalent lengths to account for fitting losses. (a) Use the equal-friction method to size the ducts. (b) Use the balanced-capacity method and the duct design worksheet to size the ducts.

11-41 Design the duct system shown in Fig. 11-48 using the balanced-capacity method and the duct design worksheet. Circular ducts are to be used and installed below

Figure 11-46. Schematic duct layout for Problem 11-39.

Figure 11-47. Layout for Problem 11-40.

Figure 11-48. Schematic for Problem 11-41.

a concrete slab. The total pressure available at the plenum is 0.18 in. H₂O, and each diffuser has a loss in total pressure of 0.025 in. H₂O. Use the equivalent-length method of accounting for losses in the fittings.

11-42 A high velocity duct layout is shown in Fig. 11-49. (a) Size the system using the static regain design method. Use maximum allowable velocities from Fig. 11-33. (b) Specify the required total pressure in the plenum. (c) Specify the fan shaft horsepower, efficiency, speed, and total pressure using Fig. 11-8.

11-43 Repeat Problem 11-42 using the velocity-reduction method.

11-44 Solve Problem 11-42 using the velocity-reduction method and SI units. Specify fan characteristics.

11-45 The schematic of a large office complex is shown in Fig. 11-50. Halls, closets, and so on, have been omitted for clarity. The air quantities shown are the maximum that each space will require, but the spaces will not all peak at the same time. (a) Lay out a duct system starting from the equipment room running due west and turning into two trunk ducts running north and south between the three rows of spaces. Each space requires a terminal box connected to the trunk duct with circular duct. (b) Estimate the maximum flow for the system and each section of the duct using the following guidelines: The east rooms will peak about 10:00 A.M. and remain constant until about 3:00 P.M. The interior rooms peak at 9:00 P.M., and remain constant until 4:00 P.M. The north and south rooms in the center peak at noon and demand gradually decreases after that. The west rooms reach one-half capacity at noon and peak at 5:00 P.M. The minimum overall load occurs at 5:00 A.M. and is about 20 percent of the maximum. (c) Size the ducts using a maximum velocity of about 2500 ft/min and the velocity-reduction method. Use a velocity of about 1000 ft/min for the branch connections

Figure 11-49. High-velocity duct system for Problem 11-42.

|← ——— 120 ft ——— →|

1000 cfm	600 cfm	500 cfm
1000 cfm	700 cfm	500 cfm
1000 cfm	700 cfm	500 cfm
1000 cfm	700 cfm	500 cfm
1000 cfm	600 cfm	500 cfm

200 ft

N

Equip. room

Figure 11-50. Schematic for Problem 11-45.

to the terminal boxes. (d) Estimate the total pressure requirements of the system at maximum flow. (e) Locate the static pressure sensor for the fan speed control. (f) Refer to Fig. 11-22 and locate operating points 1 and 2. Scale the size of the fan up or down as required by changing the capacity values along the abscissa. (g) How much power does the system require at the minimum and maximum operating points?

11-46 Lay out and size the ducts for the system provided by the instructor and provide other requested information.

11-47 Refer to Fig. 11-51 and construct the energy grade line (total pressure versus length) for the system shown. The change in total pressure is shown for each part of the system. What total pressure must each fan produce?

Figure 11-51. Schematic of a makeup and exhaust air system.

11-48 Refer to Problem 11-47 and assume the return fan does not exist. Construct the energy grade lines and give the fan total pressure.

11-49 Refer to Fig. 11-51 and assume the supply fan is moved just upstream of the coil section. Construct the energy grade lines and give the total pressure for the fans.

11-50 Refer to Problem 11-49 and assume the return fan does not exist. Construct the energy grade lines and give the fan total pressure.

CHAPTER TWELVE

Mass Transfer and the Measurement of Humidity

The concept of mass transfer has already been discussed in Chapter 3 in connection with the thermodynamic processes of humidification and dehumidification. In this chapter the physical aspects of mass transfer are briefly reviewed as they relate to the typical processes encountered in HVAC systems, such as air washers and cooling coils.

It was also mentioned in Chapter 3 that to determine the state of moist air we must know two independent properties plus the barometric pressure. The dry bulb temperature and barometric pressure may be determined easily, but the measurement of the one remaining property is more difficult.

The wet bulb thermometer is the most practical device available to measure a property related to the moisture content of the air and is a reliable instrument when properly applied. An objective of this chapter is to discuss its application from the mass transfer point of view.

12-1 COMBINED HEAT AND MASS TRANSFER

Many problems in engineering are concerned with the simultaneous transfer of mass and heat. In this book these problems deal mainly with heated and cooled air–water vapor mixtures that result in the evaporation or condensation of the water.

It is well known that a link exists between the transport of momentum, heat, and mass (1). Consider the two-dimensional boundary layer equations for an incompressible, constant property fluid with a zero pressure gradient.

$$u \frac{\partial C}{\partial x} + v \frac{\partial C}{\partial y} = D \frac{\partial^2 C}{\partial y^2} \tag{12-1}$$

$$u \frac{\partial u}{\partial x} + v \frac{\partial u}{\partial y} = \nu \frac{\partial^2 u}{\partial y^2} \tag{12-2}$$

511

$$u \frac{\partial t}{\partial x} + v \frac{\partial t}{\partial y} = \alpha \frac{\partial^2 t}{\partial y^2} \tag{12-3}$$

The boundary conditions are

$$y = 0; \quad u = 0, \quad v = v_w, \quad t = t_w, \quad C = C_w \tag{12-4}$$

$$y = \infty; \quad u = u_\infty \quad\quad t = t_\infty, \quad C = C_\infty \tag{12-5}$$

The quantities D, v, and α are the diffusivities for mass, momentum, and heat, respectively. When these quantities are equal and the wall velocity v_w is zero, the solutions to Eqs. 12-1 through 12-3 are identical and the concentration C, velocity u, and temperature profiles t are similar. In this case the Prandtl number Pr and Schmidt number Sc are both equal to one, where

$$\mathrm{Pr} = \frac{v}{\alpha} \tag{12-6}$$

and

$$\mathrm{Sc} = \frac{v}{D} \tag{12-7}$$

The ratio of the Schmidt number to the Prandtl number is the Lewis number

$$\mathrm{Le} = \frac{\mathrm{Sc}}{\mathrm{Pr}} = \frac{\alpha}{D} \tag{12-8}$$

which is also equal to one in this case.

To clearly show the connection between heat, mass, and momentum transfer consider an energy balance on a surface

$$h(t_\infty - t_w) = k \left(\frac{\partial t}{\partial y} \right)_{y=0} \tag{12-9}$$

where h and k are the convective heat-transfer coefficient and the fluid thermal conductivity, respectively. When the solution to Eq. 12-3 for t and the distance y are nondimensionalized

$$t' = \frac{t - t_w}{t_\infty - t_w} \quad\quad y' = \frac{y}{L} \tag{12-10}$$

and inserted in Eq. 12-9, the following result is obtained:

$$\mathrm{Nu} = \frac{hL}{k} = \left(\frac{\partial t'}{\partial y'} \right)_{y=0} \tag{12-11}$$

where Nu is the Nusselt number, a dimensionless heat transfer coefficient.

Fick's law may be written as

$$\dot{m}_w = -DA \left(\frac{\partial C}{\partial y} \right)_{y=0} \tag{12-12}$$

where $\dot{m}_w$ is the mass-transfer rate, mass per unit time. A mass-transfer coefficient is introduced:

$$\dot{m}_w = h_m A(C_w - C_\infty) \tag{12-12a}$$

where h_m has dimensions of length per unit time and C is the concentration, mass per unit volume. Combination of Eqs. 12-12 and 12-12a and the use of the dimensionless concentration and length from the solution to Eq. 12-1

$$C' = \frac{(C_w - C)}{(C_w - C_\infty)} \qquad y' = \frac{y}{L} \tag{12-13}$$

yields

$$\text{Sh} = \frac{h_m L}{D} = \left(\frac{\partial C'}{\partial y'}\right)_{y=0} \tag{12-14}$$

where Sh is the Sherwood number, a dimensionless mass transfer coefficient.

When the temperature and concentration profiles are similar, the dimensionless profiles C' and t' are identical. As a result the dimensionless heat- and mass-transfer coefficients given by Eqs. 12-11 and 12-14 are also identical. The link to momentum transfer or friction can be shown in a similar way. A similar development may be carried out to show the same general relationship between heat, mass, and momentum transfer in turbulent flow.

It is well known that expressions for Nusselt and Sherwood numbers have the form

$$\text{Nu} = f(\text{Re}, \text{Pr}) \tag{12-15}$$
$$\text{Sh} = f(\text{Re}, \text{Sc})$$

where the functional relations are given by

$$\text{Nu} = C_1 \text{Re}^a \text{Pr}^b \tag{12-15a}$$
$$\text{Sh} = C_1 \text{Re}^a \text{Sc}^b$$

Reynolds analogy was first used to show the connection between heat and momentum transfer and appears as

$$\frac{h}{\rho c_p \overline{V}} = \frac{f}{2} \tag{12-16}$$

where f is the Fanning friction factor. The analogy was later extended to mass transfer:

$$\frac{h_m}{\overline{V}} = \frac{h}{\rho c_p \overline{V}} = \frac{f}{2} \tag{12-16a}$$

The Reynolds analogy has long been recognized as giving reasonable results when $\text{Pr} \approx \text{Sc} \approx 1$ and the temperature potential is moderate. Many other analogies have been proposed to account for the effect of Prandtl number. Chilton and Colburn (2) have proposed one of the most widely accepted analogies called the j-factor analogy:

$$j = j_m = \frac{f}{2} \tag{12-17}$$

where

$$j = \frac{h}{\rho_a c_{pa} \overline{V}} \Pr^{2/3} = \frac{h}{Gc_{pa}} \Pr^{2/3} \tag{12-18}$$

and

$$j_m = \frac{h_m}{\overline{V}} \mathrm{Sc}^{2/3} \tag{12-19}$$

From Eqs. 12-18 and 12-19

$$\frac{h}{\rho_a c_{pa} h_m} = \left(\frac{\mathrm{Sc}}{\Pr}\right)^{2/3} = \mathrm{Le}^{2/3} \tag{12-20}$$

where $\mathrm{Le}^{2/3}$ is approximately 1.0 for moist air at usual conditions. In air-conditioning calculations it is generally more convenient to use the concentration in the form of the humidity ratio W rather than C in mass of water per unit volume. The relation between the two is

$$C = W\rho_a \tag{12-21}$$

where ρ_a is the mass density of the dry air, mass per unit volume. Equation 12-12a then becomes

$$\dot{m}_w = h_m A \rho_a (W_w - W_\infty) \tag{12-12a}$$

or

$$\dot{m}_w = h_d A (W_w - W_\infty) \tag{12-12b}$$

where

$$h_d = \rho_a h_m \tag{12-22}$$

The dimension of h_d is mass of dry air per unit area and time. The analogy of Eq. 12-20 then becomes

$$\frac{h}{c_{pa} h_d} = \mathrm{Le}^{2/3} \tag{12-20a}$$

and the j factor of Eq. 12-19 becomes

$$j_m = \frac{h_d}{\rho_a \overline{V}} \mathrm{Sc}^{2/3} \tag{12-19a}$$

The use of the foregoing analogy in moist-air problems requires caution. Over the range of temperatures of 50 to 140 F or 10 to 60 C and from completely dry to saturated air, the Lewis number ranges from about 0.81 to 0.86 (3), whereas the Schmidt and Prandtl numbers have values of about 0.6 and 0.7. Therefore, the theoretical basis for the analogy is not entirely satisfied. The analogy is also based on ideal surface and flow field conditions. For example, there is undisputed evidence that the water deposited on a surface during dehumidification roughens the surface and upsets the analogy because h is for a smooth dry surface (4). There is also evidence that disturbances in the flow field may influence the transfer phenomena and sometimes render the analogy invalid (5). The analogy seems to be most valid when there is direct contact between the air and water. In situations such as a dehumidifying heat exchanger, the condensate that collects on the surface upsets the fundamental basis for the analogy. Research recently completed related to this phenomenon will be discussed in Chapter 14. We will use the j-factor analogy most extensively for direct contact processes.

12-2   THE PSYCHROMETER

The wet bulb thermometer or a variation of it is the most convenient and simplest means of measuring a state property related to the moisture content of an air–water vapor mixture, and practically all computational methods and charts are based on its use. Figures 3-2 and 3-3 show typical arrangements of the wet bulb and dry bulb thermometers that together make up the psychrometer.

Figure 12-1 shows a schematic arrangement of a psychrometer. Although mercury thermometers are shown, thermocouples or other temperature-sensing devices can be used. To carry out an analysis of the psychrometer the following assumptions are made.

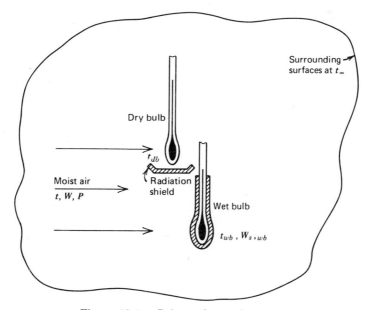

Figure 12-1.   Schematic psychrometer.

1. Heat conduction along the stem of the thermometer is negligible.
2. The air velocity is low enough for the static and stagnation air temperatures to be essentially equal.

Notice that each thermometer may exchange heat with the air by convection and with the surroundings by radiation. In many cases radiation effects are very small for the dry bulb thermometer. Because the wet bulb thermometer is generally at a different temperature than the dry bulb thermometer and the surroundings, radiation effects must usually be considered. It should always be kept in mind that a temperature-sensing device indicates its own temperature, which may not be the temperature of the surrounding medium.

If we assume that the psychrometer has reached a steady state condition, an energy balance on the dry bulb thermometer yields

$$h(t_{db} - t) = h_r(t_\infty - t_{db}) \tag{12-23}$$

where

t = actual air dry bulb temperature, F or C
t_{db} = dry bulb thermometer temperature, F or C
t_∞ = temperature of surrounding enclosure, F or C
h = convective film coefficient for the thermometer, Btu/(hr-ft²-F) or W/(m²C)
h_r = radiation heat-transfer coefficient, Btu/(hr-ft²-F) or W/(m²-C)

Solving for the air dry bulb temperatures gives

$$t = t_{db} - \left(\frac{h_r}{h}\right)_{db} (t_\infty - t_{db}) \tag{12-24}$$

A method of handling the (h_r/h) term will be described later.

The energy balance on the wet bulb thermometer must account for the evaporation of water from the wick.

$$h_d(W_{s,wb} - W)i_{fg,wb} = h(t - t_{wb}) + h_r(t_\infty - t_{wb}) \tag{12-25}$$

where

$W_{s,wb}$ = humidity ratio for saturated air at the temperature indicated by the wet bulb thermometer, lbmw/lbma or kgw/kga
h_d = mass-transfer coefficient, lbma/(ft²-hr) or kga/(m²-s)
$i_{f,g,wb}$ = enthalpy of vaporization of water at the indicated wet bulb temperature, Btu/lbmw or J/kgw
t_{wb} = indicated wet bulb temperature, F or C

One may solve for the humidity ratio of air

$$W = W_{s,wb} - B(t - t_{wb}) \tag{12-26}$$

where B is the *wet bulb coefficient* defined by

$$B = \frac{Le^{2/3}c_{pa}}{i_{fg,wb}} \left| 1 + \left(\frac{h_r}{h}\right) \frac{(t_\infty - t_{wb})}{(t - t_{wb})} \right| \tag{12-27}$$

where the analogy of Eq. 12.20a has been used to obtain the mass-transfer coefficient h_d. The specific heat of the dry air, c_{pa} is 0.24 Btu/(lbma-F) or 1004 J/(kg-C). Equations 12-26 and 12-27 permit calculation of the air–humidity ratio W for the general case. The humidity ratio of saturated air at the wet bulb temperature may be obtained from Table A-2.

For the usual situation, where the surrounding surfaces are at the dry bulb temperature ($t = t_\infty$), Eq. 12-27 reduces to

$$B = \frac{c_{pa}Le^{2/3}\left(1 + \dfrac{h_r}{h}\right)}{i_{fg,wb}} \tag{12-28}$$

When the web bulb is perfectly shielded from radiation effects, Eq. 12-27 reduces to

$$B = \frac{c_{pa}Le^{2/3}}{i_{fg,wb}} \tag{12-29}$$

The determination of the wet bulb coefficient B is the main problem in finding the humidity ratio W. Although the Lewis number is often assumed to be unity, for dry bulb temperatures of 50 to 100 F or 10 to 38 C and for completely dry to saturated air a value of 0.85 is more accurate. The radiation heat transfer coefficient h_r may be estimated from the relation

$$h_r = \sigma\varepsilon \frac{[(T_\infty)^4 - (T_{wb})^4]}{(t_\infty - t_{wb})} \tag{12-30}$$

where ε is the emittance of the wick covering the wet bulb and the Stefan-Boltzman constant σ is 0.1712×10^{-8} Btu/(hr-ft^2-R^4) or 5.673×10^{-8} W/(m^2-K^4) depending on the system of units utilized. Equation 12-30 may be expressed in a more convenient form by multiplying and dividing by 10^8. Thus

$$h_r = \sigma'\varepsilon \frac{\left|\left(\dfrac{T_\infty}{100}\right)^4 - \left(\dfrac{T_{wb}}{100}\right)^4\right|}{(t_\infty - t_{wb})} \tag{12-30a}$$

where $\sigma' = \sigma \times 10^8$.

The convection heat-transfer coefficient h may be computed from the relation

$$\mathrm{Nu}_d = \frac{hd}{k} = C\,(\mathrm{Re}_d)^n \tag{12-31}$$

Table 12-1 COEFFICIENTS FOR EQUATION 12-33

$\mathrm{Re}_d = \bar{V}d/\nu$	C	n
4–40	0.821	0.385
40–4,000	0.615	0.466
4,000–40,000	0.174	0.618

where the Prandtl number has been assumed equal to one and the coefficients C and n are given in Table 12-1. The fluid properties k and v should be evaluated at the *mean film temperature, t_f.*

$$t_f = \frac{t + t_{wb}}{2} \tag{12-32}$$

Table B-4 gives the properties of air. For convenience in computing the wet bulb coefficient B, Figures 12-2 and 12-3 have been constructed using Eqs. 12-30 and 12-31, *for the special case of the surrounding temperature t_∞ equal to the air dry bulb temperature t.* A value of $\varepsilon_{wb} = 0.9$ was used and a wet bulb thermometer diameter of 0.3 in. or 7.5 mm applies to Fig. 12-2, whereas a diameter of 0.1 in. or 2.5 mm was used in Fig. 12-3. Figures 12-2 and 12-3 apply only to Eq. 12-28 where $t = t_\infty$. For the general case a new expression must be derived for the wet bulb coefficient B if Figs. 12-2 and 12-3 are to be used or Eq. 12-27 may be used with h_r and h computed from Eqs. 12-30 and 12-31.

Example 12-1.   A wet bulb psychrometer is inserted in a 10-in. diameter, insulated air duct. The volume flow rate of the air is 450 cfm and it is thoroughly mixed. The dry bulb thermometer indicates 60 F, whereas the wet bulb reading is 50 F. The thermometers are unshielded and have a diameter of about 0.3 in. Determine the humidity ratio W.

Solution.   Because the duct is insulated, the inside surface will approach the air dry bulb temperature very closely:

$$t_\infty = t_{db} = t$$

Equation 12-26 will be used to compute W, whereas $W_{s,wb}$ may be read from Table A-2 at $t_{wb} = 50$ F, as 0.007658 lbmw/lbma. The wet bulb coefficient B is given by Eq. 12-28 for this case with h_r/h read from Fig. 12-2. The enthalpy of vaporization $i_{fg,wb}$ is obtained from Table A-1 as 1065 Btu/lbmw and the Lewis number is taken as 0.85.
  The average air velocity in the duct is given by

$$\bar{V} = \frac{\dot{Q}}{A} = \frac{450}{(\pi/4)(10/12)^2} = 825 \text{ ft/min}$$

By interpolating between curves 7 and 8 in Fig. 12-2, $h_r/h = 0.067$. By using Eq. 12-28 we get

$$B = \frac{0.240 \ (0.85)^{2/3} \ (1 + 0.067)}{1065} = 2.157 \times 10^{-4}$$

Then by Eq. 12-26

$$W = 0.007658 - (2.157)10^{-4}(60 - 50) = 0.0055 \text{ lbmw/lbma}$$

Figure 12-2. Ratio h_r/h for a wet bulb diameter of 0.3 in. (7.5 mm). (James L. Threlkeld, *Thermal Environmental Engineering*, 2nd ed., 1970, p. 202. Reprinted by permission of Prentice-Hall, Inc., Englewood Cliffs, N.J.)

Curve no.	t F	t_{wb} F	t C	t_{wb} C
1	120	100	50	38
2	120	85	50	30
3	120	70	50	21
4	90	80	33	27
5	90	70	33	21
6	90	60	33	16
7	60	55	16	13
8	60	45	16	7
9	30	28	−1	−2
10	30	20	−1	−7
11	0	−2	−17	−19

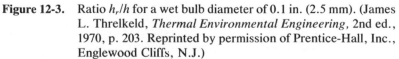

Figure 12-3. Ratio h_r/h for a wet bulb diameter of 0.1 in. (2.5 mm). (James L. Threlkeld, *Thermal Environmental Engineering*, 2nd ed., 1970, p. 203. Reprinted by permission of Prentice-Hall, Inc., Englewood Cliffs, N.J.)

Example 12-2.   An uninsulated metal duct has air flowing at 600 ft/min. An inserted thermometer indicates an air dry bulb temperature of 120 F, whereas the pipe wall temperature is estimated to be 90 F. The thermometer is not shielded and its diameter is about 0.3 in. Find the actual air dry bulb temperature.

Solution.   Equation 12-24 applies here. The ratio h_r/h must be determined analytically using Eqs. 12-30 and 12-31. The Reynolds number will first be computed as

$$\mathrm{Re}_d = \frac{\bar{V}d}{\nu} = \frac{600}{60}\frac{0.3}{12}\frac{10^4}{1.8} = 1389$$

where the viscosity ν was obtained from Table B-4 at 105 F. Then using Eq. 12-31 with coefficients from Table 12-1, we find that

$$h = \frac{k}{d}(0.615)(\mathrm{Re}_d)^{0.466} = \frac{(15.6)10^{-3}}{(0.3/12)}(0.615)(1389)^{0.466}$$

$$h = 11.2 \text{ Btu/(hr-ft}^2\text{-F)}$$

where k was obtained from Table B-4 at 105 F. Eq. 12-30 with $\varepsilon = \varepsilon_{db} = 0.8$, $t_\infty = 90$ F, and t_{wb} replaced with $t_{db} = 120$ F yields

$$h_r = 0.1712(0.8)\frac{\left[\left(\frac{550}{100}\right)^4 - \left(\frac{580}{100}\right)^4\right]}{(90 - 120)} = 1.0 \text{ Btu/(hr-ft}^2\text{-F)}$$

Then

$$\frac{h_r}{h} = \frac{1.0}{11.2} = 0.089$$

Substitution in Eq. 12-24 then gives

$$t = 120 - (0.089)(90 - 120) = 122.7 \text{ F}$$

It is evident that radiation can introduce significant errors in the dry bulb temperature. The error is quite dependent on the emittance of the thermometer bulb. If the bulb of Example 12-2 had an emittance of about 0.2, the error in the measurement is less than one degree F.

Example 12-3.   An unshielded wet bulb thermometer is placed in the duct used in Example 12-2 and indicates a temperature of 100 F. Find the humidity ratio W.

Solution.   It will be necessary to estimate the wet bulb coefficient using Eq. 12-27, and Eq. 12-26 will give W. In this case

$$h_r = 0.1712(0.9)\frac{\left[\left(\frac{550}{100}\right)^4 - \left(\frac{560}{100}\right)^4\right]}{(90 - 100)} = 1.05 \text{ Btu/(hr-ft}^2\text{-F)}$$

and h will be assumed equal to that calculated in Example 12-2, $h = 11.2$ Btu/(hr-ft^2-F). Other data required are: $W_{s,wb} = 0.0432$ lbmw/lbma; $i_{fg,wb} = 1036.7$ Btu/lbmw; Le = 0.85; and $t = 122.7$ F from Example 12-2. Then

$$B = \frac{(0.240)(0.85)^{2/3} \left| 1 + \dfrac{1.05}{11.2} \dfrac{(90 - 100)}{(122.7 - 100)} \right|}{1036.7}$$

$$B = 1.991 \times 10^{-4} \text{ lbmw/(lbma-F)}$$

and

$$W = 0.0432 - (1.991)10^{-4}(122.7 - 100) = 0.0387 \text{ lbmw/lbma}$$

In summary, the most reliable results are obtained with the wet bulb psychrometer when the air velocity is relatively high and radiation effects are small. There will always be some uncertainty in computing the wet bulb coefficient B; this effect will be greatest when W is small. Almost all humidity-measuring devices have the same limitation.

12-3 ADIABATIC SATURATION TEMPERATURE VERSUS WET BULB TEMPERATURE

In the previous section the humidity ratio W was determined by measurements of wet and dry bulb temperatures with heat and mass transfer theory. The wet bulb temperature was not considered as a thermodynamic state property and indeed it should not be, because it depends on air velocity and the bulb geometry, in addition to the state of the moist air. It should be emphasized that the wet bulb temperature and the adiabatic saturation temperature are distinctly different. The adiabatic saturation temperature is hypothetical and can only be approached in practice. When the wet bulb temperature appears in psychrometric equations and charts, it is really the adiabatic saturation temperature that is being considered.

Threlkeld (3) has analyzed the problem and correlated wet bulb temperature with the adiabatic saturation temperature (sometimes called thermodynamic wet bulb temperature). The results are shown in Figs. 12-4 and 12-5 for thermometers with diameters of 0.3 and 0.1 in. or 7.5 and 2.5 mm. The data are shown as percent deviation of $(t_{wb} - t^*)$ from the wet bulb depression $(t - t_{wb})$ for shielded and unshielded wet bulbs. Three different temperature combinations are given and several general conclusions may be drawn. It appears that the unshielded wet bulb will generally more closely approximate the adiabatic saturation temperature. Temperatures taken with air velocities greater than about 200 ft/min or 1 m/s generally have the least deviation. Lower air velocities may be used with smaller wet bulbs. Threlkeld (3) draws the following general conclusion: For atmospheric temperatures above freezing, where the wet bulb depression does not exceed about 20 F or 11 C, and where no unusual

Figure 12-4. Deviation ($t_{wb} - t^*$) in percent of the wet bulb depression ($t - t_{wb}$) for a wet bulb diameter of 0.3 in. (7.5 mm) and a barometric pressure of 14.696 psia (101.325 kPa). (James L. Threlkeld, *Thermal Environmental Engineering*, 2nd ed., 1970, p. 208. Reprinted by permission of Prentice-Hall, Inc., Englewood Cliffs, N.J.)

radiation circumstances exist, t_{wb} should differ from t^* by less than about 0.5 F or 0.27 C for an unshielded mercury-in-glass thermometer as long as the air velocity exceeds about 100 ft/min or 0.5 m/s.

Thus, for most engineering problems the wet bulb temperature obtained from a properly operated unshielded psychrometer may be used directly as the adiabatic saturation temperature. When wet bulb depression is large or radiation effects are unusual, the procedure in Sec. 12-2 should be used.

The wick on the wet bulb should be kept clean and thoroughly moistened with distilled water. Air velocity over the bulb must be maintained to obtain reliable results.

12-4   OTHER METHODS OF MEASURING HUMIDITY

In addition to the wet bulb psychrometer there are other devices that measure humidity. Most of these devices are more complicated and less reliable than the psychrometer. One device that gives the relative humidity as a direct reading is a *hygrometer*. Several general types of hygrometers exist.

Figure 12-5. Deviation $(t_{wb} - t^*)$ in percent of the wet bulb depression $(t - t_{wb})$ for a wet bulb diameter of 0.1 in. (2.5 mm) and a barometric pressure of 14.696 psia (101.325 kPa). (James L. Threlkeld, *Thermal Environmental Engineering*, 2nd ed., 1970, p. 209. Reprinted by permission of Prentice-Hall, Inc., Englewood Cliffs, N.J.)

Dew-point Hygrometer

In the most common form of this hygrometer, means are provided for cooling and observing the temperature of a surface exposed to moist air. The highest temperature at which detectable condensation occurs on the surface is the dew-point temperature. Charts and tables are used with the dew-point temperature to find the relative humidity. A bright metallic surface cooled by ether or some other refrigerant is usually employed. There are many variations of the dew-point hygrometer that are more fully discussed in reference 6.

Dimensional Change Hygrometer

Changes in humidity cause many organic materials to shrink or expand; this action may be tied to a suitable linkage to actuate a dial or switch. Materials commonly used are human hair, nylon, dacron, wood, and paper. Unfortunately these materials do not consistently have a reproducible behavior, which causes problems in calibration. The most common application of this type of hygrometer is in humidistats used to control humidifers.

Electrical Impedance Hygrometer

This type of hygrometer correlates the changing moisture content of a substance with a change in electrical impedance. An electrical or electronic signal processing system is therefore required. The advantage of the impedance hygrometer is that it provides an electrical signal for recording that may be used in telemetering data from remote locations. Reference 6 provides additional details.

Electrolytic Hygrometer

In this device a highly effective desiccant absorbs moisture from a regulated stream of moist air. The absorbed moisture is electrolyzed, and the required current is related to the humidity level. The desiccant is usually phosphorous pentoxide and the apparatus is used at moisture–air ratios of 1 to 1000 parts per million.

Gravimetric Hygrometer

This hygrometer removes water vapor from a known quantity of moist air and weighs it. Obviously precise laboratory equipment is required. Powerful desiccants or freezing processes are used to remove the water vapor from the air.

All of the hygrometers discussed must be calibrated. This may be done by maintaining an atmosphere at a known temperature and humidity and exposing the hygrometer to it. Complete calibration requires comparison at several different conditions. The National Bureau of Standards maintains a primary reference standard. Additional information and reference material on calibration and standards can be found in reference 6.

REFERENCES

1. J. D. Parker, J. H. Boggs, and E. F. Blick, *Introduction to Fluid Mechanics and Heat Transfer,* Addison-Wesley, Reading, Mass., 1969.

2. T. H. Chilton and A. P. Colburn, "Mass Transfer (Absorption) Coefficients—Prediction from Data on Heat Transfer and Fluid Friction," *Industrial and Engineering Chemistry,* November 1934.

3. James L. Threlkeld, *Thermal Environmental Engineering,* 2nd ed., Prentice-Hall, Englewood Cliffs, N.J., 1970.

4. J. L. Guillory and F. C. McQuiston, "An Experimental Investigation of Air Dehumidification in a Parallel Plate Exchanger," *ASHRAE Transactions, Vol. 29,* June 1973.

5. Wayne A. Helmer, "Condensing Water Vapor—Air Flow in a Parallel Plate Heat Exchanger," Ph.D. Thesis, Purdue University, May 1974.

6. *ASHRAE Handbook, Fundamentals Volume,* American Society of Heating, Refrigerating and Air-Conditioning Engineers, Atlanta, Ga., 1985.

PROBLEMS

12-1   Compute the heat-transfer coefficient using the *j*-factor where air is flowing over a surface at a velocity of 1200 ft/min and has a mean temperature of 100 F. The *j*-factor is known to be 0.008.

12-2   Compute the heat-transfer coefficient for air flowing at a temperature of 40 C and a velocity of 6 m/s. The *j*-factor is 0.01.

12-3   It is necessary to estimate the rate at which water is removed from an air stream by a cooling coil. The mass-transfer coefficient is not known but the coil was tested at the same Reynolds number under sensible heat-transfer conditons and the heat-transfer coefficient was 10 Btu/(hr-ft²-F) [56.8 W/(m²-c)]. Estimate the mass-transfer coefficient assuming a mean air temperature of about 50 F (10 C).

12-4   Use Eqs. 12-21 and 12-2a to show that the units of the mass-transfer coefficient h_d are mass of dry air per unit area and time. Further show that c_{pa} in Eqs. 12-20 must be on the basis of unit mass of dry air.

12-5   The Nusselt number for flow in a tube is given by Eq. 12-15 where the constants C_1, *a*, and *b* are 0.023, 0.8, and 0.3, respectively. Estimate a mass-transfer coefficient for miost air flowing in a 12-in. (30-cm) tube at the rate of 600 cfm (0.3 m³/s). Assume standard air.

12-6   Estimate the mass-transfer coefficient for moist air flowing over a 1-in. (25-mm) diameter tube starting with Eq. 12-31 for flow normal to a tube. The air velocity is 100 ft/min (0.5 m/s). Assume standard air.

12-7   Estimate the rate at which water is evaporated from a 1000-acre lake on an August day when the dry bulb and wet bulb temperatures are 100 and 75 F, respectively (43,560 ft² = 1 acre). Assume a heat-transfer coefficient *h* of 5 Btu/(hr-ft²-F) between the moist air and the lake surface and a water surface temperature of 80 F.

12-8   The sensible heat-transfer coefficient for a dry surface has been determined to be 9 Btu/(hr-ft²-F) at a certain Reynolds number. Estimate the total heat transfer to the surface per square foot at a location where the wall temperature is 50 F and the state of the moist air flowing over the surface is given by 75 F dry bulb and 65 F wet bulb. Assume that the Reynolds number does not change and the Lewis number is 0.82. The condensate present on the surface will increase the transfer coefficients by about 15 percent.

12-9   Estimate the total heat-transfer rate per square meter to a surface where the surface temperature is 9 C and the moist air flowing over the surface is at 25 C dry bulb and 19 C wet bulb. The sensible heat-transfer coefficient for the same surface under dry conditions and at the same Reynolds number is known to be 55 W/(m²-C). Assume that the effect of the wet surface is to increase the heat transfer coefficient by 25 percent.

12-10 A housekeeper hangs a wet blanket out to dry in a high wind. The blanket weighs 4 lb dry and 16 lb wet and has dimensions of 7 ft by 8 ft. Assume outdoor conditions of 90 F db and 50 percent relative humidity and that the blanket is at a temperature of 90 F. Estimate the time required for the blanket to become dry if the average heat-transfer coefficient on both sides of the blanket is 4 Btu/(hr-ft²-F).

12-11 Moist air flows over a wet bulb thermometer, which is cylindrical in shape with a diameter of 0.25 in. (including the wick). If the air velocity is 1000 ft/min, the dry bulb air temperature is 80 F, and the wet bulb temperature is 70 F, compute the (a) sensible heat-transfer coefficient h, and (b) the mass-transfer coefficient h_d.

12-12 A psychrometer is mounted in a 6-in. diameter insulated duct. Forty cfm of air is flowing in the duct. The thermometers indicate 90 and 70 F and are unshielded. The bulb diameters are approximately $\frac{1}{4}$ in. Determine (a) the humidity ratio using Chart 1 and (b) the humidity ratio by the procedures of Sec. 12-2. (c) Compare the results of (a) and (b). (d) Using Fig. 12-4, select an optimum air velocity for this situation. How could this be achieved without changing the flow rate?

12-13 Consider a wet bulb psychrometer installed in an insulated air duct, 100 mm in diameter. The thermometers are indicating temperatures of 16 and 13 C and are not shielded. The bulb diameters are about 8 mm. The air flow rate is 0.02 m³/s. (a) Compute the humidity ratio. (b) What minimum air velocity should be used to obtain a constant error in the adiabatic saturation temperature?

12-14 A psychrometer is installed in a duct where the air velocity is 700 ft/min. The thermometers are unshielded. The dry bulb thermometer indicates a temperature of 110 F and the duct wall temperature is estimated to be 85 F. The thermometer has a diameter of 0.15 in. Determine the actual air dry bulb temperature.

12-15 The wet bulb thermometer of Problem 12-14 indicates a temperature of 75 F. Determine the humidity ratio W.

12-16 The air temperature in an uninsulated metal duct is to be measured with an unshielded thermometer about 0.20 in. in diameter ($\epsilon = 0.8$) located on the duct center line where the air velocity is 1000 ft/min. The air pressure is near standard and the thermometer reads 150 F. The air surrounding the duct is at 50 F, and the ratio of the inside to outside heat-transfer coefficients for the duct is 5. Estimate the actual air temperature in the duct.

12-17 A psychrometer using two thermocouples is suspended in a room where the mean air velocity is 0.25 m/s. The dry bulb thermocouple is 1 mm in diameter, whereas the wet bulb temperature is measured with a thermocouple wick 2 mm in diameter. The thermocouples are unshielded and the surrounding surfaces are at a temperature about 2 C greater than the dry bulb. If the indicated dry bulb and wet bulb temperatures are 25 C and 19 C, estimate (a) the true dry bulb temperature and (b) the humidity ratio.

12-18 Suppose the psychrometer of Problem 12-17 is used in an uninsulated duct where moist air is flowing at an indicated dry bulb and wet bulb of 50 C and 35 C. The heat-transfer coefficients on the inside and outside of the duct have a ratio of

7 and the ambient temperature is 28 C. The mean air velocity in the duct is 1.0 m/s. Estimate (a) the dry bulb temperature of the air and (b) the humidity ratio of the air.

12-19   An unshielded psychrometer installed in an insulated duct indicates temperatures of 60 and 45 F. The bulb diameters are about 0.10 in. and the air velocity is about 800 ft/min. Estimate the adiabatic saturation temperature of the air.

12-20   An unshielded psychrometer installed in an insulated duct indicates temperatures of 33 and 21 C. Estimate the adiabatic saturation temperature if the air velocity is 6 m/s and the diameters of the temperature probes are about 2 mm.

12-21   An unshielded psychrometer is to be used where the temperatures are about 105 and 85 F. Bulb diameters are about 0.1 in. The difference between the wet bulb and adiabatic saturation temperature must be relatively constant. (a) What minimum air velocity should be used? (b) Give the approximate difference.

CHAPTER THIRTEEN

Direct Contact Heat and Mass Transfer

The process of air humidification was discussed in Chapter 3 on the basis of thermodynamics. In that case the condition line on the psychrometric chart was found to be solely a function of the water enthalpy, and all of the water mixed with the airstream was assumed to evaporate and become a part of the air–water vapor mixture.

In this chapter we will consider problems in which the quantity of water in contact with the air is much larger than the quantity added or withdrawn from the airstream. Depending on the moist air state and the water temperature a variety of results are possible. The air may be cooled and humidified or dehumidified, or heated and humidified by direct contact with water.

Only heat and mass transfer will be considered. As discussed in Chapter 4, the air may be cleansed of dust and water-soluble vapors by contact with water.

13-1 BASIC RELATIONS FOR THE AIR WASHER

The basic relations for all types of direct contact equipment are quite similar. The air washer is considered first; we shall introduce modifications later for spray dehumidifiers and cooling towers.

The primary reason for treating direct contact equipment as a separate group arises from the difficulty in evaluating the heat and mass transfer areas. For the air washer or any spray-type device that does not have packing materials, the heat and mass transfer areas are approximately equal. For a cooling tower the difference may be considerable.

The basic relations may be written as follows; see Fig. 13-1. The mass transfer is given by

$$-dG_l = G_a\, dW = h_d a_m (W_i - W)\, dL \tag{13-1}$$

where W_i is the humidity ratio at the interface between the water and moist

529

Figure 13-1.   Schematic of a spray chamber.

air. The quantity a_m is the mass-transfer area per unit volume of the chamber. G_l and G_a are the water and air mass velocities $\dot{m}_l/A_c$ and $\dot{m}_a/A_c$ in mass per unit time and area where A_c is the cross-sectional area of the chamber. The water evaporated equals the increase in moisture of the air, which must equal the mass-transfer rate. The sensible heat transfer to the air is given by

$$G_a c_{pa} \, dt_a = h_a a_h (t_i - t_a) \, dL \tag{13-2}$$

where t_i is the interface temperature and a_h is the heat-transfer area per unit volume. The total energy transfer to the air is

$$G_a(c_{pa} \, dt_a + i_{fg} \, dW) = [h_d a_m (W_i - W) i_{fg} + h_a a_h (t_i - t_a)] \, dL \tag{13-3}$$

The concept of enthalpy potential, which is discussed in detail in Chapter 14, may be used to simplify Eq. 13-3. The term in brackets on the left-hand side of Eq. 13-3 is di, where i is the enthalpy of the moist air in Btu/lbma. If it is assumed that $a_h = a_m$ and the analogy of Eq. 12-20a is used to relate the heat- and mass-transfer coefficients h and h_d with $L_e = 1$, Eq. 13-3 becomes

$$G_a di = h_d a_m (i_i - i) \, dL \tag{13-3a}$$

where $(i_i - i)$ is the enthalpy potential. An energy balance yields

$$G_a di = \pm G_l c_l \, dt_l + c_l t_l dG_l \tag{13-4}$$

The negative sign refers to parallel flow of air and water, whereas the positive sign refers to counterflow. The last term in Eq. 13-4 is very small and will be neglected in the following development. The heat transfer to the water may be expressed as

$$G_l c_l \, dt_l = h_l a_h (t_l - t_i) \, dL \tag{13-5}$$

Equations 13-1 through 13-5 are the basic relations for direct-contact equipment with the possible exception of chambers that contain packing, such as cooling towers.

Various relations useful in equipment design may be derived from the basic relations. Combining Eqs. 13-3a, 13-4, and 13-5 gives

$$\frac{i - i_i}{t_l - t_i} = -\frac{h_l a_h}{h_d a_m} = -\frac{h_l}{h_d} \tag{13-6}$$

This shows that the ratio of driving potentials for total heat transfer through the air and liquid films is equal to the ratio of film resistances for the gas and liquid film. Combining Eq. 13-2 and 13-3a yields

$$\frac{di}{dt_a} = \frac{i - i_i}{t_a - t_i} \tag{13-7}$$

while use of Eqs. 13-1 and 13-2 gives

$$\frac{dW}{dt_a} = \frac{W - W_i}{t_a - t_i} \tag{13-8}$$

Equation 13-8 indicates that at any cross section in the spray chamber the instantaneous slope of the process path on the psychrometric chart is a straight line. This is illustrated in Fig. 13-2, where state 1 represents the state of the moist air entering the chamber and point 1_i represents the interface saturation state. The initial path of the process is then in the direction of the line connecting points 1 and 1_i. As the air is heated and humidified, the water is cooled and

Figure 13-2.   Air washer humidification process.

the interface state changes. In the case shown the interface state gradually moves downward along the saturation curve. Points 2 and 3 represent other states along the process curve. Note that the path is directed toward the interface in each case. The interface states are defined by Eqs. 13-4 and 13-6. Equation 13-4 relates the air enthalpy change to the water temperature change, whereas Eq. 13-6 describes the way in which the interface state changes to accommodate the transport coefficients and the air state. Solution of Eqs. 13-4 and 13-6 for the interface state is rather complex but can be done by trial and error or by the use of a complex graphical procedure. A simpler method utilizes a psychrometric chart with enthalpy and temperature as coordinates (1). The use of these coordinates makes it possible to plot Eqs. 13-4 and 13-6 for easy graphical solution. The following example illustrates the procedure.

Example 13-1. A parallel-flow air washer is to be designed as shown in Fig. 13-1. The design conditions are as follows.

Water temperature at the inlet,	$t_{l1} = 90$ F
Water temperature at the outlet,	$t_{l2} = 75$ F
Air dry bulb temperature at the inlet,	$t_{a1} = 60$ F
Air wet bulb temperature at the inlet,	$t_{wb1} = 42$ F
Air mass flow rate per unit area,	$G_a = 1250$ lbm/(hr-ft²)
Spray ratio	$G_l/G_a = 0.75$
Air heat transfer coefficient per unit volume,	$h_a a_h = 50$ Btu/(hr-F-ft³)
Liquid heat-transfer coefficient per unit volume,	$h_l a_h = 666$ Btu/(hr-F-ft³)
Air volume flow rate,	$\dot{Q} = 7000$ cfm

Solution. The mass flow rate of the dry air is given by

$$\dot{m}_a = \frac{\dot{Q}}{v_1} = \frac{7000}{13.1} = 534 \text{ lbm/min}$$

Then the spray chamber must have a cross-sectional area of

$$A_c = \frac{\dot{m}_a}{G_a} = \frac{534(60)}{1250} = 25.6 \text{ ft}^2$$

The Colburn analogy of Eq. 12-20a with Le = 1 will be used to obtain the mass-transfer coefficient (assuming $a_m = a_h$).

$$h_d a_m = \frac{h_a a_h}{c_{pa}} - \frac{50}{0.24} = 208 \text{ lbm/(hr-ft}^3)$$

In parallel flow the air entering at 60 F is in contact with the water at 90 F, and the air leaving at state 2 is in contact with the water at 75 F. This helps to understand the construction of Fig. 13-3, which shows the

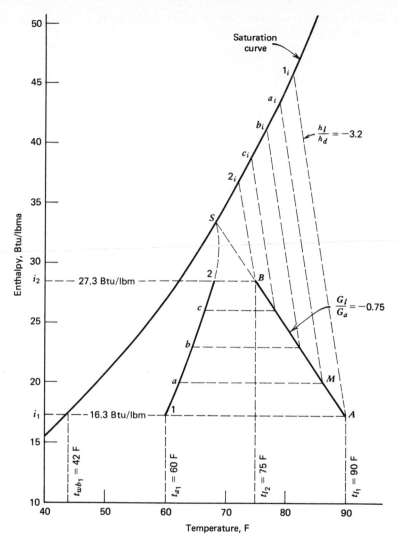

Figure 13-3. Graphical solution for Example 13-1.

graphical solution for the interface states and the process path for the air passing through the air washer. The solution is carried out as follows:

1. Locate state 1 as shown at the intersection of t_{a1} and i_1. Point A is a construction point defined by the entering water temperature t_{l1} and i_1. Note that the temperature scale is used for both the air and water.

2. The energy balance line is constructed from point A to point B and is defined by Eq. 13-4:

$$\frac{di}{dt_l} = -\frac{G_l c_l}{G_a} = -\frac{G_l}{G_a} = -0.75$$

since $c_l = 1$ Btu/(lbmw-F). Point B is determined by the temperature of leaving water $t_{l2} = 75$ F. The negative slope is a consequence of parallel flow. The line AB has no physical significance.

3. The line $A1_i$ is called a *tie line* and is defined by Eq. 13-6:

$$\frac{i - i_i}{t_l - t_i} = \frac{-h_l a_h}{h_d a_m} = \frac{-666}{208} = -3.2$$

The intersection of the tie line having slope of -3.2 with the saturation curve defines the interface state 1_i. The combination of the line AB and line $A1_i$ represents graphical solutions of Eqs. 13-4 and 13-6.

4. The initial slope of the air process path is then given by a line from state 1 to state 1_i. The length of the line $1a$ depends on the required accuracy of the solution and the rate at which the curvature of the path is changing.

5. The procedure is repeated by constructing the line aM and the tie line Ma_i, which has the same slope as $A1_i$. The path segment ab is on a line from a to a_i. Continue in the same manner until the final state of the air point 2 is reached. State point 2 is on a horizontal line passing through point B. The final state of the air is defined by $t_{a2} = 68$ F, $i_2 = 28.3$ Btu/lbma, and $t_{wb2} = 62.3$ F.

6. To complete the solution, it is necessary to determine the length of the air washer. Equation 13-3a gives

$$dL = \frac{G_a}{h_d a_m} \frac{di}{(i_i - i)} \tag{13-3b}$$

or

$$L = \frac{G_a}{h_d a_m} \int_1^2 \frac{di}{(i_i - i)} \tag{13-3c}$$

Equation 13-3c can be evaluated graphically or numerically. A plot of $1/(i_i - i)$ versus i is shown in Fig. 13-4 and the area under the curve represents the value of the integral. Using Simpson's rule with four equal increments yields

$$y = \int_1^2 \frac{di}{(i_i - i)} \approx \frac{\Delta i}{3} (y_1 + 4y_2 + 2y_3 + 4y_4 + y_5)$$

With $\Delta i = 2.75$

$$y = \left(\frac{275}{3}\right)(0.036 + 4(0.044) + 2(0.057) + 4(0.076) + 0.12) = 0.688$$

The design length is then

$$L = \frac{1250}{208}(0.688) = 6.0 \text{ ft}$$

The method used in the foregoing example can be used to predict the performance of existing direct-contact equipment. Furthermore, test data from a unit may be used in a trial-and-error procedure to determine the transport coefficients.

The graphical procedure outlined can be programmed rather easily for a small digital computer. If more than occasional calculations are needed, this would be worthwhile. Use of a computer would also permit refinements and variation of design parameters.

13-2   THE SPRAY DEHUMIDIFIER

The spray dehumidifier can often be used to advantage when a source of cold water is available. To be effective the spray dehumidifier must be used in counterflow, which can best be achieved as shown in Fig. 13-5. The following example illustrates the design procedure for this case.

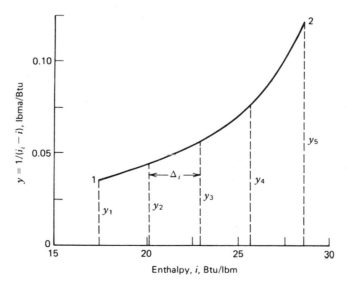

Figure 13-4.   Graphical solution of $\int di/(i_i - i)$.

Figure 13-5. Schematic of counterflow spray chamber.

Example 13-2. A counterflow spray dehumifidier is to be designed as shown schematically in Fig. 13-5. These are the design conditions:

Water temperature at the inlet,	$t_{l2} = 7$ C
Water temperature at the outlet,	$t_{l1} = 15$ C
Air dry bulb temperature at the inlet,	$t_{a1} = 28$ C
Air wet bulb temperature at the inlet,	$t_{wb1} = 22$ C
Air mass flow rate per unit area,	$G_a = 1.36$ kg/(s-m^2)
Spray ratio,	$\dfrac{G_l c_l}{G_a} = 3.25$
Air heat-transfer coefficient per unit volume,	$h_a a_h = 1210$ W/(C-m^3)
Liquid heat-transfer coefficient per unit volume,	$h_l a_h = 14700$ W/(C-m^3)
Air volume flow rate,	$\dot{Q} = 2.83$ m^3/s

Find the cross-sectional area and final state of the air.

Solution.   The mass flow rate of the dry air is

$$\dot{m}_a = \frac{\dot{Q}}{v_1} = \frac{2.83}{0.875} = 3.24 \text{ kg/s}$$

The chamber cross-sectional area is then

$$A_c = \frac{\dot{m}_a}{G_a} = \frac{3.24}{1.36} = 2.38 \text{ m}^2$$

Using the Colburn analogy and assuming that $a_m = a_h$

$$h_d a_m = \frac{h_a a_m}{c_{pa}} = \frac{1210}{1000} = 1.21 \text{ kg/(s-m}^3)$$

The graphical solution for the interface states and the process path is shown in Fig. 13-6. In the case of counterflow, the air entering at 28 C is in contact with the water at 15 C, and the air leaving is in contact with the water at 7 C. The procedure is quite similar to that given in Example 13-1. Note, however, that the energy balance line *AB* has a positive slope because of the counterflow arrangement. Notice also that because of counterflow, point *A* corresponds to t_{l1} and point *B* corresponds to t_{l2}. The final state of the air is defined by $t_{a2} = 14$ C and $t_{wb2} = 13.5$ C.

The height of the spray chamber is determined by using Eq. 13-3a with the procedure given in Example 13-1.

The use of spray dehumidifiers is less frequent than extended surface heat exchangers, which are discussed in Chapter 14.

13-3   COOLING TOWERS

A typical cooling tower used in HVAC applications is shown in Fig. 2-11. The particular model shown is a packaged mechanical draft unit.

The function of the cooling tower is to reject heat to the atmosphere by reducing the temperature of water circulated through condensers or other heat-rejection equipment. For this reason the state of the air and its path on the psychrometer chart is of little interest. The cooling range for the water (decrease in temperature) and the approach of the leaving water temperature to the ambient wet bulb temperature are most important in determining the size and cost of a cooling tower. This will be evident in the analysis to follow.

The general equations derived in Sec. 13-2 are still valid. To facilitate calculations, however, the procedure is somewhat different in this case. Equation 13-3a for the total energy transfer to the air is recalled.

$$G_a \, di = G_l c_l \, dt_l = h_d a_m (i_i - i) \, dL \qquad (13\text{-}3a)$$

To avoid consideration of the interfacial conditions, an overall coefficient U_i

Figure 13-6.   Graphical solution for Example 13-2.

is adopted that relates the driving potential to the enthalpy i_l at the bulk water
temperature t_l. Equation 13-3a then becomes

$$G_a \, di = G_l c_l \, dt_l = U_i a_m (i_l - i) \, dL \tag{13-9}$$

and, when Eq. 13-9 is integrated,

$$\frac{U_i a_m L}{G_l c_l} = \int \frac{dt_l}{(i_l - i)} \tag{13-9a}$$

Now

$$\dot{m}_l = G_l A_c \qquad \text{and} \qquad V = A_c L$$

then

$$\frac{U_i a_m V}{\dot{m}_l c_l} = \int \frac{dt_l}{(i_l - i)} \tag{13-10}$$

To review:

U_i = overall mass-transfer coefficient between the water and air, lbm/(hr-ft^2) or kg/(s-m^2)

a_m = mass-transfer surface area per unit volume associated with U_i, ft^2/ft^3 or m^2/m^3

V = total cooling tower volume, ft^3 or m^3

$\dot{m}_l$ = mass flow rate of water through the tower, lbm/hr or kg/s

c_l = specific heat of the water, Btu/(lbm-F) or kJ/(kg-C)

t_l = water temperature at a particular location in the tower, F or C

i_l = enthalpy of saturated moist air at t_l, Btu/lbm or kJ/kg

i = enthalpy of the moist air at temperature t, Btu/lbm or kJ/kg

The left-hand side of Eq. 13-10 is a measure of the cooling tower size and has the familiar form of the NTU parameter used in heat-exchanger design.

Equation 13-10 cannot be integrated in a straightforward mathematical way; however, a step-by-step finite-element approach can be used. The following example illustrates the procedure. The remainder of the design procedure is considered after the example.

Example 13-3. Water is to be cooled from 100 to 85 F in a counterflow cooling tower when the outside air has a 75 F wet bulb temperature. The water-to-air flow ratio ($\dot{m}_l/\dot{m}_a$) is 1.0. Calculate the transfer units as defined by Eq. 13-10.

Solution. Figure 13-7 is the cooling diagram for the given conditions. As the water is cooled from t_{l1} to t_{l2}, the enthalpy of the saturated air i_l follows the saturation curve from A to B. The air entering at wet bulb temperature t_{wb1} has enthalpy i_1. (This assumes the air enthalpy is only a function of wet bulb temperature.) The leaving water temperature t_{l2} and the enthalpy i_1 define point C and the initial driving potential is represented by the distance BC. The enthalpy increase of the air is a straight-line function with respect to the water temperature as defined by Eq. 13-9. The slope of the air operating line CD is therefore ($c_l \dot{m}_l / \dot{m}_a$).

Point C represents the air conditions at the inlet and point D represents the air conditions leaving the tower. Note that the driving potential gradually increases from the bottom to the top of the tower. Counterflow integration calculations start at the bottom of the tower where the air conditions are known. Evaluation of the integral of Eq. 13-10 may be carried out in a manner similar to that described in Example 13-1 by plotting t_l versus $1/(i_l - i)$; however, another method will be used here (2). The step-by-step procedure is shown in Table 13-1. Water temperatures are

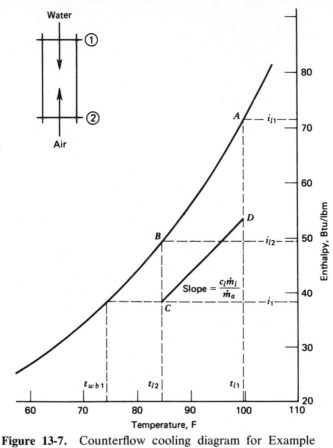

Figure 13-7. Counterflow cooling diagram for Example 13-3.

listed in column 1 in increments of one to two degrees. Although smaller increments will give greater accuracy, this must be balanced against an increase in calculation time. The film enthalpies shown in column 2 are obtained from Table A-2 as the enthalpy of saturated air at the water temperatures. Column 3 shows the air enthalpy, which is determined from

$$\Delta i = \frac{c_l \dot{m}_l}{\dot{m}_a} \Delta t_l \tag{13-9b}$$

where the initial air enthalpy i_1 is 38.5 Btu/lbma, $c_l = 1.0$ Btu/(lbmw-F), $\dot{m}_l/\dot{m}_a = 1.0$, and Δt is read in column 1 of Table 13-1. The data of columns 4 and 5 are obtained from columns 2 and 3. Column 6 is the average of two steps from column 5 multiplied by the water temperature increment (column 1) for the same step. The number of transfer units are then given in column 7 as the summation of column 6. Column 8 gives the temperature range over which the water has been cooled. The last entry in column 7

Table 13-1   COUNTERFLOW COOLING TOWER INTEGRATION CALCULATIONS

1	2	3	4	5	6	7	8
Water Temperature t_t degrees F	Enthalpy of Film i_t Btu/lbma	Enthalpy of Air i Btu/lbma	Enthalpy Difference $(i_t - i)$ Btu/lbma	Reciprocal of Enthalpy Difference $\dfrac{1}{(i_t - i)}$ lbma/Btu	Average $\dfrac{\Delta t_t}{(i_t - i)}$ F-lbma/Btu	Summation $\dfrac{\Delta t_t}{(i_t - i)}$ F-lbma/Btu	Cooling Range degrees F
85	49.4	38.5	10.9	0.0917			
					0.0905	0.0905	1
86	50.7	39.5	11.2	0.0893			
					0.1748	0.2653	3
88	53.2	41.5	11.7	0.0855			
					0.1661	0.4314	5
90	55.9	43.5	12.4	0.0806			
					0.1558	0.5872	7
92	58.8	45.5	13.3	0.0752			
					0.1451	0.7323	9
94	61.8	47.5	14.3	0.0699			
					0.1348	0.8671	11
96	64.9	49.5	15.4	0.0649			
					0.1248	0.9919	13
98	68.2	51.5	16.7	0.0599			
					0.1148	1.1067	15
100	71.7	53.5	18.2	0.0549			

is the number of transfer units required for this problem. It is evident from Table 13-1 that either an increase in the cooling range or a decrease in the leaving water temperature will increase the number of transfer units. As mentioned earlier, these two factors are quite important in cooling-tower design. The heat exchangers with which the cooling tower is connected should be designed with the cooling tower in mind. It may be more economical to enlarge the heat exchangers and/or increase the flow rate of the water than increase the size of the cooling tower. Obviously, this becomes an optimization problem where the digital computer would be useful.

To continue the problem of tower design, we need information on the overall mass-transfer coefficient per unit volume, $U_i a_m$. There is little theory to predict this coefficient; therefore, we must rely on experiments. The tower characteristic may generally be represented as a log-log plot of $U_i a_m V / \dot{m}_a$ for a constant air mass flow rate and water inlet temperature t_{l1} (Fig. 13-8). Although different air flow rates and water temperatures cause the characteristic to move up or down, variations in air flow rate of about ± 20 percent and water temperature variations of ± 10 F do not move the characteristic significantly. The slope of the characteristic varies from about -0.4 to -1.0. The required transfer units for a given set of water and air conditions are also shown in Fig. 13-8. The intersection of the two curves defines the operating point for the tower.

Assuming that the tower characteristics are known, we may determine the mass-transfer coefficient per unit volume $U_i a_m$. After many tests have been made on towers of a similar type, it is possible to predict $U_i a_m$ with reasonable accuracy. Then the volume of the tower required for a given set of conditions is given by

$$V = \frac{N \dot{m}_l c_l}{U_i a_m} \tag{13-11}$$

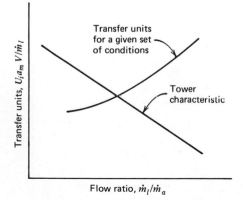

Transfer units for a given set of conditions

Tower characteristic

Transfer units, $U_i a_m V / \dot{m}_l$

Flow ratio, $\dot{m}_l / \dot{m}_a$

Figure 13-8.   Tower characteristic.

where N is the number of transfer units given by Eq. 13-10. The cross-sectional area of the tower is defined by

$$A_c = \frac{\dot{m}_a}{G_a} = \frac{\dot{m}_l}{G_l} \qquad (13\text{-}12)$$

and the height of the tower is given by

$$L = \frac{V}{A_c} \qquad (13\text{-}13)$$

Example 13-4.   Suppose the cooling tower of Example 13-3 must handle 1000 gpm of water. It has been determined that an air mass velocity of 1500 lbma/(hr-ft^2) is acceptable without excessive water carry-over (drift). The overall mass-transfer coefficient per unit volume $U_i a_m$ is estimated to be 120 lbm/(hr-ft^3) for the type of tower to be used. Estimate the tower dimensions for the required duty.

Solution.   The transfer units N required for the tower were found to be 1.107 in Example 13-3. Then the total volume of the tower is given by Eq. 13-11 as

$$V = \frac{1.107(1000)8.33(60)(1.0)}{120} = 4611 \text{ ft}^3$$

The cross-sectional area of the tower may be determined from Eq. 13-12 using the mass velocity of the air and the water-to-air ratio.

$$A_c = \frac{\dot{m}_a}{G_a} = \left(\frac{\dot{m}_l}{\dot{m}_a}\right)\frac{\dot{m}_a}{G_a} = \frac{\dot{m}_l}{G_a}$$

$$A_c = \frac{1000(8.33)(60)}{1500} = 333 \text{ ft}^2$$

which is equivalent to an 18 × 18 ft cross section. Then from Eq. 13-13

$$L = \frac{V}{A_c} = \frac{4611}{333} = 13.8 \text{ ft}$$

Caution must be exercised in using mass-transfer data from the literature for tower design. There are many variations in construction that affect the transport coefficients dramatically. The scale of the tower also seems to be important because of the ratio of wall surface area to total volume.

For many HVAC applications, relatively small factory-assembled cooling towers are used. Performance data are usually presented in a form such that a certain standard size may be selected. Figure 13-9 and Table 13-2 are typical of what might be furnished by a manufacturer for a line of towers. The entering water temperature, air wet bulb temperature, and the water flow rate determine the model to be selected for a fixed leaving water temperature of 85 F. The

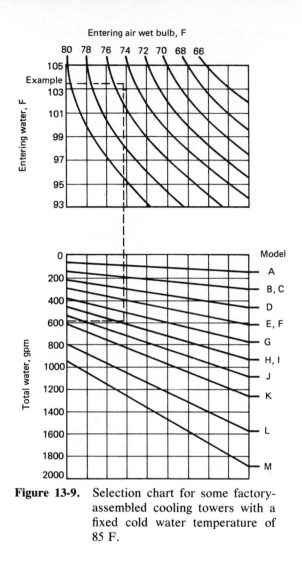

Figure 13-9.   Selection chart for some factory-assembled cooling towers with a fixed cold water temperature of 85 F.

example shown as a dotted line illustrates use of the chart. This procedure usually causes the tower to be slightly oversized. The cooling range may be computed from the entering water temperature and flow rate, and the tower capacity.

In Table 13-2 the nominal rating in tons refers to a typical refrigeration system with which the cooling tower may be used and includes only the heat transferred to the evaporator.

Example 13-5   Select a cooling tower using Fig. 13-9 and Table 13-2 for the conditions of examples 13-3 and 13-4. Compute the cooling range, approach, and heat transfer rate.

Table 13-2   PERFORMANCE DATA FOR SOME FACTORY-
ASSEMBLED COOLING TOWERS

Model	Nominal Rating		No. of Cells	No. of Fans	cfm	Motor hp
	Tons	gpm				
A	50	120	1	1	10,500	5
B	100	240	1	2	21,000	10
C	100	240	1	2	21,000	2–5
D	150	360	1	3	31,500	15
E	200	480	1	4	42,000	20
F	200	480	1	4	42,000	2–10
G	250	600	1	5	52,500	25
H	300	720	1	6	63,000	30
I	300	720	1	6	63,000	2–15
J	350	840	1	8	84,840	2–20
K	400	960	1	8	84,000	2–20
L	500	1200	1	10	105,000	2–25
M	600	1440	1	12	126,000	2–30

Solution.   The entering water temperature and air wet bulb temperature
are 100 F and 75 F, respectively, with a water flow rate of 1000 gpm.
Referring to Fig. 13-9, a model L would be the obvious choice; however,
it is somewhat over-sized. Suppose that the heat exchangers (condensers)
in the circuit could be changed so that the water would enter the tower at
103 F with a flow rate of 835 gpm. This is still the same duty of about
7.5×10^6 Btu/hr. Referring back to Fig. 12-9, a model K tower fits this
situation. The cooling range is

$$t_{l1} - t_{l2} = 103 - 85 = 18 \text{ F}$$

and the approach is

$$t_{l2} - t_{wb1} = 85 - 75 = 10 \text{ F}$$

the heat-transfer rate from the water is

$$\dot{q} = 835(60)8.33(103 - 85) = 7.5 \times 10^6 \text{ Btu/hr}$$

The reader is referred to the Equipment Volume of the *ASHRAE Handbook*
(2), which has a great deal of information on cooling tower performance and
selection.

REFERENCES

1. *ASHRAE Handbook, Fundamentals Volume,* American Society of Heating, Refrigerating and Air-Conditioning Engineers, Atlanta, GA, 1985.

2. *ASHRAE Handbook, Equipment Volume,* American Society of Heating, Refrigeraing and Air-Conditioning Engineers, Atlanta, GA, 1983.

PROBLEMS

13-1   Determine the final state of the air, the cross-sectional area, and the height of a counterflow spray dehumidifier that operates as follows:

$$t_{l1} = 50 \text{ F} \qquad t_{a1} = 95 \text{ F}$$
$$t_{l2} = 60 \text{ F} \qquad t_{wb1} = 82 \text{ F}$$
$$G_a = 1200 \text{ lbm/(hr-ft}^2) \qquad \dot{Q}_a = 5000 \text{ cfm}$$
$$\dot{Q}_w = 55 \text{ gpm} \qquad h_a a_h = 60 \text{ Btu/(hr-F-ft}^3)$$
$$h_l a_h = 800 \text{ Btu/(hr-F-ft}^3)$$

13-2   A counterflow spray dehumidifier operates as stated below. Determine (a) the final state of the air, (b) the cross-sectional area, and (c) the height of the chamber.

$$t_{l1} = 10 \text{ C} \qquad t_{a1} = 35 \text{ C}$$
$$t_{l2} = 16 \text{ C} \qquad t_{wb1} = 28 \text{ C}$$
$$G_a = 1.63 \text{ kg/(s-m}^2) \qquad \dot{Q}_a = 2.36 \text{ m}^3/\text{s}$$
$$\dot{Q}_w = 3.5 \times 10^{-3} \text{ m}^3/\text{s} \qquad h_a a_h = 1040 \text{ W/(C-m}^3)$$
$$h_l a_h = 13850 \text{ W/(C-m}^3)$$

13-3   Redesign the air washer in Example 13-1 assuming counterflow of the air and water.

13-4   Solve Problem 13-1 assuming parallel flow of the air and water spray.

13-5   Solve Problem 13-2 assuming parallel flow of the air and water.

13-6   A parallel-flow air washer is to be used as an evaporative cooler. Air will enter at 100 F (38 C) dry bulb and 62 F (17 C) wet bulb and leave at 75 F (24 C) dry bulb. Other operating conditions are as follows:

$$t_{l1} = 80 \text{ F} \ (27 \text{ C})$$
$$G_a = 1000 \text{ lbm/(hr-ft}^2) \ [1.36 \text{ kg/(m}^2\text{-s)}]$$
$$h_l a_h = 700 \text{ Btu/(hr-ft}^3\text{-F)} \ [13 \text{ kW/(m}^3\text{-C)}]$$
$$\dot{Q}_a = 4000 \text{ ft}^3/\text{min} \ (1.9 \text{ m}^3/\text{s})$$
$$h_a a_h = 55 \text{ Btu/(hr-ft}^3\text{-F)} \ [1 \text{ kW/(m}^3\text{-C)}]$$
$$G_l c_l / G_l = 0.7 \text{Btu/(lbm-F)} \ [2.93 \text{ kJ/(kg-C)}]$$

Determine the final state of the air, and the chamber cross-sectional area and length.

13-7   A counterflow cooling tower cools water from 104 to 85 F when the outside air has a wet bulb temperature of 76 F. The water flow rate is 2000 gpm and the air flow rate is 210,000 cfm. Calculate the transfer units for the tower.

13-8 A counterflow cooling tower cools water from 44 to 30 C. The outdoor air has a wet bulb temperature of 22 C. Water flows at the rate of 0.32m³/s and the water-to-air mass flow ratio is 1.0. Estimate the transfer units for the tower.

13-9 Estimate the tower dimensions for Problem 13-7. The air mass velocity may be assumed to be 1800 lbma/(hr-ft²) and the overall mass-transfer coefficient per unit volume $U_i a_m$ is about 125 lbm/(hr-ft³).

13-10 Estimate the tower dimensions for Problem 13-8. Assume a mass velocity for the air of 2.7 kg/(s-m³).

13-11 Use Fig. 13-9 and Table 13-2 to select a suitable tower(s) for the conditions of Problem 13-7.

13-12 Complete Example 13-2 to find the height of the chamber.

13-13 Repeat Example 13-3 changing the entering air wet bulb temperature to 79 F. Compare the transfer units with those of Table 13-1. How will this affect the tower dimensions?

13-14 Repeat Example 13-3, changing the entering water temperature to 105 F. Compare the transfer units with those of Table 13-1. How will this affect the tower dimensions?

13-15 The condensers for a centrifugal chiller plant require 200 gpm with water entering at 85 F and leaving at 100 F, the outdoor ambient air wet bulb temperature is 76 F. (a) Select a suitable cooling tower using Fig. 13-9 and Table 13-2 and (b) compute the cooling range, approach, and the tower capacity.

CHAPTER FOURTEEN

Extended Surface
Heat Exchangers

The term *heat exchanger* is usually applied to a device in which two fluid streams separated by a solid surface exchange heat energy. These devices may take many forms. However, ordinary metal tubes are the main components of many types. The heat exchanger is the most widely used device in HVAC applications. Although it is only a part of the overall system, a heat exchanger is required in every heating and cooling design problem that the engineer solves.

The major applications in the HVAC field are as follows: refrigerant-to-water in the case of chillers and water-cooled condensers where shell and tube configurations are often used; water-to-air and refrigerant-to-air where finned tubes are often used; and air-to-air in the case of heat recovery applications where plate-fin type surfaces or rotating heat exchangers may be used.

The heat-transfer surface may operate under conditions where only sensible heat transfer occurs or where latent and sensible heat transfer occur simultaneously as with dehumidifying coils. Both of these cases are treated in this chapter. Pressure loss is an important consideration in the design of heat exchangers; this will be considered in conjunction with the heat transfer in the discussions that follow.

In general, complications arise when one attempts to describe the rate of heat transfer from one fluid to the other in a heat exchanger. For practical purposes

$$\dot{q} = UA\Delta t_m \qquad (14\text{-}1)$$

where U is the familiar overall heat-transfer coefficient and A is the surface area associated with U. The mean temperature difference between the streams Δt_m must be used because Δt varies continuously throughout the heat exchanger. The overall heat-transfer coefficient U is also a variable resulting from changing physical properties and hydrodynamic characteristics from one part of the exchanger to another. In addition, the calculation of U is difficult and subject to large errors in some cases.

The use of Eq. 14-1 requires the assumption of an average value for U and the determination of a suitable mean temperature difference. An alternate approach to heat-exchanger problems involves the use of an average value of U and the concept of an effectiveness.

14-1   THE LMTD METHOD

With suitable assumptions it is possible to derive an expression for the mean temperature difference required in Eq. 14-1 for parallel and counterflow. The assumptions are as follows:

1. The overall heat transfer coefficient U, the mass flow rates $\dot{m}_c$ and $\dot{m}_h$, and the specific heats c_c and c_h are all constants where the subscripts c and h refer to the cold and hot streams.
2. There is no heat loss or gain external to the heat exchanger and there is no axial conduction in the heat exchanger.
3. A single bulk temperature applies to each stream at a given cross section.

Figures 14-1 and 14-2 show counterflow and parallel flow heat exchangers that represent the simplest types. The subscripts i and o refer to inlet and outlet, respectively. For both counterflow and parallel flow the appropriate mean temperature is given by (1)

$$\Delta t_m = \frac{\Delta t_1 - \Delta t_2}{\ln(\Delta t_1/\Delta t_2)} \tag{14-2}$$

Figure 14-1.   Counterflow heat exchanger.

Figure 14-2. Parallel flow heat exchanger.

where Δt_2 and Δt_1 are defined in Figs. 14-1 and 14-2. This particular mean temperature difference is the *log mean temperature difference* and is designated LMTD.

Complex Flow Patterns

In many cases the flow paths in the heat exchanger are quite complex and in some cases expressions may be developed for Δt_m; however, they are generally so complicated that charts have been developed to replace the equations. The concept of a correction factor F is used where

$$\dot{q} = UA(F)(\Delta t_m) \tag{14-3}$$

where Δt_m is computed in the same manner as the LMTD for an equivalent counterflow exchanger. Figures 14-3 and 14-4 show charts for two common flow configurations encountered in HVAC applications. The shell and tube configuration shown in Fig. 14-3 is often used for chillers and water-cooled condensers, whereas the cross-flow configuration occurs with many heating and cooling applications where air flows normal to a bank of finned tubes. The parameters P and R are defined as

$$P = \frac{t_{co} - t_{ci}}{t_{hi} - t_{ci}} \quad \text{and} \quad R = \frac{t_{hi} - t_{ho}}{t_{co} - t_{ci}} \tag{14-4}$$

P may be thought of as an effectiveness of the heat exchanger when $(\dot{m}c_p)_c < (\dot{m}c_p)_h$ where $(\dot{m}c_p)$ is the *fluid capacity rate* C. If the numerator and denom-

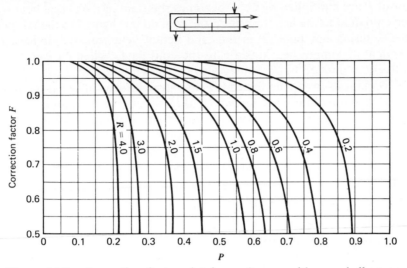

Figure 14-3. Correction factor plot for exchanger with one shell pass and two, four, or any multiple of tube passes.

Figure 14-4. Correction factor plot for single-pass cross-flow exchanger, both fluids unmixed.

inator of P are multiplied by C_c, P becomes the ratio of the actual heat transfer to the cold fluid to the heat transfer if the cold fluid is heated to the temperature of the entering hot fluid. P is thus the actual heat transfer divided by the theoretical maximum heat transfer to the cold fluid. When $C_h < C_c$, P is the effectiveness divided by R. It may also be shown that $R = C_c/C_h$.

Calculations involving the LMTD are straightforward when the fluid inlet and outlet temperatures are known. If three of the temperatures are known, together with the fluid capacity rates of the streams, the fourth temperature may be calculated by a simple energy balance. When two of the temperatures are unknown, however, trial-and-error procedures are required because of the form of the equation for the LMTD, Eq. 14-2. In the design of heat exchangers this is often true.

14-2   THE NTU METHOD

The NTU method has the advantage of eliminating the trial-and-error procedure of the LMTD method for many practical problems when only the inlet fluid temperatures are known. NTU stands for the "Number of Transfer Units."

Heat exchanger effectiveness was mentioned in the previous article as

$$\varepsilon = \frac{\text{Actual heat-transfer rate}}{\text{Maximum possible heat-transfer rate}}$$

The actual heat-transfer rate is given by

$$\dot{q} = C_h(t_{hi} - t_{ho}) = C_c(t_{co} - t_{ci}) \tag{14-5}$$

The maximum possible heat-transfer rate is expressed by

$$\dot{q}_{max} = C_{min}(t_{hi} - t_{ci}) \tag{14-6}$$

This is true because the maximum heat transfer will occur when one of the fluids undergoes a temperature change equal to the maximum in the heat exchanger, which is $(t_{hi} - t_{ci})$. The fluid experiencing the maximum temperature change must be the one with the minimum value of C to satisfy the energy balance.

The fluid with the minimum value of C may be the hot or the cold fluid. For $C_h = C_{min}$, using Eqs. 14-5 and 14-6

$$\varepsilon = \frac{\dot{q}}{\dot{q}_{max}} = \frac{C_h(t_{hi} - t_{ho})}{C_{min}(t_{hi} - t_{ci})} = \frac{(t_{hi} - t_{ho})}{(t_{hi} - t_{ci})} \tag{14-7}$$

For $C_c = C_{min}$

$$\varepsilon = \frac{\dot{q}}{\dot{q}_{max}} = \frac{C_c(t_{co} - t_{ci})}{C_{min}(t_{hi} - t_{ci})} = \frac{(t_{co} - t_{ci})}{(t_{hi} - t_{ci})} \tag{14-8}$$

It is therefore necessary to have two expressions for the effectiveness (Eqs.

14-7 and 14-8). When effectiveness is known, the outlet temperature may be easily computed. For example, when $C_h < C_c$

$$t_{ho} = \varepsilon(t_{ci} - t_{hi}) + t_{hi} \qquad (14\text{-}7a)$$

Also

$$t_{co} = \frac{\dot{q}}{C_c} + t_{ci} = \frac{C_h}{C_c}(t_{hi} - t_{ho}) + t_{ci} \qquad (14\text{-}9)$$

or

$$t_{co} = \frac{C_h}{C_c}\varepsilon(t_{hi} - t_{ci}) + t_{ci} \qquad (14\text{-}9a)$$

Expressions for ε for several flow configurations are shown in Table 14-1. Figures 14-5, 14-6, and 14-7 show graphical representations of the effectiveness for three common cases. The NTU parameter is defined as UA/C_{min} and may be thought of as a heat-transfer size factor. It may also be observed that flow configuration is unimportant when $C_{min}/C_{max} = 0$. This corresponds to the situation of one fluid undergoing a phase change where c_p may be thought of as being infinite. Evaporating or condensing refrigerants as well as condensing water vapor are examples where $C_{min}/C_{max} = 0$.

Table 14-1 THERMAL EFFECTIVENESS OF HEAT EXCHANGERS WITH VARIOUS FLOW ARRANGEMENTS

Parallel flow:	$\varepsilon = \dfrac{1 - \exp[-\mathrm{NTU}(1 + C)]}{1 + C}$
Counterflow:	$\varepsilon = \dfrac{1 - \exp[-\mathrm{NTU}(1 - C)]}{1 - C\exp[-\mathrm{NTU}(1 - C)]}$
Cross-flow (both streams are unmixed):[a]	$\varepsilon = 1 - \exp\left\{\dfrac{1}{C\eta}[\exp((-\mathrm{NTU})(C)(\eta)) - 1]\right\}$ where $\eta = \mathrm{NTU}^{-0.22}$
Cross-flow (both streams are mixed):	$\varepsilon = \mathrm{NTU}\left\{\dfrac{\mathrm{NTU}}{1 - \exp(-\mathrm{NTU})} + \dfrac{(\mathrm{NTU})(C)}{1 - \exp[-(\mathrm{NTU})(C)]} - 1\right\}^{-1}$
Cross-flow (stream C_{min} is unmixed):	$\varepsilon = \dfrac{1}{C}\{1 - \exp[-C(1 - \exp(-\mathrm{NTU}))]\}$
Cross-flow (stream C_{max} is unmixed):	$\varepsilon = 1 - \exp\left\{-\dfrac{1}{C}[1 - \exp(-(\mathrm{NTU})(C))]\right\}$
1-2 Parallel counterflow:	$\varepsilon = 2\left\{1 + C + \dfrac{1 + \exp[-\mathrm{NTU}(1 + C^2)^{1/2}]}{1 - \exp[-\mathrm{NTU}(1 + C^2)^{1/2}]}(1 + C^2)^{1/2}\right\}^{-1}$ where $\mathrm{NTU} = (UA/C_{min})$, and $C = C_{min}/C_{max}$

[a]This is an approximate expression.

Figure 14-5.   Effectiveness for counterflow exchanger performance.

The NTU method has gained greatest acceptance in connection with design of *compact heat exchangers* where a large surface area per unit volume exists. An arbitrary definition has been proposed that states that the ratio of surface area to volume in a compact exchanger is greater than 200 ft^2/ft^3 or 656 m^2/m^3. The common finned-tube exchangers used extensively in HVAC systems generally fall into this category. Compact heat exchangers are used in applications where at least one fluid is a gas. Because the thermal resistance associated

Figure 14-6.   Effectiveness for parallel flow exchanger performance.

Figure 14-7.   Effectiveness for cross-flow exchanger with fluids unmixed.

with the gas film is high compared with a fluid such as water, fins are used to increase the heat-transfer area and decrease the thermal resistance.

14-3   HEAT TRANSFER—SINGLE-COMPONENT FLUIDS

The heat transfer rate from one fluid to the other in a heat exchanger is expressed by Eq. 14-3:

$$\dot{q} = UA(F)\,\Delta t_m \tag{14-3}$$

when the LMTD method is used. When the NTU method is utilized, the heat-transfer rate is generally computed from the temperature change for either fluid. For example

$$\dot{q} = (\dot{m}c_p)_h(t_{hi} - t_{ho}) \tag{14-10}$$

since

$$\text{NTU} = \frac{UA}{(\dot{m}c_p)_{\min}} \tag{14-11}$$

It is evident that an average value of the overall coefficient U must be known for both design methods.

The concept of overall thermal resistance and the overall heat-transfer coefficient was discussed in Chapter 4. The general procedure is the same for heat exchangers. For a simple heat exchanger without fins the overall coefficient U is given by

$$\frac{1}{UA} = \frac{1}{h_o A_o} + \frac{\Delta x}{k A_m} + \frac{1}{h_i A_i} + \frac{R_{fi}}{A_i} + \frac{R_{fo}}{A_o} \tag{14-12}$$

where

h_o = heat-transfer coefficient on the outside, Btu/(hr-ft²-F) or W/(m²-C)
h_i = heat-transfer coefficient on the inside, Btu/(hr-ft²-F) or W/(m²-C)
Δx = thickness of the separating wall, ft or m
k = thermal conductivity of the separating wall, (Btu-ft)/(ft²-hr-F) or (W-m)/(m²-C)
A = area, ft² or m² where o, m, and i refer to outside, mean, and inside, respectively
R_f = fouling factor, (hr-ft²-F)/Btu or (m²-C)/W

In general the areas A_o, A_m, A_i are not equal and U may be referenced to any one of the three. Let $A = A_o$, then

$$\frac{1}{U_o} = \frac{1}{h_o} + \frac{\Delta x}{k(A_m/A_o)} + \frac{1}{h_i(A_i/A_o)} + \frac{R_{fi}}{(A_i/A_o)} + R_{fo} \qquad (14\text{-}12a)$$

Fin Efficiency

Many of the heat exchangers used in HVAC systems have fins on one or both sides. Because the fins do not have a uniform temperature, the fin efficiency η is used to describe the heat-transfer rate.

$$\eta = \frac{\text{Actual heat transfer}}{(\text{Heat transfer with fin all at the base temperature } t_b)}$$

Figure 14-8 shows a simple fin. Since the base on which the fin is mounted also transfers heat, another parameter similar to fin efficiency is defined, called the surface effectiveness η_s:

$$\eta_s = \frac{\text{Actual heat transfer for fin and base}}{(\text{Heat transfer for fin and base when the fin is at the base temperature } t_b)}$$

Figure 14-8.   A simple fin of uniform cross section.

If we assume that h is uniform over the fin and base surface, the actual heat-transfer rate is then given by

$$\dot{q} = hA\eta_s(t_b - t_\infty) \tag{14-13}$$

where A is the total surface area of the fin and base. The thermal resistance is given by

$$R' = \frac{1}{hA\eta_s} \tag{14-14}$$

For a case where both sides of the heat exchanger have fins, the overall coefficient U is, assuming no fouling

$$\frac{1}{UA} = \frac{1}{h_o A_o \eta_{so}} + \frac{\Delta x}{kA_m} + \frac{1}{h_i A_i \eta_{si}} \tag{14-15}$$

If $A = A_o$

$$\frac{1}{U_o} = \frac{1}{h_o \eta_{so}} + \frac{\Delta x}{k(A_m/A_o)} + \frac{1}{h_i \eta_{si}(A_i/A_o)} \tag{14-15a}$$

The second term on the right-hand side of Eqs. 14-15 and 14-15a represents the thermal resistance of the base and is often negligible.

Extended or finned surfaces may take on many forms ranging from the simple plate of uniform cross section shown in Fig. 14-8 to complex patterns attached to tubes. Several common configurations are considered in this section starting with the fin of uniform cross section. The heat transfer rate for the fin may be shown to be (1)

$$\dot{q} = (t_b - t_\infty)mkA_c \tanh(ml) \tag{14-16}$$

when heat transfer from the tip is zero (1). The parameter m is given by

$$m = \left[\frac{hP}{kA_c}\right]^{1/2} \tag{14-17}$$

where

k = thermal conductivity of the fin material, (Btu-ft)/(ft²-hr-F) or (W/m)/(m²-C)
P = perimeter of the fin $2(L + y)$, ft or m
A_c = cross-sectional area of the fin $(L)(Y)$, ft² or m²
h = heat-transfer coefficient, Btu/(hr-ft²-F) or W/(m²-C)

If we use the definition of fin efficiency given,

$$\eta = \frac{(t_b - t_\infty)mkA_c \tanh(ml)}{hA(t_b - t_\infty)} \tag{14-18}$$

or

$$\eta = \frac{\tanh(ml)}{(ml)} \tag{14-19}$$

The heat-transfer coefficient used to define m is assumed to be a constant. In a practical heat exchanger h will vary over the surface of the fins and will probably change between the inlet and outlet of the exchanger. The only practical solution is to use an average value of h for the complete surface. Equation 14-19 may be applied to a surface such as that shown in Fig. 14-25.

Most fins are very thin and $L \gg y$. In this case the parameter m defined by Eq. 14-17 may be simplified by setting $P = 2L$. Then

$$m = \left[\frac{2hL}{kLy}\right]^{1/2} = \left[\frac{2h}{ky}\right]^{1/2} \tag{14-20}$$

This approximation is often applied without explanation.

Finned tube heat exchangers are very popular for water-to-air or refrigerant-to-air applications. Figure 14-9 shows a sketch of a tube with circular fins. The diagram is somewhat idealized, since in practice the fin is usually wound on the tube in a helix from one continuous strip of material. A typical circular finned tube water coil is shown in Fig. 14-10. Typically the fin will be quite thin. In the case of the circular fin the solution for the fin efficiency is very complex and is not generally used for practical problems; however, Fig. 14-11 shows a plot of the solution. An approximate but quite accurate method of predicting η for a circular fin has been developed by Schmidt (2). The method is largely empirical but has many advantages when an analytical expression is required. The method is summarized as follows:

$$\eta = \frac{\tanh(mr\phi)}{(mr\phi)} \tag{14-21}$$

where m is defined by Eq. 14-20 and

$$\phi = \left(\frac{R}{r} - 1\right)[1 + 0.35 \ln(R/r)] \tag{14-22}$$

When R/r is between 1.0 and 8 and η falls between 0.5 and 1.0, the error is less than one percent of the value of the fin efficiency taken from Fig. 14-11.

Figure 14-9.   Tube with circular fins.

Figure 14-10.   Circular finned-tube water coil. (Courtesy Thermal Corporation, Houston, Tex.)

Continuous-plate fins are also used extensively in finned-tube exchangers. In this case each fin extends from tube to tube. Figure 14-12 shows such an arrangement. It is not possible to obtain a closed analytical solution for this type of fin and approximate methods are necessary. Consider the rectangular tube array of Fig. 14-13 with continuous-plate fins. When it is assumed that the heat-transfer coefficient is constant over the fin surface, an imaginary rectangular fin may be defined as shown. The outline of the fin is an equipotential line where the temperature gradient is zero. The problem is then to find η for a rectangular fin. Zabronsky (3) has suggested that a circular fin of equal area be substituted for purposes of calculating η; however, Carrier and Anderson (4) have shown that the efficiency of a circular fin of equal area is not accurate; they recommend the sector method. Rich (6) developed charts shown in reference 5 to facilitate use of the sector method. Schmidt (2) describes an approach to this problem that is nearly as accurate as the sector method and has the advantage of simplicity. Again the procedure is empirical; however, Schmidt tested the method statistically using maximum and minimum values of η that must bracket the actual fin efficiency. The method is based on the selection of a circular fin with a radius R_e that has the same fin efficiency as the rectangular

Figure 14-11. Performance of circumferential fins of rectangular cross section. (Reprinted from *ASME Transactions*, Vol. 67, 1945.)

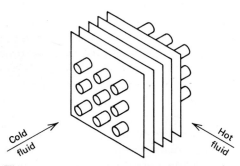

Figure 14-12. Continuous plate-fin-tube heat exchanger.

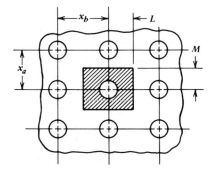

Figure 14-13. Rectangular tube array.

fin. After R_e is determined, Eq. 14-21 is used for the calculation of η. For the rectangular fin

$$\frac{R_e}{r} = 1.28\psi(\beta - 0.2)^{1/2} \tag{14-23}$$

where

$$\psi = \frac{M}{r} \quad \text{and} \quad \beta = \frac{L}{M}$$

M and L are defined in Fig. 14-13 where L is always selected to be greater than or equal to M. In other words $\beta \geq 1$. The parameter ϕ given by Eq. 14-22 is computed using R_e instead of R.

Figure 14-14 shows a triangular tube layout with continuous-plate fins. Here a hexangular fin results, which may be analyzed by the sector method (5). Schmidt (2) also analyzed this result and gives the following empirical relation, which is similar to Eq. 14-23:

$$\frac{R_e}{r} = 1.27\psi(\beta - 0.3)^{1/2} \tag{14-24}$$

where

$$\psi = \frac{M}{r} \quad \text{and} \quad \beta = \frac{L}{M}$$

M and L are defined in Fig. 14-14 where $L \geq M$. Equations 14-21 and 14-22 are used to compute η.

Special types of fins are sometimes used such as spines or fins of non-uniform cross section. Reference 5 contains data pertaining to these surfaces.

In the foregoing discussion we have assumed that the fins are rigidly attached to the base material so that zero thermal contact resistance exists. This may not always be true, particularly for plate-fin-tube surfaces. Eckels (24) has

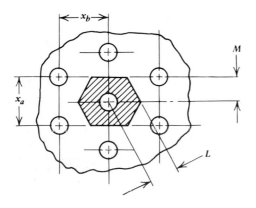

Figure 14-14. Hexangular tube array.

developed an empirical relation to predict the unit contact resistance for plate-fin-tube surfaces as follows:

$$R_{ct} = C \left[\frac{D_t \left(\frac{s}{y} - 1 \right)^2}{y} \right]^{0.6422} \tag{14-25}$$

where

R_{ct} = unit contact resistance, (hr-ft^2-F)/Btu or (m^2-C)/W
C = a constant, 2.222×10^{-6} for English units and 3.913×10^{-7} for SI units
D_t = outside tube diameter, in. or m
s = fin spacing, in. or m
y = fin thickness, in. or m

This unit contact resistance is associated with the outside tube area and added to Eq. 14-15. Because this contact resistance is undesirable as well as difficult to predict, every effort should be made to eliminate it in the manufacture of the heat exchanger. If tests are made for a surface, the contact resistance is usually reflected in the heat-transfer coefficients obtained.

Surface Effectiveness

Surface effectiveness was defined earlier as the actual heat transfer to the fin and base divided by the heat transfer to the fin and base when the fin is all at the base temperature t_b. This may be written

$$\eta_s = \frac{\dot{q}}{hA(t_b - t_\infty)} = \frac{hA_b(t_b - t_\infty) + hA_f\eta(t_b - t_\infty)}{hA(t_b - t_\infty)} \tag{14-26}$$

where $A = A_b + A_f$. Assuming h is constant over the fin and base, we see that

$$\eta_s = \frac{A_b + \eta A_f}{A} - 1 - \frac{A_f}{A}(1 - \eta) \tag{14-27}$$

and

$$\dot{q} = hA\eta_s(t_b - t_\infty) \tag{14-28}$$

14-4   TRANSPORT COEFFICIENTS INSIDE TUBES

Most HVAC heat exchanger applications of flow inside tubes and passages involve water, water vapor, and boiling or condensing refrigerants. The smooth copper tube is by far the most common geometry with these fluids. Forced convection turbulent flow is the most important mode; however, laminar flow sometimes occurs.

Turbulent Flow of Liquids Inside Tubes

Probably the most widely used heat-transfer correlation for this common case is the Dittus-Boelter equation (1)

$$\frac{\overline{h}D}{k} = 0.023(\text{Re}_D)^{0.8}(\text{Pr})^n \qquad (14\text{-}29)$$

where

$$n = 0.4 \qquad t_{\text{wall}} > t_{\text{bulk}}$$
$$n = 0.3 \qquad t_{\text{wall}} < t_{\text{bulk}}$$

Equation 14-29 applies under conditions of $\text{Re}_D > 10,000$, $0.7 < \text{Pr} < 100$, and $L/D > 60$. All fluid properties should be evaluated at the arithmetic mean bulk temperature of the fluid. Appendix B gives the thermophysical properties required in Eq. 14-29 for some common liquids and gases. Reference 5 gives other, similar correlations for special conditions. Equation 14-29 may be used for annular or noncircular cross sections for approximate calculations. In this case the tube diameter D is replaced by the hydraulic diameter D_h.

$$D_h = \frac{4(\text{Cross-sectional area})}{(\text{Wetted perimeter})} \qquad (14\text{-}30)$$

Reference 7 gives extensive data for noncircular flow channels when more accurate values are required.

Pressure drop for flow of liquids inside pipes and tubes was discussed in Chapter 9. The same procedure applies to heat-exchanger tubes; we must still take into account the considerable increase in equivalent length caused by the many U-turns, tube inlets and exits, and the headers required in most heat exchangers.

Laminar Flow of Liquids Inside Tubes

The recommended correlation for predicting the average film coefficient in laminar flow in tubes is

$$\frac{\overline{h}D}{k} = 1.86 \left[\text{Re}_D \text{Pr} \frac{D}{L} \right]^{1/3} \left(\frac{\mu}{\mu_s} \right)^{0.14} \qquad (14\text{-}31)$$

When the term in brackets is less than about 20, Eq. 14-31 becomes invalid; however, this will not occur for most heat-exchanger applications. Properties should be evaluated at the arithmetic mean bulk temperature except for μ_s, which is evaluated at the wall temperature.

A word of caution is appropriate concerning the transition from laminar to turbulent flow. This region is defined approximately by $2000 < \mathrm{Re}_D < 10{,}000$. Prediction of heat transfer and friction coefficients is uncertain during transition. The usual practice is to avoid the region by proper selection of tube size and flow rate.

Pressure drop is computed as described earlier for turbulent flow in tubes and in Chapter 9. For laminar flow the friction factor (Moody) is given by

$$f = \frac{64}{\mathrm{Re}_D} \qquad (14\text{-}32)$$

Ethylene Glycol–Water Solutions

In many systems it is necessary to add ethylene glycol to the water to prevent freezing and consequent damage to the heat exchangers and other components. The effect of the glycol on flow friction was discussed in Chapter 9, and it was shown that the lost head is generally increased when a glycol–water solution is used. The heat transfer is also adversely affected. Figures 14-15 and 14-16 give the specific heat and thermal conductivity of ethylene glycol solutions as a function of temperature and concentration. Similar data for specific gravity and viscosity are given in Chapter 9. It is very important to anticipate the use of glycol solutions during the design phase of a project

Figure 14-15.   Specific heat of aqueous solutions of ethylene glycol. (Reprinted by permission from *ASHRAE Handbook, Fundamentals Volume*, 1985.)

Figure 14-16. Thermal conductivity of aqueous solutions of ethylene glycol. (Reprinted by permission from *ASHRAE Handbook, Fundamentals Volume*, 1985.)

because the heat-transfer coefficient using a 30 percent glycol solution may be as much as 40 percent less than the coefficient using pure water. This is mainly because of the lower thermal conductivity and specific heat of the glycol solution.

Condensation and Evaporation Inside Horizontal Tubes

The prediction of heat transfer and pressure drop in two-phase flow is much more uncertain than with a single-phase flow. The mixture of vapor and liquid can vary considerably in composition and hydrodynamic behavior, and it is generally not possible to describe all conditions with one relation. Two-phase flow inside horizontal tubes is the most common situation in HVAC systems, and one or two correlations are presented for this case.

The following relations from reference 5 apply to film condensation, the dominant mode

$$\frac{\bar{h}D}{k_l} = 13.8(\mathrm{Pr})_l^{1/3} \left(\frac{i_{fg}}{c_p\,\Delta t}\right)^{1/6} \left[\frac{DG_v}{\mu_l}\left(\frac{\rho_l}{\rho_v}\right)^{1/2}\right]^{0.2} \tag{14-33}$$

where

$$\frac{DG}{\mu_l} < 5000 \quad \text{and} \quad 1000 < \frac{DG_v}{\mu_l} \left(\frac{\rho_l}{\rho_v}\right)^{1/2} < 20,000$$

The subscripts l and v refer to liquid and vapor, respectively, and Δt is the difference between the fluid saturation temperature and the wall surface temperature. When

$$20,000 < \frac{DG_v}{\mu_l} \left(\frac{\rho_l}{\rho_v}\right)^{1/2} < 100,000$$

$$\frac{\bar{h}D}{k_l} = 0.1(\text{Pr})_l^{1/3} \left(\frac{i_{fg}}{c_p \, \Delta t}\right)^{1/6} \left[\frac{DG_v}{\mu_l} \left(\frac{\rho_l}{\rho_v}\right)^{1/2}\right]^{2/3} \tag{14-34}$$

Equations 14-33 and 14-34 are for condensing saturated vapor; however, little error is introduced for superheated vapor when the wall temperature is below the saturation temperature and h is calculated for saturated vapor. Appendixes A and B give the required properties.

The average heat-transfer coefficients for evaporating R-12 and R-22 may be estimated from the following relation from reference 5:

$$\frac{\bar{h}D}{k_l} = C_1 \left[\left(\frac{GD}{\mu_l}\right)^2 \left(\frac{J \, \Delta x \, i_{fg} g_c}{Lg}\right)\right]^n \tag{14-35}$$

where

J = Joules equivalent, 778 (ft-lbf)/Btu or 1 for SI units
Δx = change in quality of the refrigerant, mass of vapor per unit mass of the mixture
i_{fg} = enthalpy of vaporization, Btu/lbm or J/kg
L = length of the tube, ft or m
C_1 = constant = 9×10^{-4} when $x_e < 0.9$, and 8.2×10^{-3} when $x_e \geq 1.0$ (x_e is the quality of the refrigerant leaving the tube)
n = constant = 0.5 when $x_e < 0.9$, and 0.4 when $x_e \geq 1.0$

The correlation was obtained from tests made using copper tubes having diameters of 0.47 and 0.71 in. and lengths from 13 to 31 ft. Evaporating temperatures varied from -4 to 32 F. Equation 14-35 is sufficient for most HVAC applications where Appendixes A and B give the required properties.

The pressure drop that occurs with a gas–liquid flow is of interest. Experience has shown that pressure drops in two-phase flow are usually much higher than would occur for either phase flowing along at the same mass rate. As in any flow, the total pressure drop along a tube depends upon three factors: (1) friction, due to viscosity, (2) change of elevation, and (3) acceleration of the fluid.

Friction is present in any flow situation, although in some cases it may contribute less than the other two factors. In horizontal flow the change in elevation is zero, and there would be no pressure drop due to this factor. Where there is a small change in gas density or little evaporation occurring, the pressure drop due to acceleration would usually be small. In flow with large changes of density or where evaporation is present, however, the acceleration pressure drop may be very significant.

There has been a variety of schemes proposed for predicting the pressure drop in two-phase flow. Some investigators have attempted to develop theories and equations valid only for a particular flow pattern. These have the disadvantage of requiring that the type of flow pattern be known, and they lead to complexities in those situations where flow patterns change along the flow. Other investigators have attempted to develop single theories valid over several other regimes. These attempts have not been completely successful.

A widely used method of predicting pressure drop in two-phase flow was developed by Lockhart and Martinelli [27]. Their method, which is essentially empirical, is based on measurements of isothermal flow of air and several liquids under essentially incompressible conditions. *The pressure drop predicted is that due to friction only.*

Extensive work has been devoted to the two-phase pressure-loss problem, but available methods remain very complex and impractical for general use. Therefore, the manufacturers of coils have resorted to experimental data for specific coils and refrigerants. Figure 14-17 is an example of such data for $\frac{1}{2}$-inch OD tubes. Trial and error may be required to obtain suitable circuiting of the coil. Circuiting will be discussed later.

14-5   TRANSPORT COEFFICIENTS OUTSIDE TUBES AND COMPACT SURFACES

Air is the most common flow medium in this case except for shell and tube evaporators and condensers where heat is transferred between a refrigerant inside the tubes to water outside the tubes. Water-to-water applications may also occur. In the case of bare tube bundles the flow may be across the tubes, parallel to the tubes, or a combination of the two when baffles are used. Compact surfaces such as finned tubes or plate fins will usually have air flowing parallel to the fins and normal to the tubes.

Bare Tubes with Parallel Flow

Kays (8) discusses both laminar and turbulent heat transfer with flow parallel to the tubes and presents results for constant heat flux per unit length of tube for fully developed velocity and temperature profiles. References 9 and 10 also pertain to this problem as well as pressure loss with flow parallel to tubes.

Figure 14-17.   Refrigerant pressure loss in ½-inch OD tubes.

Bare Tubes with a Combination of Parallel and Cross Flow

This flow pattern occurs when baffles are used in a tube and shell ex-changer. Empirical data are usually employed because of the complex nature of the flow pattern. The most successful of these methods developed from the work of Tinker and was clarified by Devore (11); the shell side flow was split into the basic components of pure cross-flow, baffle-to-baffle leakage flow, tube-to-baffle leakage flow, and bundle-to-shell bypass flow. Tinker made gross simplifications to describe the individual flow streams and developed a set of equations that could be solved simultaneously. Tinker's method was utilized in part to develop the Delaware method reported by Bell (12). This method avoided some of the complexities of Tinker's method by assuming independent correction factors for the individual streams. A study by Taborek and Palen

(13) using extensive data banks and the digital computer has led to a method that improves upon the older ones.

Bare Tubes in Cross Flow

The most common application of bare tubes in pure cross-flow involves air. Although this application is rapidly going out of style in favor of finned tubes, considerable data are available for tubes in cross-flow as shown in Fig. 14-18 and reference 7. The manner of presentation is quite typical of that used for all types of compact heat-exchanger surfaces where the j-factor, introduced in Chapter 12

$$j = \frac{\overline{h}}{Gc_p} \, \text{Pr}^{2/3} \tag{14-36}$$

and the Fanning friction factor f are plotted versus the Reynolds number

$$\text{Re} = \frac{GD_h}{\mu} \tag{14-37}$$

Figure 14-18.   Heat-transfer and flow-friction data for a staggered tube bank, 4 rows of tubes (23) (Reprinted by permission from *ASHRAE Transactions,* Vol. 79, Part II, 1973.)

The number of rows of tubes in the flow direction has an effect on the j-factor and the heat-transfer coefficient h. The data of Fig. 14-18 are applicable to an exchanger with four rows of tubes (23). For bare tubes in cross-flow, the relation between the heat-transfer coefficient for a finite number of tube rows N, and that for an infinite number of tube rows is given approximately by

$$\frac{\overline{h}}{h_\infty} = 1 - 0.32e^{-0.15N_r} \tag{14-38}$$

when $2 < N_r < 10$. One might expect the friction factor to also depend on the number of tube rows; however, this does not seem to hold true. The assumption is that since a contraction and expansion occur for each row, the friction factor is the same for each row.

The mechanical energy equation, Eq. 9-1c, with the elevation and work terms zero, expresses the lost head for a bank of tubes oriented as shown in Fig. 14-19.

$$\frac{(P_{01} - P_{02})g_c}{\rho_m g} = l_h \tag{14-39}$$

where l_h is made up losses resulting from a change in momentum, friction, and entrance and exit contraction and expansion losses. Integration of the momentum equation through the heat exchanger core, Fig. 14-19, yields (7)

$$\frac{\Delta P_0 g_c}{\rho_m g} = \frac{G_c^2}{2g\rho_m\rho_1} \left[(K_i + 1 - \sigma^2) \right. $$
$$\left. + 2\left(\frac{\rho_1}{\rho_2} - 1\right) + f\frac{A}{A_c}\frac{\rho_1}{\rho_m} - (1 - \sigma^2 - K_e)\frac{\rho_1}{\rho_2} \right] \tag{14-40}$$

where K_i and K_e are entrance and exit loss coefficients that will be discussed in the next section and σ is the ratio of minimum flow-to-frontal area of the exchanger. It may be shown that

$$\frac{A}{A_c} = \frac{4L}{D_h} \tag{14-41}$$

which is a result of the hydraulic diameter concept.

$$A = \text{total heat transfer area, ft}^2 \text{ or m}^2$$
$$A_c = \text{flow cross-sectional area, ft}^2 \text{ or m}^2$$

Referring to Eq. 14-40 the first and last terms in the brackets account for entrance and exit losses, whereas the second and third terms account for flow

Figure 14-19.   Flow normal to a tube bank.

acceleration and friction, respectively. In the case of tube bundles the entrance and exit effects are included in the friction term, that is, $K_i = K_e = 0$. Equation 14-40 then becomes

$$l_h = \frac{G_c^2}{2g\rho_m\rho_1}\left[(1 + \sigma^2)\left(\frac{\rho_1}{\rho_2} - 1\right) + f\frac{A}{A_c}\frac{\rho_1}{\rho_m}\right] \tag{14-42}$$

G_c is based on the minimum flow area and ρ_m is the mean density between inlet and outlet given by

$$\rho_m = \frac{1}{A}\int_A \rho \, dA \tag{14-43}$$

Equation 14-43 is difficult to evaluate. An arithmetic average is usually a good approximation except for parallel flow.

$$\rho_m \approx \frac{\rho_1 + \rho_2}{2} \tag{14-44}$$

A useful nondimensional form of Eq. 14-42 is given by

$$\frac{\Delta P_0}{P_{01}} = \frac{G_c^2}{2g_c\rho_1 P_{01}}\left[(1 + \sigma^2)\left(\frac{\rho_1}{\rho_2} - 1\right) + f\frac{A}{A_c}\frac{\rho_1}{\rho_m}\right] \tag{14-45}$$

where ΔP_0 and P_{01} have units of lbf/ft^2 or Pa. Equations 14-42 and 14-45 are also valid for finned tubes or any other surface that does not have abrupt contractions or expansions.

Finned-tube Heat-transfer Surfaces

Heat-transfer surfaces may take many forms; however, finned tubes are the most popular in HVAC applications. The manner in which data are presented is the same as that shown for bare tubes in Fig. 14-18 and the lost head may be computed using Eq. 14-45. Rich (14, 23) has studied the effect of both fin spacing and tube rows for the plate-fin-tube geometry. Both the j-factor and friction factor decrease as the fin spacing is decreased. The decrease in j-factor was about 50 percent and the decrease in friction factor was about 75 percent as fin pitch was increased from 3 to 20 over the Reynolds number (G_cD_h/μ) range of 500 to 1500. Figure 14-20 shows how the data correlated when the Reynolds number was based on the tube row spacing χ_b. For a given fin pitch it was found that the j-factors decreased as the number of tube rows was increased from 1 to 6 in the useful Reynolds number range. This is contrary to the behavior of bare tubes and results from the difference in the flow fields in each case. The one-row coil had significantly higher j-factors than the others, as was known from previous work done by Shepherd (15). Figure 14-21 shows the j-factor data for the coils with various numbers of tube rows. Note that the Reynolds number is based on the tube row spacing. The combination of Figs. 14-20 and 14-21 therefore gives performance data for all heat exchangers of this one tube diameter and tube pattern with variable fin pitch and number of

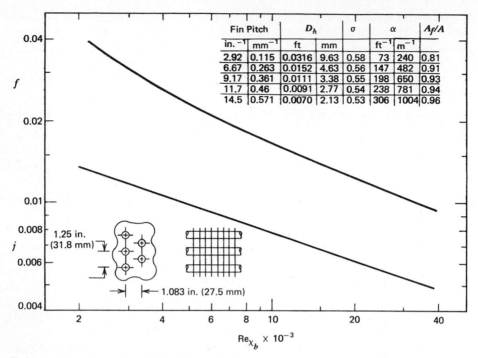

Tube outside diameter, 0.525 in. or 13.3 mm
Fin thickness, 0.006 in. or 0.152 mm

Figure 14-20.   Heat-transfer and friction data for a plate-fin-tube coil with various fin spacings and 5 rows of tubes. (Reprinted by permission from *ASHRAE Transactions*, Vol. 79, Part II, 1973.)

tube rows. Other surfaces with tube diameters and patterns in the same range will behave similarly. The study of tube-row effect (14) also showed that all rows in a plate-fin-tube coil do not have the same heat-transfer rate (Fig. 14-22). The *j*-factors are less for each successive row in the useful (low) Reynolds number range.

The friction factors behave in a manner similar to that discussed before for bare tube banks; therefore, it is assumed there is no tube-row effect.

Caution should be exercised in using published data for plate-fin-tube heat exchangers, especially if the number of rows is not given.

Research by McQuiston (25, 26) has resulted in the correlation of plate-fin-tube transport data that account for geometric variables as well as hydrodynamic effects. Figure 14-23 shows *j*-factors plotted versus the parameter *JP*, which is defined as

$$JP = \text{Re}_D^{-0.4} \left(\frac{A}{A_t} \right)^{-0.15} \tag{14-46}$$

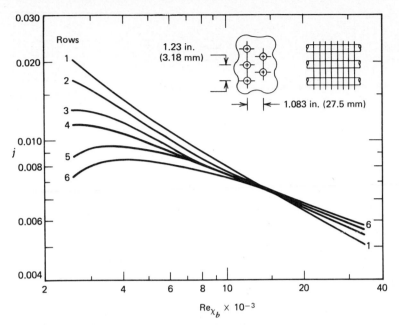

Tube outside diameter, 0.521 in. or 13.2 mm
Fin pitch, 14.5 fins/in. or 0.571 fins/mm
Fin thickness, 0.006 in. or 0.152 mm
Free flow area/frontal area, 0.536

Figure 14-21.   Heat-transfer data for plate-fin-tube coils with various number of tube rows. (Reprinted with permission from *ASHRAE Transactions*, Vol. 81, Part I, 1975.)

where

$$\mathrm{Re}_D = \frac{G_c D}{\mu} \tag{14-47}$$

and

$$\frac{A}{A_t} = \frac{4}{\pi} \frac{\chi_b}{D_h} \frac{\chi_a}{D} \sigma \tag{14-48}$$

In this case the Reynolds number is based on the outside tube diameter and A/A_t is the ratio of the total heat-transfer area to the area of the bare tubes without fins. Note that A/A_t becomes 1.0 for a bare tube bank and the correlation takes a familiar form. The tube-row effect is not accounted for in Fig. 14-23 and must be done separately using Fig. 14-21 which is well described by

$$j_n/j_1 = 1 - 1280\, N_r \mathrm{Re}_{\chi_b}^{-1.2} \tag{14-49}$$

where the subscripts n and 1 pertain to the number of tube rows. Because Fig. 14-23 is for 4 rows of tubes, it is more convenient to write

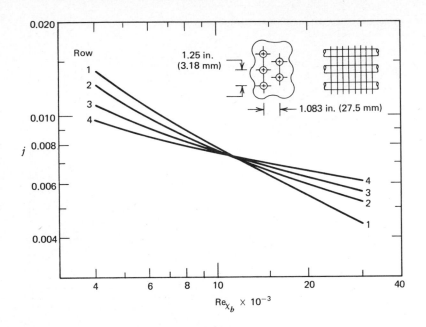

Tube outside diameter, 0.521 in. or 13.2 mm
Fin pitch, 14.5 fins/in. or 0.571 fins/mm
Fin thickness, 0.006 in. or 0.152 mm
Free flow area/frontal area, 0.536

Figure 14-22. Heat-transfer data for each row of a 4-row plate-fin-tube coil. (Reprinted with permission from *ASHRAE Transactions,* Vol. 81, Part I, 1975.)

$$j_n/j_4 = \frac{1 - 1280 \, N_r \text{Re}_{X_b}^{-1.2}}{1 - 5120 \, \text{Re}_{X_b}^{-1.2}} \tag{14-50}$$

where j_4 is read from Fig. 14-23.

Generalized correlation of friction data is more involved than that for heat-transfer data. Figure 14-24 shows such a correlation using a parameter FP defined as

$$FP = \text{Re}_D^{-0.25} \left(\frac{D}{D^*}\right)^{0.25} \left[\frac{X_a - D}{4(s - y)}\right]^{-0.4} \left[\frac{X_a}{D^*} - 1\right]^{-0.5} \tag{14-51}$$

where D^* is a hydraulic diameter defined by

$$D^* = \frac{D\left(\dfrac{A}{A_t}\right)}{1 + \dfrac{(X_a - D)}{s}} \tag{14-52}$$

The correlating parameters of Eqs. 14-51 and 15-52 have evolved over a long period of time from observations of experimental data. The friction data scatter

Figure 14-23. Heat-transfer correlation for smooth plate-fin-tube coils with 4 rows of tubes. (Reprinted with permission from *ASHRAE Transactions,* Vol. 84, Part I, 1978.)

Figure 14-24. Correlation of friction data for smooth plate-fin-coils. (Reprinted with permission from *ASHRAE Transactions,* Vol. 84, Part 1, 1978.)

more than the heat-transfer data of Fig. 14-23, which are typical. Note that the data from reference 25 are much more consistent than some of the other data that date back more than 20 years.

The data presentations of Figs. 14-23 and 14-24 have the advantage of generality and are also adaptable to the situation where moisture is condensing on the surface. This will be discussed later in this chapter. These same types of correlations may be used for other types of finned surfaces such as circular and wavy fins.

Plate Fin Heat Transfer Surfaces

Figure 14-25 illustrates the plate-fin heat-transfer surface. The fins may have several variations such as louvers, strips, or waves. Plain smooth fins are generally not used because of the low heat-transfer coefficients that arise when the flow length becomes long. The types mentioned earlier disturb the boundary layer so that the length does not influence the heat-transfer or flow-friction coefficients. Figure 14-25 is an example of data for a louvered plate-fin surface.

In computing the lost head for these surfaces, one must consider the entrance and exit losses resulting from abrupt contraction and exit expansion. The entrance and exit losses are expressed in terms of a loss coefficient K and the velocity head inside the heat exchanger core. Thus for the entrance

$$\Delta P_{0i} = K_i \frac{G_c^2}{2\rho_i g_c} \tag{14-53}$$

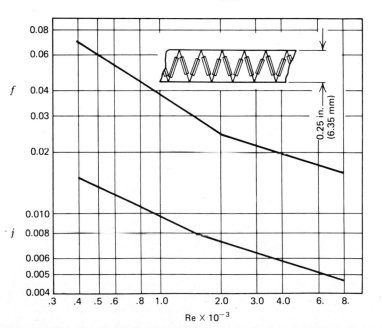

Figure 14-25.   Heat-transfer and friction data for a louvered plate-fin surface.

and for the exit

$$\Delta P_{0e} = K_e \frac{G_c^2}{2\rho_e g_c} \tag{14-54}$$

Equations 14-53 and 14-54 are included in Eq. 14-40:

$$\frac{\Delta P_0}{P_{01}} = \frac{G_c^2}{2g_c P_{01}\rho_1} \left[(K_i + 1 - \sigma^2) \right.$$

$$\left. + 2\left(\frac{\rho_1}{\rho_2} - 1\right) + f\frac{A}{A_c}\frac{\rho_1}{\rho_m} - (1 - \sigma^2 - K_e)\frac{\rho_1}{\rho_2} \right] \tag{14-55}$$

The entrance and exit loss coefficients depend on the type of surface, the contraction ratio, and the Reynolds number $G_c D_h/\mu$. The degree to which the velocity profile has developed is also important. Reference 7 gives entrance and exit loss coefficients that apply to surfaces such as that shown in Fig. 14-25. The loss coefficients are a function of Reynolds number. However, most plate-fin surfaces have flow interruptions that cause continual redevelopment of the boundary layer, which is equivalent to a very high Reynolds number condition. Figure 14-26 gives loss coefficients applicable to plate-fin type surfaces with flow interruptions such as that of Fig. 14-25.

14-6 DESIGN PROCEDURES FOR SENSIBLE HEAT TRANSFER

It is difficult to devise one procedure for designing all heat exchangers because the given parameters vary from situation to situation. All of the terminal temperatures may be known or only the inlet temperatures may be given. The mass flow rates may be fixed in some cases and a variable in others. Usually the surface area is not given.

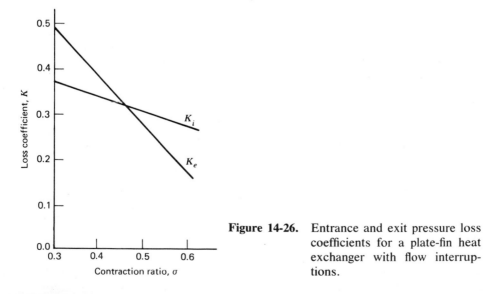

Figure 14-26. Entrance and exit pressure loss coefficients for a plate-fin heat exchanger with flow interruptions.

Earlier in the chapter the LMTD and NTU methods were described as the two general heat-exchanger design procedures. Either method may be used, but the NTU method has certain advantages. Consider the design problem where t_{hi}, t_{ho}, t_{ci}, $\dot{m}_c$, and $\dot{m}_h$ are known and the surface area A is to be determined. With either approach the heat-transfer coefficients must be determined as previously discussed so that the overall coefficient U can be computed. The NTU approach then proceeds as follows:

1. Compute the effectiveness ε and C_{min}/C_{max} from the given data.
2. Determine the NTU for the particular flow arrangement from the ε-NTU curve, such as Fig. 14-5 or Table 14-1.
3. Compute A from $A = NTU(C_{min}/U)$.

The LMTD approach is as follows:

1. Compute P and R from the given terminal temperatures.
2. Determine the correction factor F from the appropriate curve, such as Fig. 14-3.
3. Calculate the LMTD for an equivalent counterflow exchanger using Eq. 14-2.
4. Calculate A from $A = \dot{q}/U(F)(LMTD)$ where

$$\dot{q} = C_c(t_{co} - t_{ci}) = C_h(t_{hi} - t_{ho})$$

The NTU approach requires somewhat less effort in this case.

Consider the design problem where A, U, $\dot{m}_c$, $\dot{m}_h$, t_{hi}, and t_{ci} are given, and it is necessary to find the outlet temperatures t_{ho}, t_{co}. The NTU approach is as follows:

1. Calculate the NTU $= UA/C_{min}$ from given data.
2. Find ε from the appropriate curve for the flow arrangement using NTU and C_{min}/C_{max}.
3. Compute one outlet temperature from Eq. 14-7 or 14-8.
4. Compute the other outlet temperature from

$$\dot{q} = C_c(t_{co} - t_{ci}) = C_h(t_{hi} - t_{ho})$$

The LMTD approach requires iteration as follows:

1. Calculate R from $R = C_c/C_h$.
2. Assume one outlet temperature in order to compute P (first approximation).
3. Find F from the appropriate curve (first approximation).
4. Evaluate LMTD (first approximation).
5. Determine $\dot{q} = UAF(LMTD)$ (first approximation).
6. Calculate outlet temperature to compare with the assumption of step 2.
7. Repeat steps 2 through 6 until satisfactory agreement is obtained.

It is obvious that the NTU method is much more straightforward in this later problem. A complete design problem is illustrated in the following example.

Example 14-1. Design a water-to-air heating coil of the continuous plate-fin-tube type. The required duty for the coil is as follows:

Heat outdoor air from 50 to about 100 F
Air flow rate of 2000 cfm
Entering water temperature, 150 F
Leaving water temperature, 140 F
Air face velocity should not exceed 1000 fpm
Water-side head loss should not exceed 10 ft H_2O
Water connections must be on the same end of the coil
Air-side pressure drop should not exceed 1.2 in. H_2O

Solution. Figure 14-27 is a schematic of a typical water-to-air heating coil that has multiple rows of tubes. Although the water may be routed through the tubes in many different ways, the circuiting is usually such that counterflow will be approached as shown. Counterflow can usually be assumed when three or more rows are used. Because the water inlet and outlet connections must be on the same end of the coil in this case, a multiple of two rows is used if possible.

The first step in the solution will be to compute the overall heat-transfer coefficient U, which will be based on the air-side area. Equation 14-15a applies where η_{si} is equal to one and the wall thermal resistance is negligible.

$$\frac{1}{U_o} = \frac{1}{h_o\eta_{so}} + \frac{1}{h_i(A_i/A_o)}$$

The subscript o refers to the air side and the i refers to the water side. Equation 14-29 will be used to find the coefficient h_i assuming a water velocity of 4 ft/sec. Experience has shown that velocities greater than 5

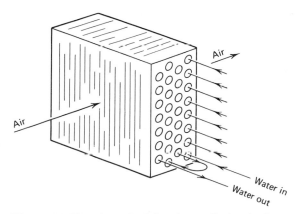

Figure 14-27. A typical heating coil circuited to approach counterflow.

ft/sec or 1.5 m/s result in very high lost head. Since at this point the tube diameter must be established, a surface geometry must be selected.

One standard plate-fin-tube surface uses $\frac{1}{2}$-in. tubes in a triangular layout as shown in Fig. 14-14 with χ_a of 1.25 in. and χ_b of 1.083 in. The fin pitch is 8 fins/in. and the fin thickness is 0.006 in. As a result of fabrication of the coil, the final tube outside diameter is 0.525 in. with a wall thickness of 0.015 in. Other geometric data will be given as required and the j-factor and friction factor will be obtained from Figs. 14-23 and 14-24. The Reynolds number based on the tube inside diameter is then

$$\text{Re}_D = \frac{\rho \bar{V} D}{\mu} = \frac{61.5(4)(0.4831/12)}{(1.04/3600)} = 34,275$$

where ρ and μ are evaluated at 145 F. The Prandtl number is

$$\text{Pr} = \frac{\mu c_p}{k} = \frac{(1.04)(1.0)}{(0.38)} = 2.74$$

Then using Eq. 14-29

$$\bar{h}_i = 0.023 \frac{k}{D} (\text{Re}_D)^{0.8}(\text{Pr})^{0.3}$$

$$\bar{h}_i = 0.023 \frac{0.38}{(0.483)/12} (34,275)^{0.8}(2.74)^{0.3}$$

$$\bar{h}_i = 1250 \text{ Btu/(hr-ft}^2\text{-F)}$$

where the exponent on the Prandtl number is for $t_{\text{wall}} < t_{\text{bulk}}$ and L/D has been assumed to be larger than 60.

To compute the air-side heat-transfer coefficient it is necessary to know the air velocity or air mass velocity inside the core. Because the coil face velocity cannot exceed 1000 ft/min, a face velocity of 900 ft/min will be assumed. Then

$$\dot{m}_a = G_{fr}A_{fr} = G_c A_c$$

and

$$G_c = G_{fr}\frac{A_{fr}}{A_c} = \frac{G_{fr}}{\sigma}$$

where the subscript fr refers to the face of the coil and c refers to the minimum flow area inside the coil. The ratio of minimum flow-to-frontal area for this case has been calculated to be 0.555.

$$G_{fr} = \rho_{fr}\bar{V}_{fr} = \frac{14.7(144)(900)60}{53.35(510)} = 4200 \text{ lbm/(hr-ft}^2\text{)}$$

and

$$G_c = \frac{4200}{(0.555)} = 7569 \text{ lbm/(hr-ft}^2\text{)}$$

The j-factor correlation of Fig. 14-23 is based on the parameter JP, which is defined by Eq. 14-46. The Reynolds number is then

$$\mathrm{Re}_D = \frac{G_c D}{\mu} = \frac{7569(0.525/12)}{0.044} = 7578$$

and the parameter A/A_t defined by Eq. 14-48 is

$$A/A_t = \frac{4(1.083)1.25(0.555)}{\pi(0.01312)12(0.525)} = 11.6$$

where the hydraulic diameter is another known dimension of the coil. The parameter JP is

$$JP = (7578)^{-0.4}(11.6)^{-0.15} = 0.0194$$

The j-factor is now read from Fig. 14-23 as 0.0066. Then

$$\mathrm{StPr}^{2/3} = \left(\frac{\bar{h}_o}{G_c c_p}\right)\left(\frac{\mu c_p}{k}\right)^{2/3} = 0.0066$$

or

$$\bar{h}_o = 0.0066(7569)(0.24)(0.71)^{-2/3} = 15.1 \text{ Btu/(hr-ft}^2\text{-F)}$$

The next step is to compute the fin efficiency and the surface effectiveness. Equations 14-20, 14-21, 14-22, and 14-24 will be used. The equivalent fin radius R_e is first computed from Eq. 14-24. The dimensions L and M are found as follows by referring to Fig. 14-14.

$$\mathrm{Dim}_1 = \frac{\chi_a}{2} = \frac{1.25}{2} = 0.625 \text{ in.}$$

$$\mathrm{Dim}_2 = \frac{[(\chi_a/2)^2 + \chi_b^2]^{1/2}}{2}$$

$$= \frac{[(0.625)^2 + (1.083)^2]^{1/2}}{2} = 0.625 \text{ in.}$$

Because Dim_1 is equal to Dim_2 in this case

$$L = M = 0.625 \text{ in.}$$

Then

$$\psi = \frac{M}{r} = \frac{0.625}{(0.525/2)} = 2.38$$

$$\beta = \frac{L}{M} = \frac{0.625}{0.625} - 1.0$$

and

$$\frac{R_e}{r} = 1.27(2.38)(1.0 - 0.3)^{1/2} = 2.53$$

From Eq. 14-22

$$\phi = (2.53 - 1)(1 + 0.35 \ln 2.53) = 2.03$$

and using Eq. 14-20

$$m = \left[\frac{2(15.1)}{100(0.006/12)} \right]^{1/2} = 24.6 \text{ ft}^{-1}$$

where the thermal conductivity k of the fin material has been assumed equal to 100 (Btu-ft)/(ft²-hr-F), which is typical of aluminum fins. Then from Eq. 14-21,

$$\eta = \frac{\tanh [(24.6)(0.525/24)(2.03)]}{(0.525/24)(2.03)(24.6)} = 0.73$$

The surface effectiveness η_{so} is then computed using Eq. 14-26 where A_f/A is 0.919.

$$\eta_{so} = 1 - 0.919(1 - 0.73) = 0.75$$

The ratio of the water-side to air-side heat transfer areas must finally be determined. The ratio of the total air-side heat transfer area to the total volume (A_o/V) or α is given as 170 ft^{-1}. The ratio of the water-side heat transfer area to the total volume (A_i/V) is closely approximated by

$$\frac{A_i}{V} = \frac{D_i \pi}{\chi_a \chi_b}$$

$$\frac{A_i}{A_o} = \left(\frac{A_i}{V} \right) \Big/ \left(\frac{A_o}{V} \right) = \frac{\pi D_i}{\chi_a \chi_b \alpha} \tag{14-56}$$

$$\frac{A_i}{A_o} = \frac{\pi(0.483/12)}{(1.25/12)(1.083/12)(170)} = 0.079$$

The overall coefficient U is then given by

$$\frac{1}{U_o} = \frac{1}{15.1(0.75)} + \frac{1}{1248(0.079)} = 0.098$$

and

$$U_o = 10.2 \text{ Btu/(hr-ft}^2\text{-F)}$$

The fluid capacity rates will now be computed. For the air

$$\dot{m} = \rho \dot{Q} = \frac{14.7(144)}{53.35(510)} (2000)(60) = 9{,}336 \text{ lbm/hr}$$

and

$$C_{air} = C_c = 0.24(9336) = 2241 \text{ Btu/(hr-F)}$$

For the water

$$\dot{q} = C_w(t_{wi} - t_{wo}) = C_{air}(t_{ao} - t_{ai})$$

and

$$C_w = C_h = C_{air} \frac{(t_{ao} - t_{ai})}{(t_{wi} - t_{wo})}$$

$$C_w = 2241 \frac{(100 - 50)}{(150 - 140)} = 11,205 \text{ Btu/(hr-F)}$$

Since $C_w > C_{air}$, $C_{air} = C_{min} = C_c$, $C_w = C_h = C_{max}$, and

$$\frac{C_{min}}{C_{max}} = \frac{2241}{11,205} = 0.20$$

The effectiveness ε is given by Eq. 14-8:

$$\varepsilon = \frac{t_{co} - t_{ci}}{t_{hi} - t_{ci}} = \frac{100 - 50}{150 - 50} = 0.50$$

Now assuming that the flow arrangement is counterflow, the NTU is read from Fig. 14-5 at $\varepsilon = 0.5$ and $C_{min}/C_{max} = 0.2$ as 0.74. Then

$$\text{NTU} = \frac{U_o A_o}{C_{min}}$$

$$A_o = \frac{0.74(2241)}{10.2} = 163 \text{ ft}^2$$

The total volume of the heat exchanger is given by

$$V = \frac{A_o}{\alpha} = \frac{163}{170} = 0.96 \text{ ft}^3$$

Since a face velocity of 900 ft/min was assumed, the face area is

$$A_{fr} = \frac{\dot{Q}}{\bar{V}_{fr}} = \frac{2000}{900} = 2.22 \text{ ft}^2$$

and the depth is

$$L = \frac{V}{A_{fr}} = 0.96/2.22 = 0.43 \text{ ft} = 5.18 \text{ in.}$$

The number of rows of tubes N_r will then be

$$N_r = \frac{L}{X_b} = 5.18/1.083 = 4.78$$

Since N_r must be an integer and a multiple of two for the flow arrangement of Fig. 14-27, six rows must be used. This will overdesign the heat exchanger. Another possibility is to use five rows with a different circuiting arrangement so that the water connections are on the same end of the coil. This will be considered later when the lost head on the water side is computed.

The lost head on the air side of the exchanger is given by Eq. 14-45, where the ratio A/A_c is given by

$$\frac{A}{A_c} = \frac{(\alpha V)}{(\sigma A_{fr})} = \frac{170(0.96)}{0.555(2.22)} = 132$$

The mass velocity G_c was previously computed as 7569 lbm/(hr-ft²) and the mean density ρ_m is approximately

$$\rho_m = \frac{P}{2R}\left(\frac{1}{T_{ci}} + \frac{1}{T_{co}}\right)$$

$$\rho_m = \frac{14.7(144)}{2(53.35)}\left[\frac{1}{510} + \frac{1}{560}\right] = 0.074 \text{ lbm/ft}^3$$

The friction factor is read from Fig. 14-24 with FP computed from Eq. 14-51. Using Eq. 14-52

$$\frac{D^*}{D} = \frac{11.6}{1 + (1.25 - 0.525)/0.125} = 1.71$$

and

$$FP = (7578)^{-0.25}(1.71)^{-0.25}\left[\frac{1.25 - 0.525}{4(0.125 - 0.006)}\right]^{-0.4}\left[\frac{1.25}{0.898} - 1\right]^{-0.5}$$

$$= 0.130$$

then from Fig. 14-24, $f = 0.027$.

$$\Delta P_0 = \frac{(7569)^2}{2(0.078)(32.2)(3600)^2}\left\{[1 + (0.555)^2]\left(\frac{0.078}{0.071} - 1\right)\right.$$

$$\left. + 0.027(132)\left(\frac{0.078}{0.074}\right)\right\}$$

$$\Delta P_0 = 3.42 \text{ lbf/ft}^2$$

and

$$l_h = \frac{\Delta P_0}{\rho_w} = \frac{3.42}{62.4} = 0.055 \text{ ft H}_2\text{O} = 0.66 \text{ in. H}_2\text{O}$$

The lost head on the tube side of the exchanger must now be computed. Recall that a velocity of 4 ft/sec was assumed to compute the heat-transfer coefficient h_i. It has also been determined that at least five rows of tubes are required and the water connections must be on the same end of the exchanger. Therefore, consider the arrangement shown in Fig. 14-28. If we use two passes per row of tubes, the water enters and leaves the same end of the coil. For the coil shown there are five separate water circuits.

Figure 14-28.   Five-row coil with two passes per row.

The flow cross-sectional area for the water may be determined from the fluid capacity rate for the water and the continuity equation.

$$\dot{m}_w = \bar{V} A \rho = \frac{C_h}{c_p}$$

and

$$A = \frac{C_h}{\bar{V} \rho c_p} = \left(\frac{11,205}{3600}\right) \frac{1}{(4)(61.5)(1.0)}$$

$$A = 0.01265 \text{ ft}^2$$

For N number of tubes

$$A = N \frac{\pi}{4} D_i^2$$

and

$$N = \frac{4A}{\pi D_i^2} = \frac{4(0.01265)(144)}{\pi(0.483)^2} = 9.94$$

Since N must be an integer, 10 tubes are required and the water velocity is reduced somewhat. This reduction in velocity will not significantly reduce the heat-transfer coefficient h_i. To adapt to the flow arrangement of Fig. 14-28 a coil that is 20 tubes high must be used. Then the height H becomes

$$H = 20X_a = 20(1.25) = 25 \text{ in.}$$

The frontal area A_{fr} was previously found to be 2.22 ft². Then the width W is

$$W = \frac{A_{fr}}{H} = \frac{2.22}{(25/144)} = 12.8 \text{ in.}$$

This arrangement will meet all of the design requirements; however, the shape of the coil, height of 25 in. and width of 12.8 in., may be unacceptable. If so, another alternative must be sought such as using six rows of tubes, placing the headers on opposite ends, or some other modification.

The lost head l_{fw} will be computed using Eq. 9-6. Lost head in the return bends will be accounted for by assuming a loss coefficient of 2 for each bend. The flow length L_w is

$$L_w = 2(5)(12.8/12) = 10.7 \text{ ft}$$

and the Moody friction factor is 0.023 from Fig. 9-1 at a Reynolds number of 34,275, which takes into account the lower water velocity. There are nine return bends in each circuit. Then

$$l_{fw} = 0.023 \frac{(10.7)}{(0.483/12)} \frac{(4)^2}{(64.4)} + 2(9) \frac{(4)^2}{(64.4)}$$

$$l_{fw} = 6 \text{ ft } H_2O$$

Assuming the 5-row configuration is not satisfactory, reconsider the circuiting of Fig. 14-27 and a 6-row coil. The coil will then have 10 tubes per row and

$$H = 10X_a = 10(1.25) = 12.5 \text{ in.}$$

The width will then be

$$W = \frac{2.22}{(12.5/144)} = 25.6 \text{ in.}$$

and

$$L_w = 2(6)(25.6/12) = 25.6 \text{ ft}$$

Assuming the friction factor is unchanged

$$l_{fw} = 0.023 \frac{(25.6)(4)^2}{(0.483/12)64.4} + 2(5) \frac{(4)^2}{64.4} = 6.2 \text{ ft } H_2O$$

The geometry of the 6-row coil is more reasonable. This value of the lost head does not include the losses in the inlet and outlet headers. Header losses may be substantial, depending on the design and fabrication, and may be equal to the losses in the tubes and return bends.

There are many different ways the heat-exchanger design problem may be posed. About the same amount of work is involved in every case, however.

The previous example shows that the process is laborious and time consuming. Therefore, almost all manufacturers have devised computer programs that carry out the design process quickly and accurately. Because of the speed of a computer, a simulation rather than a design approach may be used where the performance of a given configuration is determined.

14-7   COMBINED HEAT AND MASS TRANSFER

When the heat exchanger surface in contact with moist air is at a temperature below the dew-point temperature for the air, condensation of vapor will occur. Typically the air dry bulb temperature and the humidity ratio both decrease as the air flows through the exchanger. Therefore, sensible and latent heat transfer occur simultaneously. This process is similar to that occurring in the spray dehumidifier discussed in Chapter 13 and can be analyzed using the same procedure; however, this is not generally done.

The problem of cooling coil analysis and design is complicated by the uncertainty in determining the transport coefficients h, h_d, and f. It would be very convenient if heat transfer and friction data for dry heating coils, such as that shown in Figs. 14-23 and 14-24, could be used with the Colburn analogy of Eq. 12-20a to obtain the mass-transfer coefficients. But it has been known for some time that this approach is not reliable, and recent work (17, 18, 25, 26) has shown that the analogy does not always hold true. Figure 14-29 shows j-factors for a simple parallel plate exchanger that were obtained for different surface conditions. Although these particular j-factors are for the sensible heat transfer, the mass transfer j-factors and the friction factors exhibit the same behavior. Note that the dry surface j-factors fall below those obtained under dehumidifying conditions with the surface wet. The converging-diverging nature of the curves of Fig. 14-29 can be explained by the roughness introduced by the water on the surface and the nature of the boundary layers at different Reynolds numbers. The velocity, temperature, and concentration boundary layer thicknesses can all be approximated by

$$\frac{\delta}{x} = \frac{5}{\sqrt{Re_x}} \tag{14-57}$$

where

δ = boundary layer thickness
x = distance from inlet, measured in the same units as δ
Re_x = Reynolds number based on x

Equation 14-57 shows that at low Reynolds numbers the boundary layer grows quickly; the droplets are soon covered and have little effect on the flow field. As the Reynolds number is increased, the boundary layer becomes thin and more of the total flow field is exposed to the droplets. The roughness caused by the droplets induces mixing and larger j-factors. The data of Fig. 14-29

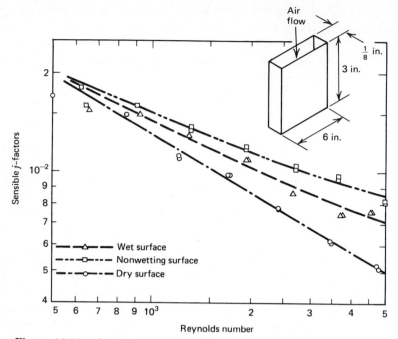

Figure 14-29. Sensible heat transfer j-factors for a parallel plate exchanger (18). (Reprinted with permission from *ASH-RAE Transactions*, Vol. 82, Part II, 1976.)

cannot be applied to all surfaces because the length of the flow channel is also an important variable. It seems certain, however, that the water collecting on the surface is responsible for breakdown of the j-factor analogy. The j-factor analogy is approximately true when the surface conditions are identical (18). That is, when the surface is wetted, the sensible and mass transfer j-factors are in close agreement. This is of little use, however, because a wet test must be made to obtain this information. Under some conditions it is possible to obtain a film of condensate on the surface instead of droplets. For example, aluminum when thoroughly degreased and cleaned with a harsh detergent in hot water experiences filmwise condensation (25). Figure 14-30 shows j-factor and friction data for a plate-fin-tube surface under dry conditions and with filmwise and dropwise condensation. The trends are the same as those shown in Fig. 14-29. The friction factors are influenced by the water on the surface over the complete Reynolds number range, whereas the j-factors are affected only at the higher Reynolds numbers. The data shown correspond to face velocities of 200 to 800 ft/min with air at standard conditions. Although not shown, the mass transfer j-factors show the same trends and are in reasonable agreement with the wet surface j-factors shown in Fig. 14-30.

Recent research involving plate-fin-tube surfaces (26) has resulted in cor-

Figure 14-30. Heat-transfer and friction data with mass transfer for a plate-fin-tube surface, 4 rows of tubes.

relations that relate dry sensible j- and f-factors to those for wetted dehumidifying surfaces. Expressions were developed that modify the parameters JP and FP of Figs. 14-23 and 14-24 for wet surface conditions. In developing these functions, it was found that a Reynolds number based on fin spacing and the ratio of fin spacing to space between the fins were useful. For film-type condensation, the modifying functions are:

Sensible j-factor

$$J(s) = 0.84 + 4 \times 10^{-5}(\text{Re}_s)^{1.25} \qquad (14\text{-}58)$$

Total j-factor

$$J_i(s) = (0.95 + 4 \times 10^{-5}(\text{Re}_s)^{1.25}) \left(\frac{s}{s-y}\right)^2 \tag{14-59}$$

Friction factor

$$F(s) = (1 + (\text{Re}_s)^{-0.4}) \left(\frac{s}{s-y}\right)^{1.5} \tag{14-60}$$

For dehumidifying conditions, the abscissa of Fig. 14-23 is changed to $J(s)JP$ and $J_i(s)JP$. The abscissa of Fig. 14-24 is changed to $F(s)FP$.

Enthalpy Potential

The enthalpy potential was mentioned in Chapter 13 and will be more fully justified here. The heat transfer from moist air to a surface at a temperature below the air dew point may be expressed as

$$\frac{\dot{q}}{A} = h(t_w - t_\infty) + h_d(W_w - W_\infty)i_{fg} \tag{14-61}$$

Using the analogy of Eq. 12-20a with Le = 1 we see that

$$\frac{\dot{q}}{A} = h_d[c_{pa}(t_w - t_\infty) + (W_w - W_\infty)i_{fg}] \tag{14-62}$$

The enthalpy of vaporization i_{fg} is evaluated at the wall temperature. Even though the Colburn analogy is not always precise, there is a proportionality between h and h_d, which is all that is required here. The enthalpy of the saturated moist air at the wall is given by

$$i_w = c_{pa}t_w + W_w(i_f + i_{fg}) \tag{14-63}$$

where i_f and i_{fg} are evaluated at the wall temperature. For the moist air in the free stream

$$i_\infty = c_{pa}t + W[i_f + i_{fg} + c_{ps}(t - t_w)] \tag{14-64}$$

The temperature t_w should be the dew-point temperature and i_f and i_{fg} should be evaluated at the dew-point temperature. However, the errors tend to compensate and Eq. 14-64 is a very good approximation. The difference in enthalpy between the surface i_w and the free stream i_∞ is then

$$i_w - i_\infty = c_{pa}(t_w - t_\infty) + i_{fg}(W_w - W_\infty) + i_f(W_w - W_\infty) \tag{14-65}$$

Comparison of Eqs. 14-62 and 14-65 then yields

$$\frac{\dot{q}}{A} = h_d[(i_w - i_\infty) - i_f(W_w - W_\infty)] \tag{14-66}$$

The last term is typically about 0.5 percent of $(i_w - i_\infty)$ and can be neglected.

Thus, the driving potential for simultaneous transfer of heat and mass is enthalpy to a close approximation, whereas temperature and concentration are the driving potentials for sensible heat and mass, respectively.

Equation 14-66 expresses the total heat transfer at a particular location in the heat exchanger; however, the moist air enthalpy at the surface i_w and in the free stream i_∞ vary throughout the exchanger, as shown in Fig. 14-31 for counterflow. In addition, most coils will have fins that must be accounted for. Then

$$\dot{q} = h_d A \eta_{ms} \Delta i_m \tag{14-67}$$

where η_{ms} is the surface effectiveness with combined heat and mass transfer and Δi_m is some mean enthalpy difference. With suitable assumptions it can be shown that Δi_m has the same form as the LMTD for counterflow:

$$\Delta i_m = \frac{\Delta i_1 - \Delta i_2}{\ln \dfrac{\Delta i_1}{\Delta i_2}} \tag{14-68}$$

This is true because i_w is directly proportional to t_c, the refrigerant temperature. Equation 14-67 expresses the total heat-transfer rate from the wall to the airstream where the wall temperature is not known explicitly. However, the heat-transfer rate from the refrigerant to the wall is given by

$$\dot{q} = h_i A_i (\Delta t_m)_i \tag{14-69}$$

where $(\Delta t_m)_i$ expresses the mean temperature difference between the refrigerant and the wall and the thermal resistance of the thin wall has been neglected. A simple iterative procedure is then necessary to solve Eqs. 14-67 and 14-69 for the total heat-transfer rate.

It was mentioned earlier in this chapter that the heat-transfer coefficient decreases from the inlet to the exit of the coil as shown in Fig. 14-22. This has a direct effect on total heat-transfer calculation because the coil surface temperature is higher than expected at the inlet due to a higher heat-transfer rate

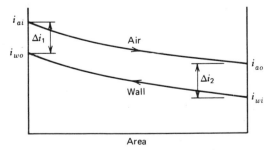

Figure 14-31. Enthalpy difference in a counterflow dehumidifying coil.

there. This should be taken into account because a portion of the coil near the air inlet may be at a temperature greater than the dew point with no mass transfer occurring.

The sensible heat transfer from the moist air to the refrigerant is computed for counterflow by Eq. 14-3:

$$\dot{q}_s = UA(\text{LMTD}) \tag{14-3}$$

where LMTD is given by Eq. 14-2 and U is given by Eq. 14-15a with η_s equal to η_{ms}.

The latent heat transfer is then easily computed from

$$\dot{q}_l = \dot{q} - \dot{q}_s \tag{14-70}$$

It is also true that

$$\dot{q}_l = \dot{m}_a(W_i - W_o)i_{fg} \tag{14-71}$$

Fin Efficiency with Mass Transfer

The fin efficiency with combined heat and mass transfer is lower than the value obtained with only sensible heat transfer. Although the basic definition is unchanged from that given in Sec. 14-3, the analysis is more complex and not exact. Threlkeld (20) has solved the problem; his solution, however, depends on knowing the water film thickness on the fin. This is a very nebulous quantity and impractical to use. An accepted method is outlined in reference 19 and is an adaptation of the work of Ware and Hacha (21) and others. Although this method appears to give good results for performance and rating purposes, it has some undesirable features since the coil surface temperature is assumed to be the only parameter affecting the fin efficiency regardless of the moist air conditions. Another disturbing feature is failure of the solution to reduce to the dry coil case when the surface and moist air conditions warrant this. These inconsistencies are troublesome when making general coil studies.

A fin of uniform cross section as shown in Fig. 14-8 has been analyzed by McQuiston (22). The method is approximate but reduces to the case of zero mass transfer and is adaptable to circular and plate-fin-tube surfaces. The analysis is outlined as follows: An energy balance on an elemental volume yields the following differential equation, assuming one-dimensional heat transfer and constant properties:

$$\frac{d^2t}{dx^2} = \frac{P}{kA_c}[h(t - t_\infty) + h_d i_{fg}(W - W_\infty)] \tag{14-72}$$

where

$$t = \text{temperature of the element, F or C}$$
$$x = \text{distance measured from base of fin, ft or m}$$
$$P = \text{circumference of the fin, ft or m}$$
$$k = \text{thermal conductivity of the fin material, (Btu-ft)/(hr-ft}^2\text{-F) or (W-m)/(m}^2\text{-C)}$$

A_c = cross-sectional area of the fin, ft² or m²

h = convective heat-transfer coefficient, Btu/(hr-ft²-F) or W/(m²-C)

t_∞ = temperature of the air–vapor mixture flowing around the fin, F or C

h_d = convective mass-transfer coefficient, lbm/(ft²-hr) or kg/(m²-s)

i_{fg} = latent heat of vaporization of water, Btu/lbm or J/kg

W = specific humidity of saturated air at temperature t, lbmw/lbma or kgw/kga

W_∞ = specific humidity of the air–vapor mixture, lbmw/lbma or kgw/kga

The analogy of Eq. 4-20a will be used to obtain the mass-transfer coefficient h_d with Le = 1

$$h_d = \frac{h}{c_{pa}} \tag{14-73}$$

where

c_{pa} = specific heat capacity of the dry air, Btu/(lbm-F) or W/(kg-C)

As suggested, the coefficient h should be for a wet surface. Other correlations may also be used. Combining Eqs. 14-72 and 14-73 gives

$$\frac{d^2t}{dx^2} = \frac{hP}{kA_c}\left[(t - t_\infty) + \frac{i_{fg}}{c_{pa}}(W - W_\infty)\right] \tag{14-72a}$$

Now, if $(W - W_\infty)$ is simply related to $(t - t_\infty)$, Eq. 14-72a can be easily solved for the temperature distribution in the fin. To justify such a simplification consider the physical aspects of a typical cooling and dehumidifying coil.

Let the air–vapor mixture enter an exchanger at a fixed condition designated by point 1 on the psychrometric chart of Fig. 14-32. Consider an evaporator with a constant temperature refrigerant operating such that the moist air very near the wall is at a temperature designated by point $w2$. The humidity ratio of the leaving air W_2 will approach W_{w2} as shown in Fig. 14-33. The process line 1-2 in Fig. 14-32 can be approximated by a straight line and a simple relationship between $(W - W_\infty)$ and $(t - t_\infty)$ exists. In fact,

$$(W_{w2} - W_1) = C(t_{w2} - t_1) \tag{14-74}$$

where C is a constant. An examination of data for many coils and various operating conditions shows that C will typically vary less than 10 percent from inlet to exit. It then seems reasonable to use an average value such as

$$C_{avg} = \frac{(C_1 + C_2)}{2} \tag{14-75}$$

Due to the shape of the saturation curve, precise location of the point $w2$ on Fig. 14-32 does not greatly affect the value of C for a particular coil condition. On the other hand, C is very sensitive to the location of point 1. For example, for $t_w = 45$ F and $t_1 = 80$ F, C varies from 1.4×10^{-4} to 0.0 as ϕ_1 varies from

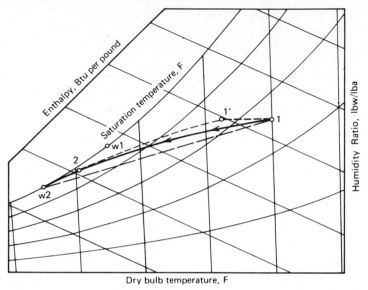

Figure 14-32.   Cooling and dehumidifying processes. (Reprinted by permission from *ASHRAE Transactions,* Vol. 81, Part I, 1975.)

50 to 29 percent. For the last condition there will be no condensation on the surface and only sensible heat transfer will occur.

When chilled water is used as a cooling medium in a counterflow arrangement, Fig. 14-34 applies. In this case the wall temperature is somewhat higher where the air enters the exchanger. Typical conditions for the moist air very near the wall are shown on Fig. 14-32 as points $w1$ and $w2$. Here the surface is completely wetted.

Figure 14-33.   Specific humidity for a constant-temperature refrigerant.

Figure 14-34. Specific humidity for chilled water
as the refrigerant.

At the inlet to the exchanger the value of C is given by

$$C_1 = \frac{W_{w1} - W_1}{t_{w1} - t_1} \qquad (14\text{-}76)$$

whereas at the exit

$$C_2 = \frac{W_{w2} - W_2}{t_{w2} - t_2} \qquad (14\text{-}76a)$$

Again C_1 is less than C_2 and an average value should be used. In most cases
C will change less than 10 percent from inlet to outlet.

Process 1-1'-2 approximates a situation where the coil is partially dry. In
this case C_1 is zero until the air reaches the location in the coil where the surface
temperature is below the air dew-point temperature. C then increases to the
value of C_2 at the exit. Again an average value of C may be used or C_2 may
be used to obtain a conservative solution.

The differential equation describing the temperature distribution in a thin
fin of uniform cross section thus becomes

$$\frac{d^2(t - t_\infty)}{dx^2} = M^2(t - t_\infty) \qquad (14\text{-}77)$$

where

$$M^2 = \frac{hP}{kA_c}\left[1 + \frac{Ci_{fg}}{c_{pa}}\right] \qquad (14\text{-}78)$$

When there is no condensation $C - 0$ and

$$M^2 = \frac{hP}{kA_c} = m^2 \qquad (14\text{-}78a)$$

The fin efficiency derived from the well-known solution of Eq. 14-77 where t_∞ and M are constants and the following boundary conditions are used:

$$x = 0 \qquad t = t_w \tag{14-79}$$

$$x = l \qquad \frac{dt}{dx} = 0$$

becomes

$$\eta_m = \frac{\tanh(Ml)}{(Ml)} \tag{14-80}$$

The approximation of Eq. 14-20 is also used here

$$\frac{hP}{kA_c} \approx \frac{2h}{ky} \tag{14-81}$$

Equation 14-80 is identical in form to the equation for the fin efficiency with no mass transfer (Eq. 14-19).

The solution may be applied to circular fins on a tube (Eq. 14-21), or to the case of plate-fin-tube heat transfer surfaces (Eqs. 14-23 and 14-24). Figure 14-11 may also be used with m replaced by M.

The surface effectiveness has the same form as Eq. 14-27

$$\eta_{ms} = 1 - \frac{A_f}{A}(1 - \eta_m) \tag{14-82}$$

The method presented is thought to be the most accurate available and is simple and straightforward to use. The method is readily adapted to the computer and is easy to use with hand calculations.

Transport Coefficients

The heat transfer and friction coefficients on the refrigerant side of the exchanger are determined by the methods discussed in Sec. 14-4. Chilled water and evaporating refrigerants are the usual cases.

The heat, mass, and friction coefficients on the air side of the exchanger should be obtained from correlations based on test data, if at all possible, since the analogy method is unreliable. The correlations of Figs. 14-23 and 14-24 as modified using $J(s)$, $J_i(s)$, and $F(s)$ (discussed at the beginning of this section), are recommended for plate-fin-tube coils. Other finned tube surfaces have similar behavior. For example, the dry-surface heat-transfer coefficients, for circular finned tubes in a staggered tube pattern, are well correlated by

$$j = 0.38 \, JP \tag{14-83}$$

and the friction factors are given by

$$f = 1.53 \, (FP)^2 \tag{14-84}$$

JP and *FP* may then be modified for a wet surface using Eqs. 14-58, 14-59, and 14-60. No information is available on the effect of tube rows on mass transfer coefficients; however, it should be similar to that for sensible heat transfer for a dry surface (Fig. 14-21).

Example 14-2. Estimate the heat, mass, and friction coefficients for a 4-row cooling coil that has the geometry of Fig. 14-20 with 12 fins per inch. The face velocity of the air is 600 ft/min and has an entering temperature of 80 F. The air leaves the coil at a temperature of 60 F. The air is at standard barometric pressure.

Solution. The correlations of Figs. 14-23 and 14-24 will be used with the parameters *JP*, *FP*, *J(s)*, and *F(s)* computed from Eqs. 14-46 through 14-60. The mass velocity is

$$G_c = \frac{60(600)14.7(144)}{0.54(53.35)540} = 4,900 \text{ lbm/(hr-ft}^2)$$

The Reynolds number based on tube diameter is then

$$\frac{G_c D}{\mu} = \frac{4900(0.525/12)}{0.044} = 4,870$$

where data from Fig. 14-20 and Table B-4 are used. To compute the parameter A/A_t, assume a 4-row coil, 1 ft in length, with 10 tubes in the face. This coil has a volume of

$$V = \frac{10(1.25)(12)4(1.083)}{1728} = 0.376 \text{ ft}^3$$

and a total of 40 tubes. The total outside surface area of the tubes is

$$A_t = 40 \pi D L = 40 \pi (0.525/12)(1) = 5.498 \text{ ft}^2$$

From Fig. 14-20, $A/V = \alpha = 238 \text{ ft}^2/\text{ft}^3$, then

$$\frac{A}{A_t} = \frac{238(0.376)}{5.498} = 16.28$$

Using Eq. 14-46

$$JP = (4870)^{-0.4}(16.28)^{-0.15} = 0.022$$

The Reynolds number based on the fin spacing is

$$\text{Re}_s = \text{Re}_D \left(\frac{s}{D}\right) = 4870 \left(\frac{0.0833}{0.525}\right) = 773$$

then

$$J(s) = 0.84 + (4 \times 10^{-5})(773)^{1.25} = 1.003$$

and

$$J_i(s) = [0.95 + (4 \times 10^{-5})(773)^{1.25}] \left[\frac{0.0833}{0.0833 - 0.006} \right]^2 = 1.293$$

then

$$JP \, J(s) = 0.0220(1.003) = 0.022$$

and

$$JP \, J_i(s) = 0.0220(1.293) = 0.028$$

Using Fig. 14-23, $j = 0.0071$ and $j_i = 0.0088$
    From Eq. 14-36

$$h = \frac{j \, G_c \, c_p}{Pr^{2/3}} = \frac{0.0071(4900)0.24}{(0.7)^{2/3}} = 10.5 \text{ Btu/(hr-ft}^2 \text{ F)}$$

and from Eq. 12-19a

$$h_d = \frac{j_i \, G_c}{Sc^{2/3}} = \frac{0.0088(4900)}{(0.6)^{2/3}} = 60.6 \text{ lbma/(ft}^2\text{-hr)}$$

The friction factor will be determined from Fig. 14-24.
Using Eqs. 14-51 and 14-52

$$D^* = \frac{0.525(16.28)}{1 + (1.25 - 0.525)/0.0833} = 0.881$$

and

$$FP = (4870)^{-0.25} \left(\frac{0.525}{0.881} \right)^{0.25} \left[\frac{1.25 - 0.525}{4(0.0833 - 0.006)} \right]^{-0.4}$$
$$\times \left[\frac{1.25}{0.881} - 1 \right] = 0.116$$

Using Eq. 14-60

$$F(s) = (1 + 773^{-0.4}) \left(\frac{0.0833}{0.0833 - 0.006} \right)^{1.5} = 1.20$$

Then

$$FP \, J(s) = 0.116(1.20) = 0.14$$

and from Fig. 14-24, $f = 0.031$.
    Note that the presence of moisture on the surface increases h, h_d, and f.

Figure 14-35. Flow diagram for heat-exchanger computer program.

Design Calculations

The design of a heat exchanger for simultaneous transfer of heat and mass is more complicated than for sensible heat transfer because the total energy transfer from the air has two components:

$$\dot{q} = \dot{m}_c c_{pc}(t_{co} - t_{ci}) = \dot{m}_a c_p(t_{ai} - t_{ao}) + \dot{m}_a i_{fg}(W_{ai} - W_{ao}) \quad (14\text{-}85)$$

The simplest case occurs when the terminal conditions are all known because

$$\dot{q} = \dot{m}_a(i_{ai} - i_{ao}) = \dot{m}_c c_{pc}(t_{co} - t_{ci}) \quad (14\text{-}86)$$

and

$$\dot{q}_s = \dot{m}_a c_p(t_{ai} - t_{ao}) = UA(\text{LMTD}) \quad (14\text{-}87)$$

When only inlet conditions are known, a double-iteration loop is required and hand calculations become impractical. The digital computer may be used to great advantage in this case. Figure 14-35 shows a flow diagram for a successful computer program.

The method discussed in Chapter 13 for the spray dehumidifier may be used to predict performance of a dehumidifying coil when hand calculations are necessary. One modification is necessary in Eq. 13-6, where the liquid film coefficient h_l must be replaced by an overall coefficient defined by

$$\frac{1}{UA} = \frac{1}{h_c A_c} + \frac{\Delta x}{k A_m} \quad (14\text{-}88)$$

The first term on the right is the refrigerant-side film resistance and the second term is the wall thermal resistance. Otherwise the procedure is identical; however, the results will be approximate.

The American Refrigeration Institute has a certification program for heating and cooling coils (19) that embodies a calculation procedure for dehumidifying coils. The standard is intended to standardize the rating of coils by setting up detailed procedures to calculate performance. Hand calculations are feasible through the use of charts and graphs. The method has many disadvantages when used for coil design, however.

REFERENCES

1. J. D. Parker, J. H. Boggs, and E. F. Blick, *Introduction to Fluid Mechanics and Heat Transfer,* Addison-Wesley, Reading, Mass., 1969.

2. T. E. Schmidt, "La Production Calorifique des Surfaces Munies D'ail-ettes," *Annexe Du Bulletin De L'Institut International Du Froid,* Annexe G-5, 1945–46.

3. H. Zabronsky, "Temperature Distribution and Efficiency of a Heat Exchanger Using Square Fins on Round Tubes," *ASME Transactions, Journal of Applied Mechanics,* Vol. 77, December 1955.

4. W. H. Carrier and S. W. Anderson, "The Resistance to Heat Flow Through Finned Tubing," *Heating, Piping, and Air Conditioning*, May 1944.

5. *ASHRAE Handbook, Fundamentals Volume,* American Society of Heating, Refrigerating and Air-Conditioning Engineers, Atlanta, Ga., 1985.

6. D. G. Rich, "The Efficiency and Thermal Resistance of Annular and Rectangular Fins," Presented at the 1966 International Heat Transfer Conference.

7. W. M. Kays and A. L. London, *Compact Heat Exchangers,* 2nd ed., McGraw-Hill, San Francisco, 1964.

8. W. M. Kays and M. E. Crawford, *Convective Heat and Mass Transfer,* 2nd ed., McGraw-Hill, San Francisco, 1980.

9. E. M. Sparrow, A. L. Loeffler, Jr., and H. A. Hubbard, *Transactions ASME, Journal of Heat Transfer,* November 1961.

10. R. G. Diessler and M. F. Taylor, *NACA TN 2351,* Washington, D.C., 1955.

11. A. Devore, *Petroleum Refiner,* Vol. 40, No. 5, May 1961 and Vol. 41, No. 12, December 1962.

12. K. J. Bell, "Bulletin No. 5," University of Delaware Engineering Experimental Station, 1963.

13. J. W. Palen and Jerry Taborek, "Solution of Shell Side Flow Pressure Drop and Heat Transfer by Stream Analysis Method," Chemical Engineering Progress Symposium Series, Vol. 65, No. 92, 1969.

14. Donald G. Rich, "The Effect of the Number of Tube Rows on Heat Transfer Performance of Smooth Plate Fin-and-Tube Heat Exchangers," *ASHRAE Transactions,* Vol. 81, Part I, 1975.

15. D. G. Shepherd, "Performance of One-Row Tube Coils with Thin-Plate Fins, Low Velocity Forced Connection," *Heating, Piping, and Air Conditioning,* April 1956.

16. F. C. McQuiston, "Heat Transfer and Flow Friction Data for Two Fin-Tube Surfaces," *ASME Journal of Heat Transfer,* Vol. 93, May 1971.

17. J. L. Guillory and F. C. McQuiston, "An Experimental Investigation of Air Dehumidification in a Parallel Plate Exchanger," *ASHRAE Transactions,* Vol. 29, June 1973.

18. F. C. McQuiston, "Heat, Mass, and Momentum Transfer in a Parallel Plate Dehumidifying Exchanger," *ASHRAE Transactions,* Vol. 82, Part II, 1976.

19. *ARI Standard 410-72,* Air Conditioning and Refrigeration Institute, Arlington, Va., 1972.

20. J. L. Threlkeld, *Thermal Environmental Engineering*, 2nd ed., Prentice-Hall, Englewood Cliffs, N.J., 1970.

21. C. D. Ware and T. H. Hacha, "Heat Transfer From Humid Air to Fin and Tube Extended Surface Cooling Coils," ASME Paper, No. 60-HT-17, August 1960.

22. F. C. McQuiston, "Fin Efficiency with Combined Heat and Mass Transfer," *ASHRAE Transactions,* Vol. 81, Part I, 1975.

23. D. G. Rich, "The Effect of Fin Spacing on the Heat Transfer and Friction Performance of Multirow Smooth Plate Fin-and-Tube Heat Exchangers," *ASHRAE Transactions,* Vol. 73, June 1973.

24. R. W. Eckels, "Contact Conductance of Mechanically Expanded Plate Finned Tube Heat Exchangers," Westinghouse Research Laboratories, Scientific Paper No. 77-1E9-SURCO-P1, March 1977.

25. F. C. McQuiston, "Heat, Mass, and Momentum Transfer Data for Five Plate-Fin-Tube Heat Transfer Surfaces," *ASHRAE Transactions,* Vol. 84, Part 1, 1978.

26. F.C. McQuiston, "Correlation of Heat, Mass, and Momentum Transport Coefficients for Plate-Fin-Tube Heat Transfer Surfaces," *ASHRAE Transactions,* Vol. 84, Part 1, 1978.

27. R. W. Lockhart, and Martinelli, R. D., "Proposed Correlation of Data for Isothermal, Two-Phase, Two-Component Flow in Pipes," *Chemical Engineering Progress,* Vol. 45, No. 39, 1949.

28. K. Ishihara, J. W. Palen, and J. Taborek, "Critical Review of Correlations for Predicting Two-Phase Flow Pressure Drop Across Tube Banks," *Heat Transfer Engineering,* Vol. 1, No. 3, Jan.–Mar. 1980.

PROBLEMS

14-1 Five thousand cfm of standard air (60 F, 29.92 in. mercury) are heated to 120 F in a cross-flow heat exchanger. The heating medium is water at 200 F, which is cooled to 180 F. Determine (a) the LMTD and correction factor F, (b) the fluid capacity rates C_c and C_h, (c) the flow rate of the water in gpm, (d) the overall conductance UA, (e) the NTU based on C_{min}, and (f) the effectiveness ε.

14-2 One m³/s of standard air is heated to 45 C in a cross-flow exchanger. Water is the hot fluid and is cooled from 90 to 85 C. Find (a) the LMTD and correction factor F, (b) the fluid capacity rates C_c and C_h, (c) the flow rate of the water in m³/s, (d) the overall conductance UA, (e) the NTU based on C_{min}, and (f) the effectiveness.

14-3 A heat exchanger is to be designed to heat 4000 cfm (1.9 m³/s) of air from 50 F (10 C) to 110 F (43 C) using hot water at 180 F (82 C) in a cross-flow arrangement.

The flow rate of the hot water is 25 gpm (1.6 L/s). Assume the overall heat-transfer coefficient based on the air side is 10 Btu/(hr-ft²F) [57 W/(m²-C)] and determine the air-side surface area using (a) the LMTD method and (b) the NTU method.

14-4 An air-cooled condenser operates in cross-flow with a constant refrigerant temperature of 125 F (52 C). The air enters at 95 F (35 C) at a flow rate of 3200 cfm (1.5 m³/s). The air-side surface area is 300 ft² (28 m²) and the overall heat-transfer coefficient is 10 Btu/(hr-ft²-F) [57 W/(m²-C)]. (a) What is the temperature of the air leaving the condenser? (b) What is the flow rate for refrigerant 22 (R-22), assuming a change in state from saturated vapor to saturated liquid?

14-5 A tube has circular fins as shown in Fig. 14-10; R is 1.0 in., whereas r is 0.5 in. The fin thickness y is 0.008 in. and the fin pitch is 10 fins/in. The average heat-transfer coefficient for the fin and tube is 10 Btu/(hr-ft²-F), whereas the thermal conductivity k of the fin and tube material is 90 (Btu-ft)/(ft²-hr-F). Determine (a) the fin efficiency η using Fig. 14-11, (b) the fin efficiency using Eq. 14-21, and (c) compare (a) and (b).

14-6 Find the surface effectiveness for the finned tube in Problem 14-5 assuming the ratio of fin area to total area is 0.9.

14-7 The finned tube of Problem 14-6 has a refrigerant flowing on the inside with an average heat-transfer coefficient of 200 Btu/(hr-ft²-F). The tube wall has a thickness of 0.015 in. Compute the overall heat-transfer coefficient based on the air-side surface area. Assume the ratio of air-side to inside surface area is 9.

14-8 Water flows through the finned tube in Problem 14-6. Find the overall heat-transfer coefficient assuming a water heat-transfer coefficient of 1100 Btu/(hr-ft²-F).

14-9 A plate-fin heat exchanger has fins that are 6 mm high, 0.16 m thick, with a fin pitch of 0.47 fins/mm. The average heat-transfer coefficient is 57 W/(m²-C) and the thermal conductivity of the fin material is 173 W/(m-C). Find the fin efficiency.

14-10 Assume that the ratio of fin area to total area in Problem 14-9 is 0.85. Find the surface effectiveness.

14-11 Refrigerant 12 flows on the opposite side of the surface in Problem 14-9. Determine the overall heat-transfer coefficient if the refrigerant heat-transfer coefficient is 1.4 kW/(m²-C). Neglect the thermal resistance of the plate.

14-12 Find the fin efficiency for a plate-fin-tube surface like that shown in Fig. 14-14 using the data shown in Table 14-2.

Table 14-2   DATA FOR PROBLEM 14-12

	a	b	y	D	h	k
(a)	1.12 in.	1.35 in.	0.010 in.	0.64 in.	10 Btu/(hr-ft²-F)	90 (Btu-ft)/(hr-ft²-F)
(b)	25 mm	22 mm	0.18 mm	10 mm	68 W/(m²-C)	170 (W-m)/(m²-C)

14-13 Assume water is flowing in the finned tubes of Problem 14-12 with a heat-transfer coefficient of 600 Btu/(hr-ft²-F) [3.4 kW/(m²-C)] and compute the overall heat-transfer coefficient. Assume the ratio of fin area to total air-side area is 0.9 and the ratio of air-side area to inside area is 10.

14-14 Compute the contact resistance for the surface of Problem 14-12. Assume a fin pitch of 12 fins per in.

14-15 Determine the average heat-transfer coefficient for water flowing in a ½-in. OD copper tube that has a wall 0.016-in. thick. The water flows at a rate of 2½ gpm and undergoes a temperature change from 180 to 150 F. The tube is over 4 ft. long.

14-16 Repeat Problem 14-15 for a 30 percent ethylene glycol solution.

14-17 Rework Problem 14-15 for 60 percent ethylene glycol solution.

14-18 Estimate the heat-transfer coefficient for water flowing with an average velocity of 4 ft/sec in a long rectangular 3 × 1 in. channel. The bulk temperature of the water is 45 F. Assume (a) cooling and (b) heating.

14-19 Repeat Problem 14-18 for 30 percent ethylene glycol instead of water.

14-20 Coolant flows through a long 12-mm ID tube with an average velocity of 1.5 m/s. The coolant undergoes a temperature change from 40 to 50 C. Compute the average heat-transfer coefficient for (a) water and (b) 30 percent ethylene glycol solution.

14-21 Water flows through a 0.34-in. ID tube with an average velocity of ½ ft/sec. The mean bulk temperature of the water is 45 F and the length of the tube is 10 ft. Estimate the average heat-transfer coefficient for (a) water and (b) ethylene glycol (30 percent solution).

14-22 Estimate the average heat-transfer coefficient for water flowing in 10-mm ID tubes at a velocity of 0.10 m/s. The length of the tubes is 3 m and the mean bulk temperature of the water is 40 C.

14-23 Estimate the average heat-transfer coefficient for condensing water vapor in ⅝-in. OD tubes that have walls 0.018 in. thick. The vapor is saturated at 5 psia and flows at the rate of 1 lbm/hr at the tube inlet. The quality of the vapor is 10 percent when it exits from the tube.

14-24 Water vapor condenses in 15-mm ID tubes. The flow rate is 0.126×10^{-3} kg/s per tube at an absolute pressure of 35 kPa. The vapor leaving the tubes has a quality of 12 percent. Estimate the average heat-transfer coefficient.

14-25 Refrigerant 22 enters the ⅜-in. OD tubes of an evaporator at the rate of 80 lbm/hr per tube. The refrigerant enters at 70 psia with a quality of 20 percent and leaves with 10 F of superheat. The effective tube length is 5 ft. Estimate the average heat-transfer coefficient.

14-26 Estimate the average heat-transfer coefficient for evaporating refrigerant 22 in 8.5-mm ID tubes that are 2 m in length. The mass velocity of the refrigerant is 200 kg/(s-m²) and the pressure and quality at the inlet are 210 kPa and 30 percent, respectively. Saturated vapor exits from the tubes.

14-27   Refer to Problem 14-15 and compute the lost head for a circuit that consists of 6 tubes, 6 ft in length connected by 5 U-bends. The equivalent length of each U-bend is 1 ft.

14-28   Refer to Problem 14-21 and compute the lost head for a circuit made up of 10 tubes, 10 ft in length, connected by 9 U-bends. The U-bends have an equivalent length of 1.5 each.

14-29   A 10-ton cooling coil uses volatile refrigerant on the tube side. There are 10 tube circuits made up of 6 tubes, $\frac{1}{2}$ in. in diameter, connected by 5 U-bends. The evaporating temperature is 30 F. Estimate the pressure loss for each circuit for (a) refrigerant 12 and (b) refrigerant 22.

14-30   For the surface shown in Fig. 14-20 with 6.7 fins per in., the air mass velocity G_{fr} is 1800 lbm/(hr-ft^2) and the air is heated from 70 to 120 F. Compute the heat-transfer coefficient and friction factor using (a) Fig. 14-20, (b) Figs. 14-23 and 14-24.

14-31   For the surface of Fig. 14-20, with 0.263 fins per mm, the air mass velocity is 4.5 kg/(m^2-s) and the mean bulk temperature is 20 C. Estimate the heat-transfer coefficient and friction factor using (a) Fig. 14-20, (b) Figs. 14-23 and 14-24.

14-32   Compute the lost pressure for Problem 14-30b assuming standard atmospheric pressure at the inlet.

14-33   Compute the lost pressure for Problem 14-31b assuming standard atmospheric pressure at the inlet.

14-34   Find the heat-transfer coefficient and friction factor for the surface of Fig. 14-25 assuming a mass velocity of 2700 lbm/(hr-ft^2). The air enters at 75 F and leaves at 55 F.

14-35   Compute the lost pressure for Problem 14-34 assuming a pressure of 14.6 lbf/in.^{2} at the inlet and flow length of 4 in.

14-36   Design a condenser for the following given conditions:
Tube side: refrigerant 22, 2-phase, t = 120 F, h_i = 200 Btu/(hr-ft^2-F)
Finned side: t_{in} = 100 F, heat transfer rate = 40,000 Btu/hr, maximum head loss = 0.3 in. H$_2$O
Construction: all aluminum, k = 100 Btu/(hr-ft^2-F).
Use the surface of Fig. 14-20 with 8 fins per in. (a) Carry out the design by hand calculations. (b) Use a computer program to find a design.

14-37   Design a hot water heating coil for a capacity of 200,000 Btu/hr. Use the surface of Fig. 14-20 with two rows of tubes and 6.7 fins per in. Air enters the coil at 70 F and water enters at 150 F. Sketch the circuiting arrangement for the tube side. Compute the head loss for both the water and air. Make the width of the exchanger about twice the height. (a) Use hand calculation method. (b) Use a computer program.

14-38   Air enters a fin-tube heat exchanger described by Fig. 14-20 (6.7 fins per in.) at the rate of 2000 cfm at 14.696 psia and with a face velocity of 1000 ft/min. The entering and leaving air temperatures are 40 and 110 F, respectively. Hot water flow through the tubes with an average velocity of 3.5 ft/sec, entering at 200 F and leaving at 190 F. Assume counterflow and use the NTU method or a computer program to determine (a) the heat-transfer area required and (b) the heat-exchanger dimensions and the number of tube rows in the air flow direction. (c) Compute the lost head on the air side in in. H_2O, and (d) select circuiting for the tubes and compute the head loss for the water side.

14-39 The surface of Problem 14-30 has 8 rows of tubes. Compute the heat-transfer coefficient.

14-40 The surface of Problem 14-31 has 6 rows of tubes. Compute the heat-transfer coefficient.

14-41 The surface of Fig. 14-20 is to be sized for a condenser in an air-conditioning unit. The condenser must reject 4.4 kW. Assume refrigerant 22 enters the condenser at 51 C as saturated vapor; it leaves as saturated liquid at the same temperature. Air enters the condenser at 35 C. Use a 5-row coil with aluminum fins with 0.571 fins per mm. Compute the air-side head loss in Pa. Use hand or computer methods to solve.

14-42 Compute the surface effectiveness for the surface of Fig. 14-25 under dehumidifying conditions. Moist air enters at 78 F dry bulb and 67 F wet bulb. Assume a base plate temperature of 47 F. The face velocity of the entering air is 600 ft/min at 14.7 psia.

14-43 Compute the surface effectiveness for the surface of Fig. 14-20 with a fin pitch of 11.7 fins per inch. The face velocity of the entering air is 500 ft/min, 85 F dry bulb and 70 F wet bulb. The surface temperatures where the air enters and leaves are 55 F and 45 F, respectively.

14-44 Air enters a finned coil at 80 F db, 67 F wb, and standard atmospheric pressure. The chilled water has a temperature of 50 F at this position in the coil. Will moisture condense from the air for the following assumed conditions? Copper tubes with wall thickness of 0.018 in. and thermal conductivity of 190 Btu/ (ft-hr-F). The heat-transfer coefficient on the inside of the tubes is 1000 Btu/ (hr-ft²-F). The product of total heat-transfer coefficient and surface effectiveness $(h_a\eta_o)$ is 60 lbma/(ft²-hr) and the ratio of total surface area to inside tube surface area is 12.

14-45 Consider a cooling coil with air entering at 27 C db, 19 C wb, and standard atmospheric pressure. The coil is all-aluminum construction with tube wall thickness of 0.50 mm and thermal conductivity of 58 W/(m-C). The heat-transfer coefficient on the inside of the tubes is 53 W/(m²-C). The product of total heat-transfer coefficient and surface effectiveness $(h_a\eta_o)$ is 2.5 kg/(m²-hr). The ratio of total surface area to inside tube surface area is 14. Will condensation occur on the surface at this location if the chilled water has a temperature of 14.3 C at this location in the coil?

14-46 A heat exchanger surface like Fig. 14-20 has a face area of 10 ft², 5 rows of tubes, and aluminum fins (11.7 fins/in.). Moist air enters the exchanger at 80 F dry bulb and 68 F wet bulb, at standard atmospheric pressure and with a face velocity of 650 ft/min. Use a coil simulation program to compute the (a) leaving air conditions, (b) total heat-transfer rate, (c) sensible heat-transfer rate, and (d) head loss for the air. The refrigerant is water that undergoes a temperature change from 45 to 50 F in a counterflow arrangement.

14-47 A heat exchanger made like the surface of Fig. 14-20 has a face area of 1 m² and 4 rows of tubes and aluminum fins (0.461 fins/mm). Air enters the coil at 26 C dry bulb and 20 C wet bulb with a face velocity of 3.5 m/s and a pressure of 100 kPa absolute pressure. Use a coil simulation program to determine the (a) leaving air conditions, (b) total heat-transfer rate, (c) sensible heat-transfer rate, (d) head loss for the air. The refrigerant is water that undergoes a change in temperature from 7 to 10 C in counterflow.

14-48 Suppose the coil of Example 14-2 had 6 rows of tubes instead of 4 and estimate the heat, mass, and friction coefficients.

14-49 Use the method described in Chapter 13 for the spray dehumidifier to solve Problem 14-46.

14-50 Use the method described in Chapter 13 for the spray dehumidifier to solve Problem 14-47.

14-51 Rework Problem 14-46 assuming barometric pressure at 5000 ft elevation.

CHAPTER FIFTEEN

Refrigeration

The refrigerator or heat pump as it is sometimes called is basic to the heating, ventilating, and air-conditioning industry. Applications vary from food processing, packaging, storing, and transportation to environmental control and comfort, the latter being of greatest interest in this text. There are basically three refrigeration methods:

1. Vapor compression
2. Absorption
3. Thermoelectric

The first two are by far the most common. Although these systems have many complex variations, only the basic cycles will be discussed in this chapter. References 3, 4, and 5 should be consulted for a more in-depth treatment.

The performance of a refrigerating system is expressed in terms of the *coefficient of performance* or *COP*.

$$\text{COP} = \frac{\text{Useful refrigerating effect}}{\text{Net energy input}} \tag{15-1}$$

The *refrigerating efficiency* η_R is the ratio of the coefficient of performance of a cycle or system to that of an ideal cycle or system:

$$\eta_R = \frac{COP}{(COP)_{\text{ideal}}} \tag{15-2}$$

The capacity of a refrigeration cycle can be measured in any one of several units such as Btu/hr or kW. A common unit of capacity in the United States is the *ton of refrigeration,* equal to 200 Btu/min or 12,000 Btu/hr of useful heat removal. The term is a carryover from the days when ice was used to cool. The latent enthalpy of fusion of ice is 144 Btu/lbm. One ton of ice melting per day requires the transfer of about 200 Btu/min of heat from the cold space.

The power input required in a gas or vapor compression system for a given refrigerating effect depends on the coefficient of performance. In terms of horsepower per ton the expression is

$$\text{hp/ton} = \frac{(12{,}000)\ \text{Btu/ton-hour}}{(2545)\ \text{Btu/hp-hour (COP)}} = \frac{4.72}{\text{COP}} \qquad (15\text{-}3)$$

15-1   THE CARNOT REFRIGERATION CYCLE

The reversible cycle is a useful thermodynamic concept. Two important characteristics of a reversible cycle are as follows:

1. No refrigeration cycle can have a higher coefficient of performance than that of a reversible cycle operating between the same source and sink temperatures.
2. All reversible refrigeration cycles operated between the same source and sink temperatures have identical coefficients of performance.

The most convenient reversible cycle to use as an ideal refrigerating cycle is the *Carnot refrigeration cycle* consisting of two isothermal processes and two adiabatic processes. Such a cycle is shown in Fig. 15-1 on temperature-entropy coordinates. Because of characteristic 2 listed before, no particular working medium needs to be specified.

During the isothermal process 2–3, heat is received by the working fluid (refrigerant) at the evaporator temperature T_e. At the end of the isothermal process the refrigerant is then compressed adiabatically. Because a reversible adiabatic process is also isentropic, the process 3–4 appears as a vertical line on the $T - s$ coordinates. The temperature of the refrigerant increases as the adiabatic process proceeds. By this work process the refrigerant temperature is increased from T_3 to T_4, enabling the refrigerant to give up heat to the environment or to some coolant. This cooling of the refrigerant occurs at the constant condenser temperature T_c shown as process 4–1. To complete the cycle the refrigerant is brought back to its initial condition (state 2) by an adiabatic expansion from state 1 to state 2.

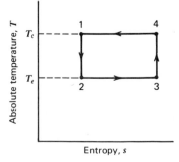

Entropy, s **Figure 15-1.**   The Carnot refrigeration cycle.

The heat absorbed by the refrigerant during the isothermal process 2–3 is

$$q_{23} = T_e(s_3 - s_2)$$

The usual thermodynamic convention is followed in which heat added to the system (the refrigerant) is positive.

The heat given up by the refrigerant during the isothermal process 4–1 is

$$q_{41} = T_c(s_1 - s_4) = T_c(s_2 - s_3) = -T_c(s_3 - s_2)$$

The net work done on the cycle is given by the first law of thermodynamics, using the convention that work done by the system is positive.

$$w_{net} = q_{net} = q_{23} + q_{41} = +T_e(s_3 - s_2) - T_c(s_3 - s_2)$$

From Eq. 15-1 the COP is

$$\text{COP} = \frac{T_e(s_3 - s_2)}{-[T_e(s_3 - s_2) - T_c(s_3 - s_2)]} = \frac{T_e}{T_c - T_e} \qquad (15\text{-}4)$$

If we attempt to actually build a working Carnot cycle, we would immediately be confronted with some serious problems. The isothermal processes, for example, might be accomplished by having a change of phase occur in the refrigerant. Three possibilities are shown in Fig. 15-2.

In case (a) the evaporation of the refrigerant occurs between points 2 and 3 and condensation occurs between points 4 and 1. In attempting to stop the evaporation at point 3 there would be difficulty in sensing the proper condition at exit since no easily measurable property (such as temperature or pressure) is changing along process 2–3. In addition, the adiabatic compression 3–4 would be complicated by the presence of moisture at the inlet. The two-phase expansion that occurs in process 1–2 is also very difficult to accomplish in a real expander.

In case (b) point 3 is easily sensed because the temperature begins to rise in a constant pressure process as soon as the saturated vapor condition is reached. The adiabatic compression 3–4 is also easily accomplished (at least

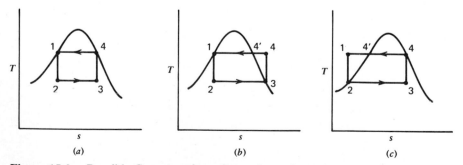

Figure 15-2.   Possible Carnot cycles using a phase change to accomplish part or all of the isothermal processes.

approximately), and no liquid will enter the compressor. The isothermal compression 4–4, however, requires an extra compressor and is difficult to approach in real equipment with significant capacity. The process 1–2 is also impractical, as was discussed in the previous case.

In case (c) point 1 is at a very high pressure relative to the pressure at 4' and would be impractical for that reason.

No practical heat engine or refrigeration device operating on the Carnot cycle has been developed, and it appears best simply to use the concept as a standard of perfection and to design the real cycle to approach this ideal as closely, as practical.

15-2   THE THEORETICAL SINGLE-STAGE CYCLE

To meet practical considerations in the adiabatic expansion and compression processes, we may replace the expander by a simple throttling valve and carry out the adiabatic compression all the way to the condenser pressure. Such a modified cycle is described in Fig. 15-3.

Refrigerant entering the compressor is assumed to be dry saturated vapor at the evaporator pressure. This is a convenient place to begin an analysis because we can easily determine all fluid properties here. The compression process 3–4 is assumed to be reversible and adiabatic and, therefore, isentropic, and is continued until the condenser pressure is reached. Point 4 is obviously

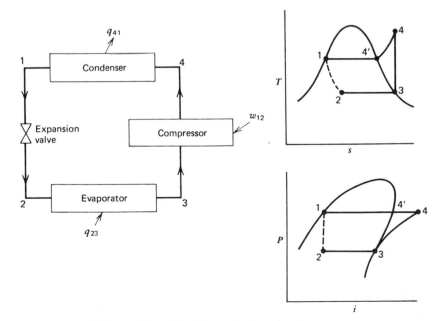

Figure 15-3.   The theoretical single-stage cycle.

in the superheated vapor region. The process 4–1 is carried out at constant pressure with the temperature of the vapor decreasing until the saturated vapor condition is reached at 4′; then the process is both at constant temperature and constant pressure during the condensation from 4′ to 1. At point 1 the refrigerant leaves the condenser as a saturated liquid. It is then expanded through a throttling valve where partial evaporation occurs as the pressure drops across the valve. The throttling process 1–2 is irreversible with an increase in entropy occurring. For this reason the process is shown as a dashed line in Fig. 15-3. For a throttling process the enthalpy at exit and at inlet are equal. To determine the coefficient of performance of this cycle the useful refrigerating effect and the net energy input must be determined. For steady flow of one pound of refrigerant:

$$i_3 - i_2 = i_3 - i_1$$

$$\text{Useful refrigerating effect} = q_{23} = i_3 - i_2 = i_3 - i_1$$

$$\text{Net energy input} = w_{34} = i_4 - i_3$$

$$\text{COP} = \frac{i_3 - i_1}{i_4 - i_3}$$

Example 15-1.   For a theoretical single, stage cycle operating with a condenser pressure of 250 psia and an evaporator pressure of 40 psia and using refrigerant 22 as the working fluid determine the (a) COP, (b) temperature at exit to compressor, (c) temperature in condenser, and (d) temperature in evaporator.

Solution.   Use the pressure-enthalpy diagram for refrigerant 22, Chart 5.

(a) $\text{COP} = \dfrac{i_3 - i_1}{i_4 - i_3} = \dfrac{104.6 - 43.1}{125.4 - 104.6} = \dfrac{61.5}{20.8} = 2.96$

(b) $t_4 = 168\ \text{F}$
(c) $t_1 = 116\ \text{F}$
(d) $t_2 = 2\ \text{F}$

The coefficient of performance is closely related to the evaporating and condensing temperatures. For the highest COP the cycle should be operated at the lowest possible condensing temperature and the highest possible evaporating temperature.

In the theoretical single-stage cycle no work is produced in the expansion process 1–2, in contrast to the adiabatic expansion in the Carnot cycle. This is lost work that cannot be put back into the system to drive the compressor; therefore, additional work is needed for the compression process in the actual cycle. In addition, extra work is required in the adiabatic compression because the enthalpy at exit from the compressor at point 4 is higher than in the Carnot cycle. On a $T - s$ diagram the area under a reversible constant-pressure process represents the change of enthalpy for that process. In Fig. 15-4, for example, the area under the constant-pressure process $4–4'–1–1'$ is equal to the change of enthalpy $(i_4 - i_{1'})$. The process $1–1'$ can be assumed to be along the saturated liquid line if subcooling of point $1'$ is not extremely large. The change of enthalpy $(i_3 - i_{1'})$ is equal to the area under the line $1'–3$. Thus, because

$$w_{34} = i_4 - i_3 = (i_4 - i_{1'}) - (i_3 - i_{1'})$$

then the enclosed area $3–4–4'–1–1'–2'–2–3$ is equal to the compressor work required per unit mass of refrigerant flow in the theoretical single-stage cycle. The additional net work above that required for the equivalent Carnot cycle w_c is equal to the sum of the two areas A_1 and A_2.

$$w_{34} = w_c + A_1 + A_2$$

The throttling process $1–2$ reduces the amount of refrigerating effect possible since point 2 is shifted to the right by the irreversible process. This loss of refrigerating capability is given by area A_3. The refrigerating effect of the theoretical cycle q_{41} is given by

$$q_{2'3} - A_3 = q_c - A_3$$

where q_c is the heat flow into the evaporator of the Carnot cycle.

The refrigerating efficiency of the theoretical single-stage cycle can be expressed in terms of the three areas described earlier and the evaporator heat flow q_c and net work w_c of an equivalent Carnot cycle.

$$\eta_R = \frac{\text{COP}}{\text{COP}_c} = \frac{(q_c - A_3)/(w_c + A_1 + A_2)}{q_c/w_c} = \frac{1 - A_3/q_c}{1 + (A_1 + A_2)/w_c}$$

The magnitude of the areas A_1 and A_2, and A_3 depends on the evaporating and condensing temperatures and on the particular refrigerant used. The vari-

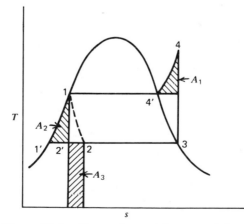

Figure 15-4.   Comparison of Carnot and theoretical cycles.

ation of η_R among the more common refrigerants is not large. Figure 15-5 shows the variation of refrigerating efficiency with evaporating temperature for several common refrigerants at a fixed condensing temperature.

Example 15-2. Compute the COP and refrigerating efficiency for a theoretical single-stage cycle operating with a condenser pressure of 1219 kPa and an evaporator pressure of 204 kPa using R-12 as the refrigerant. The work for the cycle is 31 kJ/kg of refrigerant.

Solution. From Fig. 15-3, the COP may be expressed as

$$COP = \frac{i_3 - i_1}{i_4 - i_3} = \frac{i_3 - i_1}{w_{34}}$$

And from Table A-3a

$$i_3 = 248.9 \text{ kJ/kg} \qquad i_1 = 346.3 \text{ kJ/kg}$$

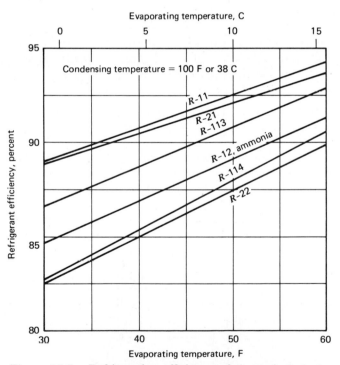

Figure 15-5. Refrigerating efficiency of theoretical single-stage cycle for various refrigerants. (James L. Threlkeld, *Thermal Environmental Engineering,* 2nd ed., 1970, p. 53, reprinted by permission of Prentice-Hall, Inc., Englewood Cliffs, N.J.)

Then

$$COP = \frac{346.3 - 248.9}{31} = 3.14$$

A Carnot cycle operating between the saturation temperatures corresponding to the given condenser and evaporator pressures has a COP given by Eq. 15-4 where from Table A-3a

$$T_e = -12 + 273 = 261 \text{ K}$$
$$T_c = 50 + 273 = 323 \text{ K}$$
$$COP_c = \frac{261}{323 - 261} = 4.2$$

The refrigerating efficiency is then

$$\eta_R = \frac{COP}{COP_d} = \frac{3.14}{4.2} = 0.75 \text{ or } 75 \text{ percent}$$

15-3   REFRIGERANTS

The working medium used in refrigeration systems is called the *refrigerant*. A refrigerant's suitability for a given application depends on many factors including its cost, and its physical, thermodynamic, and chemical properties. In this book we shall discuss three classes of compounds used as refrigerants. They are the halocarbons, the hydrocarbons, and the inorganic compounds. Table 15-1 shows some of the more common refrigerants in each of these classes. The ASHRAE designation system has been used in assigning a number to each refrigerant. In this system the halocarbons and hydrocarbons are designated as follows:

1.   The first digit on the right is the number of fluorine (F) atoms in a molecule.
2.   The second digit from the right is one more than the number of hydrogen (H) atoms in a molecule.
3.   The third digit from the right is one less than the number of carbon (C) atoms in a molecule. If this is zero, then the number is omitted.

Inorganic refrigerants are designated by adding 700 to the molecular weight of the compound. Thus, water has a molecular weight of 18 and its ASHRAE designation as a refrigerant is 718. This would commonly be referred to as R-718.

Example 15-3.   Determine the ASHRAE number designation for dichlorotetrafluoroethane, $CClF_2CClF_2$.

Solution.   There are 4 fluorine atoms, no hydrogen atoms, and 2 carbon atoms per molecule. Thus the ASHRAE designation is

Table 15-1 PROPERTIES OF COMMON REFRIGERANTS

Number	Chemical Name	Chemical Formula	Molecular Weight	Boiling Point at 1 atm		Freezing Temperature		Critical Temperature			Critical Pressure $\times 10^{-6}$ Pa	
				F	K	F	K	F	K	psia		ATM
	Halocarbons											
11	Trichloromono-fluoromethane	CCl_3	137.4	74.9	297.0	−168.0	162.0	388.4	471.2	639.5	4.409	43.52
12	Dichlorodifluoro-methane	CCl_2F_2	120.9	−21.6	243.4	−252.0	115.4	233.6	385.2	596.9	4.116	40.62
13	Monochlorotri-fluoromethane	$CClF_3$	104.5	−114.6	191.7	−294.0	92.0	83.9	302.0	561.0	3.868	38.17
14	Carbontetra-fluoride	CF_4	88.0	−198.3	145.2	−299.0	89.3	−50.2	227.5	543.0	3.737	36.88
21	Dichloromono-fluoromethane	$CHCl_2F$	102.9	47.8	282.0	−211.0	138.2	353.3	451.7	750.0	5.171	51.03
22	Monochlorodi-fluoromethane	$CHClF_2$	86.5	−41.4	232.4	−256.0	113.2	204.8	369.2	721.9	4.978	49.12
113	Trichlorotri-fluoromethane	CCl_2FCClF_2	187.4	117.6	320.7	−31.0	238.2	417.4	487.3	498.9	3.413	33.68
114	Dichlorotetra-fluoroethane	$CClF_2CClF_2$	170.9	38.8	277.0	−137.0	179.3	294.3	418.9	473.0	3.268	32.25
502	Refrigerant 22/115(48.8/51.2)	$CHClF_2/CClF_2CF_3$	111.63	−49.8	227.7	—	—	179.9	353.3	591.0	4.076	40.2
	Hydrocarbons											
50	Methane	CH_4	16.0	−258.7	111.7	−296.0	90.9	−116.5	190.7	673.1	4.641	45.80
170	Ethane	CH_3CH_3	30.0	−127.9	184.3	−297.0	90.4	90.0	305.4	709.8	4.894	48.30
290	Propane	$CH_3CH_2CH_3$	44.1	−43.7	231.1	−305.8	85.5	206.3	370.0	617.4	4.257	42.01
	Inorganic Compounds											
717	Ammonia	NH_3	17.0	−28.0	239.8	−107.9	195.4	271.4	406.2	1657.0	—	112.8
718	Water	H_2O	18.0	212.0	373.2	32.0	273.2	705.6	647.4	3208.0	—	219.5
729	Air	—	29.0	−317.8	78.8	—	—	−221.3	132.4	547.4	—	37.22
744	Carbon dioxide	CO_2	44.0	−109.2	194.7	−69.9	216.5	87.9	304.2	1070.0	—	72.88
764	Sulfur dioxide	SO_2	64.0	14.0	263.2	−103.9	197.7	315.5	430.7	1143.0	—	77.67

Source: (Abridged by permission from *ASHRAE Handbook, Fundamentals Volume,* 1985.)

$$(2 - 1)(0 + 1)(4) = 114$$

This is the designation listed in Table 15-1.

It is desirable that a refrigerant have the characteristics listed next. The significance or relative importance of each characteristic varies from one application to the next, and there is no such thing as an ideal refrigerant for all applications.

Thermodynamic Characteristics

1. *High latent enthalpy of vaporization.* This means a large refrigerating effect per unit mass of the refrigerant circulated. In small-capacity systems too low a flow rate may actually lead to problems.
2. *Low freezing temperature.* The refrigerant must not solidify during normal operating conditions.
3. *Relatively high critical temperatures.* Large amounts of power would otherwise be required for compression.
4. *Positive evaporating pressure.* Pressure in the evaporator should be above atmospheric to prevent air from leaking into the system.
5. *Relatively low condensing pressure.* Otherwise expensive piping and equipment will be required.

Physical and Chemical Characteristics

1. *High dielectric strength of vapor.* This permits the use in hermetically sealed compressors where vapor may come in contact with motor windings.
2. *Good heat-transfer characteristics.* Thermophysical properties should be such that high heat-transfer coefficients can be obtained. This covers properties such as density, specific heat, thermal conductivity, and viscosity.
3. *Satisfactory oil solubility.* Oil may be dissolved in the refrigerant or refrigerant in the oil. This can affect lubrication and heat transfer characteristics and lead to oil logging in the evaporator. A system must be designed with the oil solubility characteristics in mind.
4. *Low water solubility.* Water in a refrigerant can lead either to freeze-up in the expansion devices or corrosion.
5. *Inertness and stability.* The refrigerant must not react with materials that will contact it and its own chemical makeup must not change with time.

Safety

1. *Nonflammability.* The refrigerant should not burn or support combustion when mixed with air.

2. *Nontoxicity*. The refrigerant should not be harmful to humans directly or indirectly through foodstuffs.
3. *Nonirritability*. The refrigerant should not irritate humans (eyes, nose, lungs, or skin).

In addition to these characteristics, the refrigerant should be of relative low cost and be easy to detect in case of leaks.

15-4 COMPRESSORS

The compressor is one of the four essential parts of the vapor compression refrigeration system and may be of the reciprocating, rotary, helical rotary (screw), or centrifugal type.

Reciprocating Compressors

Most reciprocating compressors are single-acting, using pistons driven directly through a pin and connecting rod from the crankshaft. Double-acting compressors, which utilize piston rods, crossheads, stuffing boxes, and oil injection, are not extensively used.

This section covers two classifications that include most reciprocating refrigeration compressors, the halocarbon compressor is the most widely used; it is manufactured in three types of design: (1) open, (2) semi- or bolted hermetic, and (3) the welded-shell hermetic. Ammonia compressors are manufactured only in the open-type design. In the open-type compressor the shaft extends through a seal in the crankcase to an external drive.

In hermetic compressors the motor and compressor are contained within the same pressure vessel with the motor shaft as part of the compressor crankshaft, and with the motor in contact with the refrigerant. Figure 15-6a illustrates a typical hermetic compressor. A semihermetic, accessible, or serviceable hermetic compressor is of bolted construction capable of field repair (Fig. 15-6b). A welded-shell (sealed) hermetic compressor is one in which the motor-compressor is mounted inside a steel shell that in turn is sealed by welding.

The idealized reciprocating compressor is assumed to operate in a reversible adiabatic manner; pressure losses in the valves, intake, and exhaust manifolds are neglected. We will now consider the effect of some of the unavoidable realities of the compressor.

Figure 15-7 shows a schematic indicator diagram for a reciprocating compressor. The diagram is idealized because the pressure in the cylinder is assumed constant during the exhaust process $P_c = P_d$ and during the intake process P_a and P_b. However, the pressure loss in the valves is to be considered. The gas remaining in the clearance volume V_d expands in a polytropic process from state d to state a. State b is generally different from that at a because of the mixing of the expanded clearance volume vapor and the intake vapor. The vapor is compressed from state b to state c in a polytropic process. Heat transfer

Figure 15-6. Reciprocating compressors. (Courtesy of Copeland Corporation, Sidney, Ohio.)

Figure 15-7. Schematic indicator diagram for a reciprocating compressor.

may occur during the exhaust process c–d. Figure 15-8 is a pressure-specific volume diagram for the compressor of Fig. 15-7. In the state diagram of Fig. 15-8 it is assumed that

1. The same polytropic exponent n applies to the compression process b–c and the expansion process d–a.
2. Heat transfer during the exhaust process is negligible; therefore, states c and d are identical.
3. The state of the mixture of reexpanded clearance volume gas a and the intake gas 3 is the same as state b and is designated as a'. This is actually a result of assumptions 1 and 2 above. Note that the work required to compress the clearance volume vapor is just balanced by the work done in the expansion of the clearance volume vapor.

The *volumetric efficiency* η_v is defined as the ratio of the actual mass of the vapor compressed to the mass of vapor that could be compressed if the intake volume equaled the piston displacement and the state of the vapor at the beginning of the compression were equal to state 3 of Fig. 15-8. The total mass in the cylinder at the end of the intake stroke is V_b/v_b. The mass of clearance vapor is V_a/v_a. Then

$$\text{Actual mass of intake vapor} = \frac{V_b}{v_b} - \frac{V_a}{v_{a'}} = \frac{V_b - V_a}{v_b} \tag{15-5}$$

Mass of intake vapor with volume equal to piston displacement at state 3.

$$= \frac{V_b - V_d}{v_3} \tag{15-6}$$

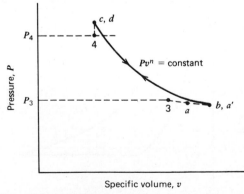

Specific volume, v

Figure 15-8. Pressure-specific volume diagram for compressor of Fig. 15-7.

The volumetric efficiency thus becomes

$$\eta_v = \frac{(V_b - V_a)v_3}{V_b - V_d)v_b}$$ (15-7)

The *clearance factor C* is defined as the ratio of the clearance volume to the piston displacement or

$$C = \frac{V_d}{V_b - V_d}$$ (15-8)

By using Eqs. 15-7 and 15-8 and the polytropic relation

$$V_a = V_d \left(\frac{P_d}{P_a}\right)^{1/n} = V_d \left(\frac{P_c}{P_b}\right)^{1/n}$$ (15-9)

It may be shown that

$$\eta_v = \left[1 + C - C \left(\frac{P_c}{P_b}\right)^{1/n}\right] \frac{v_3}{v_b}$$ (15-10)

The three main factors that influence the volumetric efficiency are accounted for in Eq. 15-10. The term within the brackets describes the effect of the clearance volume and reexpansion of the clearance volume vapor, whereas the term v_3/v_b accounts for the pressure drop and heating of the intake vapor. Obviously, it would be desirable for the clearance volume to be zero, but this is not possible due to mechanical constraints. The valve plate in the compressor must be thick enough to withstand the forces of high pressure, which causes the exhaust valve ports to have considerable volume. There must also be space for the intake valve to open. Compressor designers are constantly seeking ways to reduce the clearance volume. Figure 15-9 shows an innovative valve design to reduce the clearance volume. The exhaust valve is in the center and the intake valve is located around the circumference of the valve plate.

From Eq. 15-7 it follows that volumetric efficiency may also be expressed as

$$\eta_v = \frac{\dot{m}v_3}{PD}$$ (15-11)

where *PD* is the piston displacement in volume per unit of time. Then combining Eqs. 15-10 and 15-11 the mass flow rate of refrigerant is given by

$$\dot{m} = \left[1 + C - C \left(\frac{P_c}{P_b}\right)^{1/n}\right] \frac{PD}{v_b}$$ (15-12)

The purpose of the compressor is to move refrigerant through the system as efficiently as possible. The important effect of the clearance volume is especially evident in Eq. 15-12.

Although the polytropic exponent n must be determined experimentally,

Figure 15-9. An innovative compressor valve design. (Courtesy of Copeland Corporation, Sidney, Ohio.)

it may be approximated by the isentropic exponent k when other data are not available. Table 15-2 gives some representative values of k.

An expression for the compressor work may be derived subject to the same assumptions used in the analysis for volumetric efficiency. Referring to Fig. 15-7 we represent the work per cycle by the enclosed area of the indicator diagram.

Table 15-2   ISENTROPIC EXPONENT k FOR SOME
REFRIGERANT VAPORS

	Refrigerant		
	12	**22**	**Ammonia**
Vapor temperature, F/C	50/10	86/30	70/21
$k = c_p/c_v$	1.13	1.16	1.31

$$W = \oint V \, dP = \int_b^c V \, dP - \int_a^d V \, dP \qquad (15\text{-}13)$$

Using the realtion $PV^n = $ constant to describe the process paths

$$W = \frac{n}{(n-1)} P_b(V_b - V_a) \left[\left(\frac{P_c}{P_b}\right)^{(n-1)/n} - 1 \right] \qquad (15\text{-}14)$$

The work per unit mass of vapor compressed may be obtained by dividing Eq. 15-14 by the mass of intake vapor, Eq. 15-5.

$$w = \frac{n}{(n-1)} P_b v_b \left[\left(\frac{P_c}{P_b}\right)^{(n-1)/n} - 1 \right] \qquad (15\text{-}15)$$

The power requirement for the compressor is given by

$$\dot{W} = \frac{\dot{m}w}{\eta_m} \qquad (15\text{-}16)$$

where η_m is the compressor mechanical efficiency. Equations 15-10 through 15-16 are not intended to predict the actual performance of real compressors but rather to show the relationships among the important variables. Figure 15-10 shows representative performance data for a hermetic compressor operating with refrigerant 22. Manufacturers furnish performance data in this form or as shown in Table 15-3.

Example 15-4. Refrigerant 12 vapor enters the suction header of a single-stage reciprocating compressor at 45 psia and 40 F. The discharge pressure is 200 psia. Pressure drop in the suction valve is 2 psi and the pressure loss in the discharge valve is 4 psi. The vapor is superheated 12 F during the intake stroke. The clearance volume is 5 percent of piston displacement. Determine the (a) volumetric efficiency, (b) compressor pumping capacity if the piston displacement is 10 in.³ and the crankshaft rotates at 1725 rpm, and (c) shaft horsepower required for a mechanical efficiency of 70 percent.

Solution. The $P-v$ diagram of Fig. 15-8 will be used to aid in the problem solution. From the given data

$$P_4 = 200 \text{ psia}; \qquad P_c = P_d = P_4 + 4 = 204 \text{ psia}$$
$$P_3 = 45 \text{ psia}; \qquad P_b = P_3 - 2 = 43 \text{ psia}$$

From Chart 4, $v_3 = 0.92$ ft³/lbm and at point b, $t_b = 52$ F. $P_b = 43$ psia and $v_b = 0.98$ ft³/lbm. The clearance factor is given as 0.05, and it will be assumed that $n = k = 1.13$. Then using Eq. 15-10

$$\eta_v = \left[1 + 0.05 - (0.05) \left(\frac{204}{43}\right)^{1/1.13} \right] \frac{0.92}{0.98} = 0.8$$

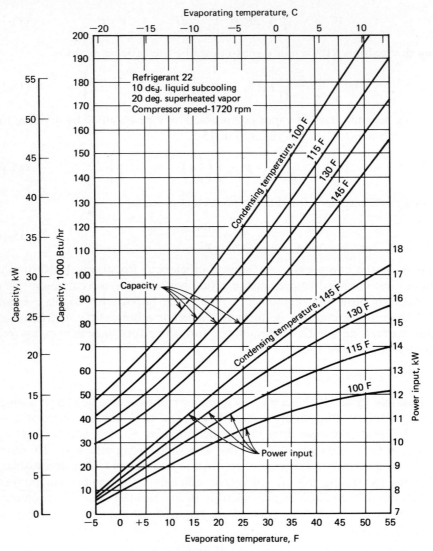

Figure 15-10. Typical capacity and power input curves for a hermetic reciprocating compressor. (Reprinted by permission from *ASHRAE Handbook, Equipment Volume*, 1975.)

Table 15-3 NOMINAL PERFORMANCE DATA FOR A SEMIHERMETIC RECIPROCATING COMPRESSOR[a]

Capacity—1000 Btu/hr

Condensing Temperature—F	Evaporating Temperature								
	10	20	25	30	35	40	45	50	55
90	154	194	219	247	278	312	349	390	433
100	141	179	203	229	258	289	324	361	401
110	129	165	187	211	238	267	299	333	370
120	118	151	171	194	218	245	274	305	338
130	107	138	156	177	199	223	247	276	306
140	96	125	142	160	180	201	224	248	274

Power Input—kW

Condensing Temperature—F	Evaporating Temperature								
	10	20	25	30	35	40	45	50	55
90	15.0	16.1	16.6	17.0	17.5	17.9	18.3	18.6	19.0
100	15.5	16.9	17.6	18.2	18.8	19.3	19.8	20.3	20.8
110	16.0	17.7	18.5	19.3	20.0	20.7	21.4	22.0	22.6
120	16.6	18.6	19.5	20.4	21.3	22.1	22.9	23.7	24.4
130	17.2	19.4	20.5	21.6	22.6	23.6	24.5	25.4	26.3
140	17.8	20.3	21.5	22.7	23.8	25.0	26.1	27.1	28.1

[a]1750 rpm, 60 Hz, refrigerant 22, 20 F superheat, 15 F subcooling, 95 F ambient.

Now from Eq. 15-11

$$\dot{m} = \frac{\eta_v(PD)}{v_3} = \frac{(0.8)}{(0.92)} \frac{(10)(1725)}{(1728)} = 8.68 \text{ lbm/min}$$

The shaft power required will be computed using Eqs. 15-15 and 15-16

$$\dot{W} = \frac{\dot{m}w}{\eta_m} = \left(\frac{\dot{m}}{\eta_m}\right) \frac{n}{(n-1)} P_b v_b \left[\left(\frac{P_c}{P_b}\right)^{(n-1)1/n} - 1\right]$$

$$\dot{W} = \left(\frac{8.68}{0.7}\right) \frac{1.13}{(1.13 - 1)} (43)(144)0.98 \left[\left(\frac{204}{43}\right)^{(1.13-1)/1.13} - 1\right]$$

$$\dot{W} = 128,298 \text{ (ft-lbf)/min} = 3.9 \text{ hp}$$

Rotary Compressors

Rotary compressors are characterized by their circular or rotary motion as opposed to reciprocating motion. Their positive displacement compression process is nonreversing and either continuous or cyclical depending on the mechanism employed. Most are direct-drive machines.

Figures 15-11 and 15-12 show two common types of rotary compressors:

Figure 15-11. Rolling piston-type rotary compressor. (Reprinted by permission from *ASHRAE Handbook, Equipment Volume*, 1975.)

Figure 15-12. Rotating vane-type rotary compressor. (Reprinted by permission from *ASHRAE Handbook, Equipment Volume*, 1975.)

the rolling piston type and the rotating vane type. These two machines are very similar with respect to size, weight, thermodynamic performance, field of greatest applications, range of sizes, durability, and sound level.

The rotary compressor performance is characterized by high volumetric efficiency because of its small clearance volume, and correspondingly low reexpansion loss. Figure 15-13 shows performance typical of compressors now in production.

Helical Rotary Compressors

The helical rotary or screw compressor also belongs to the broad class of positive-displacement compressors. Although it was developed in the early 1930s and has been used in the gas and process industries, it was first introduced to the refrigeration industry in the late 1950s.

The machine essentially consists of two mating helically grooved rotors, a male (lobes) and a female (gullies), in a stationary housing with suitable inlet and outlet gas ports (Fig. 15-14). The flow of gas in the rotors is both radial and axial. With a four-lobe male rotor rotating at 3600 rpm, the six-lobe female rotor follows at $\frac{4}{6} \times$ (3600 rpm) or 2400 rpm. The female rotor can be driven by the male rotor.

Compression in this machine is obtained by direct volume reduction with pure rotary motion. The four basic continuous phases of the working cycle are

1. *Suction.* A pair of lobes unmesh on the inlet port side and gas flows in the increasing volume formed between the lobes and the housing until the lobes are completely unmeshed.

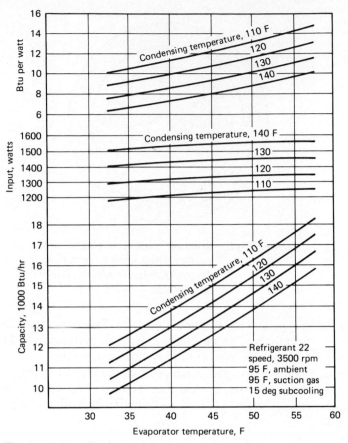

Figure 15-13. Typical performance curves for a rotary compressor. (Reprinted by permission from *ASHRAE Handbook, Equipment Volume,* 1975.

2. *Transfer.* The trapped pocket isolated from the inlet and outlet is moved circumferentially at constant suction pressure.
3. *Compression.* When remeshing starts at the inlet end, the trapped volume is reduced and the charge is gradually moved helically and compressed simultaneously toward the discharge end as the lobes' mesh point moves along axially.
4. *Discharge* starts when the compressed volume has been moved to the axial ports on the discharge end of the machine and continues until all the trapped gas is completely squeezed out.

During the remeshing period of compression and discharge, a fresh charge is drawn through the inlet on the opposite side of the meshing point. With four male lobes rotating at 3600 rpm, four interlobe volumes are filled and discharged

Port areas

Suction

Discharge

Bottom cusp

Figure 15-14.   Helical rotary (screw) compressor. (Reprinted by permission from *ASHRAE Handbook, Equipment Volume*, 1975.)

per revolution, providing 14,400 discharges per minute. Since the intake and discharge cycles overlap effectively, a smooth continuous flow of gas results.

Typical performance levels for oil-flooded screw compressors are given in Fig. 15-15.

Centrifugal Compressors

Centrifugal compressors or turbocompressors are members of a family of turbomachines that includes fans, propellors, and turbines. Such machines are characterized by a continuous exchange of angular momentum between a rotating mechanical element and a steadily flowing fluid. Because their flows are continuous, turbomachines have greater volumetric capacities, size-for-size, than do positive displacement devices. For effective momentum exchange their rotative speeds must be higher, but little vibration or wear results because of the steadiness of the motion and the absence of contacting parts.

Centrifugal compressors are used in a wide variety of refrigeration and air-conditioning applications. As many as eight or nine stages can be installed in a single casing.

Loading the Compressor

The performance diagrams of Figs. 15-10, 15-13, and 15-15 are valuable aids in designing and analyzing refrigeration systems. The compressor capacity is given in terms of the load on the evaporator, the evaporator temperature, and the condensing temperature. This makes it possible to plot curves representing the capacity of the evaporator and condenser on the same chart. Figure 15-16 is such a plot using data from Table 15-3 for a condensing temperature of 130 F. The evaporator curve is for an evaporator at fixed air flow rate and

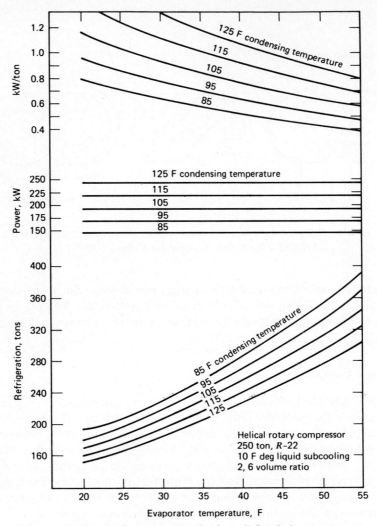

Figure 15-15. Performance levels for oil-flooded compressors. (Reprinted by permission from *ASHRAE Handbook, Equipment Volume,* 1975.)

entering air temperatures. The system must operate where the two curves cross. The curves representing the power input to the compressor and heat rejected in the condenser are also plotted.

If the load on the evaporator changes as the result of a change in air flow rate or temperature, the evaporator curve will move to the right or left as shown in Fig. 15-16, depending on whether the load decreases or increases. In all probability the condensing temperature will also change as the load changes, but we will assume it to be constant for this discussion. The important point

Figure 15-16.   Compressor and evaporator capacity from Table 15-3.

is that the curves for the evaporator and compressor define the operating condition for the compressor. Compressor power decreases as the evaporating temperature decreases and vice versa. Therefore, the evaporating temperature is an indication of compressor load. A system may be checked in the field by measuring the compressor suction pressure or power. Suction pressure is approximately equal to evaporator pressure, which is directly related to evaporator temperature. For example, a low pressure indicates the compressor is not loaded. This is an indication that the evaporator is not transferring the expected quantity of heat to the refrigerant. This may be due to any number of things, such as decreased air flow over the evaporator due to clogged filters or a loose fan belt, decreased air temperatures, and so on. The opposite case could also occur where the compressor is overloaded, as indicated by an increased suction pressure and evaporator temperature.

Control

A compressor will be at partial load much of the time. With small systems, the compressor can be started and stopped by thermostat action when the demand for cooling is satisfied. However, it is not desirable to start and stop larger compressors at frequent intervals. In this case a technique is employed to reduce the compressor capacity referred to as *compressor unloading*. This may be done by reducing the compressor speed, holding the intake valves open, or increasing the clearance volume. Let us consider the reciprocating compressor. Variable-speed compressors are available; however, unloading by holding the intake valves open is the most widely used method. Large compressors have two or more cylinders, and each cylinder can be unloaded independent of the others. For example, if one of two cylinders is unloaded, the capacity of the compressor would be decreased by about one-half. This control action is initiated by sensing the suction pressure and carried out by an electric solenoid. We noted earlier that the volumetric efficiency and capacity of a reciprocating compressor decreases rapidly as the clearance volume is increased. Therefore, capacity may be controlled by varying the clearance volume. Figure 15-17 shows a reciprocating, semihermetic compressor with this type of unloading mechanism. The spring-loaded pistons in the valve header may be positioned so that the clearance volume varies from almost zero to a value where no refrigerant is compressed at all. This system unloads all cylinders equally in a continuous rather than a stepwise manner, which results in higher operating efficiency and less vibration than the method of holding the intake valves open.

As the system becomes larger, two or more compressors may be used, each with capacity control, for further flexability in matching the refrigeration capacity to the cooling load.

Figure 15-17.   Reciprocating compressor using over reexpansion modulation control. (Courtesy of Copeland Corporation, Sidney, Ohio.)

15-5 REFRIGERATION CONTROL VALVES

Early in the development of refrigeration cycles it became evident that the system should respond to changing load and environmental conditions to better serve the system objectives and to prevent damage to the system components. The most important component in the vapor compression cycle from the standpoint of system control is the expansion valve. Proper manipulation of the expansion valve either manually or automatically can go a long way toward optimizing system performance; other devices are sometimes required, however, to control evaporator temperature or compressor load. The most common control valves are described in the following paragraphs.

Hand Expansion Valves

The earliest refrigeration systems had only hand-operated needle valves for expansion valves. The operator monitored the system and adjusted the valve according to the refrigeration load. Hand valves are seldom used today except in experimental apparatus.

Constant-pressure Expansion Valve

The constant-pressure or automatic expansion valve was the earliest type of automatic valve used in refrigeration systems. The valve acts to maintain a fixed evaporator pressure. Therefore, when evaporator pressure drops, the valve opens to allow more refrigerant flow. The opposite action results from a rise in evaporator pressure. This valve action is opposite to that desired in variable-load applications. When the refrigerating load on an evaporator increases, the pressure will increase and more refrigerant is needed to handle the increased heat-transfer rate. Conversely, when evaporator load decreases, pressure decreases, and less refrigerant is required. The automatic expansion valve is inadequate when the load is variable; it is therefore in very limited use at the present time.

Thermostatic Expansion Valve

The thermostatic expansion valve does not have the limitations of the constant-pressure valve because of a separate mechanism that positions the valve spool to admit the refrigerant as required by evaporator load (Fig. 15-18). A bulb containing a small refrigerant charge is connected by a small tube to the chamber above the valve diaphragm. The bulb is clamped to the refrigerant line where refrigerant leaves the evaporator. Spring pressure tends to close the valve, whereas bulb pressure tends to open it. The bulb is essentially a temperature-sensing element, and several degrees superheat of the refrigerant leaving the evaporator are required before the expansion valve will open. As the load on the evaporator increases, superheat will increase and the valve

Figure 15-18. Exterior and interior views of thermostatic expansion valve. (Courtesy of Sporlan Valve Company, St. Louis, Missouri.)

opens to allow a larger refrigerant flow rate. As the load decreases, the valve will close, maintaining a superheated vapor at the evaporator exit. This action protects the compressor from liquid slugging and decreases the compressor load as evaporator load decreases. Most thermostatic expansion valves are adjustable. About 5 to 10 F (3 to 5 C) superheat is usually maintained.

It is possible for the thermostatic expansion valve to overload the compressor as evaporator load continues to increase. To help prevent this condition a liquid–vapor charge may be used in the bulb. The charge is such that all the liquid is evaporated at some predetermined temperature. Very little bulb pressure rise will result with a temperature rise above this point, and the valve will not open significantly with further increase in evaporator load. Thus, the load on the compressor will be limited.

When an evaporator is subjected to a large variation in load, the pressure drop of the refrigerant as it flows through the coil will vary. The refrigerant saturation temperature will therefore vary at the coil exit, and different amounts of superheat will result at light and heavy loads. This condition is controlled by the use of an external equalizer (Fig. 15-19). The valve then senses pressure at about the same point as it senses temperature, and the superheat will be correct for whatever pressure exists in the suction line. The external equalizer must be used when one expansion valve feeds parallel circuits as shown in Fig. 15-19.

Any number of evaporators may be operated in parallel on the same compressor where a thermostatic expansion valve controls the feed to each one. Each valve controls the flow as required by that evaporator.

The thermostatic expansion valve must be selected to match the refrigeration load, the refrigerant type, the pressure difference across the valve, and the refrigerant temperature. The valves may be adjusted but do not have sufficient range to work well with different refrigerants.

Capillary Tube

The capillary tube is not a valve but effectively replaces the expansion valve in many applications. It is a long thin tube placed between the condenser and evaporator. The small diameter and relatively large length of the tube produces a large pressure drop. Effective control of the system results because the tube allows the flow of liquid more readily than vapor. Although the capillary tube operates most efficiently at one particular set of conditions, there is only a slight loss in efficiency at off-design conditions in small systems.

The main advantage of the capillary tube is its simplicity and low cost. Being a passive device it is not subject to wear. However, the very small bore of the tube is subject to plugging if precautions are not taken to maintain a clean system. Moisture can also cause plugging due to ice formation.

Almost all domestic refrigerators and room air conditioners use capillary

Figure 15-19.   External equalizer with refrigerant distributor feeding
parallel circuits.

tubes as well as domestic split system air-conditioning equipment. Extra care
must be taken to prevent tube plugging with the latter type of equipment.

Subcooled liquid refrigerant enters a properly sized capillary tube as shown
in Fig. 15-20. Pressure loss because of friction causes the refrigerant to reach
saturation at some point in the tube. At that point pressure loss becomes much
more pronounced as vapor forms and the refrigerant temperature decreases
rapidly to the tube exit. Critical flow conditions may occur at the tube exit with
a pronounced pressure loss at that point. The amount of refrigerant flowing
through the tube depends on the overall pressure difference between the condenser and the evaporator and the saturation pressure of the refrigerant entering
the tube. Therefore, a change in evaporator load or environmental conditions
that increases the pressure difference or decreases the saturation pressure will
cause an increase in refrigerant flow rate. Fortunately an increase in load
increases pressure difference and a decrease in condensing medium temperature

Figure 15-20.   Pressure and temperature of the refrigerant in a capillary tube.

decreases the pressure difference. This partially explains the favorable control characteristics of the capillary tube.

Although there are calculation procedures for the sizing of capillary tubes, the final selection must be made by trial and error. The equipment volume of the *ASHRAE Handbook* (4) contains charts for preliminary selection of capillary tubes.

The charging of capillary tube systems is rather critical. This is because part of the passive control of the system results from the flooding of the condenser at part-load conditions, which acts to decrease the condenser size, decrease subcooling, and increase the saturation pressure at the capillary inlet. If other than the proper charge is used, this important effect is lost. Too much refrigerant can also cause compressor flooding during the off cycle and result in valve damage upon startup.

Figure 15-21. Evaporator pressure regulating valve. (Courtesy of Sporlan Valve Company, St. Louis, Missouri.)

Evaporator Pressure Regulator

The purpose of an evaporator pressure regulator is to maintain a relatively constant minimum pressure in the evaporator. Because most of the evaporator surface is subjected to two-phase refrigerant, a constant minimum temperature will also be maintained. Figure 15-21 shows a typical evaporator pressure regulator. Evaporator pressure is sensed internally and balanced by a spring-loaded diaphragm. When evaporator pressure falls below a set value, the valve will close, restricting the flow of refrigerant so that evaporator pressure will rise.

The main applications of evaporator pressure regulators are to set a minimum evaporator temperature and to permit the use of different evaporators at different pressures on the same compressor.

Suction Pressure Regulator

The suction pressure regulator has the function of limiting the maximum pressure at the compressor suction. Since the compressor load is determined by suction pressure, the suction pressure regulator is a load-limiting device. This valve functions very much like the evaporator pressure regulator except that it senses compressor suction pressure (Fig. 15-22). The suction pressure regulator also reduces the compressor load during the startup period because the valve will remain closed until suction pressure is reduced to a set pressure.

15-6   THE REAL SINGLE-STAGE CYCLE

Figure 15-23 illustrates a practical single-stage cycle schematically. The practical aspects of the compressor were discussed in a previous section and are shown again as state points 3 through 4. Other factors cause the complete cycle to deviate from the theoretical. These are the pressure losses in all connecting tubing that increase the power requirement for the cycle and the heat transfer to and from the various components. Heat transfer will generally be away from the system on the high-pressure side of the cycle and will improve cycle performance. Exposed surfaces on the low-pressure side of the cycle will generally be at a lower temperature than the environment and any heat transfer will generally degrade cycle performance. As shown in Fig. 15-23, $\dot{q}_3$ will increase compressor load and $\dot{q}_4$ decreases compressor load.

It should be mentioned that the condenser and evaporator heat-transfer surfaces have pressure losses that also contribute to the overall compressor power requirement. The following example illustrates a complete cycle analysis.

Example 15-5.   A 10-ton refrigeration unit operates on R-22 and has a single-stage reciprocating compressor with a clearance volume of 3 percent

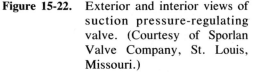

Figure 15-22.   Exterior and interior views of suction pressure-regulating valve. (Courtesy of Sporlan Valve Company, St. Louis, Missouri.)

and operates at 1725 rpm. The refrigerant leaves the evaporator at 70 psia and 40 F. Pressure loss in the evaporator is 5 psi. Pressure loss in the suction valve is 2 psi, and there is 20 F superheat at the beginning of the compression stroke. The compressor discharge pressure is 300 psia, a pressure loss of 5 psi exists in the discharge valve, and the vapor is desuperheated 10 F between the compressor discharge and the condenser inlet. Neglect pressure loss in tubing between the compressor and condenser, but the condenser pressure loss is 5 psi. The liquid has a temperature of 110 F at the expansion valve. Determine the following: (a) compressor volumetric efficiency, (b) compressor piston displacement, (c) shaft power input with mechanical efficiency of 75 percent, (d) heat rejected in the condenser, and (e) power input to compressor per ton of cooling effect.

Figure 15-23. A practical single-stage cycle.

Solution.   Most of the state points as shown in Fig. 15-23 may be located from the given data.

$$P_1 = 290 \text{ psia:} \quad T_1 = 110 \text{ F;} \quad i_2 = i_1$$
$$P_{3'} = 70 \text{ psia;} \quad t_{3'} = 40 \text{ F;} \quad P_2 = P_{3'} + 5 = 75 \text{ psia}$$
$$P_a = P_b = P_{3'} - 2 = 68 \text{ psia;} \quad t_b = 50 \text{ F}$$
$$P_c = P_d = 300 \text{ psia;} \quad P_4 = P_{4'} = P_c - 5 = 295 \text{ psia}$$

States *c*, *d*, and 4 may be completely determined following the compressor and evaporator analysis. From Chart 5, $v_3 = 0.78$ ft³/lbm and $v_b = 0.81$ ft³/lbm. Assuming that $n = k = 1.16$, the volumetric efficiency may be computed from Eq. 15-10 as

$$\eta_v = \left[1 + 0.03 - (0.03)\left(\frac{300}{68}\right)^{1/1.16} \right] \frac{0.78}{0.81} = 0.89$$

For the evaporator

$$\dot{q}_e = \dot{m}(i_{3'} - i_2) = \dot{m}(i_{3'} - i_1)$$

or

$$\dot{m} = \frac{\dot{q}_e}{(i_{3'} - i_1)}$$

From Chart 5, $i_{3'}$ = 109 Btu/lbm, i_1 = 41 Btu/lbm

$$\dot{m} = \frac{10(12,000)}{(109 - 41)} = 1765 \text{ lbm/hr}$$

From Eq. 15-11

$$\eta_v = \frac{\dot{m}v_3}{PD} \quad \text{or} \quad PD = \frac{\dot{m}v_3}{\eta_v}$$

$$PD = \frac{1765(0.78)}{0.89} = 1547 \text{ ft}^3/\text{hr}$$

or

$$\frac{PD}{\text{cycle}} = 25.8 \text{ in.}^3$$

The work per pound of refrigerant may be computed from Eq. 15-15:

$$w = \frac{n}{n - 1} P_b v_b \left[\left(\frac{P_c}{P_b} \right)^{(n-1)/n} - 1 \right]$$

$$w = \frac{1.16}{(1.16 - 1)} 68(144)(0.81) \left[\left(\frac{300}{68} \right)^{(1.16-1)/1.16} - 1 \right]$$

$$w = 13,064 \text{ (ft-lbf)/lbm}$$

$$\dot{W} = \dot{m}w = \frac{13,064(1765)}{778} = 29,638 \text{ Btu/hr}$$

$$\dot{W}_{sh} = \frac{\dot{W}}{\eta_m} = 39,517 \text{ Btu/hr} = 15.5 \text{ hp}$$

The heat rejected in the condenser is given by

$$\dot{q}_c = \dot{m}(i_1 - i_{4'})$$

However, state 4' has not been completely determined. Since the poly-tropic exponent was assumed equal to k, the isentropic compression process follows a line of constant entropy on Chart 5. Then $t_c = t_d = 188$ F, $t_4 \approx 185$ F, and $t_{4'} = t_4 - 10 = 175$ F. Then $i_{4'} = 125$ Btu/lbm.

$$\dot{q}_c = 1765(41 - 125) = -148,000 \text{ Btu/hr}$$

The total power input to the compressor per ton of cooling effect is

$$\frac{\text{hp}}{\text{ton}} = \frac{15.5}{10} = 1.55$$

Refrigerant Piping

The sizing of the piping for a refrigeration system is a very important step in the overall design process. Although the process is quite similar to that outlined in Chapter 9 for liquids, one other criterion must be considered. That is the need to ensure transport of oil from the system back to the compressor. Most of the oil resides in the compressor crankcase, but a small amount may travel out into the system, especially during startup. Where liquid is present, between the condenser and the evaporator, the oil moves along with the refrigerant quite readily. However, where only vapor is flowing, movement of the oil is dependent on the refrigerant velocity. This is especially true where the vapor is traveling vertically upward.

Refrigerant pipe sizing has become a science to itself and many manufacturers publish procedures for use of their distributors. Reference 3 should be consulted for further details.

System Control

Control of the reciprocating compressor in response to the load was discussed in Sec. 15-4. Although this is probably the most important aspect of controlling the complete system, some other factors should be mentioned. The compressor may be gradually unloaded but eventually it will have to be stopped when the load is zero. This may be done by simply turning off the power; however, with large reciprocating compressors, this will lead to problems on startup due to possible flooding of the compressor by liquid refrigerant. Therefore, shutdown is usually done on a *pump down cycle*.

Referring to Figure 15-24, when the thermostat is satisfied the solenoid valve closes, stopping the flow of refrigerant to the expansion valve. The compressor continues to run, reducing the pressure in the system between the solenoid valve and the suction side of the compressor, while compressed hot vapor is condensed and stored in the receiver. When the suction pressure reaches the set point of the low-pressure sensor, the controller disconnects the power to the compressor. Later, when the thermostat calls for cooling, the solenoid valve opens and refrigerant flows toward the compressor with a consequent rise in pressure. When the pressure reaches the low-pressure switch set point, the controller starts the compressor. During the off cycle there is only a small amount of refrigerant in the system on the suction side of the compressor, and since the low-pressure switch is set for a relatively low pressure, the compressor starts under a relatively light load. After it starts, the unloading system takes over control of the compressor.

Figure 15-24.   Refrigeration system pump-down and condenser control.

Another aspect of system control relates to the air-cooled condenser. The condenser fan is controlled to start and stop as the compressor starts and stops. This is usually adequate for small systems at normal outdoor temperatures during the summer. However, when the outdoor ambient becomes low, the condensing pressure may be so low as to cause operational problems. Therefore, a system designed for low ambient operation must be provided with *head pressure control*. This simply reduces the flow of air over the condenser in response to the compressor discharge pressure. Inlet dampers may be used, which are modulated to maintain some minimum head pressure. Another approach is to use two or more condenser fans where one or more of the fans may be cycled off to maintain the head pressure. Either of these methods may be controlled as shown in Fig. 15-24.

The reader is referred to reference 6 for information relating to the control of large refrigerating systems using centrifugal or absorption machines.

The following example illustrates the use of the compressor performance curves shown in Figs. 15-10, 15-13 and 15-15.

Example 15-6.   Construct the pressure-enthalpy diagram for a system using the compressor of Fig. 15-10 with an evaporating temperature of 35 F and a condensing temperature of 115 F. Subcooling and superheat are as specified in Fig. 15-10. Neglect system pressure losses and piping heat losses and gains.

Solution.   Chart 5 is used to obtain thermodynamic properties and to locate the isotherms for 35 F and 115 F, respectively, as shown in Fig. 15-25. The subcooling is 10 F; therefore, state 1 is determined by the 115 F and 105 F isotherms at the condenser pressure. The superheat is 20 F and state 3 is located on the 55 F isotherm at the evaporator pressure. As a result of the throttling process, state 2 is located on the 35 F isotherm where $i_2 = i_1$. Since the data of Fig. 15-10 are for a real compressor, the process from 3 to 4 is not isentropic and some other method will have to be used to locate state 4. This can be done by computing the heat transfer to the refrigerant in process 2–3 and the work done on the refrigerant in process 3–4.

$$\dot{q}_{23} = \dot{m}(i_3 - i_2)$$

and

$$\dot{m} = \frac{\dot{q}_{23}}{(i_3 - i_2)}$$

Figure 15-25.   Pressure-enthalpy diagram for Example 15-6.

From Chart 5 $i_1 = i_2 = 41$ Btu/lbm, $i_3 = 111.5$ Btu/lbm, and $\dot{q}_{23} = 130{,}000$ Btu/hr from Fig. 15-10. Then

$$\dot{m} = \frac{130{,}000}{(111.5 - 41)} = 1844 \text{ lbm/hr}$$

The compressor power is 12.5 kW or 42,650 Btu/hr from Fig. 15-10. The heat rejected from the refrigerant in the condenser is then

$$\dot{q}_{41} = \dot{q}_{23} + \dot{w}_{34} = 130{,}000 + 42{,}650 = 172{,}650 \text{ Btu/hr}$$

Also

$$\dot{q}_{41} = \dot{m}(i_4 - i_1)$$

and

$$i_4 = i_1 + \frac{\dot{q}_{41}}{\dot{m}} = 41 + \left(\frac{172{,}650}{1844}\right) = 134.6 \text{ Btu/lbm}$$

The complete cycle is thus determined in Fig. 15-25. Note an increase in entropy for the compressor of about 0.016 Btu/(lbm-R).

15-7 ABSORPTION REFRIGERATION

The mechanical vapor compression refrigeration system described in the previous sections is an efficient and practical method. However, the required energy input is shaft work or power, which is high-grade energy and expensive. The relatively large amount of work required is because of the compression of vapor that has a large volume.

It is possible to replace the vapor-compression process with a series of processes where the refrigerant vapor is absorbed by a liquid and then the liquid solution is pumped to the higher pressure. Although the shaft work required is much less, large quantities of heat must be transferred to the system. Figure 15-26 illustrates the essential features of the absorption cycle. Note that the compressor of the vapor compression cycle has been replaced by the generator, absorber, and pump, and that the solution of refrigerant and absorbent circulates through these elements. The refrigerant alone flows through the condenser, expansion valve, and evaporator. Heat is transferred to the solution in the generator, and the refrigerant vapor is separated from the solution, whereas the opposite process occurs in the absorber. The ease with which these processes may be carried out in practice depends on the refrigerant–absorbent pair and the many refinements to the cycle required in actual practice. These factors will be discussed later. The overall performance of the absorption cycle in terms of refrigerating effect per unit of energy input is generally poor; however, waste heat such as that rejected from a power plant can often be used to achieve better overall energy conservation.

Figure 15-26.   Basic absorption cycle.

15-8 PROPERTIES AND PROCESSES OF HOMOGENEOUS BINARY MIXTURES

A homogeneous mixture is uniform in composition and cannot be separated into its constituents by pure mechanical means. The properties such as pressure, density, and temperature are uniform throughout the mixture. The thermodynamics of binary mixtures is extensively covered by Bosnjakovic (1).

The thermodynamic state of a mixture cannot be specified by two independent properties as may be done with a pure substance. The composition of the mixture as described by the *concentration x,* the ratio of the mass of one constituent to the mass of the mixture, is required in addition to two other independent properties such as pressure and temperature.

The miscibility of a mixture is an especially important characteristic for an absorption system. A mixture is miscible if it does not separate after mixing. A miscible mixture is homogeneous. Some mixtures may not be miscible under all conditions with temperature being the main property influencing miscibility. Oil–refrigerant 22 is miscible at higher temperatures but nonmiscible at low temperatures. Binary mixtures for an absorption refrigeration cycle must be completely miscible in both the liquid and vapor phases with no miscibility gaps.

The behavior of a binary mixture in the vicinity of the saturation region is important to absorption refrigeration. To understand this behavior the use of imaginary experiments and a temperature versus concentration diagram is helpful. Figure 15-27a shows a piston-cylinder arrangement containing a binary mixture in the liquid phase with a concentration of material B, x_1. The piston has a fixed mass and is frictionless, so that the pressure of the mixture is always constant. Figure 15-27c is the $t-x$ diagram for the mixture with the initial state shown as point 1. As heat is slowly transferred to the mixture, the temperature will rise, and at point 2 vapor will begin to form and collect under the piston as shown in Fig. 15-27b. If the experiment is stopped at some point

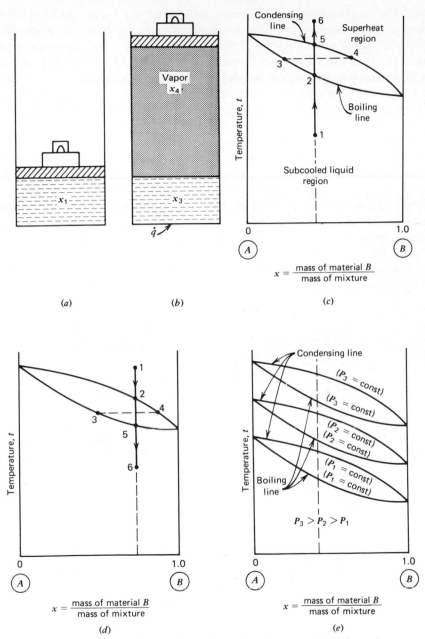

Figure 15-27. Evaporation and condensation characteristics for a homogeneous binary mixture.

above this temperature and the concentrations of the vapor and liquid determined, a rather surprising result is observed. For example, the state of the liquid is at point 3 while the state of the vapor is at point 4 and the concentration of material B in the vapor x_4 is larger than in the liquid x_3. We shall see later that this phenomenon is quite helpful in the absorption cycle.

Continued heating of the mixture will gradually vaporize all of the liquid (state point 5) with concentration $x_5 = x_1$. Further heating will superheat the vapor to point 6. When the superheated vapor is cooled at constant pressure, the entire porcess will be reversed as shown in Fig. 15-27d. Additional experiments carried out at different concentrations but at the same pressure establish the boiling and condensing lines shown in Figs. 15-27c and 15-27d. If the pressure is changed and the experiments repeated at various concentrations, the boiling and condensing lines will be displaced as shown in Fig. 15-27e. Unlike a pure substance, binary mixtures do not have a single saturation temperature for each pressure, because the saturation temperatures depend on the concentration.

A more useful representation of the properties of a binary mixture is the enthalpy-concentration diagram. Figure 15-28 is a schematic $i-x$ diagram

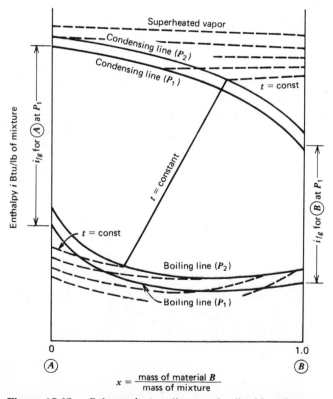

Figure 15-28. Schematic i–x diagram for liquid and vapor regions for a homogeneous binary mixture.

including the liquid and vapor regions for a homogeneous binary mixture. The condensing and boiling lines for a given pressure do not converge at $x = 0$ and $x = 1$ but are separated by a distance proportional to the enthalpy of vaporization of each constituent. Lines of constant temperature are shown in the liquid and vapor regions but not in the saturation region. When the boiling and condensing lines for more than one pressure are shown on one chart, it becomes difficult to distinguish the proper isotherms for each pressure in the saturation region. However, these isotherms may be located as needed.

Chart 2 is an enthalpy-concentration diagram for an ammonia—water mixture. Chart 2 covers the subcooled and saturation region but does not have temperature lines in the vapor region. Lines of constant temperature in the saturation region are located using the equilibrium construction lines. A vertical line is drawn upward from the saturated liquid state to the equilibrium construction line for the same pressure. From that point a horizontal line is drawn to the saturated vapor line for the same pressure. This intersection gives the saturated vapor-state point. The isotherm for the given temperature is the straight line connecting the saturated liquid and vapor states.

The most common thermodynamic processes involved in industrial systems and absorption refrigeration are described. These are the adiabatic and nonadiabatic mixing of two streams, heating and cooling including vaporization and condensation, and throttling. An understanding of these processes will help in understanding the absorption cycle, The $i - x$ diagram will be quite useful in the solution of these problems.

Adiabatic Mixing of Two Streams

Figure 15-29 shows a mixing chamber where two binary streams of different concentration and enthalpy mix in a steady-flow process. Determination of the state of the stream leaving the mixing chamber involves energy and mass balances on the control volume defined by the mixing chamber. The energy balance is

$$i_1\dot{m}_1 + i_2\dot{m}_2 = i_3\dot{m}_3 \qquad (15\text{-}17)$$

Figure 15-29.   Steady-flow adiabatic mixing process.

while the overall mass balance is given by

$$\dot{m}_1 + \dot{m}_2 = \dot{m}_3 \qquad (15\text{-}18)$$

and the mass balance for one constituent is

$$\dot{m}_1 x_1 + \dot{m}_2 x_2 = \dot{m}_3 x_3 \qquad (15\text{-}19)$$

Elimination of $\dot{m}_3$ from Eqs. 15-17, 15-18, and 15-19 gives

$$\frac{\dot{m}_1}{\dot{m}_2} = \frac{i_2 - i_3}{i_3 - i_1} = \frac{x_2 - x_3}{x_3 - x_1} \qquad (15\text{-}20)$$

Equation 15-20 defines a straight line on the $i-x$ diagram as shown in Fig. 15-29 and state 3 must lie on this line. It may be shown that

$$x_3 = x_1 + \frac{\dot{m}_2}{\dot{m}_3}(x_2 - x_1) \qquad (15\text{-}21)$$

$$i_3 = i_1 + \frac{\dot{m}_2}{\dot{m}_3}(i_2 - i_1) \qquad (15\text{-}22)$$

The $i-x$ diagram may be used to advantage to solve mixing problems, but the procedure is somewhat involved when the final state is in the mixture region.

Example 15-7. Twenty lbm per minute of liquid water–ammonia solution at 150 psia, 220 F, and concentration of 0.25 lbm ammonia per lbm of solution is mixed in a steady-flow adiabatic process with 10 lbm per minute of saturated water–ammonia solution at 150 psia and 100 F. Determine the enthalpy, concentration, and temperature of the mixture.

Solution. Chart 2 will be used to carry out the solution and Fig. 15-30 shows the procedure. State 1 is a subcooled condition located on the diagram at $t = 220$ F and $x = 0.25$. State 2 is a saturated liquid and is located at the intersection of the 150 psia boiling line and the 100 F temperature line. States 1 and 2 are joined by a straight line. State 3 is located on the connecting line and is determined by the concentration x_3 or enthalpy i_3. Using Eq. 15-21, we get

$$x_3 = x_1 + \frac{\dot{m}_2}{\dot{m}_3}(x_2 - x_1) = 0.25 + \frac{10}{30}(0.72 - 0.25)$$

$$x_3 = 0.41 \text{ lbm ammonia/lbm mixture}$$

Because the resulting mixture state lies within the saturation region, it is a mixture of vapor and liquid. To determine the temperature t_3 a graphical trial-and-error procedure shown in Fig. 15-30 is used. The fractional proportions of liquid and vapor in the mixture may also be determined from Chart 2 in a graphical manner.

$$\frac{\dot{m}_v}{\dot{m}_3} = \frac{\overline{l3}}{\overline{lv}} = 0.038 \qquad (15\text{-}23)$$

Therefore, the mixture is 3.8 percent vapor and 96.2 percent liquid.

Figure 15-30.   Schematic i–x diagram for Example 15-7.

Mixing of Two Streams with Heat Exchange

This type of process is quite common and occurs in the absorber of the absorption refrigeration device. In this case shown in Fig. 15-31 the energy balance becomes

$$\dot{m}_1 i_1 + \dot{m}_2 i_2 = \dot{m}_3 i_3 + \dot{q} \qquad (15\text{-}24)$$

The mass balance equations are identical to those for adiabatic mixing

$$\dot{m}_1 + \dot{m}_2 = \dot{m}_3 \qquad (15\text{-}18)$$

Figure 15-31.   Steady-flow mixing of two streams with heat transfer.

$$\dot{m}_1 x_1 + \dot{m}_2 x_2 = \dot{m}_3 x_3 \qquad (15\text{-}19)$$

The equation for the concentration x_3 is the same as Eq. 15-21, however, the enthalpy i_3 is given by

$$i_3 = i_1 + \frac{\dot{m}_2}{\dot{m}_3}(i_2 - i_1) - \frac{\dot{q}}{\dot{m}_3} \qquad (15\text{-}25)$$

Equation 15-25 differs from Eq. 15-19 only in the last term. The significance of this is shown in the $i-x$ diagram of Fig. 15-31. Point 3' represents a state that would occur with adiabatic mixing. Point 3 is located a distance $\dot{q}/\dot{m}_3$ directly below point 3' because $x_{3'} = x_3$ and heat is removed. If heat were added, then point 3 would be above point 3'.

Example 15-8. Five lbm per minute of saturated liquid aqua-ammonia at 100 psia and 220 F are mixed with 10 lbm/min of saturated vapor aqua-ammonia at 100 psia and 220 F. Heat transfer from the mixture is 4200 Btu/min. Find the enthalpy, temperature, and concentration of the final mixture.

Solution. The concentration of the mixture x_3 is given by Eq. 15-21. Using Chart 2, we get

$$x_3 = 0.214 + \frac{10}{15}(0.845 - 0.214) = 0.635 \ \frac{\text{lbm ammonia}}{\text{lbm of mixture}}$$

The total mass flow rate $\dot{m}_3$ is the sum of the two inlet flow rates and

$$\frac{\dot{q}}{\dot{m}_3} = \frac{4200}{15} = 280 \ \text{Btu/lbm}$$

Then using Chart 2, point 3 is located at $x_3 = x_{3'} = 0.635$ and

$$i_3 = i_{3'} - \frac{\dot{q}}{\dot{m}_3} = 578 - 280 = 298 \ \text{Btu/lbm}$$

Utilizing a straightedge and the equilibrium construction line, we find that the final temperature t_3 is about 148 F.

Heating and Cooling Processes

The absorption refrigeration cycle requires vaporization and condensation processes. For ammonia–water systems rectification is necessary to produce high-purity ammonia vapor. This is done by alternate heating and cooling processes. Figure 15-32 shows a simplified arrangement to accomplish this. To better visualize the processes, the liquid and vapor phases are separated following each heat exchanger.

For heat exchanger A

$$\dot{q}_{12} = \dot{m}_1(i_2 - i_1)$$

Figure 15-32.   Simplified apparatus for rectification of binary mixture.

$$\dot{m}_1 = \dot{m}_2$$
$$x_1 = x_2$$

For separator A,

$$\dot{m}_2 i_2 = \dot{m}_3 i_3 + \dot{m}_4 i_4$$
$$\dot{m}_2 = \dot{m}_3 + \dot{m}_4$$
$$\dot{m}_2 x_2 = \dot{m}_3 x_3 + \dot{m}_4 x_4$$

Combination of the foregoing energy- and mass-balance equations yields

$$\frac{\dot{m}_3}{\dot{m}_2} = \frac{x_4 - x_2}{x_4 - x_3} = \frac{i_4 - i_2}{i_4 - i_3} \qquad (15\text{-}26)$$

$$\frac{\dot{m}_4}{\dot{m}_2} = \frac{x_2 - x_3}{x_4 - x_3} = \frac{i_2 - i_3}{i_4 - i_3} \qquad (15\text{-}27)$$

Figure 15-33 shows the state points 1, 2, 3, and 4 on the $i-x$ diagram. The heat transfer $\dot{q}_{12}/\dot{m}_1$ may be determined graphically as shown, and the fractional components for the separator may be determined directly from the diagram.

Figure 15-33. Schematic i–x diagram for Figure 15-31.

$$\frac{\dot{m}_3}{\dot{m}_2} = \frac{\overline{24}}{\overline{34}} \quad \text{and} \quad \frac{\dot{m}_4}{\dot{m}_2} = \frac{\overline{32}}{\overline{34}}$$

The symbol $\overline{24}$ refers to the length of the line connecting points 2 and 4. Heat exchanger B and separator B may be analyzed in exactly the same way. The state points are shown in Fig. 15-33. Notice that the vapor at state 7 is almost 100 percent ammonia. This is necessary for the aqua-ammonia absorption cycle to operate efficiently. In actual practice the simple arrangement of Fig. 15-32 is inadequate for the separation of a binary mixture, and a rectifying column must be introduced between the two heat exchangers as shown in Fig. 15-34. Heat exchanger A is called the generator and heat exchanger B is referred to as the *dephlegmator*. Analysis of the dephlegmator is somewhat involved and the reader is referred to reference 2.

Throttling Process

The throttling process occurs in most refrigeration cycles. A throttling valve is shown schematically in Fig. 15-35a. Although evaporation occurs in

Figure 15-34.   The rectifying column.

throttling, and the temperature of the mixture changes, an energy balance gives $i_2 = i_1$ and the concentration remains constant $x_2 = x_1$. The state points 1 and 2 are identical on the $i-x$ diagram of Fig. 15-35b; however, it should be noted that state 1 is at pressure P_1 and state 2 is at a pressure P_2. The line $\overline{f2g}$ is located by trial and error using a straightedge and the equilibrium construction line. The temperature t_2 will generally be less than t_1. The fractional components of liquid and vapor may be determined from the line segment ratios of $\overline{f2g}$.

Example 15-9.   Saturated aqua-ammonia at 100 psia, 240 F, with a concentration of 0.3 lbm ammonia/lbm of mixture is throttled in a steady-flow process to 30 psia. Find the temperature and fractions of liquid and vapor at state 2.

Solution.   State 1 is first located using Chart 2 and is shown schematically in Fig. 15-36. The isotherm for 240 F has been located using a straightedge.

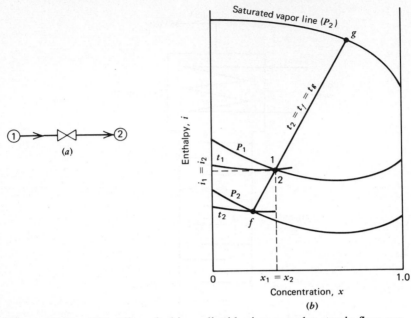

Figure 15-35. Throttling of a binary liquid mixture under steady-flow conditions.

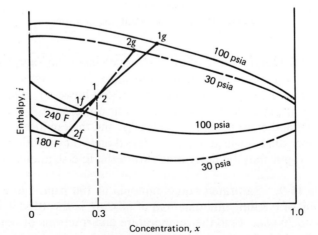

Figure 15-36. Schematic of throttling process of Example 15-9.

In this case state 1 is a two-phase mixture. State 2 is located at the same point as state 1 on the diagram. The isotherm that passes through point 2 for a pressure of 30 psia is then located by trial and error and is found to be 180 F as shown in Fig. 15-36. The fraction of vapor at state 2 is given by

$$\frac{\dot{m}_v}{\dot{m}} = \frac{\overline{22f}}{2f2g} = \frac{2.2 \text{ in.}}{7.5 \text{ in.}} = 0.289 \frac{\text{lbm vapor}}{\text{lbm mixture}}$$

$$\frac{\dot{m}_f}{\dot{m}} = 1 - 0.289 = 0.711 \frac{\text{lbm liquid}}{\text{lbm mixture}}$$

15-9 THE THEORETICAL ABSORPTION REFRIGERATION SYSTEM

Figure 15-37 shows a schematic arrangement of components for a simple theoretical absorption cycle. For simplicity we shall assume that the absorbent does not vaporize in the generator; thus only the refrigerant flows through the condenser, expansion valve, and evaporator. The vapor leaving the evaporator is absorbed by the weak liquid in the absorber as heat is transferred from the mixture. The refrigerant-enriched solution is then pumped to the pressure level in the generator where refrigerant vapor is driven off by heat transfer to the

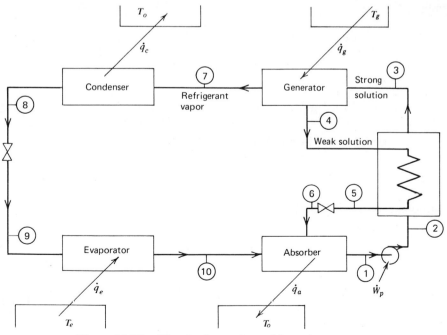

Figure 15-37. Simple absorption refrigeration system.

solution while the weak solution returns to the absorber by way of the inter-cooler.

An ideal system would be completely reversible; however, this is not theoretically possible with the system of Fig. 15-37. The Carnot cycle meets this requirement for a vapor-compression system. the maximum attainable coefficient of performance for the absorption cycle may be determined following a method developed by Bosnjakovic (1).

By the first law of thermodynamics

$$\dot{q}_a + \dot{q}_c = \dot{q}_e + \dot{q}_g + \dot{W}_p \qquad (15\text{-}28)$$

where the waste heat is

$$\dot{q}_o = \dot{q}_a + \dot{q}_c \qquad (15\text{-}29)$$

It will be assumed that the environment temperature T_o, the generator heating medium temperature T_g, and the refrigerated substance temperature T_e are all constant, absolute temperatures. The second law of thermodynamics requires that the net change in entropy for the system plus the surroundings must be greater than or equal to zero. Since the working fluid undergoes a cycle, its change in entropy is zero. Therefore,

$$\Delta S_{\text{total}} = \Delta S_g + \Delta S_e + \Delta S_o \geq 0 \qquad (15\text{-}30)$$

Because the reservoirs are internally reversible, their entropy changes may be computed as

$$\Delta S_g = -\frac{\dot{q}_g}{T_g}; \qquad \Delta S_e = -\frac{\dot{q}_e}{T_e}; \qquad \Delta S_o = \frac{\dot{q}_o}{T_o} \qquad (15\text{-}31)$$

and

$$\Delta S_{\text{total}} = -\frac{\dot{q}_g}{T_g} - \frac{\dot{q}_e}{T_e} + \frac{\dot{q}_o}{T_o} \geq 0 \qquad (15\text{-}32)$$

By combining Eqs. 15-31 and 15-32 it may be shown that

$$\dot{q}_g \frac{T_g - T_o}{T_g} \geq \dot{q}_e \frac{T_o - T_e}{T_e} - \dot{W}_p \qquad (15\text{-}33)$$

If the pump power $\dot{W}_p$ is neglected, Eq. 15-33 may be rearranged to give

$$\text{COP} = \frac{\dot{q}_e}{\dot{q}_g} \leq \frac{T_e(T_g - T_o)}{T_g(T_o - T_e)} \qquad (15\text{-}34)$$

and when all the processes are reversible

$$(\text{COP})_{\text{max}} = \frac{T_e(T_g - T_o)}{T_g(T_o - T_e)} \qquad (15\text{-}35)$$

Equation 15-35 shows that the maximum coefficient of performance for an absorption cycle is equal to the COP for a Carnot cycle operating between temperatures T_e and T_o multiplied by the efficiency of a Carnot engine operating between temperatures T_g and T_o. It is also shown that for a given environment temperature T_o the COP will increase with an increase in T_g or T_e. Unfortunately, practical absorption cycles have COPs much less than that given by Eq. 15-35.

15-10   THE AQUA-AMMONIA ABSORPTION SYSTEM

The aqua-ammonia system is one of the oldest absorption refrigeration cycles. The ammonia is the refrigerant and the water is the absorbent. Because both the water and ammonia are volatile, the generator of the simple cycle must be replaced by the combination of generator, rectifying column, and dephlegmator as shown in Fig. 15-38. This is necessary to separate almost all of the water vapor from the ammonia vapor. Note that one additional heat exchanger has also been added. A complete cycle will now be analyzed making use of the basic processes previously described.

Example 15-10   Consider the cycle of Fig. 15-38 and the following given data: condensing pressure, 200 psia; evaporating pressure, 30 psia; generator temperature, 240 F; temperature of vapor leaving dephlegmator,

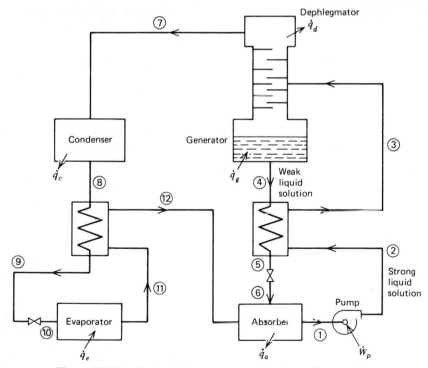

Figure 15-38.   Industrial aqua-ammonia absorption system.

130 F; and temperature of the strong solution entering the rectifying column, 200 F. The heat exchanger lowers the temperature of the liquid leaving the condenser 10 F.

States 1, 3, 4, 7, 8, and 12 are saturated. Pressure drop in components and connecting lines is negligible. The system produces 100 tons of refrigeration. Determine (a) properties P, t, x, and i for all state points of the system, (b) mass flow rate for all parts of the system, (c) horsepower required for the pump assuming 75 percent mechanical efficiency, (d) system coefficient of performance, and (e) system refrigerating efficiency.

Solution. Table 15-4 is a tabulation of thermodynamic properties and flow rates for the problem. Figure 15-39 is a schematic $i - x$ diagram showing the state points.

The given data establish all of the pressures, and the temperatures t_3, t_4, and t_7 are given. States 3, 4, and 7 are saturated states and can be located on Chart 2 and the concentration and enthalpy values read. States 7 through 12 have the same concentration because no mixing occurs. It is also true that states 1, 2, and 3 have the same concentration and $x_4 = x_5 = x_6$ as well. Now states 1, 8, and 12 are saturation states and may be located on Chart 2 making use of the known pressures and concentrations and i_1, i_8, and i_{12} read. Since $t_9 = t_8 - 10$, state 9 is determined from t_9 and x_9 and i_9 is read from the chart.

An energy balance between points 8 and 12 in Fig. 15-39 yields

$$\dot{m}_8 i_8 + \dot{q}_e = \dot{m}_{12} i_{12} = \dot{m}_8 i_{12}$$

$$\dot{m}_8 = \frac{\dot{q}_e}{i_{12} - i_8} = \frac{(100)(200)}{(631 - 148)} = 41.4 \text{ lbm/min}$$

From Fig. 15-38 it is obvious that $\dot{m}_7 = \dot{m}_8 = \dot{m}_9 = \dot{m}_{10} = \dot{m}_{11} = \dot{m}_{12}$.

Table 15-4   PROPERTIES AND FLOW RATES FOR EXAMPLE 15-10

State	P psia	t F	x lbm NH$_3$/lbm	i Btu/lbm	$\dot{m}$ lbm/min
1	30	79	0.408	− 25	262,7
2	200	79	0.408	− 24.4	262.7
3	200	200	0.408	109	262.7
4	200	240	0.298	159	221.3
5	200	97	0.298	0	221.3
6	30	97	0.298	0	221.3
7	200	130	0.996	650	41.4
8	200	97	0.996	148	41.4
9	200	86	0.996	137	41.4
10	30	0	0.996	137	41.4
11	30	40	0.996	620	41.4
12	30	57	0.996	631	41.4

Figure 15-39.   Schematic $i-x$ diagram for Example 15-10.

Mass balance on the absorber gives

$$\dot{m}_{12} + \dot{m}_6 = \dot{m}_1$$
$$m_{12}x_{12} + \dot{m}_6 x_6 = \dot{m}_1 x_1$$

and

$$\dot{m}_6 = \dot{m}_{12} \frac{(x_{12} - x_1)}{(x_1 - x_6)} = 41.4 \frac{(0.996 - 0.408)}{(0.408 - 0.298)}$$

$$\dot{m}_6 = 221.3 \text{ lbm/min}$$

$$\dot{m}_1 = \dot{m}_{12} + \dot{m}_6 = 221.3 + 41.4 = 262.7 \text{ lbm/min}$$

From Fig. 15-39, $\dot{m}_1 = \dot{m}_2 = \dot{m}_3$ and $\dot{m}_4 = \dot{m}_5 = \dot{m}_6$.

The pump work may be expressed as

$$w_{12} = i_1 - i_2 = \frac{(P_1 - P_2)v_1}{J}$$

where $J = 778$ (ft-lbf)/Btu. The specific volume v_1 is an empirically de-termined quantity and is given in Table 33, Chapter 17 of reference 3. For $x_1 = 0.408$ and $t_1 = 79$ F, $v_1 = 0.0187$ ft³/lbm. Then

$$w_{12} = \frac{144(30 - 200)(0.0187)}{778}$$

$$w_{12} = -0.588 \text{ Btu/lbm}$$
$$\dot{W}_p = w_{12}\dot{m}_1 = (-0.588)(262.7) = -154.6 \text{ Btu/min}$$
$$i_2 = i_1 - w_{12} = -25 - (-0.588) = -24.4 \text{ Btu/lbm}$$

The enthalpy, pressure, and concentration at state 2 then establish t_2.

The energy balance on the solution heat exchanger yields

$$\dot{m}_2 i_2 + \dot{m}_4 i_4 = \dot{m}_3 i_3 + \dot{m}_5 i_5$$
$$\dot{m}_4 = \dot{m}_5 \quad \text{and} \quad \dot{m}_2 = \dot{m}_3$$

Then

$$i_5 = i_4 - \frac{\dot{m}_2}{\dot{m}_4}(i_3 - i_2) = 159 - \frac{262.7}{221.3}[109 - (-24.4)]$$

$$i_5 = 0.64 \approx 0 \text{ Btu/lbm}$$

The temperature t_5 is then read from Chart 2 as a function of P_5, x_5, and i_5. The enthalpy $i_6 = i_5$ are coincident on Chart 2; however, state 5 is at 200 psia, whereas state 6 is at 30 psia. Since state 6 is subcooled, $t_6 = t_5$.

States 9 and 10 are coincident on Chart 2 since the process is one of throttling. Temperature t_{10} may be found using the method described in Example 15-8.

The energy balance on the evaporator gives

$$\dot{m}_{10}i_{10} + \dot{q}_e = \dot{m}_{11}i_{11}$$
$$\dot{m}_{10} = \dot{m}_{11}$$

$$i_{11} = \frac{\dot{q}_e}{\dot{m}_{10}} + i_{10}$$

$$i_{11} = \frac{(200)(100)}{41.4} + 137 = 620 \text{ Btu/lbm}$$

State 11 is a mixture of liquid and vapor and its temperature is found from Chart 2 to be 40 F.

The pump horsepower requirement may be computed from W_p above.

$$\text{Power} = \frac{\dot{W}_p}{\eta_m} = \frac{154.6(778)}{0.75(33,000)} = 4.86 \text{ hp}$$

To compute the coefficient of performance it is necessary to establish the generator heat-transfer rate. However, this cannot be done without an analysis of the complete generator–rectifier column dephlegmator unit. To make a reasonable analysis, details of the column design and experimental data are required. For purposes of simplicity it will be assumed that the generator heat-transfer rate is 115 percent of the net heat transfer for the complete rectifying unit. An energy balance gives

$$\dot{m}_3 i_3 + \dot{q}_g = \dot{m}_4 i_4 + \dot{m}_7 i_7 + \dot{q}_d$$
$$\dot{q}_{net} = (\dot{q}_g - \dot{q}_d) = \dot{m}_4 i_4 + \dot{m}_7 i_7 - \dot{m}_3 i_3$$
$$= 221.3(159) + 41.4(650) - 262.7(109)$$
$$\dot{q}_{net} = 33,462.4 \text{ Btu/min}$$

Then

$$\dot{q}_g = 1.15 \, \dot{q}_{net} = 38,480 \text{ Btu/min}$$
$$\dot{q}_d = 38,480 - 33,462 = 5020 \text{ Btu/min}$$

Neglecting the pump work

$$\text{COP} = \frac{\dot{q}_e}{\dot{q}_g} = \frac{100(200)}{38,480} = 0.52$$

The maximum COP may be calculated from Eq. 15-35. It will be assumed that $t_o = 80$ F and $t_g = 240$ F and $t_e = 0$ F were given.

$$(\text{COP})_{max} = \frac{460(700 - 540)}{700(540 - 460)} = 1.31$$

and

$$\eta_R = \frac{\text{COP}}{(\text{COP})_{max}} = \frac{0.52}{1.31} = 0.40$$

15-11   THE LITHIUM BROMIDE–WATER ABSORPTION SYSTEM

In recent times the lithium bromide–water absorption system has become popular for air-conditioning applications. Water is the refrigerant, whereas the lithium bromide is the absorbent. Although lithium bromide is normally solid, when mixed with water a liquid solution may be formed.

The greatest advantage of the system is the nonvolatility of the lithium bromide. Only water vapor is driven off in the generator. The lithium bromide–water system is simpler and operates with a higher COP than the water–ammonia system. The main disadvantage is the relatively high evaporating temperatures and very low system pressures. Here is an example of a typical system analysis.

Example 15-11   The following data are given for a lithium bromide–water system of the type shown in Fig. 15-37: condensing temperature, 100 F; evaporating temperature, 40 F; temperature of strong solution leaving absorber, 100 F; temperature of strong solution entering generator, 180 F; and generator temperature, 200 F.

Saturated conditions exist for states 3, 4, 8, and 10. Pressure drop in components and lines is negligible. The system cooling capacity is 10 tons. Determine (a) the thermodynamic properties of P, t, x, and i for all necessary state points, (b) the mass flow rate for each part of the system, (c) the system coefficient of performance, (d) the system refrigerating efficiency, and (e) the steam rate for the generator if saturated steam at 220 F is used.

Solution.   Parts (a) and (b) will be solved concurrently. Table 15-5 shows a tabulation of data. Properties of pure water were obtained from Table A-1, whereas solution properties were obtained from Chart 3. Note that Chart 3 expresses concentration in terms of the absorbent lithium bromide.

The high- and low-side pressures are found from the temperatures t_8 and t_{10}, and i_8 and i_{10} are found in Table A-1. The enthalpy of the pure water vapor at state 7 may be computed by Eq. 2-18 because it behaves as an ideal gas at the low pressure. States 3 and 4 are saturated conditions and may be located on Chart 3 as shown in Fig. 15-40. Enthalpy and concentration values may then be read for states 3 and 4.

For the evaporator

$$\dot{m}_9 = \frac{\dot{q}_e}{(i_{10} - i_9)} = \frac{10(200)}{1011} = 1.98 \text{ lb/min}$$

Table 15-5   PROPERTIES AND FLOW RATES FOR EXAMPLE 15-11

State	P mm/Hg	t F	x lb LiBr/lbm	i Btu/lbm	$\dot{m}$ lbm/min
1	6.3	100	0.6	—	25.8
2	49.1	—	0.6	—	25.8
3	49.1	180	0.6	−35	25.8
4	49.1	200	0.65	−27	23.8
5	49.1	—	0.65	—	23.8
6	6.3	—	0.65	—	23.8
7	49.1	200	0	1151	1.98
8	49.1	100	0	68	1.98
9	6.3	40	0	68	1.98
10	6.3	40	0	1079	1.98

Figure 15-40.   Schematic i–x diagram for Example 15-11.

For the absorber

$$\dot{m}_6 = \dot{m}_{10} \frac{(x_1 - x_{10})}{(x_6 - x_1)} = \frac{1.98(0.60)}{0.05} = 23.8 \text{ lb/min}$$

$$\dot{m}_1 = \dot{m}_6 + \dot{m}_{10} = 25.8 \text{ lb/min}$$

For the generator

$$\dot{q}_g = \dot{m}_4 i_4 + \dot{m}_7 i_7 - \dot{m}_3 i_3$$
$$= 23.8(-27) + 1.98(1151) - 25.8(-35)$$
$$\dot{q}_g = 2538 \text{ Btu/min}$$

Neglecting the pump work

$$\text{COP} = \frac{\dot{q}_e}{\dot{q}_g} = \frac{200(10)}{2538} = 0.79$$

For a generator source temperature of 220 F, refrigerated medium temperature of 45 F, and assuming an environment temperature of 100 F the maximum COP becomes

$$(\text{COP})_{\text{max}} = \frac{505(680 - 560)}{680(560 - 505)} = 1.62$$

and

$$\eta_R = \frac{0.79}{1.62} = 0.49$$

Assuming saturated water at 220 F leaves the steam coil

$$\dot{m}_s = \frac{60\dot{q}_g}{i_{fg}} = \frac{60(2538)}{965} = 158 \text{ lb/hr}$$

REFERENCES

1. Fran Bosnjakovic, *Technical Thermodynamics*, translated by Perry L. Blackshear, Jr., Holt, Rinehart, and Winston, New York, 1965.

2. James L. Threlkeld, *Thermal Environmental Engineering*, 2nd ed., Prentice-Hall, Englewood Cliffs, N.J., 1970.

3. *ASHRAE Handbook, Fundamentals Volume*, American Society of Heating, Refrigerating and Air-Conditioning Engineers, Atlanta, Ga., 1985.

4. *ASHRAE Handbook, Equipment Volume*, American Society of Heating, Refrigerating and Air-Conditioning Engineers, Atlanta, Ga., 1983.

5. W. F. Stoecker, *Refrigeration and Air Conditioning*, McGraw-Hill, New York, 1958.

6. *ASHRAE Handbook, Refrigeration Volume*, American Society of Heating, Refrigerating and Air-Conditioning Engineers, Atlanta, Ga., 1986.

PROBLEMS

15-1 A vapor compression refrigeration cycle operates on R-22 and follows the theoretical single-stage cycle. The condensing temperature is 110 F and the evaporating temperature is 40 F. The system produces 10 tons of cooling effect. Determine the (a) coefficient of performance, (b) refrigerating efficiency, (c) hp/ton of refrigeration, (d) mass flow rate of the refrigerant, (e) theoretical hp input to the compressor, and (f) theoretical piston displacement of the compressor in ft^3/min.

15-2 A vapor compression refrigeration cycle uses R-22 and follows the theoretical single-stage cycle. The condensing temperature is 48 C and the evaporating temperature is -18 C. Power input to the cycle is 2.5 kW and the mass flow rate of refrigerant is 0.05 kg/s. Determine (a) heat rejected from the condenser, (b) the coefficient of performance, (c) the enthalpy at the compressor exit, and (d) the refrigerating efficiency.

15-3 An R-12 system is arranged as shown in Fig. 15-41. Compression is isentropic. Assume frictionless flow. Find the system hp/ton.

15-4 Consider a reciprocating compressor operating with R-12. Refrigerant enters the cylinder at 20 psia (138 kPa) and 20 F (-7 C), but leaves the evaporator saturated

Figure 15-41. Schematic for Problem 15-3.

at 0.5 F (− 18 C). The vapor is discharged from the cylinder at 180 psia (1.24 MPa). Compute the volumetric efficiency for (a) a clearance factor of 0.03; (b) a clearance factor of 0.15; (c) compare the mass flow rates for (a) and (b); and (d) compare the power input to the compressor in (a) and (b).

15-5 Consider a 4-cylinder, 3-in. bore by 4-in. stroke, 800 rpm, single-acting compressor for use with R-12. The proposed operating condition for the compressor is 100 F condensing temperature and 40 F evaporating temperature. It is estimated that the refrigerant will enter the expansion valve as a saturated liquid, that vapor will leave the evaporator at a temperature of 45 F, and that vapor will enter the compressor at a temperature of 55 F. Assume a compressor volumetric efficiency of 70 percent and frictionless flow. Calculate the refrigeration capacity in tons.

15-6 Consider the compressor of Fig. 15-10.
  (a) Construct the pressure-enthalpy diagram for a condensing temperature of 130 F (54 C) and an evaporating temperature of 45 F (7 C).
  (b) What is the heat-transfer rate in the evaporator and the power input?
  (c) Suppose the load on the evaporator decreases to 110,000 Btu/hr (32 kW) and find the evaporating temperature and power input. Assume that the condensing temperature remains constant.

15-7 Refer to Problem 15-6a and sketch the capacity and power curves for 130 F (54 C) from Fig. 15-10. (a) Sketch the capacity curve for the evaporator assuming its capacity is proportional to the evaporating temperature in the ratio 4000 Btu/(hr-F) (2.1 kW/C). (b) The evaporator load decreases to 130,000 Btu/hr (38 kW) while the condensing temperature decreases to 115 F (46 C). Sketch the new evaporator-capacity curve. (c) Suppose in part (a) that the evaporator operating conditions remain fixed but the condensing temperature increases to 145 F (63 C). What will the capacity and evaporating temperature be?

15-8 A refrigeration system using the compressor described in Table 15-3 is designed to operate with a condensing temperature of 120 F with an evaporating temperature of 50 F when the outdoor ambient is 95 F. After the system is put into operation, the evaporator pressure is measured to be 69 psig and the power to the compressor is 20.7 kW. Superheat and subcooling are assumed to be as given in Table 15-3 and the ambient is about 95 F. (a) Sketch the capacity and power curves for the compressor and estimate the condensing and evaporating temperatures, (b) compare the actual and expected performance, and (c) suggest what might be done to obtain the design conditions. Show with your sketch.

15-9 Saturated R-22 at 45 F enters the compressor of a single-stage system. Discharge pressure is 275 psia. Suction valve pressure drop is 2 psi. Discharge valve pressure drop is 4 psi. Assume the vapor is superheated 10 F in the cylinder during the intake stroke. Piston clearance is 5 percent. Determine the (a) volumetric efficiency, (b) pumping capacity in lbm/min for 20 ft³/min piston displacement, and (c) horsepower requirement if mechanical efficiency is 80 percent.

15-10 Consider the single-stage vapor-compression cycle shown in Fig. 15-42. Design conditions using R-12 are:

$$\dot{q}_L = 30{,}000 \text{ Btu/hr} \qquad P_3 = 200 \text{ psia}$$
$$P_1 = 60 \text{ psia saturated} \qquad P_3 - P_4 = 2 \text{ psi}$$
$$P_2 = 55 \text{ psia} \qquad C = 0.04$$
$$T_2 = 60 \text{ F} \qquad \eta_m = 0.90$$
$$PD = 9.4 \text{ cfm}$$

(a) Determine $\dot{W}$, $\dot{q}_H$, $\dot{m}_{12}$, and sketch the cycle on a P vs. i diagram. If the load $\dot{q}_L$ decreases to 24,000 Btu/hr and the system comes to equilibrium with $P_2 = 50$ psia and $T_2 = 50$ F, (b) determine $\dot{W}$, $\dot{q}_H$, $\dot{m}$ and locate the cycle on a P vs. i diagram.

15-11 Consider an ordinary single-stage vapor-compression air-conditioning system. Because of clogged filters the air flow over the evaporator is gradually reduced to a very low level. Explain how the evaporator and compressor will be affected if the system continues to operate.

Figure 15-42. Schematic for Problem 15-10.

15-12 A vapor-compression cycle is subject to short periods of very light load; it is not practical to shut the system down. During these periods of light load, moisture condenses from the air flowing over the evaporator and freezes. Suggest a modification to the system to prevent this condition.

15-13 A vapor-compression cycle is subject to occasional overload that leads to the tripping of circuit breakers. Explain how the system can be modified to prevent compressor overload without shutting the system off.

15-14 Saturated water vapor at 50 F is mixed in a steady-flow process with a saturated lithium bromide–water solution having a concentration of 0.60 lbm LiBr per lbm of mixture. The mass of the liquid solution is five times the mass of the water vapor mixed. The mixing process occurs at constant pressure. Find (a) the concentration of the resulting mixture, and (b) the heat that must be removed in Btu/lbm of the final mixture if a saturated liquid solution is produced.

15-15 A saturated liquid aqua-ammonia solution at 220 F and 200 psia is throttled to a pressure of 10 psia. Find (a) the temperature after the throttling process, and (b) the relative portions of liquid and vapor in the mixture after throttling.

15-16 A solution of ammonia and water at 180 F, 100 psia, and with a concentration of 0.25 lbm ammonia per lbm of solution is heated at constant pressure to a temperature of 280 F. The vapor is then separated from the liquid and cooled to a saturated liquid at 100 psia. What are the temperature and concentration of the saturated liquid?

15-17 It is proposed to use hot water at 180 F or 82 C from a solar collector system to operate a simple absorption cycle. Compute the maximum possible coefficient of performance assuming an environment temperature of 100 F or 38 C and an air-cooled evaporator with the air temperature at 75 F or 24 C.

15-18 Solve Example 15-11 using the following conditions:

Condensing temperature, 95 F
Evaporating temperature, 45 F
Strong solution temperature leaving absorber, 105 F
Strong solution temperature entering generator, 180 F
Generator temperature, 200 F

15-19 Solar energy is to be used to operate a lithium bromide–water absorption system. The cooling capacity is 3 tons. Analyze the cycle following the approach used in Example 15-11 and the following conditions:

Condensing temperature, 100 F
Evaporating temperature, 40 F
Strong solution temperature leaving absorber, 105 F
Strong solution temperature entering generator, 140 F
Generator temperature, 160 F

Thermodynamic Properties

Table A-1 WATER AT SATURATION—ENGLISH UNITS

Fahrenheit Temperature	Pressure P psia	Specific Volume ft³/lbm		Enthalpy Btu/lbm			Entropy Btu/(lbm-R)		
		Saturated Solid v_i	Saturated Vapor $v_g \times 10^{-3}$	Saturated Solid i_i	Sublimation i_{ig}	Saturated Vapor i_g	Saturated Solid s_i	Sublimation s_{ig}	Saturated Vapor s_g
32.018	0.0887	0.01747	3.302	−143.34	1218.7	1075.4	−0.292	2.479	2.187
30	0.0808	0.01747	3.607	−144.35	1218.9	1074.5	−0.294	2.489	2.195
25	0.0641	0.01746	4.506	−146.84	1219.1	1072.3	−0.299	2.515	2.216
20	0.0505	0.01745	5.655	−149.31	1219.4	1070.1	−0.304	2.542	2.238
15	0.0396	0.01745	7.13	−151.75	1219.7	1067.9	−0.309	2.569	2.260
10	0.0309	0.01744	9.04	−154.17	1219.9	1065.7	−0.314	2.597	2.283
5	0.0240	0.01743	11.52	−156.56	1220.1	1063.5	−0.320	2.626	2.306
0	0.0185	0.01743	14.77	−158.93	1220.2	1061.2	−0.325	2.655	2.330

Fahrenheit Temperature	Pressure P psia	Specific Volume ft³/lbm		Enthalpy Btu/lbm			Entropy Btu/(lbm-R)		
		Saturated Liquid v_f	Saturated Vapor v_g	Saturated Liquid i_f	Evaporation i_{fg}	Saturated Vapor i_g	Saturated Liquid s_f	Evaporation s_{fg}	Saturated Vapor s_g
32.018	0.08866	0.016022	3302	0.01	1075.4	1075.4	0.00000	2.1869	2.1869
36	0.10397	0.016021	2839	4.00	1073.1	1077.1	0.00810	2.1648	2.1729
40	0.12166	0.016020	2445	8.02	1070.9	1078.9	0.01617	2.1430	2.1592
44	0.14196	0.016021	2112	12.03	1068.6	1080.7	0.02418	2.1215	2.1457
48	0.16520	0.016023	1829	16.05	1066.4	1082.4	0.03212	2.1003	2.1324
52	0.19173	0.016026	1588.6	20.07	1064.1	1084.2	0.04000	2.0795	2.1195
56	0.2219	0.016030	1382.9	24.08	1061.9	1085.9	0.04781	2.0590	2.1068
60	0.2563	0.016035	1206.9	28.08	1059.6	1087.7	0.05555	2.0388	2.0943
64	0.2952	0.016040	1055.8	32.09	1057.3	1089.4	0.06323	2.0189	2.0821
68	0.3391	0.016047	925.8	36.09	1055.1	1091.2	0.07084	1.9993	2.0701
72	0.3887	0.016055	813.7	40.09	1052.8	1092.9	0.07839	1.9800	2.0584

76	0.4446	0.016063	716.8	44.09	1050.6	1094.7	0.08589	1.9610	2.0469
80	0.5073	0.016073	632.8	48.09	1048.3	1096.4	0.09332	1.9423	2.0356
84	0.5776	0.016083	559.8	52.08	1046.0	1098.1	0.10069	1.9238	2.0245
88	0.6562	0.016094	496.3	56.07	1043.8	1099.9	0.10801	1.9056	2.0136
92	0.7439	0.016105	440.9	60.06	1041.5	1101.6	0.11527	1.8877	2.0030
96	0.8416	0.016117	392.4	64.06	1039.2	1103.3	0.12248	1.8700	1.9925
100	0.9503	0.016130	350.0	68.05	1037.0	1105.0	0.12963	1.8526	1.9822
104	1.0708	0.016144	312.8	72.04	1034.7	1106.7	0.13674	1.8354	1.9722
108	1.2044	0.016158	280.0	76.03	1032.4	1108.4	0.14379	1.8185	1.9623
112	1.3520	0.016173	251.1	80.02	1030.1	1110.2	0.15079	1.8017	1.9525
116	1.5150	0.016189	225.6	84.01	1027.8	1111.9	0.15775	1.7852	1.9430
120	1.6945	0.016205	203.0	88.00	1025.5	1113.5	0.16465	1.7690	1.9336
124	1.8921	0.016221	183.1	91.99	1023.2	1115.2	0.17152	1.7529	1.9244
128	2.1090	0.016239	165.30	95.98	1020.9	1116.9	0.17833	1.7370	1.9154
132	2.347	0.016256	149.51	99.98	1018.6	1118.6	0.18510	1.7214	1.9065
136	2.607	0.016275	135.44	103.97	1016.3	1120.3	0.19183	1.7059	1.8978
140	2.892	0.016293	122.88	107.96	1014.0	1121.9	0.19851	1.6907	1.8892
144	3.203	0.016313	111.66	111.96	1011.6	1123.6	0.20515	1.6756	1.8807
148	3.541	0.016333	101.61	115.96	1009.3	1125.2	0.21175	1.6607	1.8725
152	3.910	0.016353	92.60	119.96	1006.9	1126.9	0.21831	1.6460	1.8643
156	4.310	0.016374	84.51	123.96	1004.6	1128.5	0.22482	1.6315	1.8563
160	4.745	0.016395	77.23	127.96	1002.2	1130.1	0.23130	1.6171	1.8484
164	5.216	0.016417	70.67	131.96	999.8	1131.8	0.23774	1.6029	1.8407
168	5.726	0.016439	64.76	135.97	997.4	1133.4	0.24414	1.5889	1.8330
172	6.277	0.016462	59.42	139.97	995.0	1135.0	0.25050	1.5750	1.8255
176	6.873	0.016485	54.58	143.98	992.6	1136.6	0.25682	1.5613	1.8181
180	7.515	0.016509	50.20	147.99	990.2	1138.2	0.26311	1.5478	1.8109
184	8.206	0.016533	46.23	152.01	987.7	1139.7	0.26936	1.5344	1.8037
188	8.951	0.016558	42.62	156.02	985.3	1141.3	0.27557	1.5211	1.7967
192	9.750	0.016583	39.34	160.04	982.8	1142.9	0.28175	1.5080	1.7898
196	10.609	0.016608	36.36	164.05	980.4	1144.4	0.28789	1.4951	1.7830
200	11.529	0.016634	33.63	168.07	977.9	1145.9	0.29400	1.4822	1.7762
204	12.515	0.016661	31.15	172.10	975.4	1147.5	0.30008	1.4695	1.7696
208	13.570	0.016688	28.88	176.13	972.9	1149.0	0.30612	1.4570	1.7631
212	14.698	0.016716	26.80	180.16	970.3	1150.5	0.31213	1.4446	1.7567

Table A-1 WATER AT SATURATION—ENGLISH UNITS *(continued)*

Fahrenheit Temperature	Pressure P psia	Specific Volume ft³/lbm		Enthalpy Btu/lbm			Entropy Btu/(lbm-R)		
		Saturated Liquid v_f	Saturated Vapor v_g	Saturated Liquid i_f	Evaporation i_{fg}	Saturated Vapor i_g	Saturated Liquid s_f	Evaporation s_{fg}	Saturated Vapor s_g
216	15.903	0.016743	24.90	184.18	967.8	1152.0	0.31811	1.4322	1.7504
220	17.188	0.016772	23.15	188.22	965.3	1153.5	0.32406	1.4201	1.7441
224	18.557	0.016801	21.55	192.26	962.7	1154.9	0.32998	1.4080	1.7380
228	20.015	0.016830	20.08	196.30	960.1	1156.4	0.33586	1.3961	1.7319
232	21.57	0.016860	18.723	200.34	957.5	1157.9	0.34172	1.3842	1.7260
236	23.22	0.016891	17.476	204.39	954.9	1159.3	0.34755	1.3725	1.7201
240	24.97	0.016922	16.327	208.44	952.3	1160.7	0.35335	1.3609	1.7143
244	26.82	0.016953	15.267	212.49	949.6	1162.1	0.35912	1.3494	1.7085
248	28.79	0.016985	14.287	216.55	947.0	1163.5	0.36486	1.3380	1.7029
248	28.79	0.016985	14.287	216.55	947.0	1163.5	0.36486	1.3380	1.7029
252	30.88	0.017017	13.382	220.62	944.3	1164.9	0.37058	1.3267	1.6973

Source: Abridged from, "Steam Tables (English Units)," by Joseph H. Keenan, F. G. Keyes, P. G. Hill, and J. G. Moore. Copyright © 1969 by John Wiley & Sons, Inc., New York. Reprinted by permission of John Wiley & Sons, Inc.

Table A-1a WATER AT SATURATION—SI UNITS

Temperature C	K	Pressure Pa $P \times 10^{-5}$	Specific Volume m³/kg Saturated Solid $v_i \times 10^3$	Saturated Vapor v_g	Enthalpy kJ/kg Saturated Solid i_i	Sublimation i_{ig}	Saturated Vapor i_z	Entropy kJ/(kg-K) Saturated Solid s_i	Sublimation s_{ig}	Saturated Vapor s_g
0.01	273.16	0.006113	1.0908	2.061	−333.40	2834.8	2501.4	−1.221	10.378	9.156
−2	271.15	0.005176	1.0904	2.417	−337.62	2835.3	2497.7	−1.237	10.456	9.219
−4	269.15	0.004375	1.0901	2.838	−341.78	2835.7	2494.0	−1.253	10.536	9.283
−6	267.15	0.003689	1.0898	3.342	−345.91	2836.2	2490.3	−1.268	10.616	9.348
−8	265.15	0.003102	1.0894	3.944	−350.02	2836.6	2486.6	−1.284	10.698	9.414
−10	263.15	0.002602	1.0891	4.667	−354.09	2837.0	2482.9	−1.299	10.781	9.481
−12	261.15	0.002176	1.0888	5.537	−358.14	2837.3	2479.2	−1.315	10.865	9.550
−14	259.15	0.001815	1.0884	6.588	−362.15	2837.6	2475.5	−1.331	10.950	9.619
−16	257.15	0.001510	1.0881	7.860	−366.14	2837.9	2471.8	−1.346	11.036	9.690

Temperature C	K	Pressure Pa $P \times 10^{-5}$	Specific Volume m³/kg Saturated Liquid $v_f \times 10^3$	Saturated Vapor $v_g \times 10^3$	Enthalpy kJ/kg Saturated Liquid i_f	Evaporation i_{fg}	Saturated Vapor i_g	Entropy kJ/(kg-K) Saturated Liquid s_f	Evaporation s_{fg}	Saturated Vapor s_g
0.01	273.16	0.006113	1.0002	206 136	0.01	2501.3	2501.4	0.0000	9.1562	9.1562
2	275.15	0.007056	1.0001	179 889	8.37	2496.7	2505.0	0.0305	9.0730	9.1035
4	277.15	0.008131	1.0001	157 232	16.78	2491.9	2508.7	0.0610	8.9904	9.0514
6	279.15	0.009349	1.0001	137 734	25.20	2487.2	2512.4	0.0912	8.9090	9.0003
8	281.15	0.010724	1.0002	120 917	33.60	2482.5	2516.1	0.1212	8.8289	8.9501
10	283.15	0.012276	1.0004	106 379	42.01	2477.7	2519.8	0.1510	8.7498	8.9008
12	285.15	0.014022	1.0005	93 784	50.41	2473.0	2523.4	0.1806	8.6718	8.8524
14	287.15	0.015983	1.0008	82 848	58.80	2468.3	2527.1	0.2099	8.5949	8.8048

Table A-1a WATER AT SATURATION—SI UNITS (*continued*)

Temperature		Pressure Pa	Specific Volume m³/kg		Enthalpy kJ/kg			Entropy kJ/(kg-K)		
			Saturated Liquid	Saturated Vapor	Saturated Liquid	Evaporation	Saturated Vapor	Saturated Liquid	Evaporation	Saturated Vapor
C	K	$P \times 10^{-5}$	$v_f \times 10^3$	$v_g \times 10^3$	i_f	i_{fg}	i_g	s_f	s_{fg}	s_g
16	289.15	0.018181	1.0011	73 333	67.19	2463.6	2530.8	0.2390	8.5191	8.7582
18	291.15	0.020640	1.0014	65 038	75.58	2458.8	2534.4	0.2679	8.4443	8.7123
20	293.15	0.02339	1.0018	57 791	83.96	2454.1	2538.1	0.2966	8.3706	8.6672
22	295.15	0.02645	1.0022	51 447	92.33	2449.4	2541.7	0.3251	8.2979	8.6229
24	297.15	0.02985	1.0027	45 883	100.70	2444.7	2545.4	0.3534	8.2261	8.5794
26	299.15	0.03363	1.0032	40 994	109.07	2439.9	2549.0	0.3814	8.1552	8.5367
28	301.15	0.03782	1.0037	36 690	117.43	2435.2	2552.6	0.4093	8.0854	8.4946
30	303.15	0.04246	1.0043	32 894	125.79	2430.5	2556.3	0.4369	8.0164	8.4533
32	305.15	0.04759	1.0050	29 540	134.15	2425.7	2559.9	0.4644	7.9483	8.4127
34	307.15	0.05324	1.0056	26 571	142.50	2421.0	2563.5	0.4917	7.8811	8.3728
36	309.15	0.05947	1.0063	23 940	150.86	2416.2	2567.1	0.5188	7.8147	8.3336
38	311.15	0.06632	1.0071	21 602	159.21	2411.5	2570.7	0.5458	7.7492	8.2950
40	313.15	0.07384	1.0078	19 523	167.57	2406.7	2574.3	0.5725	7.6845	8.2570
42	315.15	0.08208	1.0086	17 671	175.91	2401.9	2577.9	0.5991	7.6206	8.2197
44	317.15	0.09111	1.0095	16 018	184.27	2397.2	2581.4	0.6255	7.5574	8.1829
46	319.15	0.10098	1.0103	14 540	192.62	2392.4	2585.0	0.6518	7.4950	8.1468
48	321.15	0.11175	1.0112	13 218	200.97	2387.6	2588.5	0.6779	7.4334	8.1113
50	323.15	0.12349	1.0121	12 032	209.33	2382.7	2592.1	0.7038	7.3725	8.0763
52	325.15	0.13628	1.0131	10 968	217.69	2377.9	2595.6	0.7296	7.3123	8.0419
54	327.15	0.15019	1.0141	10 011	226.04	2373.1	2599.1	0.7552	7.2528	8.0080
56	329.15	0.16529	1.0151	9149	234.41	2368.2	2602.6	0.7807	7.1940	7.9747
58	331.15	0.18166	1.0161	8372	242.77	2363.4	2606.1	0.8060	7.1359	7.9419
60	333.15	0.19940	1.0172	7671	251.13	2358.5	2609.6	0.8312	7.0784	7.9096
62	335.15	0.21860	1.0182	7037	259.49	2353.6	2613.1	0.8562	7.0216	7.8778

64	337.15	0.23934	1.0194	6463	267.86	2348.7	2616.5	0.8811	6.9654	7.8465
66	339.15	0.2617	1.0205	5943	276.23	2343.7	2620.0	0.9058	6.9098	7.8156
68	341.15	0.2859	1.0217	5471	284.61	2338.8	2623.4	0.9304	6.8548	7.7852
70	343.15	0.3119	1.0228	5042	292.98	2333.8	2626.8	0.9549	6.8004	7.7553
72	345.15	0.3399	1.0240	4652	301.36	2328.9	2630.2	0.9792	6.7466	7.7258
74	347.15	0.3699	1.0253	4297	309.74	2323.9	2633.6	1.0034	6.6934	7.6968
76	349.15	0.4022	1.0265	3973	318.13	2318.9	2637.0	1.0275	6.6407	7.6682
78	351.15	0.4368	1.0278	3677	326.51	2313.8	2640.3	1.0515	6.5885	7.6400
80	353.15	0.4739	1.0291	3407	334.91	2308.8	2643.7	1.0753	6.5369	7.6122
82	355.15	0.5136	1.0305	3160	343.30	2303.7	2647.0	1.0990	6.4858	7.5848
84	357.15	0.5560	1.0318	2934	351.70	2298.6	2650.3	1.1225	6.4353	7.5578
86	359.15	0.6014	1.0332	2726	360.10	2293.5	2653.6	1.1460	6.3852	7.5312
88	361.15	0.6498	1.0346	2536	368.51	2288.3	2656.6	1.1693	6.3356	7.5050
90	363.15	0.7014	1.0360	2361	376.92	2283.2	2660.1	1.1925	6.2866	7.4791
92	365.15	0.7564	1.0375	2200	385.33	2278.0	2663.3	1.2156	6.2379	7.4536
94	367.15	0.8149	1.0389	2052	393.75	2272.8	2666.5	1.2386	6.1898	7.4284
96	369.15	0.8771	1.0404	1915.0	402.17	2267.6	2669.7	1.2615	6.1421	7.4036
98	371.15	0.9433	1.0420	1789.1	410.61	2262.3	2672.9	1.2842	6.0948	7.3791
100	373.15	1.0135	1.0435	1672.9	419.04	2257.0	2676.1	1.3069	6.0480	7.3549
102	375.15	1.0880	1.0451	1565.5	427.48	2251.7	2679.2	1.3294	6.0016	7.3310
104	377.15	1.1669	1.0467	1466.2	435.92	2246.4	2682.3	1.3518	5.9557	7.3075
106	379.15	1.2506	1.0483	1374.3	444.37	2241.0	2685.4	1.3741	5.9101	7.2843
108	381.15	1.3391	1.0499	1289.1	452.83	2235.6	2688.4	1.3964	5.8650	7.2613
110	383.15	1.4327	1.0516	1210.2	461.30	2230.2	2691.5	1.4185	5.8202	7.2387
112	385.15	1.5317	1.0533	1136.9	469.76	2224.7	2694.5	1.4405	5.7758	7.2163
114	387.15	1.6362	1.0550	1068.8	478.24	2219.2	2697.5	1.4624	5.7318	7.1942
116	389.15	1.7465	1.0568	1005.5	486.72	2213.7	2700.5	1.4842	5.6882	7.1724
118	391.15	1.8627	1.0585	946.7	495.21	2208.2	2703.4	1.5060	5.6449	7.1509
120	393.15	1.9853	1.0603	891.9	503.71	2202.6	2706.3	1.5276	5.6020	7.1296
122	395.15	2.1143	1.0622	840.8	512.21	2197.0	2709.2	1.5492	5.5594	7.1086

Source: Abridged from, "Steam Tables (International Edition—Metric Units)," by Joseph H. Keenan, F. G. Keyes, P. G. Hill, and J. G. Moore. Copyright © 1969 by John Wiley & Sons, Inc., New York. Reprinted by permission of John Wiley & Sons, Inc.

Table A-2 *MOIST AIR,* STANDARD ATMOSPHERIC PRESSURE, 14.696 psi—ENGLISH UNITS

Temp. F	Humidity Ratio lb_w/lb_a W_s	Volume ft³/lb dry air v_a	v_{as}	v_s	Enthalpy Btu/lb dry air i_a	i_{as}	i_s	Entropy Btu/(lb dry air)·F s_a	s_{as}	s_s	Condensed Water Enthalpy Btu/lb i_w	Entropy Btu/lb·F s_w	Vapor Press. in. Hg p_s	Temp F
0	0.0007875	11.579	0.015	11.594	0.0	0.835	0.835	0.00000	0.00192	0.00192	−158.89	−0.3243	0.037671	0
5	0.0010207	11.706	0.019	11.725	1.201	1.085	2.286	0.00260	0.00247	0.00506	−156.52	−0.3192	0.048814	5
10	0.0013158	11.832	0.025	11.857	2.402	1.402	3.804	0.00517	0.00315	0.00832	−154.13	−0.3141	0.062901	10
15	0.0016874	11.959	0.032	11.991	3.603	1.801	5.404	0.00771	0.00400	0.01171	−151.71	−0.3089	0.080623	15
20	0.0021531	12.085	0.042	12.127	4.804	2.303	7.107	0.01023	0.00505	0.01528	−149.27	−0.3038	0.102798	20
25	0.0027339	12.212	0.054	12.265	6.005	2.930	8.935	0.01272	0.00636	0.01908	−146.80	−0.2987	0.130413	25
30	0.0034552	12.338	0.068	12.406	7.206	3.711	10.917	0.01519	0.00796	0.02315	−144.31	−0.2936	0.164631	30
32	0.0037895	12.389	0.075	12.464	7.687	4.073	11.760	0.01617	0.00870	0.02487	−143.30	−0.2915	0.180479	32
36	0.004452	12.490	0.089	12.579	8.648	4.793	13.441	0.01811	0.01016	0.02827	4.05	0.0081	0.21181	36
40	0.005216	12.591	0.105	12.696	9.609	5.624	15.233	0.02004	0.01183	0.03187	8.07	0.0162	0.24784	40
44	0.006094	12.692	0.124	12.816	10.570	6.582	17.152	0.02196	0.01374	0.03570	12.09	0.0242	0.28918	44
48	0.007103	12.793	0.146	12.939	11.531	7.684	19.215	0.02386	0.01592	0.03978	16.10	0.0321	0.33651	48
52	0.008259	12.894	0.171	13.065	12.492	8.949	21.441	0.02575	0.01840	0.04415	20.11	0.0400	0.39054	52
56	0.009580	12.995	0.200	13.195	13.453	10.397	23.850	0.02762	0.02122	0.04884	24.11	0.0478	0.45205	56
60	0.011087	13.096	0.233	13.329	14.415	12.052	26.467	0.02947	0.02442	0.05389	28.11	0.0555	0.52193	60
64	0.012805	13.198	0.271	13.468	15.376	13.942	29.318	0.03132	0.02804	0.05936	32.11	0.0632	0.60113	64
68	0.014758	13.299	0.315	13.613	16.337	16.094	32.431	0.03315	0.03214	0.06529	36.11	0.0708	0.69065	68
72	0.016976	13.400	0.365	13.764	17.299	18.543	35.841	0.03496	0.03677	0.07173	40.11	0.0783	0.79167	72
76	0.019491	13.501	0.422	13.923	18.260	21.323	39.583	0.03676	0.04199	0.07875	44.10	0.0858	0.90533	76
80	0.022340	13.602	0.487	14.089	19.222	24.479	43.701	0.03855	0.04787	0.08642	48.10	0.0933	1.03302	80
84	0.025563	13.703	0.561	14.264	20.183	28.055	48.238	0.04033	0.05448	0.09481	52.09	0.1006	1.17608	84
88	0.029208	13.804	0.646	14.450	21.145	32.105	53.250	0.04209	0.06192	0.10401	56.09	0.1080	1.33613	88

92	0.033323	13.905	0.742	14.647	22.107	36.687	58.794	0.04384	0.07028	0.11412	60.08	0.1152	1.51471	92
96	0.037972	14.006	0.852	14.858	23.069	41.871	64.940	0.04558	0.07968	0.12525	64.07	0.1224	1.71372	96
100	0.043219	14.107	0.976	15.084	24.031	47.730	71.761	0.04730	0.09022	0.13752	68.07	0.1296	1.93492	100
104	0.049140	14.208	1.118	15.326	24.993	54.354	79.346	0.04901	0.10206	0.15108	72.06	0.1367	2.18037	104
108	0.055826	14.309	1.279	15.588	25.955	61.844	87.799	0.05071	0.11537	0.16608	76.05	0.1438	2.45232	108
112	0.063378	14.411	1.462	15.872	26.917	70.319	97.237	0.05240	0.13032	0.18272	80.05	0.1508	2.75310	112
116	0.071908	14.512	1.670	16.181	27.879	79.906	107.786	0.05408	0.14713	0.20121	84.04	0.1577	3.08488	116
120	0.081560	14.613	1.906	16.519	28.842	90.770	119.612	0.05575	0.16605	0.22180	88.04	0.1647	3.45052	120
124	0.092500	14.714	2.176	16.890	29.805	103.102	132.907	0.05740	0.18739	0.24480	92.03	0.1715	3.85298	124
128	0.104910	14.815	2.485	17.299	30.767	117.111	147.878	0.05905	0.21149	0.27054	96.03	0.1783	4.29477	128
132	0.119023	14.916	2.837	17.753	31.730	133.066	164.796	0.06068	0.23876	0.29944	100.02	0.1851	4.77919	132
136	0.135124	15.017	3.242	18.259	32.693	151.294	183.987	0.06230	0.26973	0.33203	104.02	0.1919	5.30973	136
140	0.153538	15.118	3.708	18.825	33.656	172.168	205.824	0.06391	0.30498	0.36890	108.02	0.1985	5.88945	140
144	0.174694	15.219	4.245	19.464	34.620	196.183	230.802	0.06551	0.34530	0.41081	112.02	0.2052	6.52241	144
148	0.199110	15.320	4.869	20.189	35.583	223.932	259.514	0.06710	0.39160	0.45871	116.02	0.2118	7.21239	148
152	0.227429	15.421	5.596	21.017	36.546	256.158	292.705	0.06868	0.44507	0.51375	120.02	0.2184	7.96306	152
156	0.260512	15.522	6.450	21.972	37.510	293.849	331.359	0.07025	0.50723	0.57749	124.02	0.2249	8.77915	156
160	0.29945	15.623	7.459	23.082	38.474	338.263	376.737	0.07181	0.58007	0.65188	128.02	0.2314	9.6648	160
164	0.34572	15.724	8.664	24.388	39.438	391.095	430.533	0.07337	0.66622	0.73959	132.03	0.2378	10.6250	164
168	0.40131	15.825	10.117	25.942	40.402	454.630	495.032	0.07491	0.76925	0.84415	136.03	0.2442	11.6641	168
172	0.46905	15.926	11.894	27.820	41.366	532.138	573.504	0.07644	0.89423	0.97067	140.04	0.2506	12.7880	172
176	0.55294	16.027	14.103	30.130	42.331	628.197	670.528	0.07796	1.04828	1.12624	144.05	0.2569	14.0010	176
180	0.65911	16.128	16.909	33.037	43.295	749.871	793.166	0.07947	1.24236	1.32183	148.06	0.2632	15.3097	180
184	0.79703	16.229	20.564	36.793	44.260	908.061	952.321	0.08098	1.49332	1.57430	152.07	0.2694	16.7190	184
188	0.98272	16.330	25.498	41.828	45.225	1121.174	1166.399	0.08247	1.82963	1.91210	156.08	0.2756	18.2357	188
192	1.24471	16.431	32.477	48.908	46.190	1422.047	1468.238	0.08396	2.30193	2.38589	160.10	0.2818	19.8652	192
196	1.64070	16.532	43.046	59.578	47.155	1877.032	1924.188	0.08543	3.01244	3.09787	164.12	0.2880	21.6152	196
200	2.30454	16.633	60.793	77.426	48.121	2640.084	2688.205	0.08690	4.19787	4.28477	168.13	0.2941	23.4906	200

Source: Abridged by permission from *ASHRAE Handbook, Fundamentals Volume,* 1985.

Table A-2a *MOIST AIR*, STANDARD ATMOSPHERIC PRESSURE, 101.325 kPa—SI UNITS

Temp. C	Humidity Ratio kg_w/kg_a W_s	Volume m³/kg dry air			Enthalpy kJ/kg dry air			Entropy kJ/(kg dry air·K)			Condensed Water		Vapor Press. kPa	Temp C
		v_a	v_{as}	v_s	i_a	i_{as}	i_s	s_a	s_{as}	s_s	Enthalpy kJ/kg i_w	Entropy kJ/(kg·K) s_w	p_s	
−18	0.0007711	0.7222	0.0009	0.7231	−18.103	1.902	−16.201	−0.0686	0.0079	−0.0607	−370.02	−1.3596	0.12492	−18
−16	0.0009303	0.7279	0.0011	0.7290	−16.092	2.299	−13.793	−0.0607	0.0094	−0.0513	−366.06	−1.3441	0.15068	−16
−14	0.0011191	0.7336	0.0013	0.7349	−14.080	2.769	−11.311	−0.0529	0.0113	−0.0416	−362.07	−1.3287	0.18122	−14
−12	0.0013425	0.7393	0.0016	0.7409	−12.069	3.327	−8.742	−0.0452	0.0134	−0.0318	−358.06	−1.3132	0.21732	−12
−10	0.0016062	0.7450	0.0019	0.7469	−10.057	3.986	−6.072	−0.0375	0.0160	−0.0215	−354.01	−1.2978	0.25991	−10
−8	0.0019166	0.7507	0.0023	0.7530	−8.046	4.764	−3.283	−0.0299	0.0189	−0.0110	−349.93	−1.2824	0.30999	−8
−6	0.0022811	0.7563	0.0028	0.7591	−6.035	5.677	−0.357	−0.0223	0.0224	0.0000	−345.82	−1.2669	0.36874	−6
−4	0.0027081	0.7620	0.0033	0.7653	−4.023	6.751	2.728	−0.0148	0.0264	0.0115	−341.69	−1.2515	0.43748	−4
−2	0.0032074	0.7677	0.0039	0.7717	−2.011	8.007	5.995	−0.0074	0.0310	0.0236	−337.52	−1.2361	0.51773	−2
0	0.0037895	0.7734	0.0047	0.7781	−0.000	9.473	9.473	0.0000	0.0364	0.0364	−333.32	−1.2206	0.61117	0
2	0.004381	0.7791	0.0055	0.7845	2.012	10.970	12.982	0.0073	0.0419	0.0492	8.49	0.0306	0.7060	2
4	0.005054	0.7848	0.0064	0.7911	4.024	12.672	16.696	0.0146	0.0480	0.0627	16.91	0.0611	0.8135	4
6	0.005818	0.7904	0.0074	0.7978	6.036	14.608	20.644	0.0219	0.0550	0.0769	25.32	0.0913	0.9353	6
8	0.006683	0.7961	0.0085	0.8046	8.047	16.805	24.852	0.0290	0.0628	0.0919	33.72	0.1213	1.0729	8
10	0.007661	0.8018	0.0098	0.8116	10.059	19.293	29.352	0.0362	0.0717	0.1078	42.11	0.1511	1.2280	10
12	0.008766	0.8075	0.0113	0.8188	12.071	22.107	34.179	0.0433	0.0816	0.1248	50.50	0.1806	1.4026	12
14	0.010012	0.8132	0.0131	0.8262	14.084	25.286	39.370	0.0503	0.0927	0.1430	58.88	0.2099	1.5987	14
16	0.011413	0.8188	0.0150	0.8338	16.096	28.867	44.963	0.0573	0.1051	0.1624	67.26	0.2389	1.8185	16
18	0.012989	0.8245	0.0172	0.8417	18.108	32.900	51.008	0.0642	0.1190	0.1832	75.63	0.2678	2.0643	18
20	0.014758	0.8302	0.0196	0.8498	20.121	37.434	57.555	0.0711	0.1346	0.2057	84.00	0.2965	2.3389	20
22	0.016741	0.8359	0.0224	0.8583	22.133	42.527	64.660	0.0779	0.1519	0.2298	92.36	0.3249	2.6448	22
24	0.018963	0.8416	0.0256	0.8671	24.146	48.239	72.385	0.0847	0.1712	0.2559	100.73	0.3531	2.9852	24
26	0.021448	0.8472	0.0291	0.8764	26.159	54.638	80.798	0.0915	0.1927	0.2842	109.09	0.3812	3.3633	26
28	0.024226	0.8529	0.0331	0.8860	28.172	61.804	89.976	0.0982	0.2166	0.3148	117.45	0.4090	3.7823	28
30	0.027329	0.8586	0.0376	0.8962	30.185	69.820	100.006	0.1048	0.2432	0.3481	125.81	0.4367	4.2462	30

32	0.030793	0.8643	0.0426	0.9069	32.198	78.780	110.979	0.1115	0.2728	0.3842	134.17	0.4642	4.7586	32
34	0.034660	0.8700	0.0483	0.9183	34.212	88.799	123.011	0.1180	0.3056	0.4236	142.53	0.4915	5.3242	34
36	0.038971	0.8756	0.0546	0.9303	36.226	99.983	136.209	0.1246	0.3420	0.4666	150.89	0.5186	5.9468	36
38	0.043778	0.8813	0.0618	0.9431	38.239	112.474	150.713	0.1311	0.3824	0.5135	159.25	0.5456	6.6315	38
40	0.049141	0.8870	0.0698	0.9568	40.253	126.430	166.683	0.1375	0.4273	0.5649	167.61	0.5724	7.3838	40
42	0.055119	0.8927	0.0788	0.9714	42.268	142.007	184.275	0.1439	0.4771	0.6211	175.97	0.5990	8.2081	42
44	0.061791	0.8983	0.0888	0.9872	44.282	159.417	203.699	0.1503	0.5325	0.6828	184.33	0.6254	9.1110	44
46	0.069239	0.9040	0.1002	1.0042	46.296	178.882	225.179	0.1566	0.5940	0.7507	192.69	0.6517	10.0982	46
48	0.077556	0.9097	0.1129	1.0226	48.311	200.644	248.955	0.1629	0.6624	0.8253	201.06	0.6778	11.1754	48
50	0.086858	0.9154	0.1272	1.0425	50.326	225.019	275.345	0.1692	0.7385	0.9077	209.42	0.7038	12.3503	50
52	0.097272	0.9211	0.1433	1.0643	52.341	252.340	304.682	0.1754	0.8234	0.9988	217.78	0.7296	13.6293	52
54	0.108954	0.9267	0.1614	1.0882	54.357	283.031	337.388	0.1816	0.9182	1.0998	226.15	0.7552	15.0205	54
56	0.122077	0.9324	0.1819	1.1143	56.373	317.549	373.922	0.1877	1.0243	1.2120	234.52	0.7807	16.5311	56
58	0.136851	0.9381	0.2051	1.1432	58.389	356.461	414.850	0.1938	1.1432	1.3370	242.88	0.8061	18.1691	58
60	0.15354	0.9438	0.2315	1.1752	60.405	400.458	460.863	0.1999	1.2769	1.4768	251.25	0.8313	19.9439	60
62	0.17244	0.9494	0.2614	1.2109	62.421	450.377	512.798	0.2059	1.4278	1.6337	259.62	0.8563	21.8651	62
64	0.19393	0.9551	0.2957	1.2508	64.438	507.177	571.615	0.2119	1.5985	1.8105	268.00	0.8812	23.9405	64
66	0.21848	0.9608	0.3350	1.2958	66.455	572.116	638.571	0.2179	1.7927	2.0106	276.37	0.9060	26.1810	66
68	0.24664	0.9665	0.3803	1.3467	68.472	646.724	715.196	0.2238	2.0147	2.2385	284.75	0.9306	28.5967	68
70	0.27916	0.9721	0.4328	1.4049	70.489	732.959	803.448	0.2297	2.2699	2.4996	293.13	0.9551	31.1986	70
72	0.31698	0.9778	0.4941	1.4719	72.507	833.335	905.842	0.2356	2.5655	2.8010	301.51	0.9794	33.9983	72
74	0.36130	0.9835	0.5662	1.5497	74.525	951.077	1025.603	0.2414	2.9104	3.1518	309.89	1.0037	37.0063	74
76	0.41377	0.9892	0.6519	1.6411	76.543	1090.628	1167.172	0.2472	3.3171	3.5644	318.28	1.0278	40.2369	76
78	0.47663	0.9948	0.7550	1.7498	78.562	1257.921	1336.483	0.2530	3.8023	4.0553	326.67	1.0517	43.7020	78
80	0.55295	1.0005	0.8805	1.8810	80.581	1461.200	1541.781	0.2587	4.3890	4.6477	335.06	1.0755	47.4135	80
82	0.64724	1.0062	1.0360	2.0422	82.600	1712.547	1795.148	0.2644	5.1108	5.3753	343.45	1.0993	51.3860	82
84	0.76624	1.0119	1.2328	2.2446	84.620	2029.983	2114.603	0.2701	6.0181	6.2882	351.85	1.1228	55.6337	84
86	0.92062	1.0175	1.4887	2.5062	86.640	2442.036	2528.677	0.2757	7.1901	7.4658	360.25	1.1463	60.1727	86
88	1.12800	1.0232	1.8333	2.8565	88.661	2995.890	3084.551	0.2813	8.7580	9.0393	368.65	1.1696	65.0166	88
90	1.42031	1.0289	2.3199	3.3488	90.681	3776.918	3867.599	0.2869	10.9586	11.2455	377.06	1.1928	70.1817	90

Source: Abridged by permission from *ASHRAE Handbook, Fundamentals Volume,* 1985.

Table A-3 REFRIGERANT 12 SATURATION PROPERTIES—ENGLISH UNITS

Fahrenheit Temperature	Pressure psia P	Volume ft³/lbm		Enthalpy Btu/lbm			Entropy Btu/(lbm-R)		Fahrenheit Temperature
		Liquid v_f	Vapor v_g	Liquid i_f	Latent i_{fg}	Vapor i_g	Liquid s_f	Vapor s_g	
−40	9.3076	0.010564	3.8750	0	72.913	72.913	0	0.17373	−40
−38	9.8035	0.010586	3.6922	0.4215	72.712	73.134	0.001000	0.17343	−38
−36	10.320	0.010607	3.5198	0.8434	72.511	73.354	0.001995	0.17313	−36
−34	10.858	0.010629	3.3571	1.2659	72.309	73.575	0.002988	0.17285	−34
−32	11.417	0.010651	3.2035	1.6887	72.106	73.795	0.003976	0.17257	−32
−30	11.999	0.010674	3.0585	2.1120	71.903	74.015	0.004961	0.17229	−30
−28	12.604	0.010696	2.9214	2.5358	71.698	74.234	0.005942	0.17203	−28
−26	13.233	0.010719	2.7917	2.9601	71.494	74.454	0.006919	0.17177	−26
−24	13.886	0.010741	2.6691	3.3848	71.288	74.673	0.007894	0.17151	−24
−22	14.564	0.010764	2.5529	3.8100	71.081	74.891	0.008864	0.17126	−22
−20	15.267	0.010788	2.4429	4.2357	70.874	75.110	0.009831	0.17102	−20
−18	15.996	0.010811	2.3387	4.6618	70.666	75.328	0.010795	0.17078	−18
−16	16.753	0.010834	2.2399	5.0885	70.456	75.545	0.011755	0.17055	−16
−14	17.536	0.010858	2.1461	5.5157	70.246	75.762	0.012712	0.17032	−14
−12	18.348	0.010882	2.0572	5.9434	70.036	75.979	0.013666	0.17010	−12
−10	19.189	0.010906	1.9727	6.3716	69.824	76.196	0.014617	0.16989	−10
−8	20.059	0.010931	1.8924	6.8003	69.611	76.411	0.015564	0.16967	−8
−6	20.960	0.010955	1.8161	7.2296	69.397	76.627	0.016508	0.16947	−6
−4	21.891	0.010980	1.7436	7.6594	69.183	76.842	0.017449	0.16927	−4
−2	22.854	0.011005	1.6745	8.0898	68.967	77.057	0.018388	0.16907	−2
0	23.849	0.011030	1.6089	8.5207	68.750	77.271	0.019323	0.16888	0
2	24.878	0.011056	1.5463	8.9522	68.533	77.485	0.020255	0.16869	2

n	25.939	0.011082	1.4867	9.3843	68.314	77.698	0.021184	0.16851	n
4	25.939	0.011082	1.4867	9.3843	68.314	77.698	0.021184	0.16851	4
6	27.036	0.011107	1.4299	9.8169	68.094	77.911	0.022110	0.16833	6
8	28.167	0.011134	1.3758	10.250	67.873	78.123	0.023033	0.16815	8
10	29.335	0.011160	1.3241	10.684	67.651	78.335	0.023954	0.16798	10
12	30.539	0.011187	1.2748	11.118	67.428	78.546	0.024871	0.16782	12
14	31.780	0.011214	1.2278	11.554	67.203	78.757	0.025786	0.16765	14
16	33.060	0.011241	1.128	11.989	66.977	78.966	0.026699	0.16750	16
18	34.378	0.011268	1.1399	12.426	66.750	79.176	0.027608	0.16734	18
20	35.736	0.011296	1.0988	12.863	66.522	79.385	0.028515	0.16719	20
22	37.135	0.011324	1.0596	13.300	66.293	79.593	0.029420	0.16704	22
24	38.574	0.011352	1.0220	13.739	66.061	79.800	0.030322	0.16690	24
26	40.056	0.011380	0.98612	14.178	65.829	80.007	0.031221	0.16676	26
28	41.580	0.011409	0.95173	14.618	65.596	80.214	0.032118	0.16662	28
30	43.148	0.011438	0.91880	15.058	65.361	80.419	0.033013	0.16648	30
32	44.760	0.011468	0.88725	15.500	65.124	80.624	0.033905	0.16635	32
34	46.417	0.011497	0.85702	15.942	64.886	80.828	0.034796	0.16622	34
36	48.120	0.011527	0.82803	16.384	64.647	81.031	0.035683	0.16610	36
38	49.870	0.011557	0.80023	16.828	64.406	81.234	0.036569	0.16598	38
40	51.667	0.011588	0.77357	17.273	64.163	81.436	0.037453	0.16586	40
42	53.513	0.011619	0.74798	17.718	63.919	81.637	0.038334	0.16574	42
44	55.407	0.011650	0.72341	18.164	63.673	81.837	0.039213	0.16562	44
46	57.352	0.011682	0.69982	18.611	63.426	82.037	0.040091	0.16551	46
48	59.347	0.011714	0.67715	19.059	63.177	82.236	0.040966	0.16540	48
50	61.394	0.011746	0.65537	19.507	62.926	82.433	0.041839	0.16530	50
52	63.494	0.011779	0.63444	19.957	62.673	82.630	0.042711	0.16519	52
54	65.646	0.011811	0.61431	20.408	62.418	82.826	0.043581	0.16509	54
56	67.853	0.011845	0.59495	20.859	62.162	83.021	0.044449	0.16499	56
58	70.115	0.011879	0.57632	21.312	61.903	83.215	0.045316	0.16489	58
60	72.433	0.011913	0.55839	21.766	61.643	83.409	0.046180	0.16479	60

Table A-3 REFRIGERANT 12 SATURATION PROPERTIES—ENGLISH UNITS (continued)

Fahrenheit Temperature	Pressure psia P	Volume ft³/lbm Liquid v_f	Volume ft³/lbm Vapor v_g	Enthalpy Btu/lbm Liquid i_f	Enthalpy Btu/lbm Latent i_{fg}	Enthalpy Btu/lbm Vapor i_g	Entropy Btu/(lbm-R) Liquid s_f	Entropy Btu/(lbm-R) Vapor s_g	Fahrenheit Temperature
62	74.807	0.011947	0.54112	22.221	61.380	83.601	0.047044	0.16470	62
64	77.239	0.011982	0.52450	22.676	61.116	83.792	0.047905	0.16460	64
66	79.729	0.012017	0.50848	23.133	60.849	83.982	0.048765	0.16451	66
68	82.279	0.012053	0.49305	23.591	60.580	84.171	0.049624	0.16442	68
70	84.888	0.012089	0.47818	24.050	60.309	84.359	0.050482	0.16434	70
72	87.559	0.012126	0.46383	24.511	60.035	84.546	0.051338	0.16425	72
74	90.292	0.012163	0.45000	24.973	59.759	84.732	0.052193	0.16417	74
76	93.087	0.012201	0.43666	25.435	59.481	84.916	0.053047	0.16408	76
78	95.946	0.012239	0.42378	25.899	59.201	85.100	0.053900	0.16400	78
80	98.870	0.012277	0.41135	26.365	58.917	85.282	0.054751	0.16392	80
82	101.86	0.012316	0.39935	26.832	58.631	85.463	0.055602	0.16384	82
84	104.92	0.012356	0.38776	27.300	58.343	85.643	0.056452	0.16376	84
86	108.04	0.012396	0.37657	27.769	58.052	85.821	0.057301	0.16368	86
88	111.23	0.012437	0.36575	28.241	57.757	85.998	0.058149	0.16360	88
90	114.49	0.012478	0.35529	28.713	57.461	86.174	0.058997	0.16353	90
92	117.82	0.012520	0.34518	29.187	57.161	86.348	0.059844	0.16345	92
94	121.22	0.012562	0.33540	29.663	56.858	86.521	0.060690	0.16338	94
96	124.70	0.012605	0.32594	30.140	56.551	86.691	0.061536	0.16330	96
98	128.24	0.012649	0.31679	30.619	56.242	86.861	0.062381	0.16323	98
100	131.86	0.012693	0.30794	31.100	55.929	87.029	0.063227	0.16315	100
102	135.56	0.012738	0.29937	31.583	55.613	87.196	0.064072	0.16308	102
104	139.33	0.012783	0.29106	32.067	55.293	87.360	0.064916	0.16301	104

106	143.18	0.012829	0.28303	32.553	54.970	87.523	0.065761	0.16293	106
108	147.11	0.012876	0.27524	33.041	54.643	87.684	0.066606	0.16286	108
110	151.11	0.012924	0.26769	33.531	54.313	87.844	0.067451	0.16279	110
112	155.19	0.012972	0.26037	34.023	53.978	88.001	0.068296	0.16271	112
114	159.36	0.013022	0.25328	34.517	53.639	88.156	0.069141	0.16264	114
116	163.61	0.013072	0.24641	35.014	53.296	88.310	0.069987	0.16256	116
118	167.94	0.013123	0.23974	35.512	52.949	88.461	0.070833	0.16249	118
120	172.35	0.013174	0.23326	36.013	52.597	88.610	0.071680	0.16241	120
122	176.85	0.013227	0.22698	36.516	52.241	88.757	0.072528	0.16234	122
124	181.43	0.013280	0.22089	37.021	51.881	88.902	0.073376	0.16226	124
126	186.10	0.013335	0.21497	37.529	51.515	89.044	0.074225	0.16218	126
128	190.86	0.013390	0.20922	38.040	51.144	89.184	0.075075	0.16210	128
130	195.71	0.013447	0.20364	38.553	50.768	89.321	0.075927	0.16202	130
132	200.64	0.013504	0.19821	39.069	50.387	89.456	0.076779	0.16194	132
134	205.67	0.013563	0.19294	39.588	50.000	89.588	0.077633	0.16185	134
136	210.79	0.013623	0.18782	40.110	49.608	89.718	0.078489	0.16177	136
138	216.01	0.013684	0.18283	40.634	49.210	89.844	0.079346	0.16168	138
140	221.32	0.013746	0.17799	41.162	48.805	89.967	0.080205	0.16159	140
142	226.72	0.013810	0.17327	41.693	48.394	90.087	0.081065	0.16150	142
144	232.22	0.013874	0.16868	42.227	47.977	90.204	0.081928	0.16140	144
146	237.82	0.013941	0.16422	42.765	47.553	90.318	0.082794	0.16130	146
148	243.51	0.014008	0.15987	43.306	47.122	90.428	0.083661	0.16120	148
150	249.31	0.014078	0.15564	43.850	46.684	90.534	0.084531	0.16110	150
152	255.20	0.014148	0.15151	44.399	46.238	90.637	0.085404	0.16099	152
154	261.20	0.014221	0.14750	44.951	45.784	90.735	0.086280	0.16088	154
156	267.30	0.014295	0.14358	45.508	45.322	90.830	0.087159	0.16077	156
158	273.51	0.014371	0.13976	46.068	44.852	90.920	0.088041	0.16065	158
160	279.82	0.014449	0.13604	46.633	44.373	91.006	0.088927	0.16053	160
162	286.24	0.014529	0.13241	47.202	43.885	91.087	0.089817	0.16040	162

Table A-3 REFRIGERANT 12 SATURATION PROPERTIES—ENGLISH UNITS (continued)

Fahrenheit Temperature	Pressure psia P	Volume ft³/lbm		Enthalpy Btu/lbm			Entropy Btu/(lbm-R)		Fahrenheit Temperature
		Liquid v_f	Vapor v_g	Liquid i_f	Latent i_{fg}	Vapor i_g	Liquid s_f	Vapor s_g	
164	292.77	0.014611	0.12886	47.777	43.386	91.163	0.090710	0.16027	164
166	299.40	0.014695	0.12540	48.355	42.879	91.234	0.091608	0.16014	166
168	306.15	0.014782	0.12202	48.939	42.360	91.299	0.092511	0.16000	168
170	313.00	0.014871	0.11873	49.529	41.830	91.359	0.093418	0.15985	170
172	319.97	0.014963	0.11550	50.123	41.290	91.413	0.094330	0.15969	172
174	327.06	0.015058	0.11235	50.724	40.736	91.460	0.095248	0.15953	174

Source: Copyright (1956) by the Du Pont Company. Reprinted by permission.

Table A-3a REFRIGERANT 12 SATURATION PROPERTIES—SI UNITS

Celsius Temperature	Pressure Pa $P \times 10^{-5}$	Enthalpy kJ/kg			Entropy kJ/(kg-K)		Specific Volume m³/kg	
		i_f	i_{fg}	i_g	s_f	s_g	$v_f \times 10^3$	$v_g \times 10^3$
−40	0.6412	163.949	169.601	333.551	0.85804	1.58548	0.65947	242.094
−35	0.8065	168.369	167.489	335.858	0.87675	1.58004	0.66561	195.536
−30	1.0034	172.811	165.342	338.153	0.89516	1.57516	0.67198	159.487
−28	1.0919	174.595	164.472	339.067	0.90243	1.57334	0.67459	147.373
−26	1.1864	176.381	163.596	339.978	0.90967	1.57160	0.67724	136.374
−24	1.2871	178.172	162.713	340.885	0.91686	1.56993	0.67993	126.364
−22	1.3943	179.966	161.824	341.790	0.92400	1.56833	0.68267	117.243
−20	1.5083	181.765	160.927	342.692	0.93110	1.56680	0.68544	108.915
−18	1.6294	183.568	160.022	343.590	0.93816	1.56533	0.68827	101.303
−16	1.7577	185.375	159.110	344.485	0.94518	1.56393	0.69113	94.3375
−14	1.8938	187.186	158.189	345.375	0.95216	1.56258	0.69404	87.9461
−12	2.0377	189.003	157.259	346.262	0.95911	1.56129	0.69701	82.0803
−10	2.1898	190.823	156.322	347.144	0.96601	1.56005	0.70002	76.6898
−9	2.2691	191.735	155.849	347.584	0.96945	1.55945	0.70154	74.1563
−8	2.3505	192.648	155.374	348.022	0.97288	1.55886	0.70308	71.7262
−7	2.4341	193.562	154.898	348.460	0.97630	1.55829	0.70463	69.3948
−6	2.5199	194.478	154.418	348.896	0.97971	1.55773	0.70619	67.1525
−5	2.6080	195.395	153.936	349.331	0.98311	1.55718	0.70777	65.0005
−4	2.6986	196.313	153.452	349.765	0.98651	1.55664	0.70936	62.9298
−3	2.7913	197.233	152.965	350.198	0.98989	1.55611	0.71097	60.9418
−2	2.8866	198.154	152.475	350.629	0.99327	1.55560	0.71259	59.0285
−1	2.9841	199.077	151.983	351.059	0.99664	1.55509	0.71422	57.1910
0	3.0842	200.000	151.488	351.488	1.00000	1.55460	0.71587	55.4217

Table A-3a   REFRIGERANT 12 SATURATION PROPERTIES—SI UNITS *(continued)*

Celsius Temperature	Pressure Pa	Enthalpy kJ/kg			Entropy kJ/(kg-K)		Specific Volume m³/kg	
	$P \times 10^{-5}$	i_f	i_{fg}	i_g	s_f	s_g	$v_f \times 10^3$	$v_g \times 10^3$
1	3.1869	200.926	150.990	351.916	1.00336	1.55411	0.71754	53.7176
2	3.2922	201.853	150.489	352.342	1.00670	1.55364	0.71922	52.0763
3	3.4000	202.781	149.985	352.766	1.01004	1.55317	0.72091	50.4973
4	3.5104	203.711	149.478	353.189	1.01338	1.55272	0.72262	48.9758
5	3.6236	204.642	148.969	353.611	1.01670	1.55227	0.72435	47.5095
6	3.7394	205.574	148.457	354.031	1.02002	1.55184	0.72609	46.0982
7	3.8580	206.509	147.941	354.450	1.02333	1.55140	0.72786	44.7358
8	3.9794	207.445	147.422	354.867	1.02663	1.55098	0.72963	43.4225
9	4.1035	208.383	146.900	355.283	1.02993	1.55058	0.73143	42.1578
10	4.2306	209.322	146.375	355.697	1.03322	1.55017	0.73324	40.9363
11	4.3609	210.264	145.845	356.109	1.03651	1.54977	0.73507	39.7551
12	4.4939	211.207	145.313	356.520	1.03978	1.54939	0.73692	38.6171
13	4.6298	212.152	144.777	356.929	1.04306	1.54901	0.73879	37.5192
14	4.7689	213.099	144.237	357.336	1.04632	1.54863	0.74068	36.4570
15	4.9111	214.047	143.694	357.741	1.04959	1.54826	0.74259	35.4317
16	5.0564	214.998	143.147	358.145	1.05284	1.54790	0.74451	34.4409
17	5.2051	215.951	142.596	358.546	1.05609	1.54755	0.74646	33.4816
18	5.3567	216.905	142.041	358.947	1.05934	1.54720	0.74843	32.5569
19	5.5119	217.863	141.482	359.344	1.06258	1.54685	0.75042	31.6603
20	5.6703	218.821	140.919	359.740	1.06581	1.54652	0.75243	30.7944
21	5.8319	219.782	140.352	360.134	1.06904	1.54619	0.75446	29.9570
22	5.9969	220.745	139.781	360.526	1.07227	1.54586	0.75652	29.1470
23	6.1653	221.710	139.206	360.916	1.07549	1.54554	0.75859	28.3634

24	6.3374	222.678	138.625	361.303	1.07871	1.54523	0.76069	27.6030
25	6.5131	223.649	138.040	361.688	1.08193	1.54491	0.76282	26.8672
26	6.6922	224.621	137.450	362.071	1.08514	1.54461	0.76497	26.1552
27	6.8750	225.595	136.856	362.451	1.08834	1.54430	0.76715	25.4649
28	7.0615	226.574	136.256	362.829	1.09155	1.54400	0.76935	24.7958
29	7.2516	227.554	135.652	363.206	1.09475	1.54370	0.77157	24.1481
30	7.4455	228.537	135.043	363.579	1.09795	1.54341	0.77383	23.5200
31	7.6430	229.521	134.429	363.950	1.10114	1.54312	0.77611	22.9119
32	7.8451	230.511	133.806	364.317	1.10433	1.54283	0.77842	22.3192
33	8.0507	231.502	133.181	364.683	1.10752	1.54254	0.78075	21.7461
34	8.2598	232.495	132.551	365.046	1.11071	1.54226	0.78312	21.1912
35	8.4732	233.492	131.914	365.406	1.11390	1.54198	0.78552	20.6518
36	8.6909	234.493	131.270	365.763	1.11708	1.54170	0.78794	20.1276
37	8.9123	235.496	130.622	366.117	1.12027	1.54142	0.79040	19.6197
38	9.1384	236.503	129.965	366.468	1.12345	1.54114	0.79290	19.1251
39	9.3683	237.512	129.304	366.816	1.12663	1.54087	0.79542	18.6459
40	9.6026	238.526	128.635	367.161	1.12981	1.54059	0.79798	18.1799
41	9.8410	239.541	127.961	367.502	1.13299	1.54032	0.80058	17.7276
42	10.084	240.561	127.280	367.841	1.13617	1.54004	0.80321	17.2884
43	10.331	241.585	126.592	368.177	1.13935	1.53977	0.80588	16.8612
44	10.583	242.612	125.895	368.507	1.14253	1.53949	0.80858	16.4449
45	10.839	243.643	125.193	368.835	1.14571	1.53921	0.81133	16.0413
46	11.099	244.678	124.481	369.159	1.14889	1.53893	0.81411	15.6479
47	11.364	245.715	123.765	369.480	1.15207	1.53865	0.81694	15.2665
48	11.634	246.758	123.039	369.797	1.15525	1.53837	0.81981	14.8945
49	11.909	247.805	122.303	370.108	1.15844	1.53808	0.82272	14.5319
50	12.188	248.855	121.562	370.417	1.16162	1.53780	0.82568	14.1803
52	12.761	250.969	120.053	371.022	1.16799	1.53722	0.83173	13.5040
54	13.353	253.101	118.507	371.608	1.17438	1.53662	0.83799	12.8623
56	13.966	255.252	116.924	372.176	1.18078	1.53601	0.84446	12.2532

Table A-3a   REFRIGERANT 12 SATURATION PROPERTIES—SI UNITS (continued)

Celsius Temperature	Pressure Pa $P \times 10^{-5}$	Enthalpy kJ/kg			Entropy kJ/(kg-K)		Specific Volume m³/kg	
		i_f	i_{fg}	i_g	s_f	s_g	$v_f \times 10^3$	$v_g \times 10^3$
58	14.600	257.422	115.301	372.723	1.18719	1.53537	0.85115	11.6747
60	15.253	259.611	113.639	373.250	1.19361	1.53472	0.85808	11.1253
62	15.928	261.819	111.937	373.756	1.20005	1.53404	0.86528	10.6033
64	16.625	264.051	110.187	374.239	1.20650	1.53332	0.87275	10.1061
66	17.344	266.302	108.394	374.696	1.21297	1.53258	0.88053	9.63298
68	18.086	268.577	106.552	375.129	1.21947	1.53180	0.88863	9.18216
70	18.851	270.873	104.662	373.534	1.22598	1.53098	0.89709	8.75265
75	20.867	276.701	99.726	376.428	1.24231	1.52876	0.92001	7.76538
80	23.038	282.647	94.493	377.140	1.25869	1.52626	0.94603	6.88968

Source: Abstracted by permission from "Thermodynamic Table for Refrigerant 12 in SI-Units" International Institute of Refrigeration, Paris.

Table A-4 REFRIGERANT 22 SATURATION PROPERTIES—ENGLISH UNITS

Fahrenheit Temperature	Pressure psia P	Volume ft³/lbm Liquid v_f	Volume ft³/lbm Vapor v_g	Enthalpy Btu/lbm Liquid i_f	Enthalpy Btu/lbm Latent i_{fg}	Enthalpy Btu/lbm Vapor i_g	Entropy Btu/(lbm-R) Liquid s_f	Entropy Btu/(lbm-R) Vapor s_g	Fahrenheit Temperature
−40	15.222	0.011363	3.2957	0.000	100.257	100.257	0.00000	0.23888	−40
−38	16.024	0.011389	3.1412	0.506	99.971	100.477	0.00120	0.23827	−38
−36	16.859	0.011415	2.9954	1.014	99.682	100.696	0.00240	0.23767	−36
−34	17.728	0.011442	2.8578	1.524	99.391	100.914	0.00359	0.23707	−34
−32	18.633	0.011469	2.7278	2.035	99.097	101.132	0.00479	0.23649	−32
−30	19.573	0.011495	2.6049	2.547	98.801	101.348	0.00598	0.23591	−30
−28	20.549	0.011523	2.4887	3.061	98.503	101.564	0.00716	0.23534	−28
−26	21.564	0.011550	2.3787	3.576	98.202	101.778	0.00835	0.23478	−26
−24	22.617	0.011578	2.2746	4.093	97.899	101.992	0.00953	0.23423	−24
−22	23.711	0.011606	2.1760	4.611	97.593	102.204	0.01072	0.23369	−22
−20	24.845	0.011634	2.0826	5.131	97.285	102.415	0.01189	0.23315	−20
−18	26.020	0.011662	1.9940	5.652	96.974	102.626	0.01307	0.23262	−18
−16	27.239	0.011691	1.9099	6.175	96.660	102.835	0.01425	0.23210	−16
−14	28.501	0.011720	1.8302	6.699	96.344	103.043	0.01542	0.23159	−14
−12	29.809	0.011749	1.7544	7.224	96.025	103.250	0.01659	0.23108	−12
−10	31.162	0.011778	1.6825	7.751	95.704	103.455	0.01776	0.23058	−10
−8	32.563	0.011808	1.6141	8.280	95.380	103.660	0.01892	0.23008	−8
−6	34.011	0.011838	1.5491	8.810	95.053	103.863	0.02009	0.22960	−6
−4	35.509	0.011868	1.4872	9.341	94.724	104.065	0.02125	0.22912	−4
−2	37.057	0.011899	1.4283	9.874	94.391	104.266	0.02241	0.22864	−2
0	38.657	0.011930	1.3723	10.409	94.056	104.465	0.02357	0.22817	0
2	40.309	0.011961	1.3189	10.945	93.718	104.663	0.02472	0.22771	2

Table A-4 REFRIGERANT 22 SATURATION PROPERTIES—ENGLISH UNITS (continued)

Fahrenheit Temperature	Pressure psia P	Volume ft³/lbm Liquid v_f	Volume ft³/lbm Vapor v_g	Enthalpy Btu/lbm Liquid i_f	Enthalpy Btu/lbm Latent i_{fg}	Enthalpy Btu/lbm Vapor i_g	Entropy Btu/(lbm-R) Liquid s_f	Entropy Btu/(lbm-R) Vapor s_g	Fahrenheit Temperature
4	42.014	0.011992	1.2680	11.483	93.378	104.860	0.02587	0.22725	4
6	43.775	0.012024	1.2195	12.022	93.034	105.056	0.02703	0.22680	6
8	45.591	0.012056	1.1732	12.562	92.688	105.250	0.02818	0.22636	8
10	47.464	0.012088	1.1290	13.104	92.338	105.442	0.02932	0.22592	10
12	49.396	0.012121	1.0869	13.648	91.986	105.633	0.03047	0.22548	12
14	51.387	0.012154	1.0466	14.193	91.630	105.823	0.03161	0.22505	14
16	53.438	0.012188	1.0082	14.739	91.272	106.011	0.03275	0.22463	16
18	55.551	0.012221	0.97144	15.288	90.910	106.198	0.03389	0.22421	18
20	57.727	0.012255	0.93631	15.837	90.545	106.383	0.03503	0.22379	20
22	59.967	0.012290	0.90270	16.389	90.178	106.566	0.03617	0.22338	22
24	62.272	0.012325	0.87055	16.942	89.807	106.748	0.03730	0.22297	24
26	64.644	0.012360	0.83978	17.496	89.433	106.928	0.03844	0.22257	26
28	67.083	0.012395	0.81031	18.052	89.055	107.107	0.03958	0.22217	28
30	69.591	0.012431	0.78208	18.609	88.674	107.284	0.04070	0.22178	30
32	72.169	0.012468	0.75503	19.169	88.290	107.459	0.04182	0.22139	32
34	74.818	0.012505	0.72911	19.729	87.903	107.632	0.04295	0.22100	34
36	77.540	0.012542	0.70425	20.292	87.512	107.804	0.04407	0.22062	36
38	80.336	0.012579	0.68041	20.856	87.118	107.974	0.04520	0.22024	38
40	83.206	0.012618	0.65753	21.422	86.720	108.142	0.04632	0.21986	40
42	86.153	0.012656	0.63557	21.989	86.319	108.308	0.04744	0.21949	42
44	89.177	0.012695	0.61448	22.558	85.914	108.472	0.04855	0.21912	44
46	92.280	0.012735	0.59422	23.129	85.506	108.634	0.04967	0.21876	46

48	95.463	0.012775	0.57476	23.701	85.094	108.795	0.05079	0.21839	48
50	98.727	0.012815	0.55606	24.275	84.678	108.953	0.05190	0.21803	50
52	102.07	0.012856	0.53808	24.851	84.258	109.109	0.05301	0.21768	52
54	105.50	0.012898	0.52078	25.429	83.834	109.263	0.05412	0.21732	54
56	109.02	0.012940	0.50414	26.008	83.407	109.415	0.05523	0.21697	56
58	112.62	0.012982	0.48813	26.589	82.975	109.564	0.05634	0.21662	58
60	116.31	0.013025	0.47272	27.172	82.540	109.712	0.05745	0.21627	60
62	120.09	0.013069	0.45788	27.757	82.100	109.857	0.05855	0.21592	62
64	123.96	0.013114	0.44358	28.344	81.656	110.000	0.05966	0.21558	64
66	127.92	0.013159	0.42981	28.932	81.208	110.140	0.06076	0.21524	66
68	131.97	0.013204	0.41653	29.523	80.755	110.278	0.06186	0.21490	68
70	136.12	0.013251	0.40373	30.116	80.298	110.414	0.06296	0.21456	70
72	140.37	0.013297	0.39139	30.170	79.836	110.547	0.06406	0.21422	72
74	144.71	0.013345	0.37949	31.307	79.370	110.677	0.06516	0.21388	74
76	149.15	0.013393	0.36800	31.906	78.899	110.805	0.06626	0.21355	76
78	153.69	0.013442	0.35691	32.506	78.423	110.930	0.06736	0.21321	78
80	158.33	0.013492	0.34621	33.109	77.943	111.052	0.06846	0.21288	80
82	163.07	0.013543	0.33587	33.714	77.457	111.171	0.06956	0.21255	82
84	167.92	0.013594	0.32588	34.322	76.966	111.288	0.07065	0.21222	84
86	172.87	0.013647	0.31623	34.931	76.470	111.401	0.07175	0.21188	86
88	177.93	0.013700	0.30690	35.543	75.968	111.512	0.07285	0.21155	88
90	183.09	0.013754	0.29789	36.158	75.461	111.619	0.07394	0.21122	90
92	188.37	0.013809	0.28917	36.774	75.949	111.723	0.07504	0.21089	92
94	193.76	0.013864	0.28073	37.394	74.430	111.824	0.07613	0.21056	94
96	199.26	0.013921	0.27257	38.016	73.905	111.921	0.07723	0.21023	96
98	204.87	0.013979	0.26467	38.640	74.375	112.015	0.07832	0.20989	98
100	210.60	0.014038	0.25702	39.267	72.838	112.105	0.07942	0.20956	100
102	216.45	0.014098	0.24962	39.897	72.294	112.192	0.08052	0.20923	102
104	222.42	0.014159	0.24244	40.530	71.744	112.274	0.08161	0.20889	104

Table A-4   REFRIGERANT 22 SATURATION PROPERTIES—ENGLISH UNITS (*continued*)

Fahrenheit Temperature	Pressure psia P	Volume ft³/lbm		Enthalpy Btu/lbm			Entropy Btu/(lbm-R)		Fahrenheit Temperature
		Liquid v_f	Vapor v_g	Liquid i_f	Latent i_{fg}	Vapor i_g	Liquid s_f	Vapor s_g	
106	228.50	0.014221	0.23549	41.166	71.187	112.353	0.08271	0.20855	106
108	234.71	0.014285	0.22875	41.804	70.623	112.427	0.08381	0.20821	108
110	241.04	0.014350	0.22222	42.446	70.052	112.498	0.08491	0.20787	110
112	247.50	0.014416	0.21589	43.091	69.473	112.564	0.08601	0.20753	112
114	254.08	0.014483	0.20974	43.739	68.886	112.626	0.08711	0.20718	114
116	260.79	0.014552	0.20378	44.391	68.291	112.682	0.08821	0.20684	116
118	267.63	0.014622	0.19800	45.046	67.688	112.735	0.08932	0.20649	118
120	274.60	0.014694	0.19238	45.705	67.077	112.782	0.09042	0.20613	120
122	281.71	0.014768	0.18692	46.368	66.456	112.824	0.09153	0.20578	122
124	288.95	0.014843	0.18163	47.034	65.826	112.860	0.09264	0.20542	124
126	296.33	0.014920	0.17648	47.705	65.186	112.891	0.09375	0.20505	126
128	303.84	0.014999	0.17147	48.380	64.537	112.917	0.09487	0.20468	128
130	311.50	0.015080	0.16661	49.059	63.877	112.936	0.09598	0.20431	130
132	319.29	0.015163	0.16187	49.743	63.206	112.949	0.09711	0.20393	132
134	327.23	0.015248	0.15727	50.432	62.523	112.955	0.09823	0.20354	134
136	335.32	0.015336	0.15279	51.125	61.829	112.954	0.09936	0.20315	136
138	343.56	0.015426	0.14843	51.824	61.123	112.947	0.10049	0.20275	138
140	351.94	0.015518	0.14418	52.528	60.403	112.931	0.10163	0.20235	140
142	360.48	0.015613	0.14004	53.238	59.670	112.908	0.10277	0.20194	142
144	369.17	0.015712	0.13600	53.955	59.922	112.877	0.10391	0.20152	144

146	378.02	0.015813	0.13207	54.677	58.159	112.836	0.10507	0.20109	146
148	387.03	0.015917	0.12823	55.406	57.380	112.787	0.10622	0.20065	148
150	396.19	0.016025	0.12448	56.143	56.585	112.728	0.10739	0.20020	150
152	405.52	0.016137	0.12083	56.887	55.771	112.658	0.10856	0.19974	152
154	415.02	0.016252	0.11726	57.638	54.939	112.577	0.10974	0.19926	154
156	424.68	0.016372	0.11376	58.399	54.087	112.485	0.11093	0.19878	156
158	434.52	0.016497	0.11035	59.168	53.213	112.381	0.11213	0.19828	158
160	444.53	0.016627	0.10710	59.948	52.316	112.263	0.11334	0.19776	160
162	454.71	0.016762	0.10374	60.737	51.394	112.131	0.11456	0.19723	162
164	465.07	0.016902	0.10054	61.538	50.446	111.984	0.11580	0.19668	164
166	475.61	0.017050	0.97393	62.351	49.469	111.820	0.11705	0.19611	166
168	486.34	0.017204	0.094309	63.178	48.461	111.639	0.11831	0.19552	168
170	497.26	0.017367	0.091279	64.019	47.419	111.438	0.11959	0.19490	170
172	508.37	0.017538	0.088299	64.875	46.340	111.216	0.12089	0.19425	172
174	519.67	0.017719	0.085365	65.750	45.221	110.970	0.12222	0.19358	174

Table A-4a REFRIGERANT 22 SATURATION PROPERTIES—SI UNITS

Celsius Temperature	Pressure Pa	Enthalpy kJ/kg			Entropy kJ/(kg-K)		Specific Volume m³/kg	
	$P \times 10^{-5}$	i_f	i_{fg}	i_g	s_f	s_g	$v_f \times 10^3$	$v_g \times 10^3$
−40	1.0490	155.624	233.204	388.828	0.82489	1.82512	0.70935	205.841
−35	1.3162	160.923	230.162	391.085	0.84742	1.81388	0.71679	166.470
−30	1.6340	166.291	227.008	393.299	0.86975	1.80337	0.72451	135.907
−28	1.7768	168.458	225.712	394.170	0.87863	1.79934	0.72768	125.613
−26	1.9291	170.636	224.398	395.034	0.88748	1.79542	0.73091	116.262
−24	2.0912	172.826	223.064	395.890	0.89630	1.79160	0.73418	107.747
−22	2.2639	175.027	221.709	396.736	0.90509	1.78786	0.73751	99.9741
−20	2.4472	177.239	220.335	397.574	0.91385	1.78423	0.74090	92.8825
−18	2.6418	179.463	218.939	398.402	0.92259	1.78066	0.74435	86.3881
−16	2.8481	181.698	217.522	399.220	0.93129	1.77719	0.74785	80.4427
−14	3.0666	183.945	216.083	400.028	0.93997	1.77378	0.75142	74.9848
−12	3.2976	186.204	214.622	400.826	0.94862	1.77046	0.75505	69.9736
−10	3.5416	188.473	213.140	401.613	0.95725	1.76720	0.75875	65.3646
−9	3.6686	189.612	212.390	402.002	0.96155	1.76560	0.76062	63.1994
−8	3.7992	190.755	211.634	402.388	0.96585	1.76401	0.76251	61.1184
−7	3.9331	191.900	210.873	402.772	0.97014	1.76245	0.76442	59.1227
−6	4.0708	193.049	210.104	403.153	0.97442	1.76089	0.76635	57.2017
−5	4.2118	194.199	209.332	403.531	0.97870	1.75935	0.76830	55.3612
−4	4.3568	195.354	208.551	403.905	0.98297	1.75782	0.77026	53.5871
−3	4.5054	196.511	207.765	404.277	0.98724	1.75631	0.77225	51.8828
−2	4.6577	197.671	206.974	404.645	0.99150	1.75482	0.77425	50.2452
−1	4.8138	198.834	206.177	405.011	0.99575	1.75334	0.77628	48.6700
0	4.9740	200.000	205.372	405.373	1.00000	1.75187	0.77832	47.1523

1	5.1382	201.170	204.562	405.731	1.00424	1.75041	0.78039	45.6920
2	5.3065	202.343	203.744	406.087	1.00848	1.74896	0.78248	44.2848
3	5.4786	203.517	202.922	406.439	1.01271	1.74753	0.78459	42.9319
4	5.6549	204.696	202.092	406.788	1.01694	1.74612	0.78672	41.6278
5	5.8360	205.878	201.254	407.132	1.02115	1.74470	0.78887	40.3677
6	6.0207	207.062	200.412	407.474	1.02537	1.74331	0.79105	39.1576
7	6.2100	208.250	199.562	407.812	1.02958	1.74192	0.79325	37.9890
8	6.4036	209.441	198.706	408.147	1.03378	1.74055	0.79548	36.8624
9	6.6020	210.636	197.841	408.477	1.03798	1.73918	0.79773	35.7742
10	6.8045	211.834	196.971	408.804	1.04218	1.73782	0.80001	34.7261
11	7.0123	213.036	196.090	409.126	1.04637	1.73647	0.80231	33.7109
12	7.2239	214.240	195.207	409.447	1.05056	1.73514	0.80464	32.7357
13	7.4408	215.448	194.314	409.761	1.05474	1.73380	0.80699	31.7909
14	7.6626	216.660	193.412	410.072	1.05892	1.73248	0.80937	30.8778
15	7.8890	217.874	192.504	410.378	1.06309	1.73116	0.81179	29.9967
16	8.1203	219.092	191.589	410.681	1.06726	1.72985	0.81423	29.1453
17	8.3566	220.314	190.666	410.979	1.07142	1.72855	0.81670	28.3224
18	8.5980	221.539	189.735	411.274	1.07559	1.72726	0.81920	27.5268
19	8.8446	222.768	188.795	411.563	1.07974	1.72597	0.82173	26.7567
20	9.0967	224.002	187.846	411.847	1.08390	1.72468	0.82430	26.0109
21	9.3535	225.237	186.891	412.128	1.08805	1.72341	0.82689	25.2910
22	9.6158	226.477	185.926	412.403	1.09220	1.72214	0.82953	24.5937
23	9.8840	227.722	184.951	412.673	1.09634	1.72086	0.83219	23.9174
24	10.157	228.970	183.970	412.939	1.10049	1.71960	0.83489	23.2643
25	10.436	230.222	182.978	413.200	1.10463	1.71834	0.83763	22.6307
26	10.720	231.477	181.978	413.455	1.10876	1.71708	0.84041	22.0177
27	11.011	232.738	180.967	413.705	1.11290	1.71582	0.84322	21.4229
28	11.306	234.001	179.949	413.950	1.11703	1.71457	0.84608	20.8476
29	11.608	235.270	178.920	414.190	1.12117	1.71332	0.84897	20.2890

Table A-4a REFRIGERANT 22 SATURATION PROPERTIES—SI UNITS (continued)

Celsius Temperature	Pressure Pa	Enthalpy kJ/kg			Entropy kJ/(kg-K)		Specific Volume m³/kg	
	$P \times 10^{-5}$	i_f	i_{fg}	i_g	s_f	s_g	$v_f \times 10^3$	$v_g \times 10^3$
30	11.915	236.543	177.881	414.424	1.12530	1.71207	0.85191	19.7478
31	12.229	237.820	176.831	414.651	1.12943	1.71082	0.85489	19.2222
32	12.549	239.102	175.771	414.873	1.13356	1.70957	0.85791	18.7129
33	12.874	240.389	174.700	415.089	1.13769	1.70832	0.86099	18.2183
34	13.206	241.678	173.621	415.300	1.14181	1.70708	0.86410	17.7397
35	13.544	242.976	172.527	415.502	1.14594	1.70582	0.86727	17.2732
36	13.889	244.276	171.424	415.700	1.15007	1.70457	0.87049	16.8217
37	14.239	245.580	170.310	415.891	1.15420	1.70332	0.87376	16.3832
38	14.597	246.892	169.182	416.074	1.15833	1.70206	0.87708	15.9564
39	14.960	248.207	168.044	416.251	1.16246	1.70080	0.88046	15.5425
40	15.331	249.530	166.890	416.419	1.16659	1.69953	0.88390	15.1396
41	15.708	250.857	165.725	416.581	1.17073	1.69826	0.88739	14.7482
42	16.092	252.190	164.546	416.735	1.17487	1.69699	0.89095	14.3677
43	16.483	253.528	163.355	416.882	1.17901	1.69571	0.89457	13.9980

44	16.881	254.873	162.148	417.020	1.18315	1.69442	0.89825	13.6380
45	17.286	256.223	160.927	417.150	1.18730	1.69312	0.90201	13.2880
46	17.698	257.581	159.690	417.271	1.19145	1.69181	0.90583	12.9471
47	18.117	258.944	158.440	417.384	1.19561	1.69050	0.90973	12.6157
48	18.543	260.314	157.173	417.487	1.19977	1.68918	0.91371	12.2929
49	18.977	261.692	155.889	417.581	1.20394	1.68784	0.91777	11.9784
50	19.418	263.073	154.594	417.667	1.20811	1.68650	0.92190	11.6731
52	20.323	265.866	151.938	417.804	1.21648	1.68377	0.93044	11.0839
54	21.258	268.686	149.214	417.900	1.22488	1.68099	0.93936	10.5251
56	22.227	271.546	146.400	417.946	1.23334	1.67812	0.94869	9.99235
58	23.227	274.441	143.502	417.943	1.24183	1.67518	0.95847	9.48558
60	24.260	277.373	140.514	417.887	1.25038	1.67215	0.96874	9.00332
62	25.326	280.348	137.423	417.771	1.25899	1.66903	0.97957	8.54301
64	26.428	283.373	134.216	417.589	1.26768	1.66577	0.99101	8.10286
66	27.566	286.451	130.883	417.334	1.27647	1.66238	1.00314	7.68155
68	28.740	289.585	127.417	417.002	1.28535	1.65884	1.01604	7.27841
70	29.952	292.786	123.795	416.581	1.29436	1.65512	1.02983	6.89127
75	33.153	301.133	113.925	415.058	1.31757	1.64480	1.06911	5.98528
80	36.614	310.139	102.511	412.650	1.34222	1.63249	1.11803	5.15095

Source: Abstracted by permission from "Thermodynamic Table for Refrigerant R22 in SI-Units," International Institute of Refrigeration, Paris.

Thermophysical Properties

Table B-1 REFRIGERANT 12—ENGLISH UNITS

Fahrenheit Temperature	Viscosity, μ lbm/(ft-hr)			Thermal Conductivity, k Btu/(hr-ft-F)			Specific Heat, c_p Btu/(lbm-F)			Fahrenheit Temperature
	Saturated Liquid	Saturated Vapor	Gas P = 1 atm	Saturated Liquid	Saturated Vapor	Gas P = 1 atm	Saturated Liquid	Saturated Vapor	Gas P = 0	
−140	2.44			0.0655			0.199		0.1085	−140
−120	1.95			0.0631			0.202		0.1233	−120
−100	1.593			0.0608			0.204		0.1160	−100
−80	1.331			0.0585			0.207		0.1196	−80
−60	1.132			0.0561			0.209	0.126	0.1230	−60
−40	0.978			0.0538			0.212	0.133	0.1264	−40
−20	0.856	0.0246	0.0246	0.0514	0.0040	0.00400	0.214	0.139	0.1296	−20
0	0.758	0.0262	0.0258	0.0490	0.0043	0.00431	0.217	0.145	0.1327	0
20	0.679	0.0276	0.0269	0.0467	0.0046	0.00463	0.220	0.150	0.1356	20
40	0.613	0.0287	0.0280	0.0443	0.0050	0.00495	0.224	0.157	0.1385	40
60	0.557	0.0297	0.0290	0.0420	0.0053	0.00528	0.229	0.164	0.1413	60
80	0.511	0.0308	0.0301	0.0397	0.0056	0.00561	0.234	0.174	0.1439	80

100	0.471	0.0320	0.0312	0.0373	0.0060	0.00594	0.240	0.185	0.1465	100
120	0.436	0.0335	0.0322	0.0350	0.0064	0.00627	0.251	0.199	0.1490	120
140	0.403	0.0354	0.0332	0.0326	0.0068	0.00660	0.266	0.216	0.1513	140
160	0.366	0.0380	0.0342	0.0302	0.0072	0.00694	0.288	0.235	0.1536	160
180	0.325	0.0412	0.0352	0.0276	0.0076	0.00728	0.317	0.290	0.1558	180
200	0.270	0.0453	0.0362	0.0246	0.0083	0.00763	0.356	0.362	0.1579	200
220	0.198	0.051	0.0372	0.0204	0.0093	0.00798	0.406		0.1599	220
230	0.148	0.060	0.0377	0.0161	0.0107	0.00816			0.1609	230
234	0.075	0.075	0.0379	0.0130	0.130	0.00823			0.1612	234
240			0.0382			0.00834			0.1618	240
260			0.0392			0.00871			0.1637	260
280			0.0401			0.00908			0.1654	280
300			0.0410			0.00945			0.1671	300
320			0.0420			0.00982			0.1687	320
340			0.0429			0.0101			0.1703	340
360			0.0438			0.0105			0.1718	360
380			0.0447			0.0108			0.1732	380
400			0.0456			0.0112			0.1746	400

Source: Reprinted by permission from *ASHRAE Handbook, Fundamentals Volume*, 1972.

Table B-1a REFRIGERANT 12—SI UNITS

Celsius Temperature	Viscosity, $\mu \times 10^3$ (N·s/m²)			Thermal Conductivity, k W/(m·K)			Specific Heat, c_p kJ/(kg·K)			Kelvin Temperature
	Saturated Liquid	Saturated Vapor	Gas $P = 101.33$ kPa	Saturated Liquid	Saturated Vapor	Gas $P = 101.33$ kPa	Saturated Liquid	Saturated Vapor	Gas $P = 0$	
− 95	1.01			0.113			0.833		0.4543	178
− 84	0.806			0.109			0.846		0.4702	189
− 73	0.659			0.105			0.854		0.4857	200
− 62	0.550			0.101			0.867		0.5007	211
− 51	0.468			0.097			0.875	0.528	0.5150	222
− 40	0.404			0.093			0.888	0.557	0.5292	233
− 29	0.354	0.0102	0.0102	0.089	0.0069	0.0069	0.896	0.582	0.5426	244
− 18	0.313	0.0108	0.0107	0.085	0.0074	0.0075	0.909	0.607	0.5556	255
− 7	0.281	0.0114	0.0111	0.081	0.0080	0.0080	0.921	0.628	0.5677	266
4	0.253	0.0119	0.0116	0.077	0.0087	0.0086	0.938	0.657	0.5799	277
16	0.230	0.0123	0.0120	0.073	0.0920	0.0091	0.959	0.687	0.5916	289
27	0.211	0.0127	0.0124	0.069	0.0097	0.0097	0.980	0.729	0.6025	300

704

38	0.195	0.0132	0.0129	0.065	0.0104	0.0103	1.00	0.775	0.6113	311
49	0.180	0.0138	0.0133	0.061	0.0111	0.0109	1.05	0.833	0.6238	322
60	0.167	0.0146	0.0137	0.056	0.0118	0.0114	1.11	0.904	0.6335	333
71	0.151	0.0157	0.0141	0.052	0.0125	0.0120	1.21	0.984	0.6431	344
82	0.134	0.0170	0.0146	0.048	0.0132	0.0126	1.33	1.21	0.6523	355
93	0.112	0.0187	0.0150	0.043	0.0173	0.0132	1.49	1.52	0.6611	366
104	0.082	0.021	0.0154	0.035	0.0161	0.0138	1.70		0.6695	377
110	0.061	0.025	0.0156	0.028	0.0185	0.0141			0.6737	383
112	0.031	0.031	0.0157	0.022	0.0225	0.0142			0.6749	385
116			0.0158			0.0144			0.6774	389
127			0.0162			0.0151			0.6854	400
138			0.0166			0.0157			0.6925	411
149			0.0169			0.0164			0.6996	422
160			0.0174			0.0170			0.7063	433
171			0.0177			0.0175			0.7130	444
183			0.0181			0.0182			0.7193	456
194			0.0185			0.0187			0.7252	467
205			0.0189			0.0194			0.7310	478

Source: Adapted by permission from *ASHRAE Handbook, Fundamentals Volume*, 1972.

Table B-2 REFRIGERANT 22—ENGLISH UNITS

Fahrenheit Temperature	Viscosity, μ lbm/(ft-hr)			Thermal Conductivity, k Btu/(hr-ft-F)			Specific Heat, c_p Btu/(lbm-F)			Fahrenheit Temperature
	Saturated Liquid	Saturated Vapor	Gas $P = 1$ atm	Saturated Liquid	Saturated Vapor	Gas $P = 1$ atm	Saturated Liquid	Saturated Vapor	Gas $P = 0$	
−100	1.153			0.0789			0.255		0.1260	−100
−80	1.002			0.0757			0.256		0.1292	−80
−60	0.883			0.0725			0.259	0.139	0.1324	−60
−40	0.788	0.0242	0.0242	0.0693	0.0040	0.00404	0.262	0.146	0.1356	−40
−20	0.710	0.0254	0.0254	0.0661	0.0044	0.00443	0.266	0.152	0.1388	−20
0	0.646	0.0266	0.0265	0.0630	0.0048	0.00481	0.271	0.158	0.1420	0
20	0.592	0.0279	0.0276	0.0598	0.0052	0.00520	0.276	0.165	0.1452	20
40	0.546	0.0292	0.0288	0.0566	0.0056	0.00558	0.283	0.175	0.1484	40
60	0.507	0.0306	0.0299	0.0534	0.0060	0.00597	0.291	0.187	0.1515	60
80	0.474	0.0321	0.0310	0.0502	0.0064	0.00635	0.300	0.204	0.1546	80
100	0.444	0.0339	0.0321	0.0471	0.0068	0.00674	0.313	0.226	0.1577	100
120	0.422	0.0360	0.0333	0.0439	0.0072	0.00712	0.332	0.253	0.1608	120

140	0.387	0.0383	0.0344	0.0407	0.0077	0.00751	0.357	0.288	0.1638	140
160	0.340	0.0411	0.0354	0.0371	0.0084	0.00790	0.390	0.332	0.1668	160
180	0.285	0.045	0.0365	0.0318	0.0105	0.00828	0.433		0.1697	180
190	0.244	0.049	0.0371	0.0288	0.0119	0.00848			0.1712	190
200	0.182	0.058	0.0376	0.0238	0.0140	0.00867			0.1726	200
205	0.074	0.074	0.0379	0.0177	0.0177	0.00876			0.1733	205
220			0.0387			0.00905			0.1754	220
240			0.0398			0.00944			0.1782	240
260			0.0408			0.00982			0.1810	260
280			0.0419			0.01021			0.1836	280
300			0.0429			0.01059			0.1863	300
320			0.0439			0.0110			0.1888	320
340			0.0450			0.0114			0.1913	340
360			0.0460			0.0118			0.1937	360
380			0.0470			0.0121			0.1960	380
400			0.0480			0.0125			0.1983	400
420			0.0490			0.0129			0.2005	420
440			0.0500			0.0133			0.2026	440

Source: Reprinted by permission from *ASHRAE Handbook, Fundamentals Volume, 1972.*

Table B-2a   REFRIGERANT 22—SI UNITS

Celsius Temperature	Viscosity, $\mu \times 10^3$ (N-s)/m²			Thermal Conductivity, k W/(m-K)			Specific Heat, c_p kJ/(kg-K)			Kelvin Temperature
	Saturated Liquid	Saturated Vapor	Gas $P = 101.33$ kPa	Saturated Liquid	Saturated Vapor	Gas $P = 101.33$ kPa	Saturated Liquid	Saturated Vapor	Gas $P = 0$	
−73	0.4766			0.137			1.07		0.5275	200
−62	0.4142			0.131			1.07		0.5409	211
−51	0.365			0.125			1.08	0.582	0.5543	222
−40	0.326	0.0100	0.0100	0.120	0.0069	0.00699	1.10	0.611	0.5677	233
−29	0.294	0.0105	0.0105	0.114	0.0076	0.00767	1.11	0.636	0.5811	244
−18	0.267	0.0110	0.0110	0.109	0.0083	0.00832	1.13	0.662	0.5945	255
−7	0.245	0.0115	0.0114	0.103	0.0090	0.00900	1.16	0.691	0.6079	266
4	0.226	0.0121	0.0119	0.0980	0.0097	0.00966	1.18	0.733	0.6213	277
16	0.210	0.0126	0.0124	0.0924	0.010	0.0103	1.22	0.783	0.6343	289
27	0.196	0.0133	0.0128	0.0869	0.011	0.0110	1.26	0.854	0.6473	300
38	0.184	0.0140	0.0133	0.0815	0.012	0.0117	1.31	0.946	0.6603	311
49	0.174	0.0149	0.0138	0.0760	0.012	0.0123	1.39	1.06	0.6732	322

60	0.160	0.0158	0.0142	0.0704	0.013	0.0130	1.49	1.21	0.6858	333
71	0.141	0.0170	0.0146	0.0642	0.015	0.0137	1.63	1.39	0.6984	344
82	0.118	0.019	0.0151	0.0550	0.018	0.0143	1.81		0.7105	355
88	0.101	0.020	0.0153	0.0498	0.021	0.0147			0.7168	361
94	0.075	0.024	0.0155	0.0412	0.024	0.0150			0.7226	367
96	0.031	0.031	0.0157	0.0306	0.031	0.0152			0.7256	369
105			0.0160			0.0157			0.7344	378
116			0.0165			0.0163			0.7461	389
127			0.0169			0.0170			0.7578	400
138			0.0173			0.01767			0.7687	411
149			0.0177			0.01833			0.7800	422
160			0.0181			0.0190			0.7905	433
171			0.0186			0.0197			0.8009	444
183			0.0190			0.0204			0.8110	456
194			0.0194			0.0209			0.8206	467
205			0.0198			0.0216			0.8302	478
216			0.0203			0.0223			0.8395	489
227			0.0207			0.0230			0.8482	500

Source: Adapted by permission from *ASHRAE Handbook, Fundamentals Volume*, 1972.

Table B-3 WATER—ENGLISH UNITS

Fahrenheit Temperature	Viscosity, μ lbm/(ft-hr)			Thermal Conductivity, k Btu/(hr-ft-F)			Specific Heat, c_p Btu/(lbm-F)				Fahrenheit Temperature
	Saturated Liquid	Saturated Vapor	Gas $P = 1$ atm	Saturated Liquid	Saturated Vapor	Gas $P = 1$ atm	Saturated Liquid	Saturated Vapor	Gas $P = 0$	Gas $P = 1$ atm	
32	4.23	0.0192		0.329	0.0100		1.007		0.4438		32
40	3.65	0.0197		0.334	0.0103		1.005		0.4442		40
50	3.09	0.0202		0.339	0.0105		1.003		0.4445		50
60	2.64	0.0207		0.345	0.0107		1.001	0.4446	0.4448		60
70	2.29	0.0213		0.350	0.0109		1.000	0.4454	0.4452		70
80	2.01	0.0218		0.354	0.0112		0.999	0.4464	0.4456		80
90	1.77	0.0224		0.359	0.0114		0.998	0.4475	0.4459		90
100	1.58	0.0229		0.363	0.0116		0.998	0.4488	0.4463		100
120	1.29	0.0240		0.371	0.0121		0.998	0.4522	0.4472		120
140	1.080	0.0251		0.378	0.0125		1.000	0.4567	0.4480		140
160	0.927	0.0261		0.383	0.0130		1.001	0.4626	0.4490		160
180	0.804	0.0272		0.388	0.0135		1.003	0.4699	0.4500		180

200	0.708	0.0283	0.0289	0.392	0.0140	0.0143	1.006	0.4791	0.4510	0.4892	200
212	0.660	0.0289	0.0294	0.394	0.0143	0.0144	1.008	0.4859	0.4516	0.4867	212
220	0.631	0.0292	0.0305	0.395	0.0145	0.0147	1.009	0.4902	0.4521	0.4819	220
240	0.567	0.0302	0.0315	0.397	0.0151	0.0153	1.013	0.5019	0.4532	0.4780	240
260	0.515	0.0312	0.0326	0.398	0.0158	0.0158	1.017	0.5157	0.4544	0.4780	260
280	0.471	0.0322	0.0337	0.398	0.0165	0.0164	1.022	0.5319	0.4556	0.4750	280
300	0.433	0.0331	0.0348	0.397	0.0172	0.0170	1.029	0.5508	0.4569	0.4729	300
320	0.401	0.0341	0.0359	0.396	0.0181	0.0175	1.040	0.573	0.4582	0.4714	320
340	0.374	0.0350	0.0369	0.394	0.0190	0.0181	1.053	0.598	0.4595	0.4706	340
360	0.350	0.0359	0.0391	0.391	0.0199	0.0194	1.066	0.626	0.4609	0.4703	360
400	0.311	0.0377	0.0445	0.383	0.0222	0.0226	1.085	0.695	0.4638	0.4709	400
500	0.248	0.0427	0.0499	0.349	0.0306	0.0262	1.180	0.968	0.4715	0.4745	500
600	0.197	0.0504	0.0553	0.296	0.0486	0.0298	1.528	1.71	0.4798	0.4817	600
700			0.0556	0.188	0.1203	0.0300			0.4885	0.4895	700
706	0.101	0.101	0.0580	0.139	0.139	0.0317			0.4890	0.4900	706
750			0.0607			0.0337			0.4929	0.4935	750
800			0.0715			0.0418			0.4973	0.4977	800
1000									0.5150	0.5152	1000

Source: Reprinted by permission from *ASHRAE Handbook, Fundamentals Volume,* 1972.

Table B-3a WATER—SI UNITS

Celsius Temperature	Viscosity, $\mu \times 10^3$ (N·s)/m²			Thermal Conductivity, k W/(m·K)			Specific Heat, c_p kJ/(kg·K)				Kelvin Temperature
	Saturated Liquid	Saturated Vapor	Gas $P = 101.33$ kPa	Saturated Liquid	Saturated Vapor	Gas $P = 101.33$ kPa	Saturated Liquid	Saturated Vapor	Gas $P = 0$	Gas $P = 101.33$ kPa	
0	1.75	0.00794		0.569	0.0173		4.216		1.858		273
4	1.51	0.00814		0.578	0.0178		4.208		1.860		277
10	1.28	0.00835		0.587	0.0182		4.199		1.861		283
16	1.09	0.00856		0.597	0.0185		4.191		1.862		289
21	0.947	0.00880		0.606	0.0189		4.187	1.861	1.864		294
27	0.831	0.00901		0.613	0.0194		4.183	1.865	1.866		300
32	0.732	0.00926		0.621	0.0197		4.178	1.869	1.867		305
38	0.653	0.00947		0.628	0.0201		4.178	1.874	1.869		311
49	0.533	0.00992		0.642	0.0209		4.178	1.879	1.872		322
60	0.446	0.0104		0.654	0.0216		4.187	1.893	1.876		333
71	0.383	0.0108		0.663	0.0225		4.191	1.912	1.880		344
82	0.332	0.0112		0.672	0.0234		4.199	1.937	1.884		355
								1.967			

93	0.293	0.0117		0.678	0.0242		4.212	2.006	1.888		366
100	0.273	0.0119	0.0119	0.682	0.0247	0.0247	4.220	2.034	1.891	2.048	373
104	0.261	0.121	0.0122	0.684	0.0251	0.0249	4.224	2.052	1.893	2.038	377
116	0.234	0.0125	0.0126	0.687	0.0261	0.0254	4.241	2.101	1.897	2.018	389
127	0.213	0.0129	0.0130	0.689	0.0273	0.0265	4.258	2.159	1.902	2.001	400
138	0.195	0.0133	0.0135	0.689	0.0286	0.0273	4.280	2.227	1.908	1.989	411
149	0.179	0.0137	0.0139	0.687	0.0298	0.0284	4.308	2.306	1.913	1.980	422
160	0.166	0.0141	0.0144	0.685	0.0313	0.0294	4.354	2.399	1.918	1.974	433
171	0.155	0.0145	0.0148	0.682	0.0329	0.0303	4.409	2.504	1.924	1.970	444
183	0.145	0.0148	0.0153	0.677	0.0344	0.0313	4.463	2.621	1.930	1.969	456
205	0.129	0.0156	0.0162	0.663	0.0384	0.0336	4.543	2.910	1.942	1.972	478
260	0.103	0.0177	0.0184	0.604	0.0530	0.0391	4.940	4.053	1.974	1.987	533
316	0.0814	0.208	0.0206	0.512	0.0841	0.0453	6.397	7.159	2.009	2.017	589
371			0.0229	0.325	0.208	0.0516			2.045	2.049	644
375	0.0418	0.0418	0.0230	0.241	0.241	0.0519			2.047	2.052	648
399			0.0240			0.0549			2.064	2.066	672
427			0.0251			0.0583			2.082	2.084	700
538			0.0296			0.0723			2.156	2.157	811

Source: Adapted by permission from *ASHRAE Handbook, Fundamentals Volume*, 1972.

Table B-4 AIR—ENGLISH UNITS

Fahrenheit Temperature	Viscosity, μ lbm/(ft-hr)			Thermal Conductivity, k Btu/(hr-ft-F)			Specific Heat, c_p Btu/(lbm-F)				Fahrenheit Temperature
	Saturated Liquid	Saturated Vapor	Gas P = 1 atm	Saturated Liquid	Saturated Vapor	Gas P = 1 atm	Saturated Liquid	Saturated Vapor	Gas P = 0	Gas P = 1 atm	
−352	0.7865			0.1040	0.00312						−352
−334	0.5348			0.0942	0.00370						−334
−316	0.3993	0.0133		0.0838	0.00433						−316
−298	0.3194	0.0157	0.0154	0.0740	0.00497	0.00480	0.4690		0.2394		−298
−280	0.2664	0.0182	0.0171	0.0636	0.00584	0.00532	0.4962	0.2676	0.2394	0.2456	−280
−262	0.2297	0.0208	0.0188	0.0537	0.00705	0.00589	0.5268	0.2891	0.2394	0.2442	−262
−244	0.1815	0.0247	0.0204	0.0439	0.00890	0.00641	0.5784	0.3440	0.2394	0.2430	−244
−226	0.1016	0.0346	0.0220	0.0312	0.01213	0.00693	0.6689	0.4778	0.2394	0.2422	−226
−220.6	0.05009	0.05009	0.0225	0.01964	0.01964	0.00711			0.2394	0.2420	−220.6
−208			0.0236			0.00745			0.2394	0.2418	−208
−190			0.0251			0.00797			0.2394	0.2415	−190
−172			0.0266			0.00844			0.2394	0.2411	−172

−136	0.0295	0.00948	0.2394	0.2406	−136
−100	0.0324	0.0105	0.2394	0.2403	−100
−64	0.0351	0.0114	0.2396	0.2403	−64
−28	0.0375	0.0124	0.2396	0.2401	−28
8	0.0402	0.0134	0.2396	0.2401	8
44	0.0426	0.0142	0.2399	0.2403	44
80	0.0448	0.0151	0.2401	0.2403	80
116	0.0472	0.0160	0.2403	0.2406	116
152	0.0494	0.0168	0.2406	0.2408	152
188	0.0516	0.0176	0.2411	0.2413	188
224	0.0535	0.0183	0.2415	0.2418	224
260	0.0554	0.0191	0.2420	0.2422	260
296	0.0576	0.0199	0.2427	0.2430	296
332	0.0593	0.0206	0.2434	0.2437	332
368	0.0612	0.0214	0.2442	0.2444	368
404	0.0632	0.0221	0.2449	0.2451	404
440	0.0649	0.0228	0.2458	0.2461	440
620	0.0733	0.0264	0.2511	0.2513	620

Source: Adapted by permission from *ASHRAE Handbook, Fundamentals Volume,* 1972.

Table B-4a AIR—SI UNITS

Celsius Temperature	Viscosity, $\mu \times 10^3$ (N-s/m²)			Thermal Conductivity, k W/(m-K)			Specific Heat, c_p kJ/(kg-K)				Kelvin Temperature
	Saturated Liquid	Saturated Vapor	Gas P = 101.33 kPa	Saturated Liquid	Saturated Vapor	Gas P = 101.33 kPa	Saturated Liquid	Saturated Vapor	Gas P = 0	Gas P = 101.33 kPa	
−213	0.325			0.180	0.0054						60
−203	0.221			0.163	0.0064						70
−193	0.165	0.0055		0.145	0.0075		1.963		1.002		80
−183	0.132	0.0065	0.00635	0.128	0.0086	0.0083	2.077	1.12	1.002		90
−173	0.1101	0.0075	0.00706	0.110	0.0101	0.0092	2.205	1.21	1.002	1.028	100
−163	0.0949	0.0086	0.00775	0.093	0.0122	0.0102	2.421	1.44	1.002	1.022	110
−153	0.0750	0.0102	0.00843	0.076	0.0154	0.0111	2.80	2.00	1.002	1.017	120
−143	0.0420	0.0143	0.00909	0.054	0.021	0.0120			1.002	1.014	130
−140	0.0207	0.0207	0.00929	0.034	0.034	0.0123			1.002	1.013	133
−133			0.00974			0.0129			1.002	1.012	140
−123			0.01038			0.0138			1.002	1.011	150
−113			0.0110			0.0146			1.002	1.009	160

− 93	0.0122	0.0164	1.002	1.007	180
− 73	0.0134	0.0181	1.002	1.006	200
− 53	0.0145	0.0198	1.003	1.006	220
− 33	0.0155	0.0215	1.003	1.005	240
− 13	0.0166	0.0231	1.003	1.005	260
7	0.0176	0.0246	1.004	1.006	280
27	0.0185	0.0261	1.005	1.006	300
47	0.0195	0.0276	1.006	1.007	320
67	0.0204	0.0290	1.007	1.008	340
87	0.0213	0.0304	1.009	1.010	360
107	0.0221	0.0317	1.011	1.012	380
127	0.0229	0.0331	1.013	1.014	400
147	0.0238	0.0344	1.016	1.017	420
167	0.0245	0.0357	1.019	1.020	440
187	0.0253	0.0370	1.022	1.023	460
207	0.0261	0.0383	1.025	1.026	480
227	0.0268	0.0395	1.029	1.030	500
327	0.0303	0.0456	1.051	1.052	600

Source: Reprinted by permission from *ASHRAE Handbook, Fundamentals Volume,* 1972.

Weather Data

Table C-1 CLIMATIC CONDITIONS FOR THE UNITED STATES AND CANADA

State and Station	Latitude, Degree and Minutes	Longitude, Degree and Minutes	Elevation ft	Elevation m	Winter 99 Percent F	Winter 99 Percent C	Winter 97½ Percent F	Winter 97½ Percent C	Summer Dry Bulb 2½ Percent F	Summer Dry Bulb 2½ Percent C	Summer Wet Bulb 2½ Percent F	Summer Wet Bulb 2½ Percent C	Daily Range F	Daily Range C
Alabama, Birmingham AP	33 3	86 5	610	186	17	−8	21	−6	94	34	75	24	21	12
Alaska, Anchorage AP	61 1	150 0	90	27	−23	−31	−18	−28	68	20	58	14	15	8
Arizona, Tucson AP	32 1	111 0	2584	788	28	−2	32	0	102	39	66	19	26	14
Arkansas, Little Rock AP	34 4	92 1	257	78.3	15	−9	20	−7	96	36	77	25	22	12
California, San Francisco AP	37 4	122 2	8	2.4	35	2	38	3	77	25	63	17	20	11
Colorado, Denver AP	39 5	104 5	5283	1610	−5	−21	1	−17	91	33	59	15	28	16
Connecticut, Bridgeport AP	41 1	73 1	7	2.1	−6	−21	9	−13	84	29	71	22	18	10
Delaware, Wilmington AP	39 4	75 3	78	24	10	−12	14	−10	89	32	74	23	20	11
District of Columbia, Washington National	38 5	77 0	14	4.3	14	−10	17	−8	91	33	74	23	18	10
Florida, Tallahassee AP	30 2	84 2	58	18	27	−3	30	−1	92	33	76	24	19	11
Georgia, Atlanta AP	33 4	84 3	1005	306	17	−8	22	−6	92	33	74	23	19	11
Hawaii, Honolulu AP	21 2	158 0	7	2.1	62	17	63	17	86	30	73	23	12	7
Idaho, Boise AP	43 3	116 1	2842	866	3	−16	10	−12	94	34	64	18	31	17
Illinois, Chicago O'Hare AP	42 0	87 5	658	201	−8	−22	−4	−20	89	32	74	23	20	11
Indiana, Indianapolis AP	39 4	86 2	793	242	−2	−19	2	−17	90	32	74	23	22	12
Iowa, Sioux City AP	42 2	96 2	1095	334	−11	−24	−7	−22	92	33	74	23	24	13
Kansas, Wichita AP	37 4	97 3	1321	403	3	−16	7	−14	98	37	73	23	23	13
Kentucky, Louisville AP	38 1	85 4	474	144	5	−15	10	−12	93	34	75	24	23	13
Louisiana, Shreveport AP	32 3	93 5	252	76.8	20	−7	25	−4	96	36	76	24	20	11
Maine, Caribou AP	46 5	68 0	624	190	−18	−28	−13	−25	81	27	67	19	21	12
Maryland, Baltimore AP	39 1	76 4	146	44.5	10	−12	13	−11	91	33	75	24	21	12
Massachusetts, Boston AP	42 2	71 0	15	4.6	−6	−14	9	−13	88	31	71	22	16	9
Michigan, Lansing AP	42 5	84 4	852	260	−3	−19	1	−17	87	31	72	22	24	13
Minnesota, Minneapolis/St. Paul	44 5	93 1	822	251	−16	−27	−12	−24	89	32	73	23	22	12
Mississippi, Jackson AP	32 2	90 1	330	101	21	−6	25	−4	95	35	76	24	21	12

Location	Lat	Long	Elev											
Missouri, Kansas City AP	39 1	94 4	742	226	2	−17	6	−14	96	36	74	23	20	11
Montana, Billings AP	45 4	108 3	3567	1087	−15	−26	−10	−23	91	33	64	18	31	17
Nebraska, Lincoln CO	40 5	96 5	1150	351	−5	−21	−2	−19	95	35	74	23	24	13
Nevada, Las Vegas AP	36 1	115 1	2162	659	25	−4	28	−12	106	41	65	18	30	17
New Hampshire, Concord	43 1	71 3	339	103	−8	−22	−3	−19	87	31	70	21	26	14
New Jersey, Atlantic City CO	39 3	74 3	11	3.4	10	−12	13	−11	89	32	74	23	18	10
New Mexico, Albuquerque AP	35 0	106 4	5310	1618	12	−11	16	−9	94	34	61	16	30	17
New York, Syracuse AP	43 1	76 1	424	129	−3	−19	2	−17	87	31	71	22	20	11
North Carolina, Charlotte AP	35 1	81 0	735	224	18	−8	22	−6	93	34	74	23	20	11
North Dakota, Bismarck AP	46 5	100 5	1647	502	−23	−31	−19	−28	91	33	68	20	27	15
Ohio, Cleveland AP	41 2	81 5	777	237	1	−17	5	−15	88	31	72	22	22	12
Oklahoma, Stillwater	36 1	97 1	884	269	8	−13	13	−11	96	36	74	23	24	13
Oregon, Pendleton AP	45 4	118 5	1492	455	−2	−19	5	−15	93	34	64	18	29	16
Pennsylvania, Pittsburgh AP	40 3	80 0	1137	347	1	−17	5	−15	86	30	71	22	22	12
Rhode Island, Providence AP	41 4	71 3	55	17	5	−15	9	−13	86	30	72	22	19	11
South Carolina, Charleston AFB	32 5	80 0	41	12	24	−4	27	−3	91	33	78	26	18	10
South Dakota, Rapid City	44 0	103 0	3165	965	−11	−24	−7	−22	92	33	65	18	28	16
Tennessee, Memphis AP	35 0	90 0	263	80.2	13	−11	18	−8	95	35	76	24	21	12
Texas, Dallas AP	32 5	96 5	481	147	18	−8	22	−6	100	38	75	24	20	11
Utah, Salt Lake City	40 5	112 0	4220	1286	3	−16	8	−13	95	35	62	17	32	18
Vermont, Burlington AP	44 3	73 1	331	101	−12	−24	−7	−22	85	29	70	21	23	13
Virginia, Norfolk AP	36 5	76 1	26	7.9	20	−7	22	−6	91	33	76	24	18	10
Washington, Spokane AP	47 4	117 3	2357	718	−6	−21	2	−17	90	32	63	17	28	16
West Virginia, Charleston AP	38 2	81 4	939	286	7	−14	11	−12	90	32	73	23	20	11
Wisconsin, Milwaukee AP	43 0	87 5	672	205	−8	−22	−4	−20	87	31	73	23	21	12
Wyoming, Casper AP	42 5	106 3	5319	1621	−11	−24	−5	−21	90	32	57	14	31	17
Alberta, Calgary AP	51 1	114 1	3540	1079	−27	−33	−23	−31	81	27	61	16	25	14
British Columbia, Vancouver AP	49 1	123 1	16	4.9	15	−9	19	−7	77	25	66	19	17	9
Manitoba, Winnipeg AP	49 5	97 1	786	240	−30	−34	−27	−33	86	30	71	22	22	12
New Brunswick, Fredericton	45 5	66 3	74	23	−16	−27	−11	−24	85	29	69	21	23	13
Nova Scotia, Halifax AP	44 4	63 3	136	41.5	1	−17	5	−15	76	24	65	18	16	9
Ontario, Ottawa AP	45 2	76 0	339	103	−17	−27	−13	−25	87	31	71	22	21	12
Quebec, Montreal AP	45 3	74 0	98	30	−16	−27	−10	−23	85	29	72	22	18	10
Saskatchewan, Regina AP	50 3	105 0	1884	574	−33	−36	−29	−34	88	31	68	20	27	15

Source: Abridged by permission from *ASHRAE Handbook, Fundamentals Volume,* 1977.

Table C-2 AVERAGE DEGREE DAYS[a] FOR CITIES IN THE UNITED STATES AND CANADA

State and Station	Average Winter Temperature F	C	July	August	September	October	November	December	January	February	March	April	May	June	Yearly Total
Alabama, Birmingham	54.2	12.7	0	0	6	93	363	555	592	462	363	108	9	0	2551
Alaska, Anchorage	23.0	5.0	245	291	516	930	1284	1572	1631	1316	1293	879	592	315	10864
Arizona, Tucson	58.1	14.8	0	0	0	25	231	406	471	344	242	75	6	0	1800
Arkansas, Little Rock	50.5	10.6	0	0	9	127	465	716	756	577	434	126	9	0	3219
California, San Francisco	53.4	12.2	81	78	60	143	306	462	508	395	363	279	214	126	3015
Colorado, Denver	37.6	3.44	6	9	117	428	819	1035	1132	938	887	558	288	66	6283
Connecticut, Bridgeport	39.9	4.72	0	0	66	307	615	986	1079	966	853	510	208	27	5617
Delaware, Wilmington	42.5	6.17	0	0	51	270	588	927	980	874	735	387	112	6	4930
District of Columbia, Washington	45.7	7.94	0	0	33	217	519	834	871	762	626	288	74	0	4224
Florida, Tallahassee	60.1	15.9	0	0	0	28	198	360	375	286	202	86	0	0	1485
Georgia, Atlanta	51.7	11.28	0	0	18	124	417	648	636	518	428	147	25	0	2961
Hawaii, Honolulu	74.2	23.8	0	0	0	0	0	0	0	0	0	0	0	0	0
Idaho, Boise	39.7	4.61	0	0	132	415	792	1017	1113	854	722	438	245	81	5809
Illinois, Chicago	35.8	2.44	0	12	117	381	807	1166	1265	1086	939	534	260	72	6639
Indiana, Indianapolis	39.6	4.56	0	0	90	316	723	1051	1113	949	809	432	177	39	5699
Iowa, Sioux City	43.0	1.10	0	9	108	369	867	1240	1435	1198	989	483	214	39	6951
Kansas, Wichita	44.2	7.11	0	0	33	229	618	905	1023	804	645	270	87	6	4620
Kentucky, Louisville	44.0	6.70	0	0	54	248	609	890	930	818	682	315	105	9	4660
Louisiana, Shreveport	56.2	13.8	0	0	0	47	297	477	552	426	304	81	0	0	2184
Maine, Caribou	24.4	−3.89	78	115	336	682	1044	1535	1690	1470	1308	858	468	183	9767
Maryland, Baltimore	43.7	6.83	0	0	48	264	585	905	936	820	679	327	90	0	4654
Massachusetts, Boston	40.0	4.40	0	9	60	316	603	983	1088	972	846	513	208	36	5634
Michigan, Lansing	34.8	1.89	6	22	138	431	813	1163	1262	1142	1011	579	273	69	6909
Minnesota, Minneapolis	28.3	−1.72	22	31	189	505	1014	1454	1631	1380	1166	621	288	81	8382
Mississippi, Jackson	55.7	13.5	0	0	0	65	315	502	546	414	310	87	0	0	2239
Missouri, Kansas City	43.9	6.94	0	0	39	220	612	905	1032	818	682	294	109	0	4711
Montana, Billings	34.5	1.72	6	15	186	487	897	1135	1296	1100	970	570	285	102	7049
Nebraska, Lincoln	38.8	4.11	0	6	75	301	726	1066	1237	1016	834	402	171	30	5864
Nevada, Las Vegas	53.5	12.28	0	0	0	78	387	617	688	487	335	111	6	0	2709

Location															
New Hampshire, Concord	33.0	0.60	6	50	177	505	822	1240	1358	1184	1032	636	298	75	7383
New Jersey, Atlantic City	43.2	6.56	0	0	39	251	549	880	936	848	741	420	133	15	4812
New Mexico, Albuquerque	45.0	7.20	0	0	12	229	642	868	930	703	595	288	81	0	4348
New York, Syracuse	35.2	2.11	6	28	132	415	744	1153	1271	1140	1004	570	248	45	6756
North Carolina, Charlotte	50.4	10.56	0	0	6	124	438	691	691	582	481	156	22	0	3191
North Dakota, Bismarck	26.6	−2.67	34	28	222	577	1083	1463	1708	1442	1203	645	329	117	8851
Ohio, Cleveland	37.2	3.22	9	25	105	384	738	1088	1159	1047	918	552	260	66	6351
Oklahoma, Stillwater	48.3	9.39	0	0	15	164	498	766	868	664	527	189	34	0	3725
Oregon, Pendleton	42.6	6.22	0	0	111	350	711	884	1017	773	617	396	205	63	5127
Pennsylvania, Pittsburgh	38.4	3.89	0	9	105	375	726	1063	1119	1002	874	480	195	39	5987
Rhode Island, Providence	38.8	4.11	0	16	96	372	660	1023	1110	988	868	534	236	51	5954
South Carolina, Charleston	56.4	13.9	0	0	0	59	282	471	487	389	291	54	0	0	2033
South Dakota, Rapid City	33.4	1.11	22	12	165	481	897	1172	1333	1145	1051	615	326	126	7345
Tennessee, Memphis	50.5	10.6	0	0	18	130	447	698	729	585	456	147	22	0	3232
Texas, Dallas	55.3	13.3	0	0	0	62	321	524	601	440	319	90	6	0	2363
Utah, Salt Lake City	38.4	3.89	0	0	81	419	849	1082	1172	910	763	459	233	84	6052
Vermont, Burlington	29.4	−1.11	28	65	207	539	891	1349	1513	1333	1187	714	353	90	8269
Virginia, Norfolk	49.2	9.89	0	0	0	136	408	698	738	655	533	216	37	0	3421
Washington, Spokane	36.5	2.83	9	25	168	493	879	1082	1231	980	834	531	288	135	6655
West Virginia, Charleston	44.8	7.44	0	0	63	254	591	865	880	770	648	300	96	9	4476
Wisconsin, Milwaukee	32.6	0.667	43	47	174	471	876	1252	1376	1193	1054	642	372	135	7635
Wyoming, Casper	33.4	1.11	6	16	192	524	942	1169	1290	1084	1020	657	381	129	7410
Alberta, Calgary	—	—	109	186	402	719	1110	1389	1575	1379	1268	798	477	291	9703
British Columbia, Vancouver	—	—	81	87	219	456	657	787	862	723	676	501	310	156	5515
Manitoba, Winnipeg	—	—	38	71	322	683	1251	1757	2008	1719	1465	813	405	147	10679
New Brunswick, Fredericton	—	—	78	68	234	592	915	1392	1541	1379	1172	753	406	141	8671
Nova Scotia, Halifax	—	—	58	51	180	457	710	1074	1213	1122	1030	742	487	237	7361
Ontario, Ottawa	—	—	25	81	222	567	936	1469	1624	1441	1231	708	341	90	8735
Quebec, Montreal	—	—	9	43	165	521	882	1392	1566	1381	1175	684	316	69	8203
Saskatchewan, Regina	—	—	78	93	360	741	1284	1711	1965	1687	1473	804	409	201	10806

aBased on degrees F, quantities may be converted to degree days based on degrees C by multiplying by 5/9. This assumes 18 C corresponds to 65 F.

Source. Abridged by permission from *ASHRAE Handbook, Systems Volume,* 1980.

Table C-3 ANNUAL BIN WEATHER DATA FOR OKLAHOMA CITY, OKLAHOMA
35°-24' N 97°-36' W 1285 FT ELEVATION

Temperature Bin-F	Time Group						Total
	1–4	5–8	9–12	13–16	17–20	21–24	
	Hours of Dry-bulb Occurrence: MCWB						
100/104	0: 0	0: 0	0: 0	2: 77	0: 0	0: 0	2: 77
95/ 99	0: 0	0: 0-	5: 74	70: 76	29: 76	0: 0	104: 76
90/ 94	0: 0	0: 0	55: 76	153: 74	88: 74	0: 0	296: 74
85/ 89	2: 73	0: 0	116: 74	145: 72	120: 73	24: 74	407: 73
80/ 84	20: 72	33: 73	148: 72	153: 68	168: 71	96: 72	618: 71
75/ 79	121: 71	93: 71	132: 68	115: 65	144: 67	171: 70	776: 69
70/ 74	229: 68	221: 69	138: 65	118: 58	117: 63	186: 68	1009: 66
65/ 69	161: 64	161: 65	98: 58	98: 54	93: 56	136: 62	747: 61
60/ 64	120: 58	99: 58	95: 53	135: 51	96: 51	97: 55	642: 54
55/ 59	87: 53	104: 53	105: 50	108: 47	116: 49	81: 51	601: 50
50/ 54	96: 48	103: 49	137: 46	94: 45	133: 45	121: 47	684: 46
45/ 49	98: 43	96: 43	100: 42	66: 40	87: 41	122: 42	569: 42
40/ 44	150: 39	121: 39	98: 38	67: 37	91: 38	140: 38	667: 38
35/ 39	144: 35	153: 35	89: 34	54: 34	76: 34	105: 34	621: 35
30/ 34	107: 31	140: 30	74: 30	40: 30	50: 30	93: 30	504: 30
25/ 29	63: 25	51: 26	27: 26	24: 24	24: 25	40: 25	229: 25
20/ 24	36: 20	41: 21	23: 20	17: 19	23: 19	31: 21	171: 20
15/ 19	19: 16	37: 15	17: 16	1: 12	5: 14	16: 15	95: 15
10/ 14	7: 10	7: 11	3: 9	0: 0	0: 0	1: 11	18: 10

Source: Reprinted by permission from *Bin and Degree Hour Weather Data for Simplified Energy Calculations,* ASHRAE, Inc., Atlanta, Ga., 1986.

Table C-4 ANNUAL BIN WEATHER DATA FOR CHICAGO, ILLINOIS 41°-47′ N 87°-45′ W 607 FT ELEVATION

Temperature Bin-F	Time Group						
	1–4	5–8	9–12	13–16	17–20	21–24	Total
	Hours of Dry-bulb Occurrence: MCWB						
90/ 94	0: 0	0: 0	15: 74	64: 73	18: 71	0: 0	97: 73
85/ 89	0: 0	0: 0	64: 71	113: 68	45: 70	0: 0	222: 69
80/ 84	0: 0	7: 71	131: 67	113: 65	90: 67	21: 70	362: 67
75/ 79	25: 70	51: 68	114: 64	116: 64	125: 65	81: 68	512: 65
70/ 74	120: 66	125: 65	131: 62	111: 62	151: 63	167: 65	805: 64
65/ 69	156: 62	155: 62	88: 60	65: 58	80: 59	143: 60	687: 61
60/ 64	159: 57	130: 57	78: 54	73: 53	72: 55	103: 57	615: 56
55/ 59	111: 52	112: 52	93: 49	109: 49	96: 51	101: 52	622: 51
50/ 54	112: 48	92: 48	88: 46	89: 47	100: 47	104: 47	585: 47
45/ 49	92: 44	102: 43	97: 42	96: 40	83: 42	107: 43	577: 42
40/ 44	120: 38	104: 39	72: 37	94: 37	131: 37	115: 38	636: 38
35/ 39	106: 34	105: 34	133: 33	134: 33	131: 33	111: 34	720: 33
30/ 34	182: 30	182: 30	156: 29	129: 29	135: 30	173: 30	957: 30
25/ 29	93: 25	98: 25	66: 25	68: 24	94: 25	92: 25	511: 25
20/ 24	75: 21	75: 21	60: 20	40: 20	44: 21	60: 21	354: 21
15/ 19	63: 16	56: 16	30: 16	21: 15	33: 15	40: 16	243: 16
10/ 14	25: 11	32: 11	25: 10	13: 10	11: 11	19: 11	125: 11
5/ 9	5: 7	14: 6	11: 5	8: 6	16: 6	12: 5	66: 6
0/ 4	15: 1	16: 0	7: 0	4: 1	5: 0	11: 1	58: 0
−5/− 1	1:− 4	4:− 7	1:− 6	0: 0	0: 0	0: 0	6:− 6

Source: Reprinted by permission from *Bin and Degree Hour Weather Data for Simplified Energy Calculations*, ASHRAE, Inc., Atlanta, Ga., 1986.

Table C-5 ANNUAL BIN WEATHER DATA FOR DENVER, COLORADO 39°-45' N 104°-52' W 5283 FT ELEVATION

Temperature Bin-F	Time Group						
	1–4	5–8	9–12	13–16	17–20	21–24	Total
	Hours of Dry-bulb Occurrence: MCWB						
95/ 99	0: 0	0: 0	0: 0	3: 60	0: 0	0: 0	3: 60
90/ 94	0: 0	0: 0	15: 58	88: 59	15: 58	0: 0	118: 59
85/ 89	0: 0	0: 0	72: 59	112: 58	51: 58	0: 0	235: 58
80/ 84	0: 0	1: 57	107: 58	146: 56	91: 57	3: 58	348: 57
75/ 79	0: 0	16: 58	124: 56	100: 54	120: 56	30: 57	390: 55
70/ 74	12: 55	49: 56	121: 54	95: 52	112: 53	83: 56	472: 54
65/ 69	90: 54	99: 54	103: 51	136: 49	129: 51	140: 54	697: 52
60/ 64	129: 52	118: 52	109: 48	106: 46	108: 50	129: 51	699: 50
55/ 59	133: 50	137: 49	120: 44	119: 43	106: 46	147: 47	762: 47
50/ 54	154: 44	128: 44	135: 41	138: 40	98: 41	130: 43	783: 42
45/ 49	137: 41	139: 40	118: 38	84: 38	118: 38	120: 40	716: 39
40/ 44	170: 36	138: 36	83: 34	81: 33	84: 35	109: 36	665: 35
35/ 39	148: 32	135: 32	127: 31	103: 31	120: 32	125: 32	758: 32
30/ 34	132: 29	145: 29	96: 28	71: 28	122: 28	147: 28	713: 28
25/ 29	148: 24	120: 24	59: 24	36: 25	79: 25	123: 24	565: 24
20/ 24	98: 20	116: 20	28: 19	14: 20	51: 20	92: 20	399: 20
15/ 19	35: 16	49: 15	17: 15	14: 14	20: 16	29: 16	164: 15
10/ 14	29: 10	21: 10	8: 9	9: 10	15: 10	24: 11	106: 10
5/ 9	19: 6	19: 6	10: 5	3: 4	7: 6	7: 5	65: 6
0/ 4	20: −1	23: 0	7: 0	2: 2	11: 0	17: −1	80: 0
−5/−1	6: −6	7: −6	1: −4	0: 0	3: −6	5: −5	22: −6

Source: Reprinted by permission from *Bin and Degree Hour Weather Data for Simplified Energy Calculations,* ASHRAE, Inc., Atlanta, Ga., 1986.

Table C-6   ANNUAL BIN WEATHER DATA FOR WASHINGTON, D.C. 38°-51' N 77°-02' W 14 FT ELEVATION

Temperature Bin-F	Time Group						
	1–4	5–8	9–12	13–16	17–20	21–24	Total
			Hours of Dry-bulb Occurrence: MCWB				
90/ 94	0: 0	0: 0	11: 76	60: 77	15: 77	0: 0	86: 77
85/ 89	0: 0	0: 0	54: 75	142: 72	84: 74	0: 0	280: 73
80/ 84	2: 74	11: 75	140: 72	172: 70	155: 71	51: 74	531: 71
75/ 79	89: 73	88: 73	166: 68	168: 64	169: 67	145: 71	825: 69
70/ 74	186: 69	188: 68	164: 64	135: 61	173: 63	198: 67	1044: 66
65/ 69	178: 64	156: 63	119: 59	81: 56	102: 59	178: 62	814: 61
60/ 64	133: 58	119: 58	86: 54	103: 51	73: 53	102: 57	616: 56
55/ 59	106: 53	137: 53	113: 50	97: 47	90: 48	75: 51	618: 51
50/ 54	99: 48	90: 48	95: 46	99: 44	129: 45	111: 47	623: 46
45/ 49	128: 43	110: 43	128: 41	108: 40	102: 41	135: 42	711: 41
40/ 44	137: 38	128: 38	99: 36	113: 36	130: 36	135: 38	742: 37
35/ 39	148: 33	142: 34	134: 32	107: 31	113: 32	137: 33	781: 33
30/ 34	138: 29	137: 29	97: 28	47: 29	84: 28	114: 28	617: 28
25/ 29	62: 25	96: 25	35: 24	19: 24	25: 24	52: 24	289: 24
20/ 24	39: 20	41: 20	8: 18	5: 19	10: 19	13: 19	116: 20
15/ 19	6: 15	5: 14	10: 13	4: 15	4: 15	6: 16	35: 15
10/ 14	9: 10	12: 11	1: 12	0: 0	2: 11	8: 10	32: 10

Source: Reprinted by permission from *Bin and Degree Hour Weather Data for Simplified Energy Calculations,* ASHRAE, Inc., Atlanta, Ga., 1986.

Pipe and Tube Data

Table D-1 STEEL PIPE DIMENSIONS—ENGLISH AND SI UNITS

Nominal Pipe Size in.	Schedule Number	Diameter OD		ID		Wall Thickness		Inside Cross-Sectional Area	
		in.	mm	in.	mm	in.	mm	ft^2	m$^2 \times 10^3$
$\frac{1}{4}$	40	0.540	13.7	0.364	9.25	0.088	2.23	0.00072	0.067
	80			0.302	7.67	0.119	3.02	0.00050	0.046
$\frac{3}{8}$	40	0.675	17.1	0.493	12.5	0.091	2.31	0.00133	0.124
	80			0.423	10.7	0.126	3.20	0.00098	0.091
$\frac{1}{2}$	40	0.840	21.3	0.622	15.8	0.109	2.77	0.00211	0.196
	80			0.546	13.9	0.147	3.73	0.00163	0.151
$\frac{3}{4}$	40	1.050	26.7	0.824	20.9	0.113	2.87	0.00371	0.345
	80			0.742	18.8	0.154	3.91	0.00300	0.279
1	40	1.315	33.4	1.049	26.6	0.133	3.38	0.00600	0.557
	80			0.957	24.3	0.179	4.55	0.00499	0.464
$1\frac{1}{2}$	40	1.900	48.3	1.610	40.9	0.145	3.68	0.01414	1.314

Nominal	Schedule	OD (in)	OD (mm)	ID (in)	ID (mm)	Wall (in)	Wall (mm)	Area	Weight
	80			1.500	38.1	0.200	5.08	0.01225	1.138
2	40	2.375	60.3	2.067	52.5	0.154	3.91	0.02330	2.165
	80			1.939	49.3	0.218	5.54	0.02050	1.905
2½	40	2.875	73.0	2.469	62.7	0.203	5.16	0.03322	3.086
	80			2.323	59.0	0.276	7.01	0.02942	2.733
3	40	3.500	88.9	3.068	77.9	0.216	5.49	0.05130	4.766
	80			2.900	73.7	0.300	7.62	0.04587	4.262
4	40	4.500	114.3	4.026	102.3	0.237	6.02	0.08840	8.213
	80			3.826	097.2	0.337	8.56	0.07986	7.419
5	40	5.563	141.3	5.047	128.1	0.258	6.55	0.1390	12.91
	80			4.813	122.3	0.375	9.53	0.1263	11.73
6	40	6.625	168.3	6.065	154.1	0.280	7.11	0.2006	18.64
	80			5.761	146.3	0.432	11.0	0.1810	16.82
8	40	8.625	219.1	7.981	202.7	0.322	8.18	0.3474	32.28
	80			7.625	193.7	0.500	12.7	0.3171	29.46
10	40	10.75	273.1	10.020	254.5	0.365	9.27	0.5475	50.86
	80			9.750	247.7	0.500	12.7	0.5185	48.17

Source: Adapted from A.S.A. Standards B36.10.

Table D-2   TYPE L COPPER TUBE DIMENSIONS—ENGLISH AND SI UNITS

Nominal Tube Size in.	Diameter				Wall Thickness		Inside Cross-Sectional Area	
	OD		ID					
	in.	mm	in.	mm	in.	mm	in.^{2}	m$^2 \times 10^3$
$\frac{1}{4}$	0.375	9.53	0.315	8.00	0.030	0.762	0.0779	0.0503
$\frac{3}{8}$	0.500	12.7	0.43	11	0.035	0.889	0.145	0.094
$\frac{1}{2}$	0.625	15.9	0.545	13.8	0.040	1.02	0.233	0.150
$\frac{5}{8}$	0.750	19.1	0.666	16.9	0.042	1.07	0.348	0.225
$\frac{3}{4}$	0.875	22.2	0.785	19.9	0.045	1.14	0.484	0.312
1	1.125	28.58	1.025	26.04	0.050	1.27	0.825	0.532
$1\frac{1}{4}$	1.375	34.93	1.265	32.13	0.055	1.40	1.26	0.813
$1\frac{1}{2}$	1.625	41.28	1.505	38.23	0.060	1.52	1.78	1.15
2	2.125	53.98	1.985	50.42	0.070	1.78	3.10	2.00
$2\frac{1}{2}$	2.625	66.68	2.465	62.61	0.080	2.03	4.77	3.08
3	3.125	79.38	2.945	74.80	0.090	2.29	6.81	4.39
$3\frac{1}{2}$	3.625	92.08	3.425	87.00	0.100	2.54	9.21	5.94
4	4.125	104.8	3.905	99.19	0.110	2.79	12.0	7.74

Source: Based on ASTM B-88.

Useful Data

Table E-1   CONVERSION FACTORS

Length
1 ft = 30.48 cm
1 in. = 2.54 cm
1 m = 39.37 in.
1 micron = 10^{-6} m = 3.281×10^{-6} ft
1 mile = 5280 ft

Area
1 m^2 = 1550.1472 in.^{2}
1 m^2 = 10.76392 ft^2

Volume
1 ft^3 = 7.48 U.S. gallons = 1728 in.^{3}
1 m^3 = 6.1×10^4 in.^{3}
1 m^3 = 35.3147 ft^3
1 m^3 = 264.154 U.S. gallons

Mass
1 kg = 2.20462 lbm
1 lbm = 7000 grains = 453.5924 g

Force
1 N = 0.224809 lbf
1 lbf = 4.44822 N

Table E-1   CONVERSION FACTORS (*continued*)

Energy
1 Btu = 778.28 ft-lbf
1 kilocalorie = 10^3 calories = 3.968 Btu
1 J = 9.48×10^{-4} Btu = 0.73756 ft-lbf
1 kW-hr = 3412 Btu = 2.6562×10^6 ft-lbf

Power
1 hp = 33,000 (ft-lbf)/min
1 hp = 745.7 W
1 W = 3.412 Btu/hr
1 W = 0.001341 hp
1 W = 0.0002843 tons of refrigeration

Pressure
1 atm = 14.6959 psia = 2116 lbf/ft^2 = 101325 N/m^2
1 in. of water = 249.08 Pa
1 in. of mercury = 3376.85 Pa
1 lbf/in.^{2} = 6894.76 Pa
1 Pa = 1 N/m^2 = 1.4504×10^{-4} lbf/in.^{2}

Temperature
1 degree R difference = 1 degree F difference = $\frac{5}{9}$ degree C difference = $\frac{5}{9}$ degree K difference
degree F = $\frac{9}{5}$ (degree C) + 32
degree C = $\frac{5}{9}$ (degree F − 32)

Thermal Conductivity

$$1 \frac{\text{Btu}}{\text{hr-ft-F}} = 0.004134 \frac{\text{calorie}}{\text{s-cm-C}} = 1.7307 \frac{\text{W}}{\text{m-C}}$$

$$1 \frac{\text{W}}{\text{m-C}} = 0.5778 \frac{\text{Btu}}{\text{hr-ft-F}}$$

Table E-1   CONVERSION FACTORS (*continued*)

$$1 \frac{\text{Btu-in.}}{\text{hr-ft}^2\text{-F}} = 0.1442 \frac{\text{W}}{\text{m-C}}$$

$$1 \frac{\text{W}}{\text{m-C}} = 6.933 \frac{\text{Btu-in.}}{\text{hr-ft}^2\text{-F}}$$

Heat Transfer Coefficient

$$1 \frac{\text{Btu}}{\text{hr-ft}^2\text{-F}} = 5.678 \frac{\text{W}}{\text{m}^2\text{-C}}$$

$$1 \frac{\text{W}}{\text{m}^2\text{-C}} = 0.1761 \frac{\text{Btu}}{\text{hr-ft}^2\text{-F}}$$

Viscosity, absolute
1 poise = 100 centipoises

$$1 \frac{\text{lbm}}{\text{sec-ft}} = 1490 \text{ centipoises} = 1.49 \frac{\text{N-s}}{\text{m}^2}$$

$$1 \frac{\text{lbf-sec}}{\text{ft}^2} = 47{,}800 \text{ centipoises}$$

$$1 \text{ centipoise} = 0.001 \frac{\text{N-s}}{\text{m}^2}$$

Viscosity, kinematic
1 ft^2/sec = 0.0929 m^2/s
1 m^2/s = 10.764 ft^2/sec

Specific heat

$$1 \frac{\text{calorie}}{\text{gm-C}} = 1 \frac{\text{Btu}}{\text{lbm-F}}$$

$$1 \frac{\text{Btu}}{\text{lbm-F}} = 4186.8 \frac{\text{J}}{\text{kg-C}}$$

$$1 \frac{\text{kJ}}{\text{kg-C}} = 0.2388 \frac{\text{Btu}}{\text{lbm-F}}$$

Index

SI Base Units

Dimensions	SI Units
Mass	kilogram (kg)
Length	meter (m)
Time	second (s)
Temperature	degree Kelvin (K)

Derived SI Units Frequently Used in HVAC

Dimension	Unit	Special Name and Symbol
Acceleration, angular	rad/s^2	
Acceleration, linear	m/s^2	
Area	m^2	
Density	kg/m^2	
Energy	N-m	joule (J)
Force	$(kg\text{-}m)/s^2$	newton (N)
Frequency	1/s	hertz (Hz)
Power	J/s	watt (W)
Pressure	N/m^2	pascal (Pa)
Specific heat capacity	J/(kg-C)	
Stress	N/m^2	pascal (Pa)
Thermal conductivity	W/(m-C)	
Thermal flux density	W/m^2	
Velocity, angular	rad/s	
Velocity	m/s	
Viscosity, dynamic	$(N\text{-}s)/m^2$	
Viscosity, kinematic	m/s^2	
Volume	m^3	